Radiation Safety Guide for Nuclear Medicine Professionals

Pankaj Tandon · Dibya Prakash
Subhash Chand Kheruka
Nagesh N Bhat

Radiation Safety Guide for Nuclear Medicine Professionals

 Springer

Pankaj Tandon
Atomic Energy Regulatory Board
Mumbai, India

Subhash Chand Kheruka
Medical Physics Unit, Department
of Radiology and Nuclear Medicine
Sultan Qaboos Comprehensive Cancer
Care and Research Centre (SQCCCRC)
Muscat, Oman

Dibya Prakash
Molecular Imaging & Therapy, a Unit
of Vitrana Healthcare LLP and Nuclear
Medicine Solutions
Ghaziabad, India

Nagesh N Bhat
Bhabha Atomic Research Center
Radiological Physics and Advisory
Division
Mumbai, India

ISBN 978-981-19-4520-5   ISBN 978-981-19-4518-2   (eBook)
https://doi.org/10.1007/978-981-19-4518-2

This Springer imprint is published by the registered company Springer Nature Singapore Pte Ltd. The registered company address is: 152 Beach Road, #21-01/04 Gateway East, Singapore 189721, Singapore

Preface

The origin of this book goes back to our YouTube videos, which are an effort to bring nuclear medicine out of the books, journals and classrooms and allow learning for everyone from anywhere. Along with conventional topics, we also cover practical topics that are not taught anywhere but are faced in the day-to-day life of nuclear medicine professionals. From this book medium, we cordially invite everyone from the nuclear medicine fraternity to come forward and share their knowledge to make nuclear medicine learnings easy and accessible. Since, I had already authored one book with Springer, the editorial director Dr Naren Aggarwal proposed to compile the content of videos in the form of a book which they wish to publish. However, I was unsure about how the book would shape with staggered topics. One name that came to my mind was Dr Pankaj Tandon from Atomic Energy Regulatory Board, who probably could be a co-author and provide some more topics to make a sensible book. When contacted, he said he was already working on a book and looking for a publisher. We collaborated and moved forward by adding two more co-authors for their expertise, and this book is in front of you.

The book contains years of experience from all the four authors and can work as a reference book for working nuclear medicine professionals and a textbook for Radiological Safety Officer (RSO) examination appearing students. We tried to cover all the aspects related to Radiation Safety, starting from basic radiation physics, radiation biology, quality control, design of facilities, dosimetry, mechanism of radiation detectors, emergency preparedness, transport of radioactive materials, and various regulations applied to nuclear medicine facilities and finally, the short questions for RSO examination appearing students. In many places, regulatory documents say 'contact RSO for further management', whereas there is no proper reference to which RSOs can refer and act in various radiological situations. Hopefully, this book would be helpful in all such situations.

From Chapters 1 to 5, the basics of radiation physics, radiation quantities and units, the interaction of radiation with matter, principles of radiation protection, biological considerations, dose limits and their significance, radiation hazards and their control, exposure, exposure rate, exposure rate constant, half-value layer, tenth-value layer, the build-up factor, effective half-life, the annual limit of intake (ALI), derived air concentration (DAC) and surface contamination limits have been covered.

Chapter 6 explains occupational and public exposures, types and categories of exposure, identification of exposed individuals, death of patients administered with radiopharmaceuticals, possibilities of exposure in nuclear medicine, elimination of radionuclides from internal routes, the principle behind the estimation of effective dose, among others. Chapter 7 speaks about the biological bases of radiation protection, radiation effects at the cellular level, relative biological effectiveness, deterministic and stochastic effects, acute radiation syndrome and damage to individual organs. Chapters 8, 9 and 10 explain the planning and design of nuclear medicine facilities, high-dose therapy facilities and cyclotrons, respectively. They include site selection, layout and area requirement, equipment and accessories, staff, shielding requirements and calculations for SPECT-CT and PET-CT facilities, and calculation of ceiling thickness above the PET-CT. The chapter on high-dose therapy provides details of the isolation room, shielding requirements, delay and decay tank specifications, and a discussion on the alpha therapy facility. The cyclotron chapter additionally provides shielding calculations for unshielded and self-shielded cyclotrons. Chapter 11 explains personnel monitoring, radiation protection survey, objectives and benefits of personnel monitoring, dose limits for planned and emergency exposure situations, dose limit for medical exposure of patients, comforters, carers, and volunteers of biomedical research, personnel monitoring during pregnancy, TLD overexposure investigation, situations that do not warrant personnel monitoring and the survey of nuclear medicine facility.

Chapters 12, 13 and 14 explain radiation safety consideration in nuclear medicine, medical cyclotron and radiopharmaceutical preparation. Radiation safety in nuclear medicine says about various radioisotopes used in nuclear medicine therapy, pre- and post-therapy precautions, radiation safety of nursing staff and visitors, discharge criteria for the patients in the isolation ward, optimization of radiation doses to non-target tissues and a brief about handling emergency procedures. The chapter on medical cyclotron discusses the surveillance programme, the safety of pregnant radiation workers, handling, storage, and disposal of radioactive waste, and record keeping. Chapter 14 provides detailed insight about dose calibrators, including choices and pressure of gases, calibration factors, energy response curve, probability of photoelectric and Compton effects, sources of error and quality control. The chapter discusses radiopharmaceutical doses for adults, children and obese patients with the methods to modify them, medical events or misadministration, and the control of radiation hazards in radiopharmacy.

Chapter 15 explains the working mechanism of radiation detectors and includes basic principles of radiation detector, their characteristics, types of radiation detectors, the working mechanism of gas-filled detectors, voltage-response curve, analogue pocket dosimeters, digital pocket dosimeters, gun monitor, proportional counters, Geiger-Muller (GM) counters, scintillation detectors, CT detectors, semiconductor detectors and thermo-luminescent dosimeters (TLDs).

Chapter 16 talks about planar and single-photon emission computed (SPECT) gamma camera quality control with National Electrical

Manufacturers Association (NEMA) standards and other widely accepted protocols. Many factors which contribute to the final image quality, such as uniformity, resolutions, collimation, count rate capability and the hard copy devices, have been explained. For tomographic imaging, an additional set of parameters that influence clinical images, e.g. centre of rotation, gantry and collimator hole alignment, rotational stability of detector heads and integrity of the reconstruction algorithms, have been explained. Chapter 17 shows the quality control procedures for positron emission tomography (PET) machines and includes acceptance testing and routine quality control procedures. All minor details such as instruments needed, activity needed, and step-by-step procedure with illustrative images are provided so that nuclear medicine physicists can refer to this book and do the quality control effectively on their own.

Chapter 18 enlightens about the management of radiation emergencies and their preparedness. Since the best management of radiation emergencies is to prevent the occurrence, the chapter starts with prevention first and then discusses the complete set of situations such as spillage of radiopharmaceuticals, the incidental release of radioactive dust, fumes and gases, medical events (misadministration), medical emergencies including death involving patients administered with therapeutic doses of radiopharmaceuticals, unauthorized access to nuclear medicine facility, loss or theft of radioactive source, fire, bomb threat, natural disasters and accident of the vehicle carrying radioactive material.

Chapters 19 and 20 discuss nuclear medicine and CT dose assessments considering practical dosimetry situations. The Nuclear Medicine Dose Assessment chapter explains the need, ALARA and AHASA concepts, the term 'absorbed dose', its units, the formula for calculating absorbed dose and its components, resources for raw data, the concept of equivalent dose and effective dose and their uses, various systems of dose assessment including Medical Internal Radiation Dosimetry (MIRD), the International Commission on Radiation Protection (ICRP), the Radiation Dose Assessment Resource (RADAR) methods, free resources for dose assessment of diagnostic nuclear medicine and a discussion on practical therapeutic nuclear medicine dose assessment. Chapter 20 talks about CT dose assessment and includes the principle of CT machine functioning, the various terms used in CT dose assessment, estimation of CT doses, calculation of effective doses from system-generated CT dose reports with practical examples, diagnostic reference levels (DRLs) and achievable doses.

A large number of applications of radioactive material (RAM) necessitate its transport from one place to another. Chapter 21 explains rules associated with the transport of RAM, terms used in transportation, type of packages and their requirements, category of packages, transport index, marking, labelling, placarding, etc. Chapter 22 explains the legislation and role of national regulatory authorities in Nuclear Medicine. It covers the Atomic Energy Act, various rules issued under this act, AERB safety directives, the safety code for nuclear medicine facilities and the roles of employer, licensee, the RSO, nuclear medicine physician and nuclear medicine technologist as per terms used by the AERB. Chapter 23 explains the radioactive waste disposal and safe management of disused sealed radioactive sources. It explains the fundamental principle of radioactive waste management, classification of wastes, collection and disposal, record keeping, management of cadavers containing radionuclides and disposal of disused sources.

Chapter 24 is very helpful to students appearing for the Radiological Safety Officer (RSO) examination. It contains model 250 multiple-choice questions, 100 true-and-false questions, 60 fill-in-the-blank questions and 40 match-the-following questions. Similar questions may be asked in the RSO written examination. But, for viva questions, one needs to go through the many chapters of this book, and if someone uses this book as a preparation tool, it would definitely be useful.

We hope that efforts put in to prepare this manuscript are helpful to the nuclear medicine fraternity and students. All suggestions and comments are welcomed at nuclearmedicinesolution@gmail.com.

Ghaziabad, India   Dibya Prakash

Acknowledgements

With the greatest gratitude and humility, we thank our families for sacrificing their time and allowing us to write this book. Throughout the journey of this book, they have extended unconditional love with a lot of encouragement, patience and support. We are truly indebted to them.

There are few names without whose support this book would not have been possible. The first name is Dr Michael G Stabin, who is a true academician and believes in spreading knowledge to grow science. He knows his knowledge is so ample that no one can steal it, and his importance will remain forever (this fear is seen in many professionals). He selflessly always supported us in every endeavour. The chapter 'Nuclear Medicine Dose Assessment', without his contribution, would not even have existed in this book. We sincerely thank Dr Sarika Sharma Prashar, PhD, from the Postgraduate Institute of Medical Research and Education (PGIMER), Chandigarh, who shared her valuable expertise and experience for this chapter. We thank Mr Deepak Aheer from Siemens Healthineers, Mr Pranav Ratna from GE Healthcare and Ms Stuti Saxena from Saxons International Private Limited, who have provided valuable inputs at different stages of this book.

We are also thankful to my colleagues from AERB, Ms Manju Saini, Dr Alok Pandey and Mr D.M. Rane who have given their expert opinion while writing the chapters. Thanks are due to my dear friend Dr Lalit Mohan Aggarwal, Professor in Medical Physics at BHU, Varanasi, for his constant encouragement and motivation during the writing of this book. Further, thanks are due to Dr. Sanjay Gambhir of SGPGIMS, Lucknow, my mentors Dr. B.R. Mittal of PGIMER, Chandigarh and Dr. Anshu R Sharma of Kokilaben Dhirubhai Ambani Hospital and Medical Research Institute, Mumbai and the Senior colleagues of Radiology and Nuclear Medicine department of Sultan Qaboos Comprehensive Cancer Care & Research Centre, Muscat Oman for their constant guidance and support.

We thank Dr Naren Aggarwal, Editorial Director—Books, Asia, Medicine and Life Sciences, Springer, for offering the publishing contract, encouragement and support in producing the book.

Contents

Abstract

An atom is considered to be fundamental unit of an element while molecule for a compound. The quest to find fundamental particle of matter continued in science to subatomic and sub-nuclear explorations with various research tools such as accelerators. Detection of these structures is done in indirect means as they cannot be directly visualized. As a fallout of such studies, more insights were thrown on not only subatomic and sub-nuclear structures but also on mechanistic details of emission of various radiation from extra-nuclear and nuclear origin.

1.1 Introduction

Radiation can be produced either by isotopes or by devices such as X-ray or accelerators. The interaction property of the radiation does not depend on their origin but the basic characteristics such as energy and distribution of energy do depend on their origin.

Radiation is known to deposit its energy in highly non-uniform manner when it interacts with any medium. The fact which makes it inflect severe damage or changes is that it's non-uniform deposition can lead to severe damage at microscopic levels leading to both chemical and biological damages. Extent of damage also differs from one radiation to another by virtue of their capability to ionize the media.

1.2 Chronological Events of Radiation Interaction

When radiation interacts with any medium, initial deposit of energy is carried out by physical interaction irrespective of nature of media. These interactions quickly get followed up by successive events such as chemical and biological interactions. At every step of interaction, one can find various processes and their modifying factors. The chronologies of events also vary with the process. Biological interactions are very slow in nature stretching over several seconds to several years, even to the extent of lifetime of individuals (Table 1.1).

1.2.1 Physical Interaction of Radiation

Ionizing radiation has the capability to interact with media in which they fall upon or pass-through. The mode of interaction largely depends on type of radiation itself and the type of media involved. If radiation is uncharged type such as gamma, X-rays (photons) or neutrons, usually they create secondary charged particles through various modes of primary interactions. These

Table 1.1 Chronological events of radiological interactions

Phase	Events and processes	Time scale (s)	Modifying factors
Physical	Energy absorption, ionizations and excitation	10^{-18} to 10^{-15}	Dose, distribution of ionizing events (LET, dose rate)
Physico-chemical	Rearrangement of excited ionized molecules, formation of diffusible radicals such as H^0, OH^0 and e_{aq}	10^{-15} to 10^{-8}	Free radical scavengers
Chemical	Primary lesion-bioradicals, molecular alterations, formation of bioradicals by indirect action, long lived lesions in macromolecules	10^{-8} to 10^{-3}	Radiosensitizers radioprotectors
Biochemical	Enzymatic reactions, recognition of lesions, repair, fixation of damage	10^{-3} to 10^4	Repair inhibitors energy metabolism
Cellular level	Cell death, cell loss, division kinetics, mutation	10^4 to 10^7	Modifiers of tissue repair
Systemic (multicellular organisms)	Hormonal effects, immune reactions, vascular changes, functional impairment, adaptation, carcinogenesis, ageing, death	10^8 to 10^{10} (typical lifetime of individuals)	Tissue regeneration, medical management, rehabilitation

Table 1.2 Dependence of energy on spatial distribution of ionization events

Type of radiation	Energy	Approximate distance between primary ionizations in A°
Beta	0.5 MeV	5000
	1000 eV	50
	100 eV	5
Alpha	5 MeV	8
	1 MeV	2

secondary charged particles such as electrons and protons create further ionization in the medium. Interaction probability largely depends on bulk density, effective Z and/or electron density of the medium. When nuclear interactions are involved, the interaction probability can depend on various nuclear properties of the media involved.

Mean free path between interactions defines the density of ionization or events. This can change with type of radiation, mass of the particle, Z of the particle and also properties of the media. It is customary to define density of ionization in tissue or water as a reference media while defining type of radiation. Densely ionizing radiation such as alpha particles emanated by heavy radionuclides and heavy charged particles as found in particle accelerators are typically characterized by their capability to create dense ionization, while electrons and photons are known for their sparsely distributed ionizations (Table 1.2).

Density of ionization is usually defined in the form of specific ionization or linear energy transfer (LET). By definition, LET can be defined as amount of energy deposited in reference media (water or tissue) per unit length of traverse by radiation. Since LET changes with energy of radiation, path of radiation can create a profile which is known as track with varying LET along the track. However, to avoid such complexities of micro dosimetry concepts, either initial LET or average LET can be used to understand the type of radiation involved.

Among the several theories proposed to explain the mechanism of action of radiations, Dessauer's 'point heat theory' is one of the oldest. This theory suggests that the absorption of radiation enhances the temperature of the system thereby resulting in the thermal inactivation. However, the dose of radiation sufficient to kill a man (5 Gy) can raise the body temperature by just over 10^{-3} °C, which is less than day-to-day variations.

Lethal dose $5 \text{ Gy} = 500 \times 100 \text{ ergs / g}$

$$= \frac{500 \times 100}{4.2 \times 10^7} \text{ cal / g}$$

Temperature rise $\approx 1/1000^\circ\text{C}$

On the basis of this argument, Dessauer's theory was disregarded. However, recently Norman and his colleagues have again developed the concept of 'point heat' and proposed the 'thermal spike model'. In the case of densely ionizing radiations, as a result of the high rate of energy deposition, localized heating and temperature rise can occur. Biological materials within this high-temperature cylinder (spike) can undergo thermal inactivation.

1.2.2   Chemical Steps of Interactions

Absorption of radiation by water results in excitation and ionization, which finally results in the formation of radicals such as H°, OH° and hydrated electron (e^-_{aq}). Some of the radicals recombine with each other leading to the formation of H_2 and H_2O_2. Radiolysis of water can be summarized as follows:

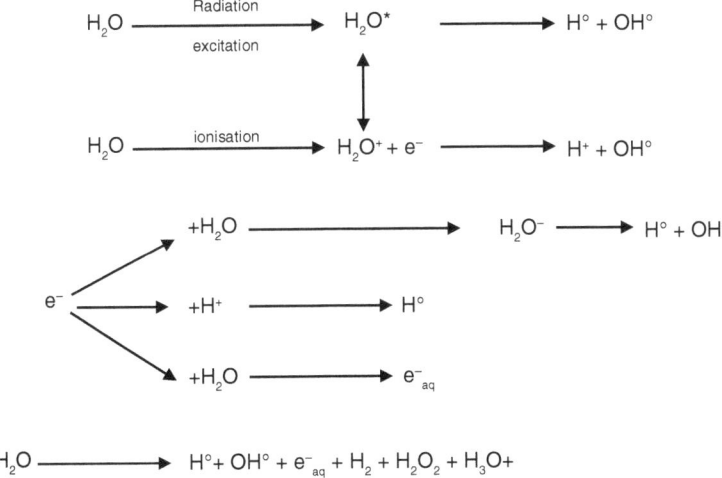

Due to the presence of unpaired electrons, the free radicals are highly reactive. Survival studies carried out in the presence of several radical scavengers such as nitric oxide, mercaptoethanol, nitrite, ethanol, etc. have led to the idea that OH° is responsible for most of the radiobiological effects in microorganisms and mammalian cells. Furthermore, the survival response of several organisms to radiations in the frozen or dry state, which prevents the indirect action, suggests that more than 50% of the lethal effects of low LET radiations are due to indirect action. However, it must be remembered that in the case of more densely ionizing radiations such as alpha rays, neutrons and accelerated charged particles, direct action dominates, and the contribution of indirect action is very limited.

1.2.3   Biological Interactions

Biological interactions of radiation are very slow in nature. It would be more precise to say *interaction of biological system with radiation* rather than interaction of radiation with biological system. In these steps, the life responds to the damage that is inflected by radiation either directly as in physical interactions or indirectly as in chemical interactions. The extent of direct and indirect interactions depends on LET of radiation with high LET radiation predominantly interacting directly and vice versa for low LET radiations. Various mechanisms and types of biological outcome are explained in a separate chapter.

1.3   Atomic Structure

This section reviews the basic composition of atoms, the small particles of which all matter is made. Atoms have a small, dense central core or nucleus containing neutrons and protons. It is proven that majority of atomic mass is concentrated in the nucleus. It is also proven that the nucleus occupies the least volume compared to the volume of atom which makes the density of the nucleus very high. Around this core circulates a cloud of electrons, which are normally considered to occupy a number of shells around the nucleus (Fig. 1.1). Neutrons and protons are called nucleons. They have about the same mass, whereas electrons are much smaller (about 1/1800th of a proton mass). Neutrons have no electric charge. Protons have a positive charge of +1, and electrons have a negative charge of −1. An atom with its complete number of electrons is therefore electrically neutral.

1.3.1   The Electrons

The electrons circulate around the nucleus in different orbits, sometimes called as 'shells'. Each

electron is bound, or held in orbit, by a fixed amount of energy.

The electrons in the inner orbits closest to the nucleus are bound tightest to the atom, whereas the electrons in the outermost orbits are loosely attached. It does not take much energy to strip outer electrons (valence electrons) off the atom; when this happens, the atom is said to be ionized. The atom in such a situation will have a net positive charge because it has lost one or more electrons (Fig. 1.2).

A special unit of energy called the electron volt (eV) is used for atomic and nuclear processes and electromagnetic radiation. [An eV is very small indeed, being the energy acquired by a sin-

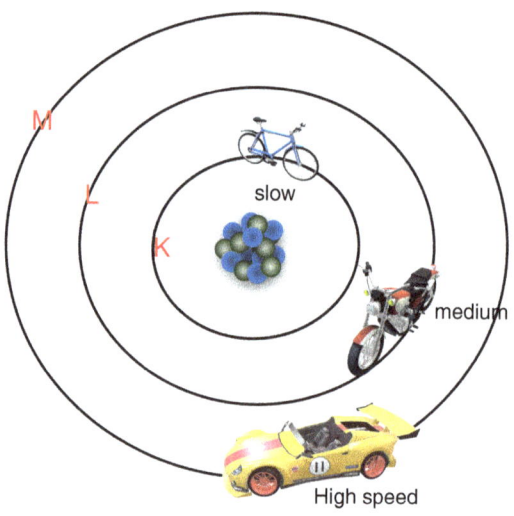

Fig. 1.2 Comparison of electron energy state principle to a vehicle speed

Fig. 1.1 Simplified model of atomic structure with central nucleus surrounded by electrons

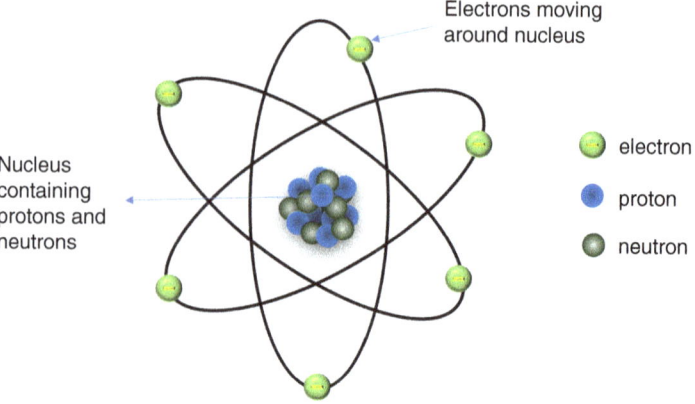

gle electron moving through an electric potential of 1 volt.] The electrons in the outermost shell of the atom are the ones that are involved in chemical reactions, including all the biochemistry in the body. These electrons have binding energies of a few electron volts (eV). In contrast, nucleons are bound together in the nucleus with much greater energies, up to thousands of electron volts (keV) or millions of electron volts (MeV).

1.3.2 The Nucleus

The number of neutrons and protons in the nucleus determines.

(a) The stability of the nucleus
(b) The identity of the material

Any combination of neutrons and protons is referred to as a nuclide. Certain combinations of neutrons and protons result in a stable condition: these correspond to the stable elements. However other combinations of neutrons and protons result in an unstable nucleus which will lose energy in the form of radiation to reach a more stable state: these are radionuclides.

The number of protons in the nucleus also determines the number of orbiting electrons, and it is the number of these electrons and their position in shells that determine the chemical properties of the substance. The number of protons is referred to as the atomic number (Z). The total number of neutrons and protons is referred to as the atomic mass number (A). Normally a nuclide or radionuclide is written in a notation which uses these numbers.

$$_Z^A X$$

where X refers to the specific chemical element.

For example, $_{53}^{131}I$ is the common radionuclide of iodine which has an atomic number 53, which defines the properties of iodine, but a total number of neutrons and protons equal to 131 mean the nucleus is unstable. Normally this is written simply as ^{131}I or Iodine-131.

A section of the periodic table is given in Fig. 1.3, to show how the elements form groups that have similar properties. The examples can be metals, rare gases, etc. which are usually arranged in columns in the periodic table.

Here we are more concerned with the properties of radionuclides, and these can also be listed in a table that demonstrates the number of pro-

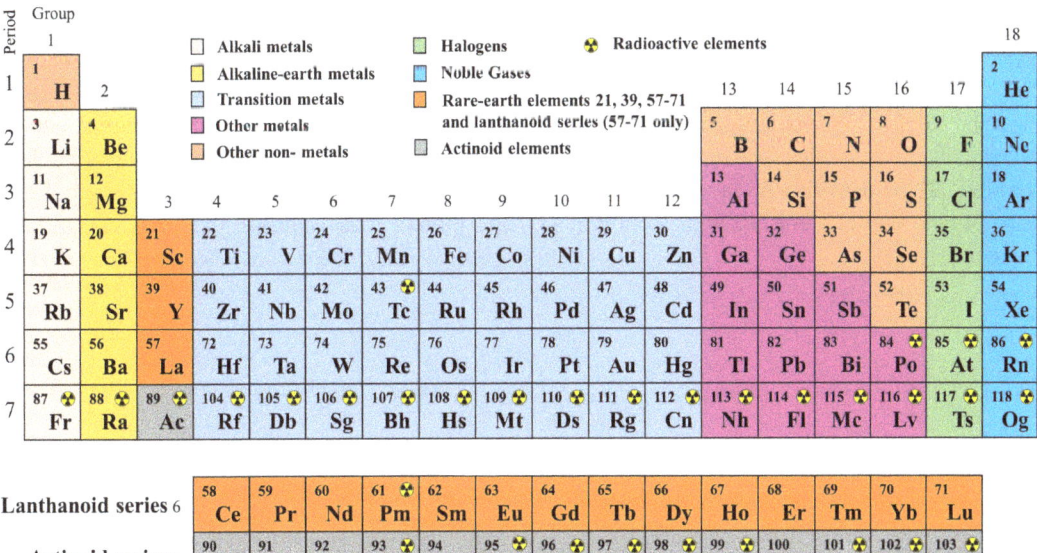

Fig. 1.3 The periodic table of elements

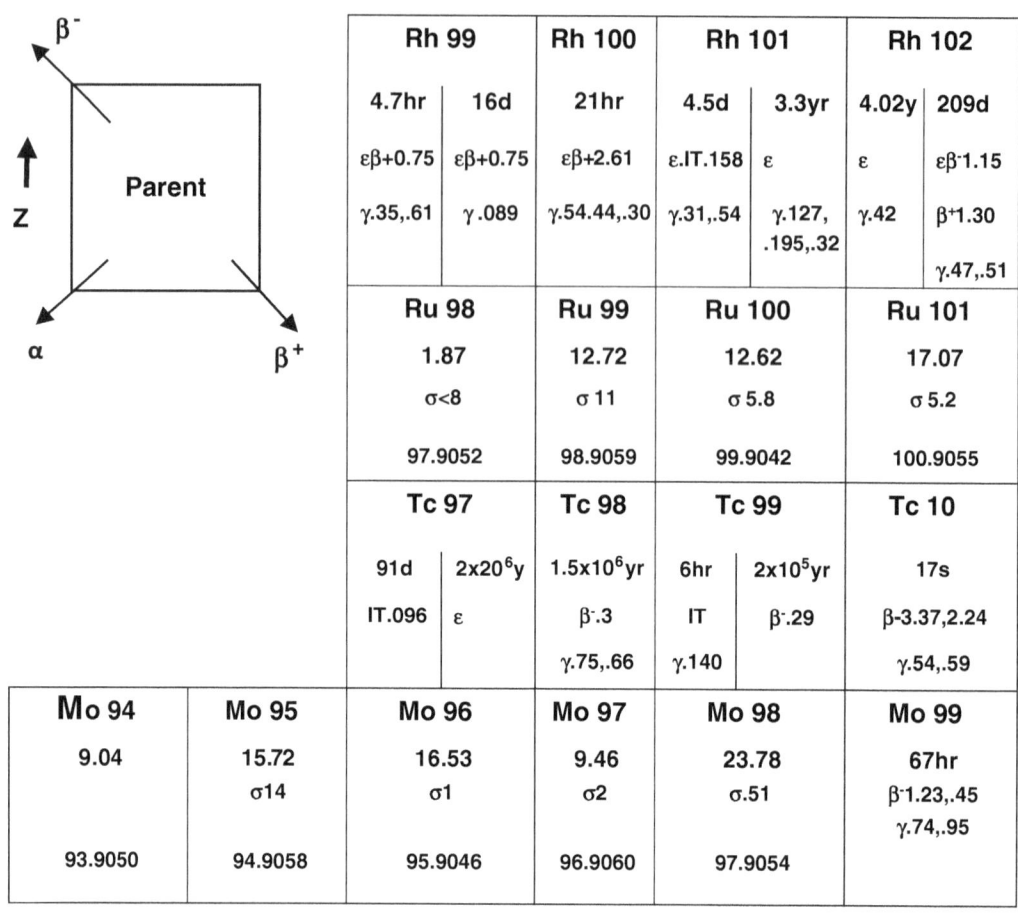

Fig. 1.4 Sample portion of chart of nuclides showing nuclides with the same atomic number with different mass number and nuclear properties

tons and neutrons in the nucleus: This table is called a chart of nuclides (see Fig. 1.4). It can be seen that all the nuclides for an element have the same atomic number, but different atomic mass number (Fig. 1.5).

Note that the Z defines the material and is independent of A. It is most common to define the specific species by the element's symbol and the atomic mass number (since the A will be fixed and known for a specific nuclide).

Fig. 1.5 N/Z ratio for
stable isotopes

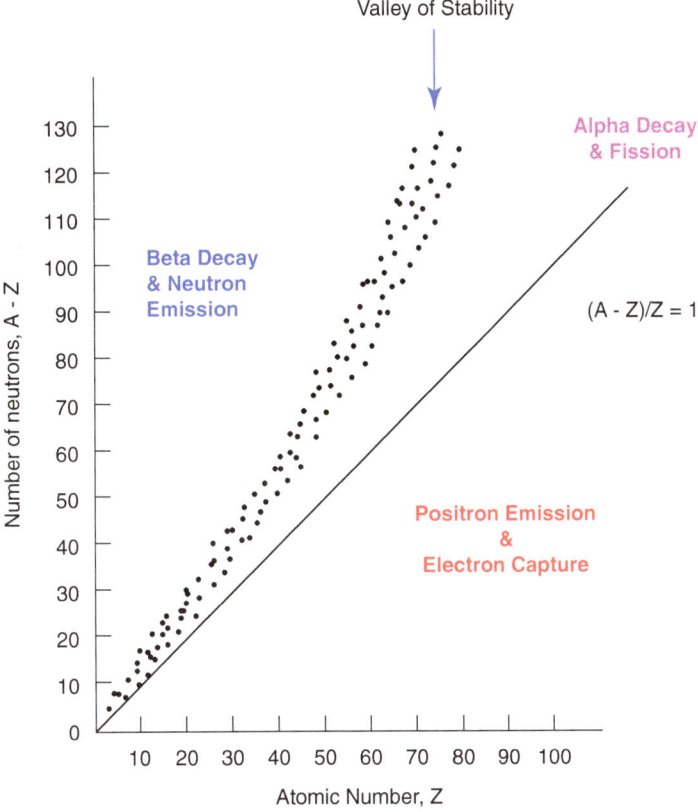

1.4   Properties of Radioactive Materials and Radiation Sources

Radiation can be created either artificially by radiation-generating machines or by isotopes. The difference is that in radiation-generating machines such as X-ray machines and accelerators, the particles derive their energy from the voltage applied or the potential generated, while in isotopes, by virtue of their nuclear property, they give out a particular type or combination of energetic particles by the process of nuclear decay. The isotopes can be naturally existing or man-made by artificial means using nuclear reactors or accelerators by bombarding particles such as neutrons and protons. Various properties of radioactive decay are explained below.

1.4.1   Stability of a Radionuclide

Every element can have different isotopes (same Z different A or number of neutrons N). However, only few combinations of Z and N can lead to stable isotope, while excess of either proton or neutron can yield unstable nucleus leading to radioactive decay [1]. By the process of decay, the radionuclide will try to stabilize itself by moving towards the line of stability as shown below. The dots in the curve show stable region of isotopes. An isotope above this curve (rich in neutrons) tries to reach the curve by β⁻ decay in which a neutron is converted into proton inside the nucleus—thereby N decreases and P increases (Z increases by +1)—while an isotope below this stability region decays by positron emission (β⁺) where a proton is converted into neutron, and the process is reverse of β⁻ decay.

Neutron-rich conditions can be achieved either by fission process wherein fissile materials will naturally have more N/Z stability ratio compared to daughter products which are lower in Z; hence, they tend to be β⁻ active. To get proton-rich conditions, generally proton accelerators are used to bombard stable isotopes and add additional protons in which case the resultant nuclei tend to become β⁺ or positron-emitting type of radionuclides.

1.4.2 Binding Energy

When two nucleons, also known as hadrons (proton and neutrons which are part of nucleus), in numbers P and N together lead to A (P + N), to keep them together in the nucleus, certain amount of binding energy (BE) is involved. Heavier nucleus has more BE, while lighter nucleus will have lesser BE as a thumb rule due to number of hadrons involved. The specific BE is the BE per nucleon, which defines the property of nucleus.

$$\text{Specific BE} = \frac{\text{Total BE}}{\text{Total number of nucleons}}$$

The specific BE curve can be used to understand most properties of nuclear reactions. The curve below depicts variation of specific BE with Z (Fig. 1.6).

The specific BE increases with increasing A initially due to increased nuclear interactions inside the nucleus between nucleons, and it peaks out around the element Fe and then starts declining due to instability of a competing proton-proton repulsion. Fusion of two lighter nuclei increases the specific BE product nuclei, while fission of heavier nuclei also increases the specific BE as the fission products move towards more a stable region. In both the processes, the excess BE will be released as reaction energy.

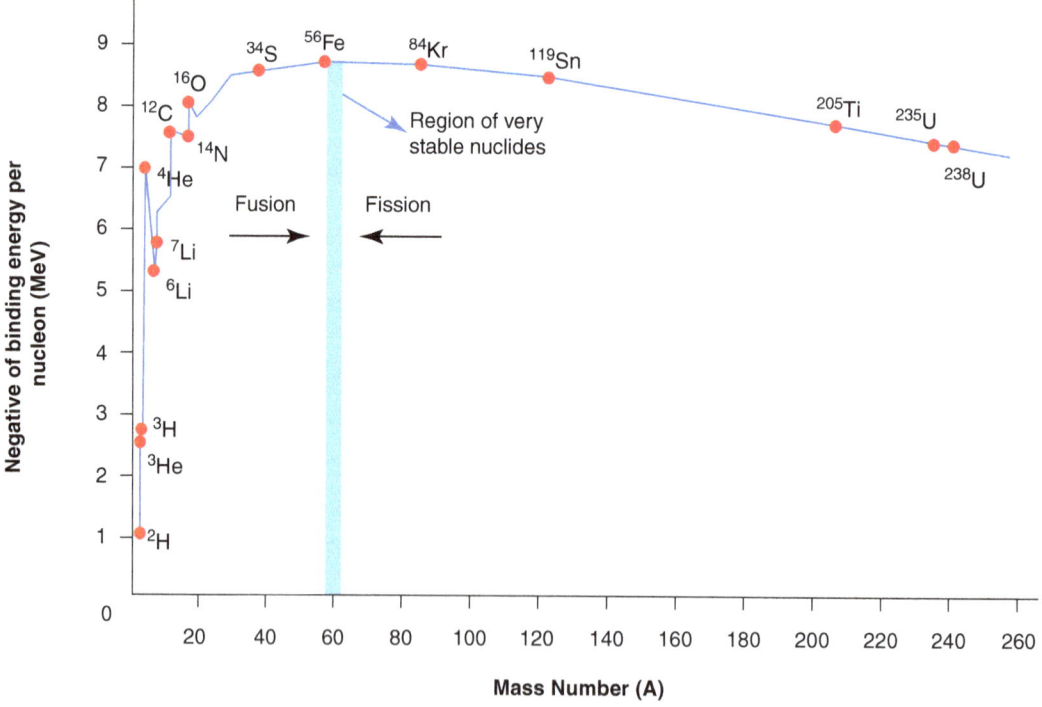

Fig. 1.6 Variation of binding energy per nucleon with atomic mass A

Fig. 1.7 Representation of radioactive decay using a decay scheme

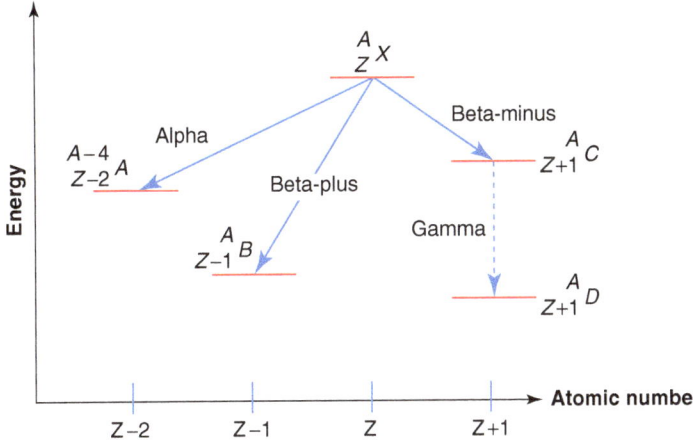

1.4.3 Radioactive Decay Scheme

The nucleus has well-defined energy levels, rather like the electron energy levels of the atom. When some radionuclides decay, the nucleus may transform straight into the lowest energy or 'ground state' of the daughter nucleus, which is stable, and that is the end of the disintegration process. But with many radionuclides, the transformation goes to one or more 'excited states' of the daughter nucleus, which then emits gamma rays of exactly the right energies to get to the ground state.

Radioactive decay can be a very complex process. One way of showing the details in a more easily understood format is to use a decay scheme (Fig. 1.7), which describes the transformation of the radioactive nucleus (parent nucleus) to another (daughter nucleus). The processes in the decay scheme are represented by arrows. If a nucleus is unstable, the nucleus will have an associated energy level, and the result of radioactive decay is that the nucleus will attempt to reach a stable lower energy, with the difference in energy being transferred to the emitted radiation. In the following diagram, horizontal lines represent the energy levels, from the highest (unstable) level of the parent nucleus to the lowest energy level of the daughter nucleus, sometimes via intermediate energy levels. Usually this results in an instantaneous (or 'prompt') further decay with release of energy in the form of a gamma photon (although this is not always the case). The intermediate level simply involves the structural organization of the nucleus so there is no further conversion of neutrons to protons or protons to neutrons. A gamma ray is simply a small burst or packet of energy rather than a physical particle.

The direction of arrows indicates the change in atomic number that occurs as a result of the decay, with either an increase, decreases or, in the case of gamma emission, no change in the atomic number (and therefore element name). The energy levels are usually also indicated so that the energy of the emission is known (being the difference between the energy levels). There may be more than one possible route for a radionuclide to reach the stable state. In this case it is normal to also indicate the probability or % chance of this happening.

1.5 Radioactive Decay and Decay Series

Radioactivity is a phenomenon in which an unstable nucleus of an element disintegrates with the emission of energy and becomes a new element. Henri Becquerel, a French physicist, discovered in 1896 that a compound of uranium emitted some invisible radiant energy. Radioactive decay involves a transition from the original nuclide to the product nuclide. The energy difference between the two states of

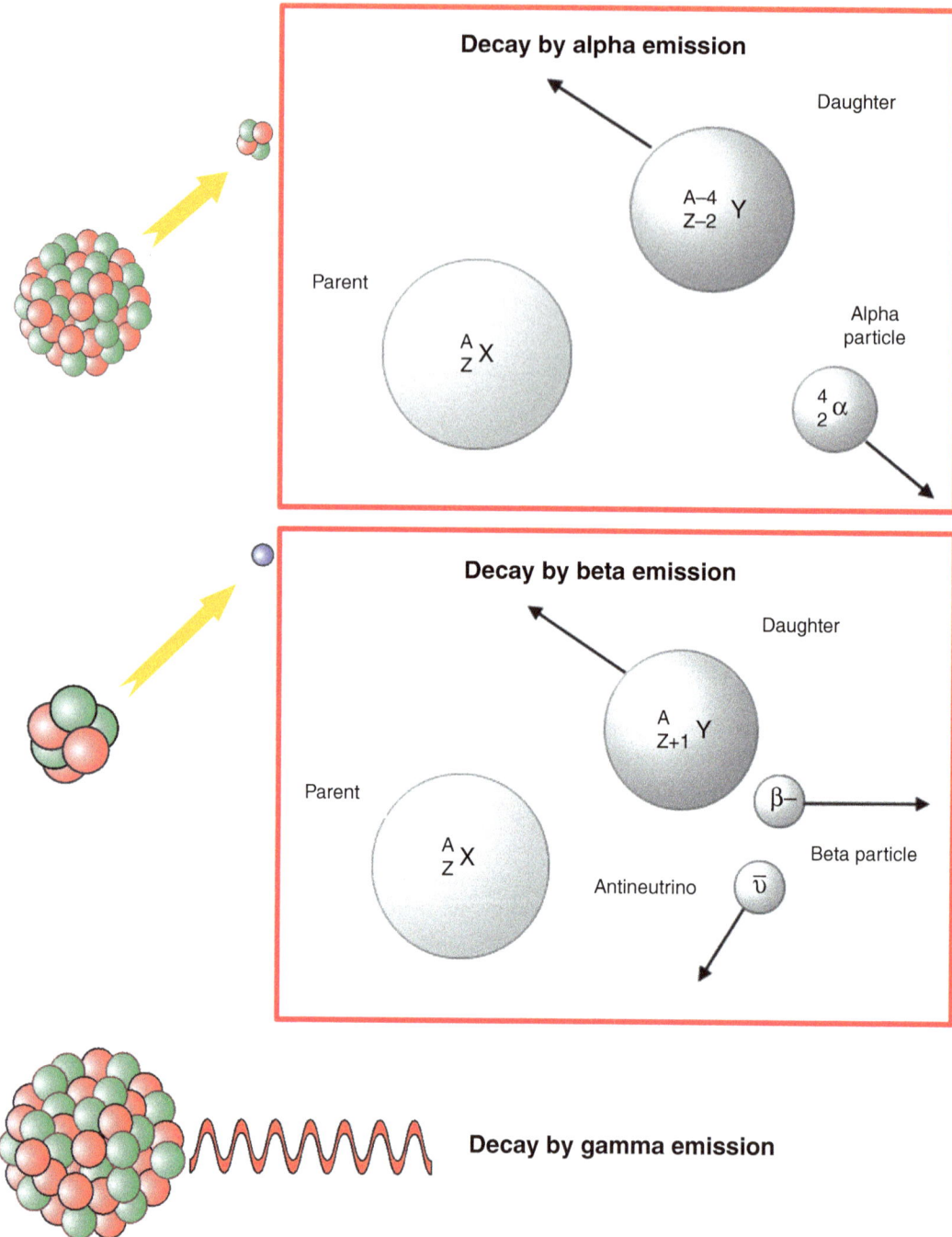

Fig. 1.8 Representative decay schemes of alpha, beta and gamma emissions

products involved in the transition corresponds to the decay energy. This decay may appear in the form of electromagnetic radiation and/or as the kinetic energy of the products involved in the reaction. The mode of decay is dependent on the particular nuclide involved. The radioactive decay can be characterized by α, β and γ radiation [2]. The following diagram summarizes properties of these decays (Fig. 1.8 and Table 1.3).

Table 1.3 Summary of properties of parent and daughter radionuclides

Decay mode	Characteristics of parent radionuclide	Change in atomic number (Z)	Change in atomic mass	Comments
Alpha	Neutron poor	−2	−4	Alphas monoenergetic
Beta	Neutron rich	+1	0	Beta energy spectrum
Positron	Neutron poor	−1	0	Positron energy spectrum
Electron capture	Neutron poor	−1	0	K-capture; characteristic X-rays emitted
Gamma	Excited energy state	None	None	Gammas monoenergetic
Internal conversion	Excited energy state	None	None	Ejects orbital electrons, characteristic X-rays and auger electrons emitted

1.5.1 Basic Concepts of Radioactivity

The activity $A(t)$ of a radioactive substance at time t is defined as number of disintegrations per second.

$$A(t) = |\,dN/dt\,| = \lambda N(t)$$

The simplest radioactive decay is characterized by a radioactive parent nucleus P decaying with a decay constant λ_p into a stable daughter nucleus D, i.e.

$$P \xrightarrow{\lambda_p} D$$

The activity of parent nuclei at time t is given as:

$$A_p = A_p(0)e^{-\lambda_p t}$$

where $A_p(0)$ is the initial activity of parent nuclei at time $t = 0$.

Thus, both the number of parent nuclei N_p and activity of parent nuclei A_p decrease exponentially with time.

1.5.2 Positron Emission

For radionuclides that are proton rich, or cyclotron produced, there are two common modes of decay: (i) By emission of positrons effectively which changes a proton into a neutron. Positrons have similar mass to electrons but a charge of +1 instead of −1. In the case of a positron, it does not last long in the real world

before it combines with an ordinary electron, annihilating them both and releasing energy as gamma rays that travel in opposite directions (Fig. 1.9).

1.5.3 Electron Capture

In electron capture the nucleus absorbs an electron from an inner electron shell, which also converts a proton to a neutron, usually with additional release of energy in the form of gamma rays. This process is usually a competing process for β + emission.

There are many electron capture radionuclides in nuclear medicine. They include ^{125}I, ^{123}I, ^{201}Tl, ^{67}Ga and ^{111}In.

1.5.4 Alpha Emission

In the case of radionuclides that have excess protons and neutrons, these usually decay by emission of alpha particles, which are like helium nuclei with mass of 4 and charge of +2. These are less common in nuclear medicine although recently have been used for therapy.

Before discussing gamma emission in more detail, it is useful to consider the decay schemes outlined below (Fig. 1.10).

When radioactive decay occurs by positron or beta emission or by electron capture, the result is the formation of a nuclide of a different element. This is demonstrated by the following diagram (Fig. 1.11) which summarizes the transitions that would occur for different modes of decay.

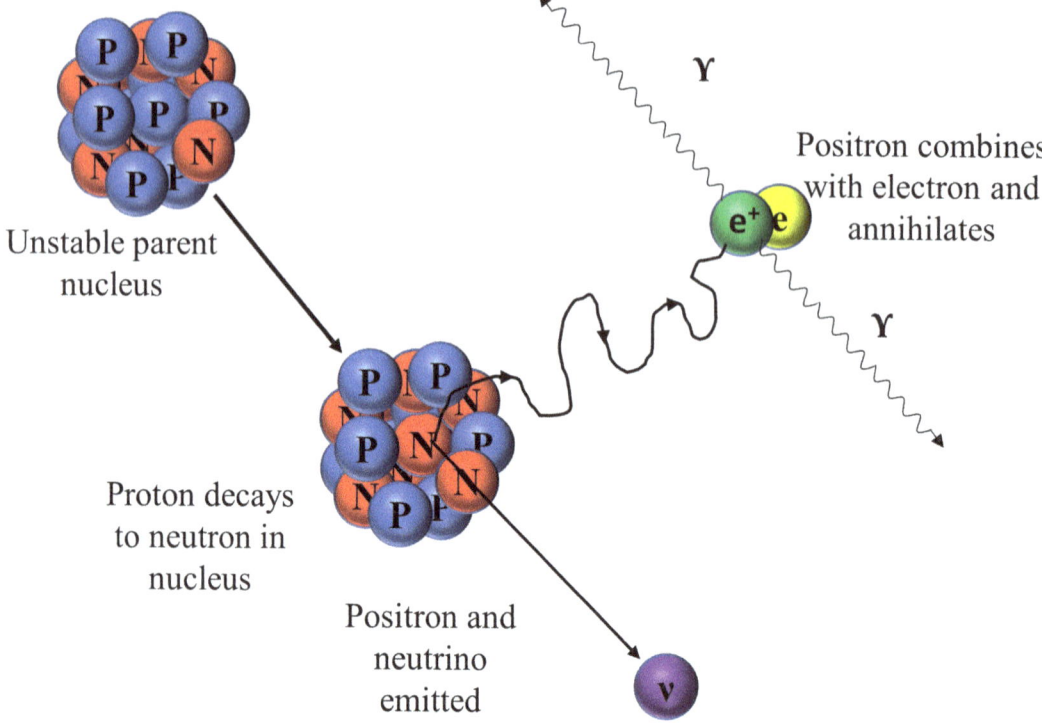

Unstable parent
nucleus

Positron combines
with electron and
annihilates

Proton decays
to neutron in
nucleus

Positron and
neutrino
emitted

Fig. 1.9 Annihilation of positron and electron produces two 511 keV gamma photons

Fig. 1.10 Decay
scheme showing the
shift on Z for the three
decay methods

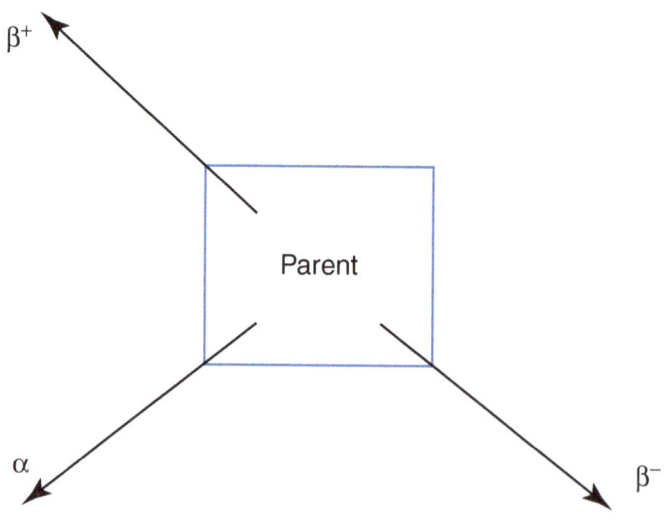

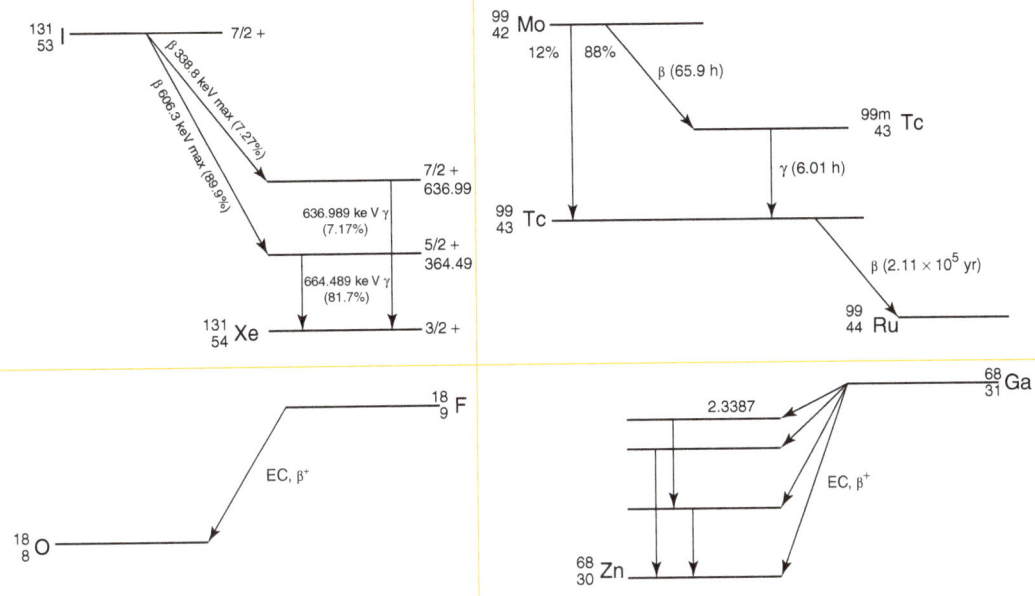

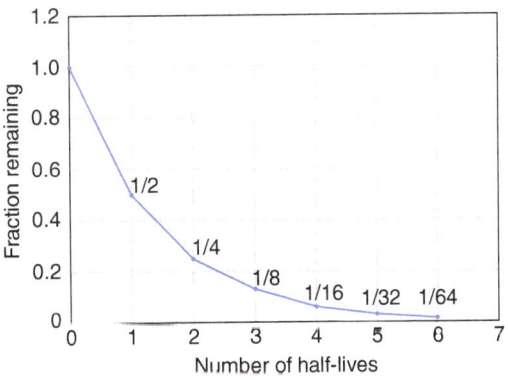

Fig. 1.11 Decay schemes of common radionuclides used in nuclear medicine

1.6 Concept of Half-Life

Half-life $t_{1/2}$ of a radioactive substance is the time during which the number of radioactive nuclei decays to half of the initial value $N(0)$ present at time $t = 0$.

After one half-life, the fraction of activity remaining is 1/2. After 'n' such half-lives ($t = n \times t_{1/2}$), the fraction of activity remaining is $(1/2)^n = 1/2^n$. Although this fraction may become very small, it can theoretically never fall to zero (Fig. 1.12).

The decay constant and half-life t are related as follows:

$$\lambda = 0.693 / t_{1/2}$$

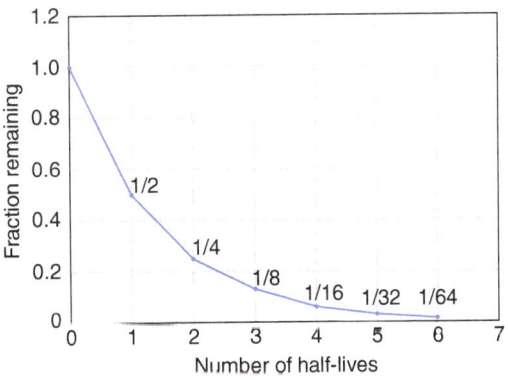

Fig. 1.12 Concept of half-life

Table 1.4 Commonly used radionuclides and their half-lives

Radionuclide	Half-life
Phosphorus-32	14.3 days
Iridium-192	74 days
Cobalt-60	5.25 years
Caesium-137	30 years
Carbon-14	5760 years
Uranium-238	4.5×10^9 years

1.7 Specific Activity

Specific activity 'a' is defined as the activity per unit mass, $a = A/m$. For short half-life radionuclides, it is easier to obtain large specific activities and vice versa (Table 1.4).

$$\text{Mean Life} = \frac{\text{Total life (time of all the radioactive atoms)}}{\text{Total number of such atoms in it}}$$

$$\text{or} \quad \tau = 1.44 t_{1/2}$$

Mean lifetime (τ) can also be defined as the average time of existence or the life expectancy of a radioactive nucleus before decay. The mean life is the time required for the number of atoms or their activity to fall to $1/e = 0.368$ of its initial value.

The unit of radioactive decay constant λ is s^{-1}. The disintegration law $N = N_o e^{-\lambda t}$ applies universally to all radioactive nuclides. But the radioactive constant λ is different for each nuclide. The λ values of the known radioactive nuclides (from the naturally occurring radioactive series of elements) extend between $\lambda = 2.31 \times 10^8$ s^{-1} (for ^{212}Po; $t_{1/2} = 3 \times 10^{-7}$ s) and $\lambda = 1.58 \times 10^{-18}$ s^{-1} (for ^{232}Th; $t_{1/2} = 1.4 \times 10^{10}$ y), a range of over 10^{24}.

1.8 Average (Mean) Life

Average or mean life 'τ' of a radioactive substance represents the average life expectancy of all radioactive atoms in the substance at time $t = 0$. The decay constant and average life are thus related as follows:

$$\tau = 1 / \lambda$$

It is also possible to determine the mean life or average life expectancy of the atoms of a radioactive species. Mean life (τ) of a radioactive substance is defined as the ratio of the total lifetime of all the radioactive atoms to the total number of such atoms in it.

1.9 Successive Radioactive Transformation: Radioactive Decay Chains

A very often encountered situation is a radioactive decay chain in which a nuclide decays to a daughter nucleus which itself disintegrates to another unstable nucleus and so on. In the simple case of three nucleus chain, i.e.

$$A \xrightarrow{\lambda_1} B \xrightarrow{\lambda_2} C \,(\text{stable})$$

where 'λ_1' and 'λ_2' are the decay constants of atoms of type A and type B and N_1 and N_2 respectively represent the number of atoms of each type that are present at any time t.

Successive radioactive decay has led to four series of decay chains. They are classified based on the initial radionuclides, i.e. uranium series with ^{232}U, thorium series with ^{232}Th, neptunium series with ^{237}Np and actinium series with Ac^{235}U. Among these, neptunium series is considered as artificial series.

1.10 Artificial Sources of Radiation

Radiation can also be generated by artificial methods using devices such as X-ray tubes and accelerators. Typically, a charged particle such as ion or electrons is used as initial source of particles and subjected to a potential difference generated by various methods. Devices generating low energy are usually simple ones with the potential directly applied between the terminals as in X-ray tubes, while higher energy ones rely upon complex principles of successive accelerations as in cyclotrons, betatron, microtron, linac and tandem pelletron. Cyclotrons are for production of positron-emitting radionuclides, while other methods such as neutron bombardment in reactor

or extraction of fission products are used in radiotherapy for artificial production of radionuclides.

1.10.1 X-Rays

Sometimes the decay of a radionuclide leaves a vacancy in an inner electron shell of the daughter nuclide. This happens after electron capture or internal conversion. The vacancy is filled by another electron from higher electron orbits, which are further away from the nucleus. This results in a release of energy in the form of characteristic X-ray as shown in Fig. 1.13.

This process has one very important practical application in nuclear medicine. A number of useful radionuclides (e.g. ^{201}Tl) decay by electron capture, which is followed by X-ray emission. (There may also be gamma rays.) X-rays usually have much lower energies than gamma rays and are generally of no use for scanning. One exception is ^{201}Tl. Because it is such a heavy element, the X-rays from its daughter, mercury-201, are about 68–80 keV. These ^{201}Hg X-rays are less than the ideal nuclear medicine energy of about 140 keV, but they are still useful for scanning

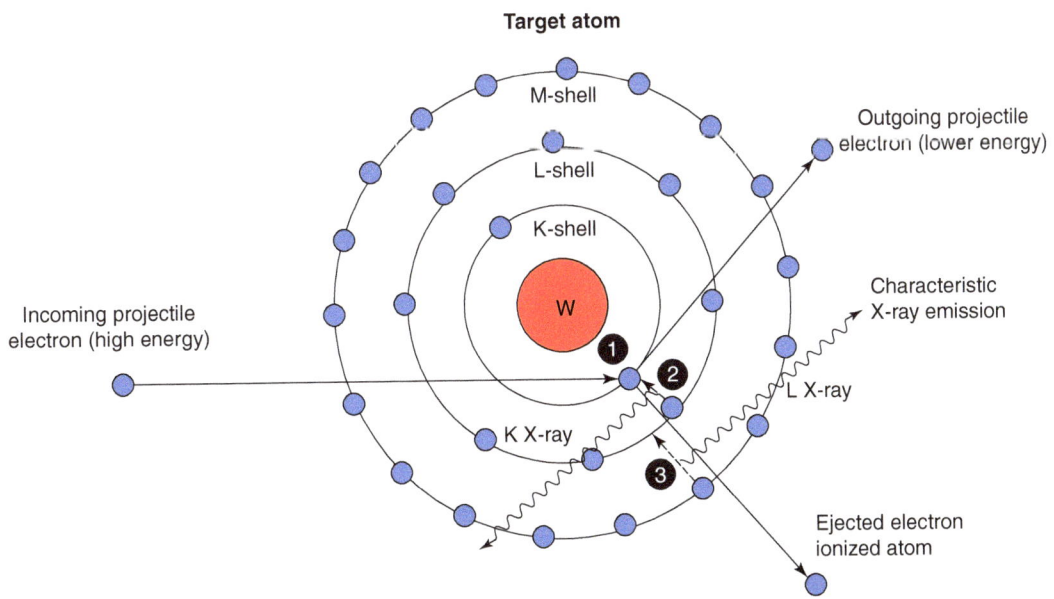

Fig. 1.13 Characteristics of X-ray

because ^{201}Tl itself emits very few gamma rays (in only 10% of disintegrations, with energy of 167 keV).

Sometimes, electrons from higher orbits may fill electron shell vacancies, and the excess energy may get transferred to another electron of the same atom leading to discrete and low-energy electron emission. The kinetic energy of these electrons corresponds to initial transition energy minus the binding energy of the electron emitted. These are called Auger electrons.

Characteristic X-rays have many applications. They are popularly used in elemental quantifications by X-ray fluorescence method. In these applications, high energy X-rays are used for bombarding the material of interest, and the emit-ted characteristic X-rays are analysed for their energy spectra. In some advanced applications, proton is used instead of X-ray to get very high signal-to-noise ratio, which facilitates detection of very low concentration of elements. This method is called proton-induced X-ray emission or PIXE. Characteristic X-rays also find applications demanding low-energy X-ray radiographs.

References

1. Evans RD. The atomic nucleus. New York: McGraw-Hill Inc.; 1955.
2. Beiser A. Concepts of modern physics. 6th ed. New York: McGraw-Hill Inc.; 2003. ISBN: 0-07-244848-2

Abstract

A unit is necessary for the measurement of any physical quantity. The International Commission on Radiation Units and Measurement (ICRU) reviews and updates, from time to time, the concepts related to quantities and their units in radiation physics, dosimetry and radiological protection.

Some of the quantities of interest are activity, air Kerma, exposure, absorbed dose, effective dose, equivalent dose, collective equivalent dose, annual limit of intake and derived air concentration. In 1980, the ICRU recommended SI units for the above quantities. These new units along with the corresponding old units are discussed in the following paragraphs.

2.1 Activity, 'A'

The activity, 'A', of radioactive material is a measure of its spontaneous transformation. It is defined as the average number of spontaneous nuclear transformations (or disintegration) taking place per unit of time.

The special name of the unit of activity is Becquerel (Bq).

1 Bq = 1 disintegration per second

The old unit of activity is Curie (Ci)

$$1 \text{Ci} = 3.7 \times 10^{10} \text{ disintegration per second}$$
$$= 3.7 \times 10^{10} \text{ Bq}$$
$$= 37 \text{ GBq}$$

Both the indirectly ionizing (photons and neutrons) and directly ionizing (charged particles) radiations transfer part or all of their energy when they interact with matter.

2.2 Kerma, 'K' (*Kinetic Energy Released per Unit Mass*)

The field of indirectly ionizing radiation at any point in matter is given by the quantity Kerma, 'K', which is defined as the sum of the initial kinetic energies of all the charged particles liberated by radiation in material of mass 1 kg.

The SI unit of Kerma is Gray and 1 Gy = 1 J kg^{-1}

When the reference material is air, the quantity is called air Kerma.

2.3   Exposure, 'X'

Exposure, 'X', is defined as the absolute value of the total charge of the ions of a single sign produced in the air when all the secondary electrons (inclusive of positrons and electrons) are liberated by photons in the air of mass Δm and are completely stopped in the air.

The unit of exposure is C kg^{-1}

With the present technique, it is difficult to measure exposure when photon energies involved lie above a few MeV or below a few keV.

The unit of exposure in use is Roentgen, 'R', which is defined as the amount of gamma radiation that would liberate 1 esu of charge of either sign in 1 cm^3 of air at STP.

$$1R = 1 \text{ esu of charge liberated per cm}^3 \text{ of air at STP}$$
$$= 2.58 \times 10^{-4} \text{C kg}^{-1}\left(\text{air}\right)$$

Except at very high energies, the exposure defined above is the ionization equivalent of the air Kerma. And by definition exposure is a quantity restricted to photons and air as the medium.

2.4   Dose, 'D'

The effects (physical, chemical and biological) of radiation depend not only on the energy transferred to the medium but also on the energy absorbed by it. The quantity absorbed dose (or simply dose) is defined as the amount of energy absorbed per unit mass of the medium at the point of interest.

The SI unit of dose is Gray (Gy) and 1 Gy = 1 J kg^{-1}

The old unit of dose is rad which is equal to energy absorption of 100 ergs per gram of material.

1 rad = 100 ergs g^{-1}
1 rad = 10^{-2} J kg^{-1}
1 rad = 10^{-2} Gy

2.5   Equivalent Dose, 'H$_T$'

Effects of radiation depend not only on the amount of energy absorbed but also on the spatial distribution of ion pairs. Hence, the biological damage caused by the same dose of different radiations may be different, if they have different rates of energy loss per unit of path length, which in other terms is referred to as linear energy transfer (LET). Alpha particles, because of their high energy, charge and mass, cause greater ionization per unit path length than gamma radiations, which mediate through singly charged electrons. One Gray of the alpha dose is found to be more effective than one Gray of the gamma dose. Hence, in radiation protection, to account for this variation in the effectiveness of different types of radiation, the radiation weighting factor (W_R) is used to multiply the absorbed dose due to each type of radiation. The weighted absorbed dose is called equivalent dose H_T.

That is, $H_T = \Sigma_R \times D_{T,R}, W_R$, where $D_{T,R}$ is the absorbed dose in tissue for radiation R of radiation weighting factor W_R.

Since W_R is a dimensionless quantity, the unit of dose equivalent is also J kg^{-1}. The radiation weighting factor was formerly called the quality factor (QF).

The special name for the unit of equivalent dose is Sievert (Sv).

Table 19.4 at Chap. 19 shows W_R values for different types of radiations as per ICRP-103. The values of W_R reflect the relative biological effectiveness (RBE), a term used in radiobiology, for different types and energies of radiation in the production of stochastic effects.

For radiation protection purposes, 20 mGy of gamma dose, 1 mGy of alpha dose and 2 mGy of fast neutron dose are equivalent. It should be noted that this is true only for low equivalent dose levels such as millisievert (mSv) and centisievert (cSv) and applicable only for radiation protection purposes.

Table 2.1 Old and new units

Quantity	Old unit	SI unit	Relationship between units
Radioactivity	Ci (curie)	Bq (becquerel)	$1\text{ Bq} = 2.7 \times 10^{-11}\text{ Ci}$
Exposure	R (roentgen)	C kg^{-1} (coulombs per kg)	$1\text{ R} = 2.58 \times 10^{-4}\text{ C kg}^{-1}$
Dose	rad	Gy (gray)	1 gray = 100 rads
Equivalent dose	rem	Sv (sievert)	1 Sv = 100 rems
Effective dose	rem	Sv (sievert)	1 Sv = 100 rems

Equivalent dose in Sv = Dose in Gy $\times W_R$

Formerly, the unit of dose equivalent was roentgen equivalent man (rem).

Dose equivalent in rem = Dose in rad $\times W_R$

In radiation protection 1 Sv is too large a quantity. Hence, dose equivalent is expressed in units of mSv (10^{-3} Sv). On the other hand, Bq is too small a quantity for many applications. Hence, the amount of radioactivity is expressed in MBq (10^6 Bq), GBq (10^9 Bq), etc. Table 2.1 gives the radiation quantities and their units.

2.6   Effective Dose, '*E*'

Radiation exposure may occur to the whole body (uniform irradiation) or to individual organs of the body (non-uniform irradiation). Non-uniform irradiation will have to be restricted to avoid not only deterministic effects but also stochastic effects. The ICRP recommends dose limits (DL) for stochastic effects and deterministic effects.

If several tissues, T_1, T_2, T_3, etc., individually receive equivalent doses, H_{T1}, H_{T2}, H_{T3}, etc., then the total risk to the individual should not exceed that resulting from the stipulated dose limit to uniform whole-body irradiation. A number of organs are considered based on their sensitivity and the seriousness of the damage. Risk factors are age- and sex-dependent. Depending on the extent to which the risk from stochastic effects in a tissue/organ may contribute to the total risk from stochastic effects, a weighting factor called the tissue weighting factor, W_T, is assigned to

each tissue/organ as given in Table 19.5 (at Chap. 19) as per ICRP-103. Thus, the effective dose (E) is defined as,

$$E = \sum_T W_T H_T$$

W_T represents the contribution of tissue and T is the total risk due to stochastic effects resulting from uniform irradiation of the whole body.

2.7   Collective Effective Doses, '*S*'

The ICRP uses further quantities related to the exposed group or population. These quantities take account of the number of people exposed to a source by multiplying the average dose to the exposed group from the source by the number of individuals in the group. The relevant quantities are the collective equivalent dose S_T, which relates to a specified tissue or organ, and the collective effective dose S. The unit of this collective quantities is the person Sievert.

2.8   Annual Limit on Intake (ALI)

ALI means the greatest value of the annual intake of the specified radionuclide which will result in a committed dose equivalent not exceeding the annual dose equivalent limit, prescribed by the competent authority, even if intake occurred every year for 50 years. Important ALI values are given in Table 2.2.

Table 2.2 ALI values of radioisotopes for inhalation

Radionuclide	ALI (MBq)
I-123	200
I-125	2
I-131	1
I-132	200
Co-57	30
Cr-51	300
Ga-67	200
Tl-201	400
Hg-197	100
Hg-203	20
Mo-99	50
Tc-99m	2000
Y-90	8
H-3	1000
In-111	100
P-32	10

Generally, ALI values for inhalation are 10 times more than that of ingestion

2.9 Derived Air Concentration (DAC)

DAC means the maximum concentration of a radionuclide in the ambient air which, if inhaled by a person for 2000 hrs in a year, at a breathing rate of 1.2 m³/h, will not result in an annual effective dose equivalent above the limits prescribed by the competent authority.

$$DAC = ALI / 2.4 \times 10^3 \, Bq / m^3$$

Reference

1. ICRP-103: The 2007 Recommendations of the International Commission on Radiological Protection.

Abstract

The kinetic energy of radiation released during radioactive decay is measured in electron volts (eV). An electron volt is the kinetic energy gained by an electron when accelerated by a 1 volt potential difference. A volt of electrons is a tiny unit. The energy of various radiations released during radioactive decay (including X-rays) will be substantially greater; hence they are measured in kilo (10^3) electron volts (keV) or million (10^6) electron volts (MeV). In contrast, visible light has an energy of 1 to 4 eV.

The radiation released by a radioisotope is invisible to the naked eye and cannot be felt by the human body. They interact with the atoms when they strike matter to cause excitations and ionizations. Excitation is raising an atom's orbital electron to a higher energy state, while ionization is the process of removing one or more electrons from an atom, resulting in the formation of an ion pair, a positive and a negative ion. Both of these processes result in energy being transferred from radiation to matter. Ionizing radiations are so named because they have the capacity to ionize materials.

The biological, chemical and physical impacts of radiation are ultimately due to ionization. This feature of radiation is utilized to detect and measure ionizing radiations. The interaction of ionizing radiation with matter is covered in Sect. 1 of this chapter.

Electromagnetic and particulate ionizing radiations are the two types of ionizing radiation where charged and uncharged particles may be found in particle radiation. In addition, Sect. 2 of this chapter briefly discusses the different methods for producing radionuclides utilized in nuclear medicine.

3.1 Section 1: Interaction of Radiation with Matter

3.1.1 Interaction of Charged Particles

Alpha and beta particles are known as charged particles. Collisions with electrons and nuclei of the atoms in the medium cause these charged particles to lose energy. Along the route of a charged particle in the medium, this causes excitation and ionization. Ionization occurs when the energy imparted to an electron in the medium is sufficient to remove it from the atom entirely. Excitation is the process of raising an electron to a higher energy level. The ionization of an alpha particle is seen schematically in Fig. 3.1.

Linear energy transfer refers to the amount of energy absorbed in the particle's medium per unit path length (LET). It's commonly measured in keV/m.

Because biological effects are dependent on the rate of energy absorption in the medium, the

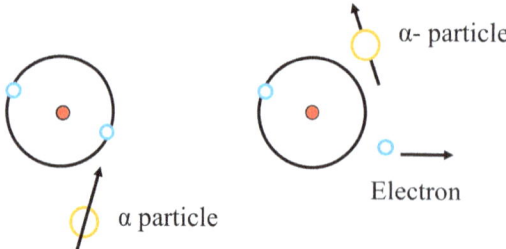

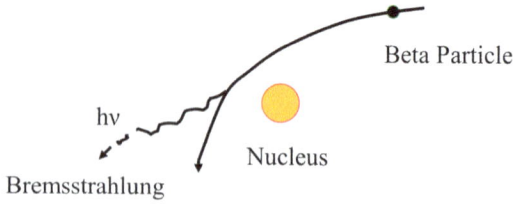

Fig. 3.2 Process of bremsstrahlung

Fig. 3.1 Ionization by alpha particles. Specific ionization is the number of ion pairs created per unit travel length of a charged particle. The specific ionization is related to the particle's mass and charge but is inversely proportional to its velocity

idea of LET is critical. The specific ionization of alpha particles is high because they are doubly charged and have a relatively large mass. As a result, these processes cause alpha particles to lose energy in the matter rather quickly. Because alpha rays from the same radionuclide release the same amount of energy, their range in a given material will be similar. The alpha particle's range is commonly measured in centimetres of air.

On the other hand, beta particles have a mass of 1/7300 that of an alpha particle and contain a single charge. As a result, when beta particles enter any substance, they disperse more and take a more circuitous route. They get deeper into the material. The LET and specific ionization of beta particles is lower than those of alpha particles. The energy of beta particles and the density of the medium determine the range of beta particles in any medium. The connection is an empirical relationship between the range (in g/cm² of aluminium) and electrons' energy (in MeV) with energies greater than 0.6 MeV.

$$R_{max}\left(g\,/\,cm^2\right)=0.53\,E_{max}-0.106$$

3.1.1.1 Radiative Collision

As seen in Fig. 3.2, when a fast-moving charged particle passes near a nucleus, it deflects and loses energy in electromagnetic radiation. Bremsstrahlung is the term for the radiation released as a result of this process. This procedure generates continuous X-rays from X-ray

equipment. In the shielding of high-energy beta particles, the formation of bremsstrahlung is of significant concern. The intensity of bremsstrahlung rises with the medium's atomic number and decreases as the particle's mass increases. As a result, radiation-induced energy loss is more critical in heavy elements than in light particles like electrons.

3.1.1.2 Range of Charged Particles in Matter

A charged particle comes to rest in the medium after losing its kinetic energy. The range of a particle is the distance it travels before coming to a stop. The energy, charge and mass of a particle and the density and atomic number of the medium determine its range. Because alpha particles lose energy quickly, they can only travel short distances, limiting their penetrating potential. A 5 MeV alpha particle has a range of less than 5 cm in the air. Even a thin sheet of paper may stop these particles and cannot penetrate human skin. Because beta particles have a smaller mass and charge, they lose less energy during each impact than alpha particles with the same starting kinetic energy, resulting in a greater range. In the air, a 1 MeV beta particle has a range of almost 40 cm and 0.5 cm in tissue. Table 3.1 shows the beta particle range released by some of the most often employed radionuclides in nuclear medicine.

3.1.2   Electromagnetic Radiations

This is a kind of wavelike disruption that occurs when electric charges vibrate. Electromagnetic radiations include radio waves, infrared radiations, visible light, ultraviolet radiation, X-rays

Table 3.1 Range of beta particles emitted by important radionuclides

Radionuclides	Maximum energy (MeV)	Range in medium	
		Tissue	Air
Phosphorus-32 (^{32}P)	1.71	0.8 cm	620 cm
Yttrium-90 (^{90}Y)	2.284	1.1 cm	1062 cm
Lutetium-177 (^{177}Lu)	0.497	1.7 mm	126.65 cm

Fig. 3.3 Illustration of photon interactions

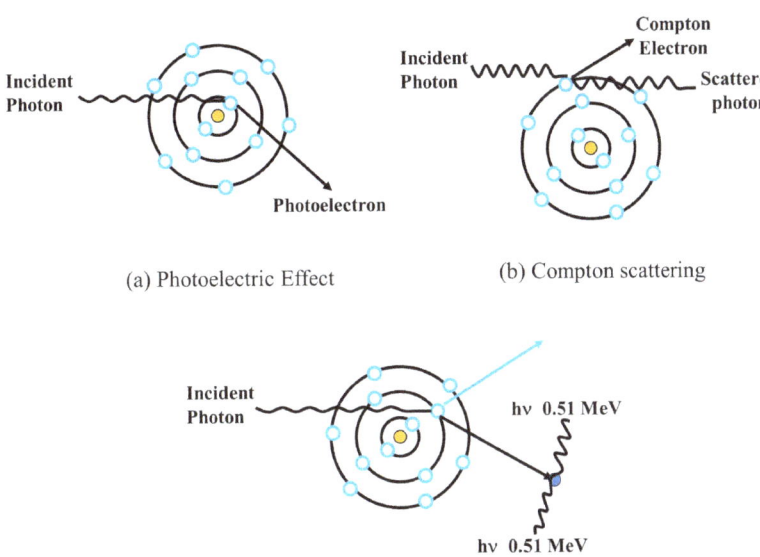

(a) Photoelectric Effect

(b) Compton scattering

(c) Pair Production

and gamma rays. They simply vary in terms of wavelength or frequency. The relation establishes a link between wavelength and frequency.

$$C = \lambda \gamma$$

where λ is the wavelength in centimetres, γ is the frequency in hertz and C is the speed of electromagnetic radiation in a vacuum (3×10^8 m/s). The relationship gives the energy of electromagnetic radiation.

$$E = h\gamma$$

where h is referred to as Plank's constant.

Only X-rays and gamma rays have enough energy to ionize matter out of all electromagnetic radiations. Photoelectric absorption, Compton scattering and pair production are the three basic mechanisms by which gamma and X-rays (also known as photons) interact with matter.

3.1.2.1 Photoelectric Absorption (Effect)

All the energy of the incoming photon is transferred to an atomic electron, which is then released from its parent atom in the photoelectric process. The photon has been absorbed fully. With the production of distinctive X-rays, an outer orbital electron may replace the vacancy formed by the expelled electron. There's also the chance of an Auger electron being generated by the atom's internal absorption of certain X-rays. The photoelectric effect's probability reduces as the photon's energy rises but increases as the medium's atomic number increases. At energies less than 100 keV, the photoelectric effect is thus the dominating interaction for high atomic number materials such as lead. The photoelectric process is shown schematically in Fig. 3.3(a).

A photon is a unit of energy. With kinetic energy, E will release an electron. $E_e = E-\phi$, where ϕ is the electron's binding energy in that orbit.

The photoelectric process has following distinguishing properties:

1. In the photoelectric process, bound electrons are involved.
2. When the photon energy is slightly greater than the electron's binding energy, the likelihood of ejection is greatest.
3. The photoelectric absorption coefficient changes as $1/E^3$ with energy.
4. The photoelectric absorption coefficient changes roughly with the absorber's atomic number (Z^3).
5. As photon energy rises, photoelectron ejection in the forward direction becomes more likely.

3.1.2.2 Compton Scattering

A portion of the energy of the incoming photon is transferred to a free electron via the Compton process. Free electrons are the atom's outermost electrons, with very low binding energies. Because free electrons are involved in Compton scattering, the process is unaffected by the atomic number of the material in which the interaction occurs. Because many materials have around the same amount of electrons per gram (3×10^{23}), absorption by this method is almost equivalent for materials of identical mass. Figure 3.3(b) shows how the photon sends just a portion of its energy to the electron and scatters with less energy. The Compton electron's energy is eventually absorbed by the medium.

Some salient features of Compton interaction are:

1. Photons and a free electron are involved in this process.
2. In this process, the mass attenuation coefficient is *independent of the Z of the medium*.
3. As the energy of the incident photon increases, the probability of interaction decreases.
4. Depending upon the scattering angle, some energy is absorbed, and the rest is scattered.

5. This interaction is more predominant in soft tissue in 100 keV to 10 MeV in the energy range.
6. The electron will be ejected more in the forward direction as the energy of an incident photon increases and carries a larger portion of the energy.

3.1.2.3 Pair Production

An energy photon may be transformed into a positron-electron pair in the strong electric field around a nucleus. This is referred to as pair production, and it is an example of energy to matter conversion, as depicted schematically in Fig. 3.3(c). The minimum photon energy necessary for this process is 1.02 MeV, with any extra photon energy being divided as kinetic energy between the electron and positron. A positron would collide with an electron after its path. The two particles annihilate, producing two photons with energies of 0.51 MeV each. Increases in photon energy above the threshold and the atomic number of the substance enhance the chance of pair creation. However, at energies below 10 MeV, it is not a significant mechanism of interaction with most materials.

The following are some of the most notable characteristics of pair production:

1. It results from a photon's interaction with the nucleus.
2. This process has a threshold energy of 1.02 MeV.
3. The likelihood of this process increases with the atomic number (Z^2) of the medium.
4. As the photon's energy increases, this process becomes more dominant.

Thus, all three interactions result in the photon energy being transferred to electrons, which subsequently loses energy to the medium.

3.1.3   Attenuation of Gamma Radiation in Matter

When gamma radiation travels through matter, it is likely to be exposed to all three interactions to

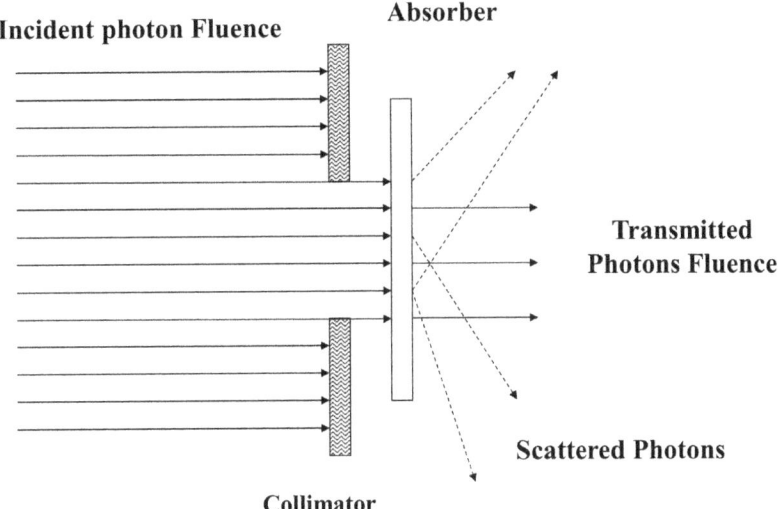

Fig. 3.4 Attenuation of electromagnetic radiation

Incident photon Fluence

Absorber

Transmitted Photons Fluence

Scattered Photons

Collimator

varying degrees. It's conceivable for parts of it to be absorbed, dispersed and transferred without coming into contact. This is shown schematically in Fig. 3.4. *The energy of the radiation and the nature of the medium define the prevailing process. Photoelectric, Compton and pair production attenuation coefficients are the probabilities of each event occurring.* The total attenuation coefficient is the sum of these three components.

The amount of radiation transmitted through a material decreases with thickness, which may be explained using the exponential relationship.

$$I = I_0 e^{-\mu_x}$$

where I_0 denotes the incoming radiation intensity, I denotes the transmitted radiation intensity, x denotes the material thickness and μ denotes the total attenuation coefficient (μ). If x is in cm, it is μ equal to 1/cm (cm^{-1}) and is known as the total linear attenuation coefficient. The natural logarithm of the proportion of incoming radiation intensity attenuated by the unit thickness of the material is represented by the numerical value μ. The mass attenuation coefficient is defined as μ/ρ, where is the density of the medium represented in (g/cm^3).

It should be noted that this rule is comparable to the one that describes the decrease of radioactivity with time and that total attenuation requires an infinite thickness. It's worth noting the distinctions between the modes of attenuation of charged

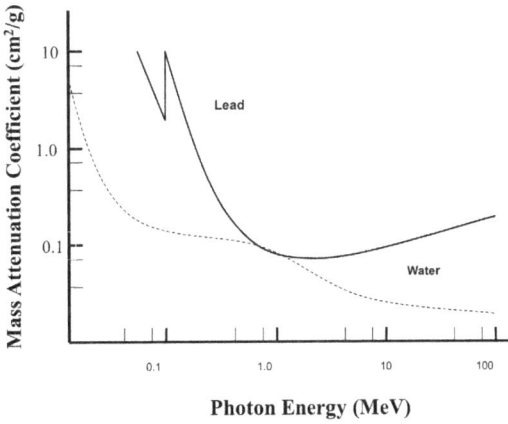

Fig. 3.5 Mass attenuation coefficient of water and lead for different photon energies

particles and photons. The overall linear attenuation coefficient is determined by the photon's energy and the attenuating medium's atomic number. Figure 3.5 depicts the variation of the total mass attenuation coefficient as a function of energy for lead and water.

Figure 3.5 depicts that:

1. The dominant interaction at low energy is photoelectric.
2. The Compton process is the most important interaction in the medium energy area.
3. Pair production begins at 1.02 MeV and grows as energy increases.

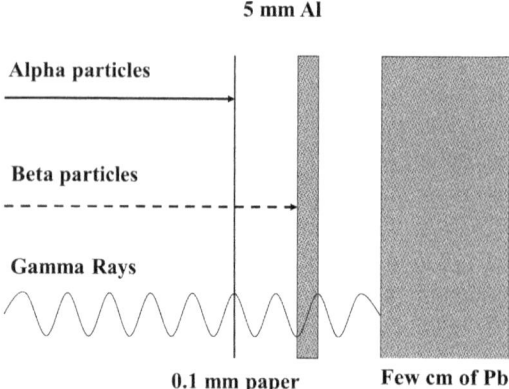

Fig. 3.6 Relative absorption of different types of radiations

4. The graph has a dip, corresponding to the shift from diminishing Compton interaction predominance to growing pair formation process predominance.

Figure 3.6 shows the relative absorption of alpha, beta and gamma radiations in various absorbers.

3.1.4　Interaction of Neutrons

Neutrons, being uncharged particles, cannot directly induce ionization. A neutron's energy determines the sort of interaction it has. Their interactions with the interacting medium's atomic nuclei are mostly atomic nuclei. Below is a list of the most important interactions in a medium.

1. **Elastic Scattering**: This interaction is similar to a billiard ball collision in which the incident neutron loses part of its energy, which is then transferred to the target nucleus as kinetic energy. In the case of lightweight nuclei like hydrogen, this is a crucial interaction.
2. **Inelastic Scattering**: The neutron is absorbed by the target nucleus and re-emitted with a loss of energy, leaving the nucleus energized. One or more gamma rays are emitted when the target nucleus decays to the ground state.
3. **Radiative Capture**: The neutron is grabbed by the target nucleus in his encounter, result-

ing in the formation of a compound nucleus in an excited state. Gamma radiation is emitted as the target nucleus decays to the ground state. Slow neutrons, i.e. neutrons with a kinetic energy of a few electron volts, are more likely to interact in this way.

4. **Neutron Capture Resulting in Other Particles**: The incoming neutron is grabbed by the target nucleus, resulting in the formation of a compound nucleus. A charged particle, similar to a proton or alpha particle, is emitted by this composite nucleus. At extremely high neutron energy, this interaction is more likely.

Table 3.2 lists the feasible types of interactions at various neutron energies.

3.1.5　Nuclear Cross-Section (σ)

In a nucleus, the probability of the process occurring due to incident neutron per cm^2 is defined as the nuclear cross-section (σ). A microscopic cross-section is what it's called. Macroscopic cross-section refers to the overall cross-section of all the nuclei in a unit volume of 1 cm^3. It's represented by Σ and written as

$$\Sigma = N_0 \rho \sigma / A \quad \text{in } \text{cm}^{-1}$$

N_0 is the Avogadro Number, ρ is the medium's density, σ is the cross-section and A is the medium's atomic weight.

Table 3.2 Type of neutron interaction with their energies

Group	Energy	Type of interaction
Neutrons with thermal energy	0–05 eV (avg. 0.025 eV)	Capture process
Neutrons with intermediate energy	0.5 eV to 10 keV	Elastic scattering and capture
Fast Neutrons	10 keV to 10 MeV	Elastic and inelastic scattering
Very fast Neutrons	Greater than 10 MeV	Inelastic scattering

An exponential rule in general regulates the transport of neutrons through materials.

This equation is identical to the exponential attenuation of photons, except that instead of μ being replaced by Σ

3.2   Section 2: Production of Radionuclides Used in Nuclear Medicine

In nuclear medicine, there are three main methods for producing radionuclides: (A) reactor based, (B) accelerator based (cyclotron) and (C) generator based.

3.2.1   Reactor Based

A target is bombarded with neutrons in a reactor to produce a radionuclide whose nuclei contain extra neutrons; the important reaction is (n,γ). The radionuclides used more commonly in SPECT studies are produced in a reactor. A target is blasted with neutrons in a reactor to create a radionuclide with additional neutrons in its nuclei; the main reaction is (n,γ). A reactor produces the radionuclides that are most typically utilized in SPECT investigations.

3.2.1.1 Nuclear Reaction (n,γ)

^{98}Mo (n,γ) -^{99}Mo to produce ^{99m}Tc

In this reaction, only a limited proportion of target nuclei are converted to product radionuclide. The product cannot be isolated from the majority of the target since it is chemically identical. This product radionuclide is of low specific activity. The fission of uranium is another process employed in reactors. This reaction is used to make a high specific activity product, which is presently employed to make ^{99}Mo/^{99m}Tc generators (column type).

^{235}U (n, fission) ^{99}Mo

The amount of the radionuclide compared to the total quantity of the stable isotope is known as a specific activity. Fission occurs when a nucleus splits, releasing energy and neutrons. Other radionuclides useful in nuclear medicine

are created in the reactor and are mentioned further down.

3.2.1.2 Standard β Emitters for Internal Radiotherapy

Palliative treatment employs the β emitters ^{32}P and ^{90}Y, which have relatively high β energies. As gels or microspheres, they are injected into joints and cavities. The radioisotope ^{32}P is also used in the labelling of nucleotides. The radiosynthesis of tumour-seeking drugs is another type of internal treatment using β emitters. ^{89}Sr, ^{153}Sm, ^{177}Lu, ^{188}Re and ^{131}I are the most regularly employed elements. Trivalent metals, in particular, are increasingly being used to treat bone metastases. ^{177}Lu linked to peptides, for example, is used to treat prostate cancer.

Similarly, ^{188}Re may produce many chelates that might be useful in medicine. Thyroid cancers are often treated using the radioisotope ^{131}I in the form of iodide. The most significant therapeutic radionuclide is ^{131}I, which has a long history of use in treating follicular thyroid cancer.

^{32}P (T½ = 14.3 days)

The neutron threshold reaction ^{32}S(n,p)^{32}P produces the radioisotope ^{32}P. This is the reaction that Chiewitz and Hevesy utilized to create the first medicinal radionuclide using a Ra/Be neutron source. The reaction cross-section averaged across the fission neutron spectrum is less than the cross-section averaged over the Ra/Be neutron spectrum. Despite this, owing to a significantly larger neutron flux in a reactor than at a Ra/Be source, ^{32}P is created in enormous amounts in a reactor through this pathway in non-carrier added (NCA) form [1, 2]. Purified ^{32}S powder is pressed into a pellet or melted together and inserted in an Al capsule as the target material. Irradiation takes place over many weeks at a rapid neutron flux of 2–3×10^{14} ncm^{-1} s^{-1}. After that, the target is melted, and ^{32}P is leached out using H_2SO_4 and ethyl alcohol in one of two ways: either ^{32}S is distilled over, and ^{32}P is leached out with dilute HCl, or ^{32}S is distilled over, and ^{32}P is leached out with dilute HCl. Cation-exchange chromatography is used to purify radiophosphorus even more. Finally, the radioisotope ^{32}P is obtained as $H_3{}^{32}PO_4$ in batch yields of about 100 GBq. It's a commercially available product.

^{89}Sr (T½ = 50.5 days)

The ^{88}Sr(n, γ)^{89}Sr reaction produces some of this radionuclide. However, ^{89}SrCl$_2$ has only been employed in the palliative treatment of malignant skeletal metastases because of its limited specific activity. A manufacturing technique employing the neutron threshold reaction ^{89}Y(n,p)^{89}Sr has been established to create radiopharmaceuticals with high specific activity. The target material is Y_2O_3 powder that has been compressed into a pellet and put within an Al capsule. A rapid neutron flux of 2–3×10^{14} ncm^{-2} s^{-1} is used for many weeks of irradiation. The chemical processing begins with the irradiation target being dissolved in HNO$_3$ and the majority of the yttrium being extracted in tributylphosphate. Following that, numerous cation-exchange chromatographic stages are used to purify ^{89}Sr. The refined product is then obtained as ^{89}SrCl$_2$ in dilute HCl, yielding roughly 20 GBq each batch.

^{90}Y (T½ = 2.7 days)

Because the target is monoisotopic and the (n,γ) cross-section with thermal neutrons is 1.28 b, this radionuclide is readily generated by the ^{89}Y(n,γ)^{90}Y reaction, even in a low-power reactor. The ^{90}Sr/^{90}Y generator system, on the other hand, is employed to produce a product with a high specific activity. The parent nuclide ^{90}Sr ($T_{1/2} = 28.9$ years) is produced with a high yield (5.77%) in the fission of ^{235}U and is isolated from the other fission products ^{90}SrCl$_2$ by a time-consuming method. Because the coproduced ^{89}Sr ($T_{1/2} = 50.5$ days) has a shorter half-life and a lower yield (4.69%), it does not pose a severe issue in the generator's preparation after many months of decay. ^{90}Y is eluted with 0.003 M ethylenediaminetetraacetic acid for normal usage after being fixed on a Dowex-50 cation-exchange chromatographic column. ^{90}Y is isolated from the parent ^{90}Sr in another system by batch extraction with di-(2-ethylhexyl) phosphoric acid [3]. Both generating systems are available for purchase.

^{131}I (T½ = 8.02 days)

The fission of ^{235}U, in which the irradiated ^{235}UAl$_3$ pellet is first treated to extract ^{99}Mo, is one method of producing this radionuclide. Following that, radioiodine is separated using anion-exchange chromatography with a distillation step.

The target material for the second pathway of ^{131}I synthesis, namely, the chemical sequence ^{130}Te (n,γ)$^{131m, \, 131g}$Te $\xrightarrow{\beta}$ ^{131}I, is either pure Te metal or TeO$_2$. With thermal neutrons, the cross-sections for the creation of ^{131m}Te ($T_{1/2} = 30$ h) and ^{131g}Te ($T_{1/2} = 25$ min) are only 12 and 192 mb, respectively, necessitating rather extended irradiations.

Irradiation is usually carried out for a few days at a neutron flux of $\sim 10^{14}$ ncm^2 s^{-1}.

The longer-lived ^{131m}Te decays 77.8% via emission to ^{131}I directly and 22.2% by IT to ^{131g}Te, which then decays to ^{131}I. The chemical procedure begins after a 2-day waiting period during which most of the ^{131m,g}Te has decomposed to ^{131}I. The radioiodine is distilled over and collected in a dilute solution of NaOH after the metallic Te target is dissolved in an oxidizing combination of chromic acid and H$_2$SO$_4$. The ^{131}I batch yield achieved utilizing the two target and processing methods is almost identical. It is, however, less than that obtained from fission. The radioiodine obtained in all three situations is iodide. The activation of ^{130}Te to ^{131m,g}Te, followed by its decay to ^{131}I, is the most common method of producing ^{131}I in small reactor labs.

^{153}Sm (T½ = 1.93 days)

The simple neutron capture reaction ^{152}Sm(n,γ)^{153}Sm [3] produces this radionuclide [3]. Although the cross-section of the reaction is rather large, the target isotope ^{152}Sm is only 26.75% in nature. As a result, many labs utilize Sm$_2$O$_3$ of natural isotopic composition as a target. However, >98% enriched ^{152}Sm$_2$O$_3$ is used as a goal to boost the product's particular activity while avoiding longer-lived radio-contaminants such as ^{145}Sm ($T_{1/2} = 340$ days) and ^{151}Sm ($T_{1/2} = 93$ a). A few milligrams of the substance are sealed in a quartz ampoule, then put in an Al capsule and irradiated for several hours in a medium flux reactor. The target material is absorbed in ethylene diamine tetra-methylene phosphoric acid (EDTMPA) dissolved in alkaline water at the end of the irradiation. At 75 °C, the solution is heated for roughly 1 h. It is then utilized as a stock solution. ^{153}Sm has a batch output of about 20 GBq and specific activity of around 18 GBq mg^{-1} of ^{152}Sm$_2$O$_3$. This precise activity is adequate for much research, and as a result, this radionuclide is finding more uses. However, a greater specific activity might be beneficial in other cases. Unfortunately, no technology has yet been identified to manufacture enough ^{153}Sm in NCA form. The two approaches that have been examined so far, ^{153}Eu(n,p)^{153}Sm and ^{140}Nd(α,n)^{153}Sm, exclusively deal with nuclear reaction cross-section measurements; no attempts at manufacturing have been recorded.

^{177}Lu ($T_{1/2}$ = 6.65 days)

This radionuclide may be produced directly via the ^{176}Lu(n,γ)^{177}Lu reaction or indirectly through the ^{176}Yb (n, γ) ^{177}Yb $\rightarrow$ ^{177}Lu reaction. The thermal neutron cross-section and the resonance integral are rather large in the direct neutron capture technique, but since the abundance of ^{176}Lu in natural Lu is only 2.6 percent, an enriched target (~43%) is often utilized [1]. Approximately 10 mg of ^{176}Lu$_2$O$_3$ is enclosed in a quartz tube and irradiated for a few days at a neutron flux of ~1 × 10^{14} ncm^{-2} s^{-1}. The target is dissolved in HCl after irradiation. A high batch yield of the final product ^{177}LuCl$_3$ is achieved. However, the approach has two main flaws: first, the product's specific activity is restricted to roughly 1600 GBq mg^{-1} of Lu after a 4 days irradiation in a reasonably high flux reactor, and second, the longer-lived ^{177m}Lu (T_{12} = 160.4 days), which is also generated during the irradiation, is present in a minor quantity. The activation of ^{176}Yb to ^{177}Yb ($T_{1/2}$ = 1.9 h) followed by decay results in a substantially lower batch yield of ^{177}Lu than the first approach. This is owing to two factors: (a) natural Yb contains just 12.7% of the target nuclide ^{176}Yb, and (b) the neutron capture cross-section and resonance integral are both rather low. As a result, 96% enriched ^{176}Yb$_2$O$_3$ is often utilized as

a target material. In a typical run, 10 mg of material is sealed in a quartz ampoule and irradiated for a few days in a nuclear reactor. The dissolving of the irradiation target in HCl begins the chemical processing of the target. Ion-exchange chromatography is then used to extract NCA (no carrier added) ^{177}Lu from the majority of Yb. When the target material is bulky, and the reaction product is at the NCA level, the separation is typical of those seen in nuclear reaction investigations. However, the purity standards are significantly more severe than in nuclear reaction investigations due to the intended medicinal usage. ^{177}LuCl$_3$ with a specific activity of about 3000 GBq mg^{-1} of Lu has been produced due to recent research [1]. This is around 2.0 times more expensive than the straightway. Another benefit of the indirect ^{177}Lu manufacturing process is the substantially lower ^{177m}Lu impurities (104%). The overall yield of ^{177}Lu obtained by the indirect technique, on the other hand, is much smaller. It's worth noting that the synthesis of ^{177}Lu through charged particle-driven reactions like Ta(p, spall), Hf(p,x) and ^{176}Yb(d,n) has also been studied.

^{188}Re ($T_{1/2}$ = 17.0 h)

The ^{187}Re(n,γ)^{188}Re reaction or the ^{188}W/^{188}Re generator system may be used to produce this radionuclide. Initially, the direct manufacturing method was used. The direct manufacturing approach is seldom employed due to the product's low specific activity and the advent of the simple ^{188}W/^{188}Re generator technology. However, since it is based on a two-step neutron capture process, the indirect method is difficult to follow in a regular reactor. The reaction ^{186}W(n,γ)^{187}W happens in the first step. The ^{187}W product is radioactive and has a half-life of 23.7 h. However, it has a very large neutron capture cross-section, causing a rivalry between radioactive decay of ^{187}W and the ^{187}W and the ^{187}W(n,γ) ^{188}W reaction, resulting in the production of ^{188}W ($T_{1/2}$ = 69.8 days). The second neutron capture phase is preferred in a reactor with a large flux. A metallic or oxide tungsten target is utilized in the manufacturing of ^{188}W. Typical neutron fluxes between 5 × 10^{14} and 2 × 10^{15} n cm^{-2} s^{-1} are used during many weeks of irradiation. The irradiation target is chemically processed by dissolving it in hot 1 M NaOH in the presence of H_2O_2 and then purifying it many times to achieve pure ^{188}W. The pure ^{188}W is normally loaded on an Al_2O_3 column to manufacture the ^{188}W/^{188}Re generator, and the ^{188}Re is eluted with saline, similar to the ^{99}Mo/^{99m}Tc generator.

3.2.1.3 Standard α Particle Emitters for Targeted Therapy

Only a few radionuclides with adequate characteristics for internal radionuclide treatment in humans are known to decay through α particle emission [3]. ^{211}At, ^{213}Bi, ^{223}Ra and ^{225}Ac are the four most widely utilized ones. The next sections go through their production processes.

^{211}At ($T_{½}$ = 7.2 h).

For over 40 years, this radionuclide has been studied. It decays to ^{207}Bi (42%) or ^{211}Po (58%) through direct α particle emission ($T_{1/2} = 0.5$ s), followed by EC emission. As a result, every ^{211}At decay is accompanied by the emission of an α particle. Because astatine is a halogen, it was expected that numerous radioiodination procedures might be applied to ^{211}At. Under some situations, though, astatine may take on a metallic appearance. Astatine has previously been used to bind antibodies, proteins and inorganic colloids. However, the compound's stability under physiological settings is the primary issue. Despite this, significant progress has been made in the field of α therapy employing ^{211}At-labelled molecules. Several nuclear procedures have been employed to produce ^{211}At, employing either a direct or indirect pathway through the decay of the parent ^{211}Rn [4]. A minor amount of parent ^{211}Rn ($T_{1/2} = 14.6$ h) is produced either by spallation of ^{232}Th or the ^{209}Bi(^{7}Li,5n)^{211}Rn reaction. Several fresh initiatives have successfully generated ^{211}At in tiny amounts for preclinical study using the indirect technique [5]. Nonetheless, the direct manufacturing reaction ^{209}Bi(^{7}Li,5n)^{211}Rn remains the preferred approach. The excitation function is very well-known [6]. To prevent the generation of ^{210}At ($T_{1/2} = 8.1$ h), which decays to the lon-

ger-lived -particle producing radionuclide ^{210}Po ($T_{1/2} = 138.0$ days) and generates a large additional radiation dose, the energy of the α particles must be maintained below 29 MeV during synthesis. The predicted yield of ^{211}At for the optimal energy range of $E_α = 28 \rightarrow 20$ MeV is 25.3 MBq μA h^{-1}, without any contamination from ^{210}At. A thin layer of Bi (melted, pressed or vacuum evaporated) on an Al backing serves as the irradiation target. Internal target systems with slanting beams have received high current irradiations [7, 8]. A batch yield of up to 7 GBq ^{211}At was recorded after a 4-hour irradiation at 55 A beam current [7]. Dry distillation at 650 °C in an Ar stream is typical for extracting ^{211}At from the irradiation target. At 77 °C, the ^{211}At withdrawn from the distillation equipment is trapped. The trapped astatine is retrieved using a tiny amount of organic solvent for following radiosynthesis activity. Recently, an automated flow system [9] with the in-line acid dissolution of irradiation bismuth metal for application in the separation of ^{211}At was developed [9]. The batch yield of ^{211}At was roughly 1 GBq after 50 min of irradiation of Bi targets (masses about 4.5 g) with 40 α particle beams.

^{213}Bi ($T_{½}$ = 45.6 min)

This radioisotope decays only 2.2% through α particle emission directly to ^{209}Tl and 97.8% by β decay to ^{213}Po ($T_{1/2} = 3.7$ μs), which decays 100% to ^{209}Pb ($T_{1/2} = 3.2$ h) via α particle emission. As a result, each decay of ^{213}Bi results in the emission of an α particle. A generator system loaded with the parent activity ^{225}Ac $T_{1/2} = 10.0$ days produces this rather short-lived α particle emitter. Because the radionuclide ^{225}Ac is difficult to make, it was questioned if a short-lived α particle emitter might be used as a therapeutic agent. Recent research on a PSMA-tagged monoclonal antibody labelled with ^{213}Bi has led to the compelling conclusion [10] that certain ‑resistant cancers may be efficiently treated with this radionuclide if several treatments are carried out. Further research and optimization work on the generator system looks beneficial to maintaining a consistent and stable supply of ^{213}Bi.

^{223}Ra ($T_½$ = 11.4 days)

This radionuclide emits only α particles at 100% efficiency. Its importance has expanded in recent years, and it is now approved for use as a drug in several countries, particularly for the treatment of blood cancer in the form of ^{223}RaCl2. It was initially discovered through the ^{226}Ra (n, γ) ^{227}Ra (42.2 min) $\overset{\beta}{\rightarrow}$ ^{227}Ac (21.8 a) $\overset{\beta}{\rightarrow}$ ^{223}Ra chain.

^{227}Ac, the parent activity, is available in a limited amount, and ^{223}Ra is withdrawn from it regularly. More ^{227}Ac must be generated due to the radionuclide's growing importance.

^{225}Ac ($T_½$ = 10.0 days)

This radionuclide decays via the series:

$$^{225}Ac \overset{\alpha}{\rightarrow} {}^{221}Fr(4.9\min) \overset{\alpha}{\rightarrow} {}^{217}At(32ms) \overset{\alpha}{\rightarrow} {}^{213}Bi(45.6\min) \overset{\beta^-}{\rightarrow}$$

$$^{213}Po(3.7\mu s) \overset{\beta^-}{\rightarrow} {}^{209}Pb(3.2h) \overset{\beta^-}{\rightarrow} {}^{209}Bi$$

Consequently, each decay of ^{225}Ac produces four α particles, making it a very effective therapeutic source. The decay chain products ^{221}Fr and ^{213}Bi, on the other hand, have detectable half-lives, prompting concerns that their release may cause excessive toxicity in normal organs. Today, this radioisotope is employed for medicinal purposes both directly and as a generator parent of ^{213}Bi.

The creation of ^{225}Ac has taken a lot of time and work. It has been optimized for separation from nuclear waste (^{229}Th) [11], and generators are now commercially available. The overall quantity of ^{225}Ac activity accessible every year, on the other hand, is restricted to around 75 GBq. Two cyclotron manufacturing procedures, ^{226}Ra(p,2n)^{225}Ac and ^{232}Th (p, spall)^{225}Ac, are being developed in response to the growing demand for this radionuclide.

3.2.2 Cyclotron Based

A target is bombarded with charged particles (protons, deuterons, etc.) in a cyclotron to generate a radionuclide with neutron-deficient nuclei. Positron emitters are radionuclides produced in a cyclotron. A cyclotron produces the radionuclides needed in positron emission tomography (PET). Table 3.3 lists the many cyclotron-produced radionuclides now in use.

3.2.3 Generator Based

A generator is a system that comprises a long-lived parent radionuclide that is created by any of the abovementioned technologies and decays to a short-lived daughter that is employed in the preparation of radiopharmaceuticals. There are many kinds of generating systems available today, briefly described here.

3.2.3.1 Generator for ^{99}Mo/^{99m}Tc

As illustrated in Fig. 3.7 (a, b), sodium ^{99m}Tc-pertechnetate is produced by elution of a ^{99}Mo/^{99m}Tc generator daily. The ^{99}Mo/^{99m}Tc-generating system separates ^{99m}Tc pertechnetate from ^{99}Mo molybdate using an alumina (aluminium oxide) column. Based on the elution process technique, there are two kinds of ^{99}Mo/^{99m}Tc-generating systems: wet column type and dry column type. Both work on the same fundamental principles, yet they operate in somewhat different ways. This section explains both the procedure and the results.

Type of Wet Column

Figure 3.7(a) shows a schematic representation of the wet column type. Replace the shipping cap with a sterile 18–20-gauge needle. To elute the generator, use a sterile isopropyl alcohol pad to wipe the septum of the protected evacuated vial, then insert the needle and open the valve. Allow

Table 3.3 Commonly cyclotron-produced radionuclides

Radionuclide	Half-life (min)	Beta (MeV)	Method of production
^{11}C	20	0.96	^{14}N (p,α)
^{13}N	10	1.19	^{16}O (p,α)
^{15}O	2	1.72	^{14}N (d,n)
^{18}F	110	0.635	^{18}O (p,n)

Fig. 3.7 (**a**, **b**): (**a**) Schematic diagram of a wet column ^{99}Mo/^{99m}Tc generator (left) that has a 0.9% NaCl reservoir and (**b**) a dry column ^{99}Mo/^{99m}Tc generator (right)

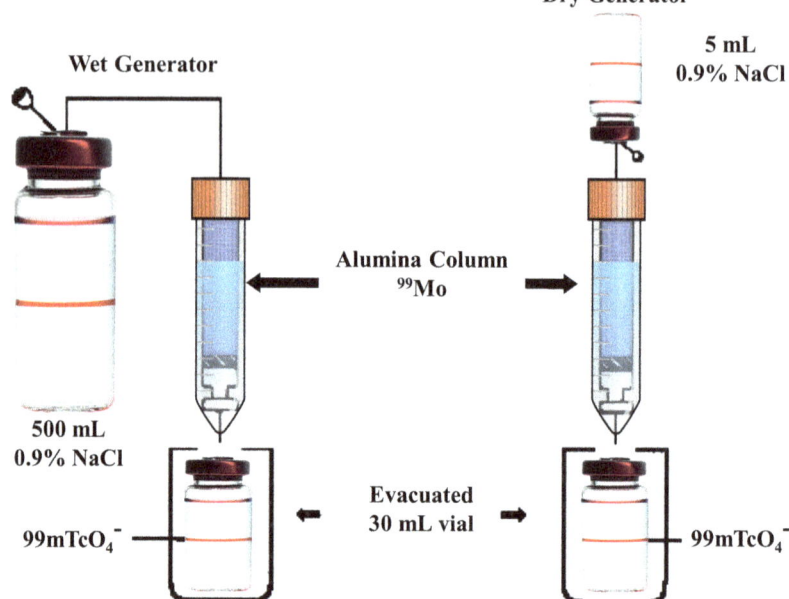

5 to 10 min for eluting. Close the valve after elution, remove the vial, cover it with a lead cap and replace the needle on the generator.

Problems with wet column generator systems include the following:

(a) If the generator does not produce any eluent, you may need to replace the elution vial (it may have lost its vacuum), the needle or both.
(b) Using a high-capacity evacuated vial to elute the generator will result in a low pertechnetate concentration, which may cause issues with certain labelling processes. As a result, the smaller the vial that should be utilized, the older the generator.
(c) If the eluent is not eluted for a long enough period, the total activity and volume of the eluent will be reduced.
(d) If a generator hasn't been eluted in 2 to 3 days, it should be eluted the day before to remove oxidizing agents and carrier 99Tc from the column.

(e) If bubbles get caught in the column, generators may elute slowly.
(f) If the generator fails to elute adequately after following all guidelines, the manufacturer may need to replace the generator.
(g) Isopropyl alcohol has an adverse effect on the generator since it does not elute with the needed efficiency.

Dry Column Type

Figure 3.7(b) shows a schematic representation of the wet column type. The following are some basic guidelines for using it:

(a) To elute the generator, you'll need a non-bacteriostatic sterile saline vial and an.
(b) empty evacuated vial.
(c) The empty evacuated vial should be put into a suitably insulated lead container before initiating the elution.
(d) Using a sterile alcohol pad, sterilize the septum of the saline vial. After removing the old vial or needle cover, place the saline vial over

the generator's input needle (saline charge needle).

(e) Wipe the emptied vial septum with an alcohol pad to sterilize it. Place the protected empty evacuated container aseptically over the collecting needle after removing the existing vial or needle cover from the generator.

(f) Allow 5–10 min for eluting. The column will be dried by drawing filtered air through the evacuated vial's surplus vacuum.

(g) After elution, remove the collecting vial and replace it with the sterile shipping vial or another sterile elution vial. The use of an empty evacuated container is favoured because it allows the column to dry more quickly, resulting in higher pertechnetate yields in successive elutions.

(h) After the saline charge vial has been eluted, it may be removed and replaced with the shipping vial.

Problems Associated with Dry Column Generators

(a) Elute for long enough (5 to 10 min) to collect the saline on the column; otherwise, the generator yield on successive elutions may suffer.

(b) Due to the loss of vacuum in the evacuated container, poor elution may ensue.

(c) If little or no saline is eluted from the column, it might be because the evacuated vial was in place before the saline vial.

Chemical separation of the daughter radionuclide from its parent is possible. ^{99m}Tc is separated from ^{99}Mo by running saline through the 99Mo/99mTc generator column. The ^{99m}Tc is eluted as sodium pertechnetate (Na^{99m}TcO$_4$) from the generator column, while the ^{99}Mo stays immobilized. Solvent extraction and sublimation are two other separating methods. Figure 3.8 shows the elution curve of the ^{99}Mo/^{99m}Tc generator.

Parent-daughter relationships may be divided into two categories. Transient equilibrium occurs when the parent's half-life is 10–100 times larger than the daughter's, for example, ^{99}Mo (66 h) → ^{99m}Tc (6.02 h). The second equilibrium is known as secular equilibrium, in which the parent's half-life is 100–1000 times the daughter's half-life, for example, ^{90}Sr (29.1 year) → ^{90}Y (64 h) or ^{68}Ge (271 days) → ^{68}Ga (68 min).

Fig. 3.8 Elution profile of ^{99}Mo/^{99m}Tc generator

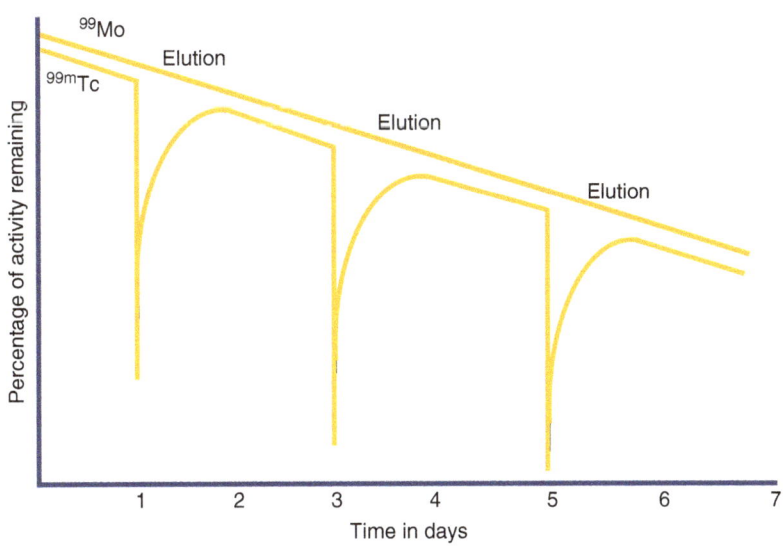

3.2.4 Generator-Produced Standard Positron Emitters

Two extensively utilized positron emitters, ^{68}Ga ($T_{1/2}$ = 1.15 h) and ^{82}Rb ($T_{1/2}$ = 1.3 min), are produced via generator systems, i.e. the decay of their parent radionuclides ^{68}Ge ($T_{1/2}$ = 271.0 days) and ^{82}Sr ($T_{1/2}$ = 25.3 days), respectively. They're typically used in PET tests at places that don't have a cyclotron. The β^+ emitter ^{68}Ga is often employed for PET attenuation correction and, in recent years, for labelling compounds that have important applications in evaluating blood-brain barrier integrity and tumour localization. Due to its good half-life and chemical characteristics, it is an excellent metallic PET radionuclide. Some radiopharmacists predict that if appropriately labelled chemicals are discovered, this radionuclide will be able to compete with, if not completely replace, ^{18}F. The β + emitter ^{82}Rb is used in cardiac PET for various purposes, including myocardial imaging, myocardial viability study and non-invasive detection of coronary artery disease.

3.2.4.1 ^{68}Ge/^{68}Ga Generator System

Nine nuclear mechanisms were examined in the generation of the parent radionuclide ^{68}Ge. Only two of these, however, were put into practice. Spallation of As or Br and natGa(p,xn)68Ge is an example. Previously, the spallation process was the technique of choice, and the radionuclide was only generated at big accelerators that allowed for lengthy irradiations in parasitic places. Wet chemical processing was used to purify the product in many steps. By irradiating a Ga$_4$Ni alloy with 20 MeV protons and separating radiogermanium by solvent extraction, the ^{nat}Ga(p,xn)^{68}Ge technique was studied. The product batch yield was very low (2 GBq) despite the comparatively high current of 45 A and irradiation period of 60 h, owing to the low energy of the protons utilized. Because both ^{69}Ga and ^{71}Ga contribute to the creation of ^{68}Ge, further cross-section measurements revealed that the usable portion of the excitation function of this nuclear process extends to energies up to 50 MeV and even beyond. As a result, with the advent of contemporary

accelerators capable of producing proton beams of 70–100 MeV, the (p,xn) reaction has gotten much interest in recent years.

Gallium metal encapsulated in niobium is often used as an irradiation target. Long irradiations with protons of 100 MeV were used. Wet chemical processing was carried out after many weeks of waiting for short-lived products to expire, and radiogermanium was extracted as GeCl$_4$ in toluene or benzene. Several extractions and back-extraction cycles were also used to purify ^{68}Ge (particularly from ^{65}Zn). The batch yield of ^{68}Ge was around 18 GBq, with a radio nuclidic purity of >99%. Stable metal impurities were often less than 1 ppm. According to the excitation function, Ep = 70 → 20 MeV looks to be a good energy range for ^{68}Ge synthesis through this approach.

The separated ^{68}Ge is usually loaded on a column packed with tin oxide, titanium oxide or an organic resin, and the positron-emitting daughter ^{68}Ga is regularly eluted with 0.01–1 M HCl to prepare the ^{68}Ge/^{68}Ga generator system. In recent years, this generator has undergone significant methodological development to increase its performance (e.g. minimization of breakthrough, improvement in elution yield and concentration of the eluate). After much effort, the ^{68}Ga eluate for medical applications can now be processed relatively quickly [12]. This technology has much potential, especially given the growing relevance of ^{68}Ga in labelling molecules for PET studies [13].

3.2.4.2 ^{82}Sr/^{82}Rb Generator System

Six nuclear procedures were researched for the synthesis of the parent radionuclide ^{82}Sr. Initially, the Mo (p, spall)^{82}Sr technique, similar to that of ^{68}Ge, was widely employed to synthesize radionuclides. Including a significant quantity of ^{85}Sr ($T_{1/2}$ = 64.9 days) in the separated ^{82}Sr was a major downside of the procedure. In subsequent years, the focus turned to the ^{85}Rb(p,4n)82Sr reaction. The ^{nat}Rb(p,xn)^{82}Sr technique, on the other hand, has shown to be more convenient and cost-effective in actual applications (by avoiding the enriched material). Ep = 70 → 50 MeV is the ideal energy range for producing ^{82}Sr. RbCl encapsulated in stainless steel is often used as an irradiation target. At 50 µA, the typical irradiation period with 70 MeV protons is roughly a week [14]. The irradiation target's wet chemical processing, which uses ion-exchange chromatography, usually begins approximately a week later. Using the $^{nat}Rb(p,xn)^{82}Sr$ nuclear process instead of the spallation approach, the ^{85}Sr impurity in ^{82}Sr is significantly decreased. Controlling the projectile's effective energy range inside the target, on the other hand, is critical. The energy limit should not be lower than 40 MeV to maintain the $^{85}Sr/^{82}Sr$ ratio below 0.25. $E_p = 70 \rightarrow 50$ MeV seems to be a very good energy range for producing ^{82}Sr through this technique.

The separated ^{82}Sr is placed into an appropriate column (e.g. Chelex 100 chelating resin, Dowex-1 anion exchanger and Purolite S950) and purged with an eluent, usually a high amount of saline, to prepare the $^{82}Sr/^{82}Rb$ generator system.

Elution with saline is used to eliminate the daughter ^{82}Rb regularly. It's worth noting that the amount of ^{85}Sr in ^{82}Sr impacts the generator's quality. Because its distinctive γ-ray peak at 514 keV cannot be separated from the 511 keV annihilation peak of the β^+ emitter ^{82}Rb in the event of strontium breakthrough from the column, the quantity of ^{85}Sr in the eluted ^{82}Rb is difficult to assess. As a result, it's important to keep an eye on the ^{85}Sr level and the column's integrity.

References

1. Vertes A, Nagy S, Klencsár Z, Lovas RG, Rösch F. Handbook of Nuclear Chemistry, vol. 4. Heidelberg, Germany: Springer; 2011; see the contribution by Mirzadeh, S., Mausner, L.F., Garland, M.A. on Reactor produced medical radionuclides, page 1857; by Qaim, S.M. on Cyclotron production of medical radionuclides, page 1903; and by Rösch, F., Knapp, F.F. on Radionuclide generators, page 1935
2. Manual for Reactor Produced Radionuclides, IAEA-TECDOC-1340, IAEA, Vienna, Austria, 2003.
3. Wilbur DS. Chemical and radiochemical considerations in radiolabeling with α-emitting radionuclides. Curr Radiopharm. 2011;4:214–47.
4. Lahiri S, Maiti M. Recent developments in nuclear data measurements and chemical separation methods in accelerator production of astatine and technetium radionuclides. Radiochim Acta. 2012;100:85–94.
5. Crawford JR, Yang H, Kunz P, Wilbur DS, Schaffer P, Ruth TJ. Development of a preclinical 211Rn/211At generator system for targeted alpha therapy research with 211At. Nucl Med Biol. 2017;48:31–5.
6. Qaim, S.M., Tárkányi, F., Capote, R. Nuclear Data for the Production of Therapeutic Radionuclides. IAEA Technical Report Series No. 473, IAEA, Vienna, Austria, 2011.
7. Zalutsky MR, Zhao XG, Alston KL, Bigner D. High-level production of alpha-particle emitting 211At and preparation of 211At-labeled antibodies for clinical use. J Med Med. 2001;42:1508–15.
8. Lebeda O, Jiran R, Rális J, Stursa J. A new internal target system for production of 211At on the cyclotron U-120M. Appl Radiat Isot. 2005;63:49–53.
9. O'Hara MJ, Krzysko AJ, Niver CM, Morrison SS, Owsley SL, Hamlin DK, Dorman EF, Wilbur DS. An automated flow system incorporating in-line acid dissolution of bismuth metal from a cyclotron irradiated target assembly for use in the isolation of astatine-211. Appl Radiat Isot. 2017;122:202–10.
10. Sathekge M, Knoesen O, Meckel M, Modiselle M, Vorster M, Marx S. 213Bi-PSMA-617 targeted alpha-radionuclide therapy in metastatic castration-resistant prostate cancer. Eur J Nucl Med Mol Imaging. 2017;44:1099–100.
11. Apostolidis C, Molinet R, Rasmussen G, Morgenstern A. Production of 225Ac from 229Th for targeted alpha therapy. Anal Chem. 2005;77:6288–91.
12. Rösch F, Riss PJ. The renaissance of the $^{68}Ge/^{68}Ga$ radionuclide generator initiates new developments in Ga-68 radiopharmaceutical chemistry. Curr Top Med Chem. 2010;10:1633–68.
13. Rösch F, Baum RP. Generator-based PET radiopharmaceuticals for molecular imaging of tumors: on the way to THERANOSTICS. Dalton Trans. 2011;40:6104–11.
14. Van der Meulen NP, van der Walt TN, Steyn GF, Raubenheimer HG. The production of ^{82}Sr using larger format RbCl targets. Appl Radiat Isot. 2013;72:96–9.

Radiation Protection Standards in Relation to ICRP Recommendations

4

Abstract

Man has always been exposed to ionizing radiation from various natural sources. The extent of this natural background exposure varies with the location. However, no ill effects have been uniquely correlated with these variations. Either no deleterious effects are produced at these levels of exposure, or their frequency is too low to be statistically observable.

The hazards of ionizing radiation became apparent in connection with exposures from manmade sources. The discovery of X-rays and the separation of radioactive substances at the end of the last century brought great benefits, albeit associated with unforeseen hazards. Within 5 years, 170 cases of radiation injuries were recorded, and by 1922, about 100 radiologists had died from radiation overexposure. These observations of radiation injuries in man and the results of radiobiological experiments improved the knowledge of the health effect of radiation, and thereby the need for protection became obvious. The International Commission on Radiological Protection (ICRP) was established in 1928 under the name of the "International X-ray and Radium Protection Committee (IXRPC)" and published its first recommendation in 1931. Later on, in 1950, it was restructured and named as the present name. The recommendations of ICRP issued from the time-to-time deal with basic principles of radiation protection based on available knowledge from radiobiological and epidemiological findings.

The chapter discusses principles behind radiation protection, biological considerations, dose limits and their significance.

4.1 Introduction

A direct source of information on radiation hazards in man is the follow-up of population groups exposed to certain radiation levels. The term 'epidemiology' covers statistical techniques and adequate identification and diagnosis of observed effects. Epidemiological information is mostly available for situations involving high doses and dose rates. On the other hand, the normal operation encounters low doses and dose rates in radiation work. Hence, any information derived from such populations is interested in assessing radiation risks to set up protection standards.

Epidemiological studies [1] do not explain how radiation produces cancer and are therefore not useful in extrapolating the results from high doses and dose rates to low-dose regions. The mechanisms of induction of radiation effects and the relationship between dose and response can be obtained from the radiobiological experiments with microorganisms, mammalian cells and experimental animals. The radiobiological information thus provides the conceptual basis for interpreting epidemiological results using the dose-response relationship based on mechanistic model(s).

There is strong evidence suggesting that most cancers originate from damage to single cells. Cancer initiation starts from the deregulation of cell growth and the loss of control of cell proliferation. Although the cells may have initiated changes, they will not lead to malignancy until they are stimulated or 'promoted' by chemicals and hormones present in their environment. Radiation exposure contributes to one relevant mutation step among the many needed to induce a cancer cell (frank malignancy). Thus, the age at which radiation-induced cancer is expressed differs from that of spontaneous cancer. In this sense, radiation replaces time, advancing slightly the occurrence of malignancy (having the same effect of reaching a higher age).

4.2   Radiation Effects

The biological effects of radiation originate from the effects which occur at the cellular level. Absorption of radiation by cells may result in cell killing or a modification without loss of viability. Both of these effects can cause harmful effects in multicellular organisms. Cell killing can result in **deterministic** effects, whereas cell-modifications may give **stochastic** effects [2, 3].

4.2.1   Deterministic Effects

Deterministic effects are generally caused by killing a significant fraction of cells in some parts of the body. The appearance of these effects in any organ depends upon the sensitivity and proliferation kinetics of cells. Organs with a large population of stem cells with high division activity, such as red bone marrow, small intestine (crypt epithelial cells), oral mucosa, colon, gonads, etc., show a higher sensitivity as compared to radioresistant organs made up of non-dividing differentiated cells (muscles, kidney, nervous system, heart and connective tissues).

The killing of a small fraction of cells by radiation does not result in functional organ impairment or clinically detectable effects. As a result, these effects manifest only when the absorbed radiation dose is adequate to kill a large fraction of cells. This is why each deterministic effect is characterized by a **threshold dose**, which depends on the cells' radiosensitivity. It is also true that the effects do not appear below this dose, but at doses well above the threshold, deterministic effects occur without exception. This means a causal relationship between radiation exposure and the effect. Another important feature of the deterministic effect is its severity increases with dose. Radiation sickness (NVD syndrome), induction of sterility, leucopenia, anaemia, reddening of the skin, depilation, skin burns, cataract (opacity of eye lens) and death are some examples of the deterministic effects. These effects generally occur when the absorbed doses are >1 Gy. An acute whole-body dose of 3–5 Gy of gamma radiation will result in the death of 50% of those exposed within 60 days due to severe haematopoietic damage. This dose is referred to as $LD_{50}(60)$.

Human beings can survive much larger doses if the exposures are non-uniform such as partial-body/localized or in situations where at least 10% of the bone marrow is spared. Doses absorbed within a short time (acute exposures) are far more damaging than protracted, fractionated or chronic exposures. This is due to the ability of cells to repair a significant fraction of radiation damage induced at low dose rates. As a result, the threshold dose for a deterministic effect increases with the decreasing dose rate. Dose limits recommended by the Commission ensure that none of the deterministic effects can appear during the normal working conditions. Only under accidental exposures or during therapeutic irradiations are deterministic effects can be seen.

4.2.2   Prenatal Effects

The human embryo/foetus is very radiosensitive, particularly during the first half of gestation. In

utero exposure entails an increased risk of malformations for exposures during the first 8 weeks.
Information derived from those exposed in utero
at Hiroshima and Nagasaki has brought to light
the high radiosensitivity of the CNS during the
8–15 weeks of pregnancy. During this period, the
risk of severe mental retardation has been estimated to be 40% Sv^{-1} for high dose rate exposure. The pregnancy period from 16 to 25 weeks
has a risk of 10% Sv^{-1}, whereas the risk of mental
retardation does not exist for the rest of the pregnancy. Based on the reanalysis of the mental
retardation data and detailed intelligence quotient
(IQ) studies among the children exposed in utero,
the ICRP has assumed that the induction of
severe mental retardation is a deterministic effect,
with a possible threshold of about 0.1 Sv. At low
doses/dose rates, the effect is best described by
an IQ shift of 30 points Sv^{-1}. In the light of this
information, for women declared to be pregnant,
the ICRP has recommended special dose limits.
Prenatal exposures are also associated with a risk
of excess childhood cancer incidence. The ICRP
now accepts a risk estimate of 10^{-1} Sv^{-1} for childhood cancers. However, it must be stressed that
the risk due to few mSv is very small compared
to the background risk of 1 in 2000 live births.

4.2.3   Stochastic Effects

Occupational exposures generally result in small
absorbed doses. These doses cannot result in
deterministic effects but can modify a few cells
leading to cell transformation or an alteration of
genetic information. The mechanism behind two
important stochastic effects of radiation, i.e.,
radiation carcinogenesis and genetic effects or
heritable effects, is cell modifications in somatic
cells constituting the body structure and germ
cells responsible for reproduction. There is no
threshold dose for stochastic effects, and the
extent of observed effects raises with dose compared to an unexposed cohort of similar characteristics. In a large group of individuals exposed
to radiation, the number of individuals who suffer
from these effects is related to per caput dose. No
human data exists at low doses <200 mSv, which
conclusively estimates the risk due to large variations in background incidences of both cancers
and heritable effects.

4.2.3.1 Radiation Carcinogenesis

Transformation and mutations in somatic cells
have a potential to induce radiation carcinogenesis after a long latent period of several years. This
phenomenon is probabilistic and does not occur
in most of those exposed even to large doses.
Radiogenic leukaemias are mostly expressed
during the first decade, but most solid cancers
appear only during the second and third decades
or even later. Radiogenic cancers do not differ
from those that occur spontaneously and are
indistinguishable. Hence, there is no possibility
to be certain that cancer is indeed caused by radiation exposure. Only the probability of cancer
induced by radiation can be calculated based on
the magnitude of radiation exposure.

Further, a very high baseline incidence of
~20% of cancers among the Western populations
poses a statistical problem to detect a small
excess attributable to any cause. The frequency
of excess cancers among radiation workers is so
small that it is difficult to detect it even by carefully designed epidemiological studies involving
several thousand workers. Studies involving
95,000 UK radiation workers showed a statistically significant leukaemia risk with a mean estimate of 0.76% Sv^{-1}. Atomic bomb survivor (ABS
cohort) study (Life Span Study-LSS) constitutes
the single largest database for the estimation of
cancer risk.

4.2.3.2 Genetic Effects

Alteration of genetic information in germ cells
(sperm/ovum) during the reproductive life has
the potential to transmit these mutations to the
future generation(s). These mutations can result
in some genetic disorders (diseases). Heritable
genetic changes can be expressed in the next generation (dominant and sex-linked diseases) or
future generations (recessive genetic disorders).
Since the human germ cells already carry many
mutations, the prevalence of genetic disorders
among the human population is very high
(>70%). Only 10–20% of the genetic diseases are

serious. Due to the large background incidence of genetic disorders, there is no possibility of detecting the enhanced incidence of genetic disorders or evaluating the genetic risk from the epidemiological studies on exposed human populations. However, offspring of mice exposed to large doses resulted in a significant increase in genetic changes. Based on these studies, it is concluded that 1 Gy per generation doubles the frequency of mutations in germ cells. This is called **doubling dose** and forms the basis for genetic risk evaluation. The ICRP 103 (2007) recommendations have suggested a risk estimate of 0.2% Sv^{-1} for the whole population, and the age-truncated estimate for the working population (18–65 years) is 0.1% Sv^{-1}. These risk figures have been revised since ICRP 60 (1991).

The main message is that at low doses of radiation of interest in risk estimation, the risk of adverse hereditary effects is small compared to the baseline frequencies of genetic diseases in the population. This is consistent with the findings from the genetic studies carried out on A-bomb survivors in Japan.

4.3 Weighing Factors Used in Radiation Protection

Two weighting factors are extensively used to estimate protection quantities. These factors often become necessary many a time due to (i) radiation exposure involving a different type of radiation with different biological efficacies and (ii) non-uniform exposures either due to the internal concentration of radionuclides or due to partial external exposures involving few or only one tissue.

4.3.1 Radiation Weighting Factor

As a thumb rule, densely ionizing radiation such as heavy charged particles and energetic neutrons are known to cause more damage at the cellular level. These radiations are also known to possess a higher hazard for stochastic effects. Relative biological effectiveness is typically more than 1 for such radiation. Though RBE and W_R are not the same quantitatively, they have many similarities

conceptually. The W_R can be considered a conservative estimation of RBE for stochastic effects by a type of radiation. While RBE is a measurable quantity, W_R is an estimate and cannot be considered for any measurement from basic principles. A type of radiation weighted for its hazard amounted to an equivalent dose and represented as H.

$$H = \sum_R DW_R$$

where

D = Physical dose (measurable)
W_R = Radiation weighting factor for the type of radiation R
H = Equivalent dose weighted for all types of radiation R

There has been a considerable revision in the values of W_R in the recent past. The Table 4.1 below summarizes the same.

4.3.2 Tissue Weighting Factor

Most of the internal exposure involving radioisotopes, which may accumulate in a tissue or in a set of tissues, pose more hazards to those tissues as doses delivered to tissues tend to be more. Besides, the radio sensitivity of these issues may not be the same. Hence, it is necessary to weigh the radiosensitivity of such tissues involved in receiving the dose for their sensitivity toward stochastic effects. This weighting factor is termed tissue weighting factor W_T. The concept of rela-

Table 4.1 Comparison of radiation weighting factor in ICRP 103 and ICRP 60

Type and energy range	Radiation weighting factor, W_R	
	ICRP 60	ICRP 103
Photons	1	1
Electrons and muons	1	1
External protons and charged pions	5	2
Alpha particles, fission fragments, heavy nuclei	20	20
Neutrons	Steps 5, 10, 20, 10, 5	A continuous curve is recommended

tive detriment is used while estimating W_T, which is described later in this chapter.

$$E = \sum_R DW_R \sum_T W_T = \sum_T H_T W_T$$

where

E = Effective dose
H_T = Equivalent dose to the tissue T
W_T = Tissue weighting factor

4.4 Risk Projection Models

The appearance of many new solid tumours among the A-bomb survivors during the last two decades suggested a relationship between the attained age and the expression of excess cancers. It was observed that the number of excess cancers continued to increase even after 40 years of exposure. It was clear from these observations that the absolute risk model, which suggests a fixed level of cancer risk for a given level of exposure, does not hold good for solid tumours. On the other hand, the excess cancer incidence was a function of the baseline cancer incidence. Since the incidence of many cancers increases with age, radiogenic excess cancers also express in larger numbers during the older age (relative risk model).

A few thousands of A-bomb survivors are still alive. It is of paramount importance to estimate the lifetime risk for the cohort under study to estimate the possible risk over the entire life time. This lifetime risk is then projected onto the population of interest, for example, occupational workers or the general public. Hence, the lifetime risk must be evaluated by projecting the existing risk over the entire lifetime using the relative risk model. The relative risk model (also called the multiplicative risk model) predicts a much higher risk than the absolute risk model used in ICRP-26. In ICRP 103, which has nearly completed the life span studies for most relevant cancers, consideration is given to both additive and relative risk models in a certain proportion. This is because, for many cancers which appear in the early years after radiation exposure, the study has already accounted the lifetime risk. Hence, the additive model will suffice, while for some cancers such as thyroid and skin cancers which may appear very slowly over the age, the relative risk model is still being considered (Table 4.2).

4.4.1 Nominal Fatality Probability Coefficients

Applying the multiplicative risk model, the ICRP has calculated the lifetime risk of mortality due to radiogenic cancer in five representative populations, China, Japan, Puerto Rico, the UK and the USA, with different spontaneous cancer rates. The risk expressed as 'nominal fatality probability coefficient' ranged from 6 to 13% per Gy and an average of 9.4% Gy^{-1} for acute exposure (see Table 4.3).

Similar calculations have been carried out by UNSCEAR and BEIR committees, about 20% higher than ICRP. Considering these, ICRP has estimated the fatality probability of 5×10^{-2} Sv^{-1} when the radiation exposures occur at low dose rates. This can be compared with the 1977 estimate of 1.25×10^{-2} Gy^{-1}. In general, females show a higher risk (~20%) than males. Age-truncated risk analysis shows a higher risk for children (a factor of 2) in 0–19 years. On the other hand, the working population aged 18–64 years has a risk coefficient of 20% lower than that of the whole population (4×10^{-2} Gy^{-1}).

Table 4.2 Proportion of additive (EAR) and multiplicative models used for estimation of cancer risk (ICRP 103)

ERR	EAR ratio used
Breast, bone marrow	0:100
Thyroid, skin	100:0
Lung	30:70
All others	50:50

Table 4.3 Fatal cancer probability (% Sv^{-1}) for different populations following acute exposure

Population type	Projection model	
	Multiplicative	NIH
Japan	10.2	9.3
USA	11.2	8.7
Puerto Rico	9.5	10.2
UK	12.9	9.7
China	6.3	6.0

4.4.2 Dose and Dose Rate Effectiveness Factor

Risk figures derived from the Japanese data pertain to high dose and acute exposure (high dose rate conditions). Most of the radiobiological data for low LET radiations suggest a reduction in the effectiveness of radiation for cancer induction with the reduction in dose and dose rate. In the absence of reliable information from human epidemiological data from low dose rate exposures, ICRP has adopted a dose and dose rate effectiveness factor (DDREF) of 2. There is no such reduction factor for high LET radiations.

4.5 Detriments

The Commission realized that it is inadequate to consider the fatal cancer risk and serious genetic effects in two generations alone for assessing the harm associated with radiation exposure. The total harm that an exposed group and its descendants may experience is now assessed in ICRP-60 for estimating 'detriment' based on (a) fatal cancer risk, (b) morbidity associated with non-fatal cancers and (c) serious as well as other genetic disorders in all generations. Relative life lost due to each type of cancer is also considered while arriving at the aggregated detriment, which is expressed as the probability of death or its equivalent. Contribution of genetic risk was also included while estimating the detriment (Table 4.4).

Detriment is estimated based on several factors such as curability, quality of life, number of fatal and non-fatal cancers associated and the average length of life lost due to cancer. These factors are estimated based on clinical experience with cancer treatment and have nothing to do with radiation exposure. However, when clubbed with the incidence rate of excess cancers based on ABS cohort, detriment becomes a measure of radiation risk.

Detriment estimated for individual tissue gives an idea of sensitivity. The relative detriment measures the fractional contribution of risk of a particular tissue compared to the whole body. This factor is used to categorize various tissues into four groups based on their sensitivity and assigned tissue weighting factor W_T. The W_T value measures the fractional contribution of tissue to stochastic risk compared to whole body. In contrast, a sum of all W_T for all tissues under consideration for stochastic risk will amount to 1 (Tables 4.5 and 4.6).

Table 4.4 Detriment for radiation exposure for the whole population and occupational workers [4]

Exposed population	Quality of life and Lethality adjusted cancer risk	Lethality adjusted heritable effects	Detriment ICRP 103[4] % Sv^{-1}
Whole population (age 0–100 year)	5.5	0.2	5.7
Adult workers (truncated age 18–65 year)	4.1	0.1	4.2

Table 4.5 Estimation of relative detriment based on ABS cohort [4]

Tissue	Nominal risk coefficient (cases per 10,000 persons Sv^{-1})	Lethal fraction	Nominal risk-adjusted for lethality and quality of life	Relative cancer-free life lost	Detriment	Relative detriment
Oesophagus	16	0.93	16	0.91	14.2	0.034
Stomach	60	0.83	58	0.89	51.8	0.123
Colon	50	0.48	38	1.13	43.0	0.102
Liver	21	0.95	21	0.93	19.7	0.047
Lung	127	0.89	126	0.96	120.7	0.286
Bone	5	0.45	3	1.00	3.4	0.008
Skin	670	0.002	3	1.00	2.7	0.006
Breast	49	0.29	27	1.20	32.6	0.077
Ovary	7	0.57	6	1.16	6.6	0.016
Bladder	42	0.29	23	0.85	19.3	0.046
Thyroid	9	0.07	3	1.19	3.4	0.008
Bone marrow	23	0.67	20	1.17	23.9	0.057
Other solid	88	0.49	67	0.97	65.4	0.155
Gonads (hereditary)	12	0.80	12	1.32	15.3	0.036
Total	*1179*		*423*		*422*	*1.000*

Table 4.6 Tissue weighting factors (W_T) are estimated by categorizing tissues based on their relative contribution to the detriment (ICRP 103) [4]

Tissue	W_T	$\Sigma\ W_T$
Bone marrow (red), colon, lung, stomach, breast	0.12	0.60
Gonads	0.08	0.08
Bladder, oesophagus, liver, thyroid	0.04	0.16
Bone surface, brain, salivary glands, skin	0.01	0.04
Remainder tissues (Nominal W_T applied to the average dose to 13 tissues: adrenals, extrathoracic region, gall bladder, heart, kidneys, lymphatic nodes, muscle, oral mucosa, pancreas, prostate (males), small intestine, spleen, thymus and uterus/cervix (females))	0.12	0.12

4.6 Dose Limits

Dose limits have evolved ever since mankind realized the importance of safety while working with radiation. The initial dose limits imposed were only to avoid deterministic effects. With the advancement of studies, it was realized that radiation could also induce stochastic effects at lower levels. While no proven threshold dose exists for stochastic effects, it is also true that at low doses (<200 mSv), no human data exists that can conclusively determine the existence of risk or otherwise.

Dose limits of ICRP apply to occupational and public exposures. The limits have evolved over the decades with refinement of scientific information collected by ragulatory communities and authorities (Table 4.7 and Fig. 4.1). At present, the dose limits are based on the evaluation of detriment resulting from a continuous exposure over a working lifetime of 47 years. Limits are chosen such that the consequence of exposure are 'just short of unacceptable' or 'just tolerable'. The ICRP assumes a maximum risk figure of 10^{-3} year^{-1} (1 in 1000 in one year) mortalities in very important industries. Based on this risk level and the variation in natural background radiation levels, the Commission confirmed that the dose limits for the public should be 1 mSv year^{-1}. The dose limit for

Table 4.7 Evolution of limits

Recommendation body and year	Limits and their basis
IXRPC 1928	1 Sv per year estimated (restriction on working hours for medical professionals to avoid serious deterministic effects such as skin burn)
IXRPC 1934	500 mSv per year estimated (threshold-based)
ICRP 1951	3 mSv per week (estimated)
ICRP 1966	50 mSv per week for workers 5 mSv for members of the public
ICRP 26 (1977)	50 mSv per year for workers 1 mSv per year for members of the public
ICRP 60 (1991)	20 mSv per year for workers 1 mSv per year for members of the public
ICRP 103 (2007)	20 mSv per year for workers 1 mSv per year for members of the public

the eye lens is 15 mSv y^{-1}, while that for the skin, hands and feet is 50 mSv year^{-1}.

The basis for occupational exposure for women who are not pregnant is the same as that of men. However, in the light of information regarding the high radiosensitivity of the human conceptus, if the woman is pregnant, the Commission has suggested a dose limit of 1 mSv to the conceptus for the duration after the declaration of pregnancy to childbirth, and intake of radionuclides during this period should not exceed 0.05 of ALI (annual limit of intake). Estimates of personnel dose should include both external and internal exposures. The limits for pregnant women intend to limit the dose to foetus to about less than 1 mSv.

To provide some flexibility, the dose limits to both occupational and public exposures are permitted to be averaged over 5 years; the new

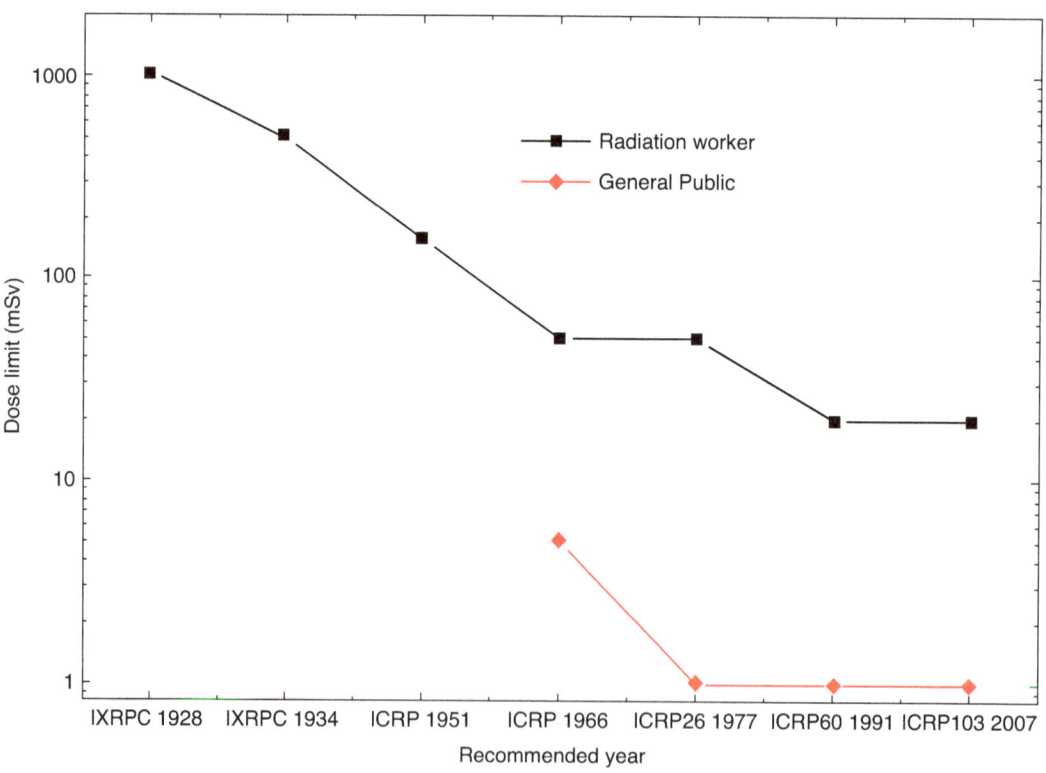

Fig. 4.1 Evolution of protection limits over years of use of radiation technology

Table 4.8 Dose limits recommended by ICRP 103 (2007)

Exposure condition	Dose limit (mSv per year)		
	Occupational	Apprentices (16–18 years)	Public
Whole body: (Effective dose)	20 mSv per year, averaged over a defined period of 5 years with no more than 50 mSv in a single year	6 mSv in a year	1 mSv in a year, averaged over 5 years
Parts of the body: (Equivalent dose) The lens of the eye Skin** Hands and feet***	150* mSv per year 500 mSv per year 500 mSv per year	50 mSv in a year 50 mSv in a year 150 mSv in a year	15 mSv in a year 50 mSv in a year –

*Recent recommendation of 20 mSv to the eye lens is not yet implemented by most countries, including India. Considerations to reduce the eye lens dose to below 1 Sv of the lifetime equivalent dose are under consideration. Protective measures are being deployed stringently
**Averaged over areas of no more than 1 cm^2 regardless of the exposed area. The nominal depth is 7.0 mg cm^{-2}
***Averaged over areas of the skin not exceeding about 100 cm^2
Note 1 Dose limits do not apply to medical exposures, natural radiation sources and under conditions resulting from accidents.

dose limits and the explanatory notes are given in Table 4.8.

It is the view of the Commission that the dose limits constitute a part of the system of protection with emphasis on the ALARA principle and not a target to be achieved. Suppose any individual exposure has crossed the limits. In that case, there need be no special restrictions on the individual's exposure: instead, the regulatory body should look into the design and operational aspects of protection in the installation.

These standards provide such a high level of protection that the radiological aspect of the working environment should not influence the administrative conditions of service of occupationally exposed persons.

Reassessment of risk of cancer, development of the concept of aggregated detriment and new information on the sensitivity of the human conceptus forms the bases for the revision of general recommendations and effective dose limits by the ICRP in 1991.

4.7 Principle of Implementation of Radiation Protection

Radiation protection, dose limits and various protection quantities are based on conservative estimation of maximum possible risk. The dose limits should not be construed as prescriptions or as annual targets that can be achieved by the end of the year. Also, the risk figures such as detriment should not be used to estimate individual risk, population risk, risk due to diagnostic exposure, risk due to exposure to background radiation, etc. The doses involved are very low or one-time incidence in nature. In a nutshell, individualization of risk figures is not feasible from the conservative estimations. These estimates are only used to set operational limits to achieve safe working conditions for occupational workers in a particular and safe environment for the mankind in general from any radiation technology procedure.

Three philosophical principles of radiation protection are:

(i) *Justification*: Use of any radiation technology needs to be justified. If any alternate method is available that derives equal or better outcomes, the use of radiation technology is not justified. The procedure involving radiation is allowed only when the outcome outweighs the risks involved. In medicine, when other methods such as ultrasound, MRI, etc. will suffice, they are usually resorted to. Even among radiation procedures, simple diagnostics such as X-rays are preferred over complex ones such as CT,

PET, etc. These procedures are allowed only when necessary to derive medical conclusions and planning of medical procedures.

(ii) *Optimization*: Once the use of radiation procedure is justified, the same should be optimized to ensure safe working for workers (occupational dose), subjects (patient dose, area exposed, number of exposures, etc.), the public and the environment (dose to the surrounding). Optimization is based on socio and economic considerations without compromising on the safety and dose limits.

(iii) *ALARA*: as low as reasonably achievable— This principle states everything about work principles, safe practices and ethics. While optimization ensures adherence to dose limits in extreme work conditions, ALARA helps to minimize the actual doses to practically up to $1/10^{th}$ of the dose limits or even lower in most cases.

The three philosophical principles of protection are achievable by three physical principles of radiation protection, not in the respective orders but by collectively implementing them to achieve safe work conditions.

(a) *Time*: This is a linear factor. Spending more time near the radiation field yields more doses. One should not spend time unnecessarily by safe work practice but only when work demands the present. For example, radiation sources should not be stored in sitting places or workers. The job of comforters should be minimized or helped by patient parties (one-time incidence) instead of workers for whom it is a regular affair.

(b) *Distance*: This is an inverse quadratic factor. The dose decreases with the inverse of the distance from the point source square. This may not be strictly true for broad source and scattered radiation but holds well as a thumb rule.

(c) *Shielding*: When time and distance cannot suffice to achieve sufficient dose reduction, shielding, which is an exponential factor, helps to reduce the doses significantly for radiation workers and the public. Thumb rules such as half-value thickness (HVT) and tenth-value thickness (TVT) for gamma and X-rays help to reduce the radiation field wherein usually high Z, high-density material is used. For certain radiation such as β, β^+, neutrons and particle rays, special considerations are needed based on their interaction with the shielding materials, and appropriate shielding material should be chosen.

4.8   Summary

- The protection limits for public and occupational workers are evaluated based on the epidemiological studies conducted on exposed populations such as atomic bomb survivors of Hiroshima and Nagasaki, the Chernobyl accident, etc.

- The philosophy of radiation protection is to completely avoid all deterministic effects by setting the limits well below the threshold dose and minimizing the risk due to stochastic effects.

- All protection limits are evaluated based on stochastic effects since deterministic effects are impossible at such low doses.

- Induction of cancer and genetic or hereditary effect are the two stochastic effects that are very important in radiation protection.

- The protection quantities such as radiation weighting factor, tissue weighting factor, equivalent dose and effective dose are based on stochastic effects.

- There is no clear evidence that shows that radiation can induce cancer at low doses. The risk estimation is based on high dose data from follow-up studies of exposed individuals.

- There is no clear evidence that radiation can induce genetic or hereditary effects in humans, even in individuals exposed to high doses, as in atomic bomb survivors of Hiroshima and Nagasaki.

- Genetic risk estimation is done using the doubling dose method using the model or experimental systems.

- Risk projection models help us arrive at the lifetime risk and estimate the risk to different types of population based on the limited available data from follow-up studies of exposed individuals.
- Nearly 1/10th of the dose limits apply to the general public compared to radiation workers. Children are more sensitive to radiation. The most important reason for setting lower limits for the general public is the presence of children in this cohort.

References

1. Status of the Dosimetry for the Radiation Effects Research Foundation (DS86) (2001).
2. Health Risks from Exposure to Low Levels of Ionizing Radiation BEIR VII (1998).
3. Recommendations of the International Commission on Radiological Protection. ICRP Publication 60. Ann. ICRP 21 (1-3) (1990).
4. The 2007 Recommendations of the International Commission on Radiological Protection. ICRP Publication 103. Ann. ICRP 37 (2-4). (2007).

Radiation Hazard Evaluation and Control in Nuclear Medicine

Abstract

The use of radioisotopes in unsealed form has potential radiation hazards to the person handling the radiation sources, the people around and the environment. Radiation hazards are classified into two categories such as 'external' and 'internal'. External radiation hazards can be controlled by using one or a combination of factors such as time, distance and shielding. The lesser the 'time' spent in the radiation field, the lesser is the radiation dose received by an individual. Increasing the distance is a very effective method of reducing the dose rate as isotropic emission of radiation follows the 'inverse square law'. The concepts of 'half-value layer (HVL)' and 'tenth-value layer (TVL)' are widely used in designing shielding and depend on the composition of the material used for shielding and the energy of the radiation. When radioactive material gets into the person's body through any means such as ingestion, inhalation or subcutaneous absorption (through wounds), it gives rise to an internal radiation hazard. The effective half-life, annual limit of intake (ALI) and derived air concentration (DAC) are different terms used to estimate internal hazards.

5.1 Introduction

With the availability of many artificially produced short-lived radioisotopes, medical diagnosis has become easier and more specific in many diseases. However, using these radioisotopes in unsealed form has potential radiation hazards to the person handling the radiation sources, the people around and the environment if they are not handled properly and the facilities provided are not adequate. The patients are also likely to be affected if a medical event (mis-administration) occurs. Commonly used radioisotopes in nuclear medicine are ^{18}F, ^{32}P, ^{67}Ga, ^{68}Ga, ^{89}Sr, ^{99m}Tc, ^{131}I, ^{201}Tl, etc. A few newer radioisotopes, such as ^{186}Re, ^{177}Lu, ^{153}Sm, ^{90}Y and ^{225}Ac, have been added to the radioisotopes used for therapeutic applications.

Radiation hazards are classified into two categories: (a) External radiation hazard which is due to the sources present outside the body i.e. in the surroundings, such as in storage, during transportation, while handling, etc., whereas (b) internal radiation hazard is due to the radioactive material which gets into the body of the person by way of ingestion, inhalation or through subcutaneous absorption (through wounds).

5.2   External Radiation Hazard

External radiation hazards are more likely to occur in routine work as most radioisotopes used in nuclear medicine emit gamma rays, X-rays and bremsstrahlung. This type of radiation may give rise to exposure to the person at a distance.

The irradiation capability of ionizing radiation is measured using the quantity 'exposure rate'.

5.2.1   The Term 'Exposure'

Radiation exposure is a measure of the ionization of air due to ionizing radiation from photons, i.e. X- or Υ-rays. It is defined as the electric charge freed by such radiation in a specified volume of air divided by the mass of that air.

IUPAC definition (1982): For X- or Υ-radiation in the air, it is the sum of the electrical charges of all the ions of one sign produced when all electrons liberated by photons in a suitably small element of volume of air are completely stopped, divided by the mass of the air in the volume element [1].

It is represented as X,

$$\text{and} \qquad X = \frac{q}{m} \qquad (5.1)$$

where q is the sum of electrical charges of ions of one sign and m is the mass of that air. The SI (*Système International*) unit is coulombs/kg.

The older definition of exposure:

It is the amount of X- or Υ-radiation that produces ionization of 1 electrostatic unit (1 esu = 0.001293 g of air) of either positive or negative charge per cubic cm (cc) of dry air at standard temperature and pressure (STP), i.e. at 273.15 K or 0 °C temperature and 10^5 Pa of pressure.

The older unit is roentgen (R).

$$1R = 2.58 \times 10^{-4}\, \text{coulombs / kg}$$

5.2.2   Exposure Rate

Exposure per unit time is called exposure rate. It is denoted as $\dot{X}$.

$$\dot{X} = \frac{\text{Exposure}}{\text{Time}} = \frac{X}{t} \qquad (5.2)$$

SI unit of exposure rate is coulombs/kg.s or ampere/kg, whereas the older and commonly used unit is roentgen/hour (R/h).

5.2.3   Exposure Rate Constant

It is a measure of exposure rate from unit point source activity at unit distance per unit time. It is also called 'gamma ray constant or gamma constant' and denoted as Γ (capital gamma) and applicable to an un-collimated point source. It allows the calculation of exposure rate:

- for a point source.
- for a given activity.
- of a gamma-emitting radionuclide, and
- at a specified distance from the source.

The SI unit is $\mu Sv.m^2/MBq$ h; however, coulomb.m^2/kg.GBq.s is also used. The older units R.cm^2/h.mCi or R.m^2/h.Ci are also used in day-to-day practice. In nuclear medicine, conventionally, it has been expressed as exposure rate (in R/h) from 1 mCi (37 MBq) point source at a distance of 1 cm. Table 5.1 gives half-life and exposure rate constants for important radioisotopes used in nuclear medicine.

Table 5.1 Exposure rate constants for radionuclides used in nuclear medicine [2]

Radionuclide	Physical half-life	Exposure ray constant ($C.m^2/kg. GBq.s$)	Exposure ray constant ($R.cm^2/mCi.h$)
^{18}F	109.74 min	1.10×10^{-9}	5.68
^{68}Ga	67.71 min	1.05×10^{-9}	5.43
^{15}O	2.04 min	1.14×10^{-9}	5.86
^{11}C	20.48 min	1.13×10^{-9}	5.86
^{13}N	9.97 min	1.13×10^{-9}	5.86
^{82}Rb	1.25 min	1.23×10^{-9}	6.33
^{99m}Tc	6.01 hours	1.54×10^{-10}	0.795
^{123}I	13.13 hours	3.46×10^{-10}	1.78
^{125}I	59.4 days	3.38×10^{-10}	1.75
^{131}I	8 days	4.26×10^{-10}	2.2
^{133}Xe	5.3 days	1.10×10^{-10}	0.568
^{201}Tl	73 hours	8.72×10^{-11}	0.45
^{67}Ga	78 hours	1.55×10^{-10}	0.803
^{111}In	68 hours	6.7×10^{-10}	3.46
^{99}Mo	65.94 hours	1.78×10^{-10}	0.917
^{57}Co	271.4 days	1.09×10^{-10}	0.563
^{60}Co	5.27 years	2.5×10^{-9}	12.9
^{137}Cs	30.16 years	6.64×10^{-10}	3.43

5.3 Control of External Radiation Hazards

External radiation hazards can be controlled by using all three factors.

1. Time	2. Distance	3. Shielding

The *dose rate* at a point away from the source depends on the last two factors. The dose received by an individual depends on both the dose rate and the time spent by the person near the point. *A fourth factor that contributes to external radiation is the strength of the source.* If the strength (activity) of the source is more, the radiation coming out of the source is more.

5.3.1 Strength of the Source

The number of emissions per unit time from a radioactive source is proportional to the source strength. Thus at a given distance from the source of the activity present, the dose rate is directly proportional to the strength of the source (activity).

The activity of the source reduces with time, whether the source is being used or just stored, and follows an exponential decay. Thus, activity (A_t) of the source at any time 't' can be represented by

$$A_t = A_0\, e^{-\mu t} = A_0\, e^{-0.693t/T^{1/2}} \tag{5.3}$$

where A_0 = initial activity of the source, t = time elapsed, λ = decay constant, T = half-life and *is a constant, unique to the radioisotope,* $e^{-0.693t/T^{1/2}}$ is called a *decay factor*.

5.3.2 Shielding

Radiation energy is absorbed (partially or fully) when radiation passes through matter (Fig. 5.1). Alpha particles are easily stopped by any matter such as a few centimetres (cm) of air, less than 1 mm of water or paper or any other material. Most beta particles are stopped by a few mm of plastic, water or glass. Only electromagnetic radiation (X-rays and gamma rays) is not easily

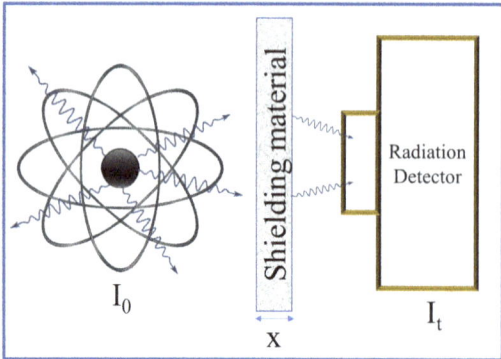

Fig. 5.1 Shielding

absorbed, and they follow the exponential attenuation law:

$$I = I_0 e^{-\mu x} = I_0 e^{-0.693x/HVT} \qquad (5.4)$$

where I = transmitted intensity of radiation (after the shielding), I_0 = incident intensity (before the shielding), e = base of the natural logarithms (2.71828…), x = thickness of the shield (say in mm), μ = linear attenuation coefficient, HVT = half-value thickness of the material (in mm), $e^{-0.693x/HVT}$ is combinedly called 'attenuation factor' and denoted as μ_L. Therefore, Eq. (5.4) may be written as:

$$I = I_0 \cdot \mu_L \qquad (5.5)$$

5.3.2.1 Half-Value Thickness (HVT) and Tenth-Value Thickness (TVT)

HVT and TVT are also called half-value layers (HVL) and tenth-value layers (TVL). These are defined as the thickness of a shield or an absorber material that reduces the radiation intensity by one-half or one-tenth of its initial value. The concepts of HVL and TVL are widely used in shielding design calculations. They are dependent on the material used for shielding and the energy of the radiation emitted by the source. In general,

1. The higher the atomic number (Z) of the material, the lower is the HVL or TVL.
2. The higher the density, the lower is the HVL or TVL.
3. The lower the energy of radiation, the lower is the HVL or TVL.

The material that requires thinner HVLs or TVLs is better shielded. Thus lead, tungsten, concrete and bricks are considered good shielding materials for gamma rays and X-rays.

Exposure rate after n HVLs

$$X' = \frac{\dot{X}}{(2)^n} \qquad (5.6)$$

Exposure rate after n TVLs

$$X' = \frac{\dot{X}}{(10)^n} \qquad (5.7)$$

where $\dot{X}$ is initial exposure rate and X' is exposure rate after n HVL or TVL.

5.3.2.2 Relationship Between HVL and TVL

We know, the intensity I_0 of a radiation source reduces to $I_0/2$ after 1 HVL and $I_0/10$ after 1 TVL. Therefore, from Eq. (5.4),

$$I_0 / 2 = I_0 e^{-\mu HVL}$$
$$\Rightarrow \quad \text{‰} = e^{-\mu HVL}$$
$$\Rightarrow \quad e^{\mu HVL} = 2$$
$$\Rightarrow \quad \mu HVL = \ln 2$$
$$\Rightarrow \quad \mu = \ln 2 / HVL$$

$$I_0 / 10 = I_0 e^{-\mu TVL}$$
$$1/10 = e^{-\mu TVL}$$
$$e^{\mu TVL} = 10 \qquad (5.8)$$
$$\mu TVL = \ln 10$$
$$\mu = \ln 10 / TVL$$

Equating both the above equations.

$$\ln 2 / HVL = \ln 10 / TVL$$

$$\Rightarrow \quad TVL / HVL = \ln 10 / \ln 2$$

$$\Rightarrow \quad TVL / HVL = 2.30258 / 0.69315 = 3.322$$

$$\therefore 1 TVL = 3.322 \, HVL$$

(5.9)

From Eq. (5.8),

$$HVL = \frac{\ln 2}{\mu} \quad \text{and} \quad TVL = \frac{\ln 10}{\mu} \quad (5.10)$$

In nuclear medicine, Eq. (5.4) is an important formula for calculating shielding. By simply placing the value of the linear attenuation coeffi-cient, one can calculate HVL or TVL. Table 5.2 gives the values of HVL and TVL of lead for some of the important radioisotopes used in nuclear medicine. David S. Smith and Michael G. Stabin [2] have compiled exposure rate constants and lead shielding values for over 1100 radionuclides.

Concrete is considered as best shielding material for the walls of any room. In the isolation ward where the patient with a large amount of radioiodine is treated or in a facility using positron-emitting radionuclides, the walls usu-ally have concrete walls. Table 5.3 gives HVL and TVL values for ^{131}I and ^{18}F in different materials such as tungsten, concrete and brick.

Table 5.2 HVL and TVL of lead (density 11.34 g/cc) for common radionuclides

Radionuclide	Major X- and gamma ray energies (keV)	Half-value layer in lead (mm)	Tenth-value layer in lead (mm)
^{18}F	511 (193%)	4.1	15.1
^{68}Ga	511 (184%)	5.12	16.0
^{15}O	511 (200%)	4.95	15.1
^{11}C	511 (200%)	4.95	15.1
^{13}N	511 (200%)	4.95	15.1
^{82}Rb	511 (190%), 777 (13%)	5.3	16.8
^{99m}Tc	140 (89%)	0.234	0.905
^{123}I	27 (71%), 159 (83%)	0.667	1.12
^{124}I	511 (50%), 603 (62%), 1693 (30%)	7.2	30.5
^{125}I	35 (100%)	0.0211	0.0623
^{131}I	364 (81%)	2.74	9.93
^{133}Xe	30 (38%), 81 (37%)	0.0379	0.4
^{201}Tl	71 (47%), 167 (11%)	0.258	0.887
^{67}Ga	93 (38%), 184 (21%), 300 (17%)	0.861	4.8
^{111}In	23 (69%), 171 (91%), 245 (94%)	0.257	1.96
^{99}Mo	740 (13.5%), 780 (5%)	5.83	23.4
^{57}Co	136.47 (99.8%), 706.42 (0.183%)	0.298	0.85
^{60}Co	1170 (100%), 1330 (100%)	15.6	45.3
^{137}Cs	662 (85%, ^{137m}Ba)	7.19	21.8
^{225}Ac	99.6 (0.7%), 99.8 (1%), 150.1 (0.6%)	0.0698	1.21

Table 5.3 HVL and TVL for ^{18}F in tungsten, concrete and brick

Sl. no.	Material	Density (g/cc)	^{131}I HVL (cm)	^{131}I TVL (cm)	^{18}F HVL (cm)	^{18}F TVL (cm)
1	Tungsten	19.25	0.2	0.6	0.295	0.98
2	Concrete	2.35	3.0	10	3.4	11.3
3	Brick	1.6	4.5	14.5	3.71 ± (0.29)	12.3

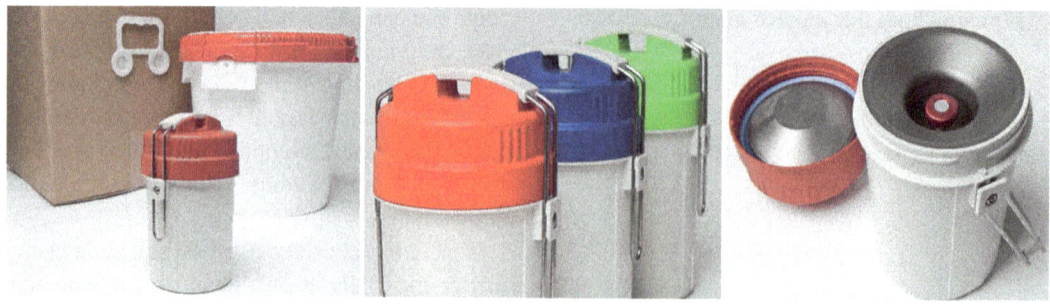

Fig. 5.2 F-18 vial transport containers. Image courtesy: www.cyclotron.nl

5.3.2.3 The Buildup Factor

The external dose rate at any distance from the source can be reduced by introducing an appropriate shielding material of appropriate thickness between the source and the measurement point. However, calculations based on just HVL and TVL values do *not* provide sufficient shielding since they neglect to scatter buildup factors (AAPM 108 [3]).

The *buildup factor* is a correction factor that considers the influence of the scattered radiation and any *secondary particles* in the medium during shielding calculations. It depends on the type and amount of shielding material and the photon's energy. The buildup factors have been calculated for many different types of shielding materials and can be found in tables of various publications, including references [4, 5]. Equation (5.4) can be modified as:

$$I = I_0 . B . e^{-\mu x} \qquad (5.11)$$

Here B is the buildup factor. Nowadays, websites and mobile applications help to calculate the dose rate and appropriate shielding for different materials and radionuclides, including buildup factors. One such website is http://www.radpro-calculator.com/. Figure 5.2 shows shielding containers for carrying ^{18}F-radiopharmaceuticals.

Based on HVL and TVL, a few examples are shown below to understand the basic calculations.

Example 5.1 Find the radiation field at a 1 m distance from 1110 MBq (30 mCi) of ^{131}I.

Solution For ^{131}I exposure rate constant = 2.2 R/h/mCi at 1 cm

$$\therefore \text{for } 30 \text{ mCi, exposure rate constant } \dot{X} = 2.2 \times 30$$
$$\therefore \text{Exposure rate constant at } 1\text{-m} \left(100 \text{ cm}\right)$$
$$\text{Distance} = \frac{2.2 \times 30}{100 \times 100}$$
$$= 66 \times 10^{-4} \text{ R / h}$$
$$= 6.6 \text{ mR / h}$$

Example 5.2 Initial © 100 mR/h (1000 µSv/h). Find exposure rate after having 5 HVTs shielding thickness.

Solution From Eq. (5.3)

$$X' = \frac{\dot{X}}{(2)^n}$$

$$X' = \frac{100}{(2)^5}$$

$$X' = \frac{100}{32}$$

$$X' = 3.1 \ \text{mR} / \text{h}$$

Solution Exposure rate constant for ^{99m}Tc = 0.8 R/h/mCi.

∴ Exposure rate of 10 mCi ^{99m}Tc = 10×0.8 R/h = 8 R/h.

From Eq. (5.3)

$$X' \text{at 5HVTs} = \frac{\dot{X}}{(2)^n}$$

$$X' = \frac{8}{(2)^5}$$

$$X' = \frac{8}{32}$$

$$X' = 0.25 \text{R} / \text{h} \quad \text{or} \quad 250 \ \text{mR} / \text{h} (2.5 \text{mSv} / \text{h})$$

Example 5.3 Find the radiation field at 10 mCi (370 MBq) of ^{99m}Tc at 5 HVTs.

Example 5.4 Find the radiation field at 1 m distance of 30 mCi (1110 MBq) of I-131.

Solution

Exposure rate constant for ^{131}I at 01 cm distance = 2.2 R / h / mCi

∴ Exposure rate of 30 mCi ^{99m}Tc at 01 cm distance = 30×2.2

∴ Exposure rate of 30 mCi ^{99m}Tc at 100 cm distance = $\dfrac{30 \times 2.2}{100 \times 100}$

$$= 66 \times 10^{-4} \text{R} / \text{h}$$

$$= 6.6 \text{mR} / \text{h} (66 \mu Sv / \text{h})$$

5.3.3   Distance

A radioactive source emits radiation in all directions (isotropically). Consider a source at the centre of a sphere of radius 'r' cm and emitting 'S' photons/s. Since all the gamma photons should pass through the surface of the sphere, the intensity on the surface is given by:

$$I = \frac{S}{\text{Surface area of the sphere}} = \frac{S}{4\pi r^2}$$

$$(5.12)$$

This property, i.e. isotropic emission of radiation from a source, follows the 'inverse square law'. If I_1 and I_2 are intensities at distances d_1 and d_2, the law can be represented by:

$$I_1 / I_2 = (d_2)^2 / (d_1)^2 \qquad (5.13)$$

This equation can also be written as:

$$I_1 \times (d_1)^2 = I_2 \times (d_2)^2 \qquad (5.14)$$

For example, if the dose rate at 1 m from a source is 4 mGy/h, and we double the distance, i.e. a distance of 2 m, the dose rate becomes one-fourth of the initial, i.e. 1 mGy/h. Hence, increasing the distance is a very good method of decreasing the dose rate.

Example 5.5 If the radiation field at 50 cm is 100 mR/h, find the radiation field at 25 cm and 100 cm.

Solution

Radiation field at 25 cm :	Radiation field at 100 cm :
$I_1 \times (d_1)^2 = I_2 \times (d_2)^2$	$I_1 \times (d_1)^2 = I_2 \times (d_2)^2$
$100 \times 50^2 = I_2 \times 25^2$	$100 \times 50^2 = I_2 \times 100^2$
$I_2 = \dfrac{100 \times 50 \times 50}{25 \times 25}$	$I_2 = \dfrac{100 \times 50 \times 50}{100 \times 100}$
$I_2 = 400 \ \text{mR}/\text{h}$	$I_2 = 25 \ \text{mR}/\text{h}$

5.3.4   Time

The lesser the time spent in the radiation field, the lesser would be the radiation dose received.

Total dose received = Dose rate × Time

Therefore, a combination of source strength, shielding, distance and time can result in increasing or decreasing the dose received from an external radiation source. The formula for total dose received from a radioactive source is,

$$\text{Total dose received} = A.e^{-0.693x/\text{HVT}}\ \Gamma.T / 4\pi r^2$$

$$(5.15)$$

where T = time spent in hours, r = distance in metres, A = source strength (activity in GBq), Γ = gamma ray constant in Gy/h at 1 m from 1 GBq, x = thickness (cm) of a shield between the person and the source, HVT = half-value thickness (cm) of the material of the shield.

5.4   Internal Radiation Hazard

Factors that affect the internal radiation hazard of a radioisotope are (1) chemical nature, (2) organs/tissues in which it is concentrated, (3) effective half-life, (4) energies of all types of radiations (alpha, beta, gamma), (5) mode of intake (inhalation, ingestion, etc.), particle size (in the case of inhalation), metabolic behaviour, etc. Various limits, constraints and investigation levels have been published in various ICRP publications.

ICRP 130 (2015) [6], ICRP 134 (2016) [7], ICRP 137 (2017) [8] and ICRP 141 (2019) [9] provide limits for occupational intakes for radionuclides in different parts of the body. They describe the assessment of biokinetic and dosimetric models, individual and workplace monitoring methods and general aspects of retrospective dose assessment. They also provide data on inhalation, ingestion and systemic biokinetics, including individual elements, their radioisotopes, information on chemical forms, their physical half-lives and decay modes.

5.4.1   Effective Half-Life

When radioactive material gets into the body through ingestion, inhalation or subcutaneous absorption (through wounds), it gives rise to an internal radiation hazard. Any activity inside the body delivers the dose to the organ (and nearby organs) if it resides in the organ until its activity reduces to the negligible value by radioactive decay or it is eliminated from the body by biological processes. Thus, the effective half-life of a radioisotope in the body results from the physical half-life of the radioisotope and the biological half-life of the chemical compound. The effective half-life (T_e) is expressed in terms of the physical half-life (T_p) and biological half-life (T_b) by the following equation:

$$T_e = \frac{Tp \times Tb}{Tp + Tb}$$

$$(5.16)$$

The biological half-life of an isotope may vary, to some extent, from person to person depending on the biokinetic behaviour of the human body. For example, the biological half-life of ^{131}I in the inorganic form (present in the blood before absorption in the thyroid) is of the order of a few hours. In contrast, it is in the range of 15–80 days when it becomes organic (in the thyroid), depending upon the individual's age [10]. However, this varies considerably among individuals and in different countries due to different levels of stable iodine in the diet [11].

5.4.2 ALI and DAC

Considering all factors mentioned in Sect. 5.4, a quantity called annual limit on intake (ALI) is calculated for each radioisotope (in a given chemical form and for each intake mode). By definition, ALI means the greatest value of the annual intake of the specified radionuclide, resulting in a committed dose equivalent not exceeding the annual dose equivalent limit prescribed by the competent authority, even if intake occurred every year for 50 years. ALI values for a few commonly used radioisotopes are given in Table 5.4. Usually, ALI value for inhalation is 10 times more than ALI for ingestion.

Thus, 2 MBq of ^{125}I intake taken today by inhalation, as a single intake (or spread over the whole year in small quantities), results in a total whole-body equivalent dose in the next 50 years not more than 20 mSv. Two MBq of ^{125}I can be permitted every year.

Another useful quantity for radioisotope laboratories is derived air concentration (DAC), expressed in Bq/m^3.

$$DAC = \frac{ALI}{2.4 \times 10^3} \qquad (5.17)$$

where 2.4×10^3 is the volume of air (m^3) intake of a standard man during working hours throughout the year.

ALI refers to intake by each worker.
DAC refers to air contamination in the working place.

5.4.3 Surface Contamination

While handling radioisotope in unsealed form, workplace contamination is the most common hazard. Good work practices and handling facilities are very important to avoid contamination in the workplace.

Any detection of contamination on clothing above background needs segregation and appropriate action. Similarly, contamination on the hands, feet or any body part should be thoroughly decontaminated. Recommended limits for contamination on work surfaces are as follows.

For alpha: 10^{-5} μCi/cm^2 (0.37 Bq/cm^2).
For beta: 10^{-4} μCi/cm^2 (3.7 Bq/cm^2).

Surface contamination limits for different areas are given in Table 5.5.

Table 5.4 ALI and DAC values for important radioisotopes used in medicine

Radioisotope	ALI (Bq)		DAC (Bq/m^3)
	Ingestion	Inhalation	
^{99m}Tc	1.0×10^9	2.0×10^9	8.3×10^5
^{123}I	9.0×10^7	2.0×10^8	8.3×10^4
^{125}I	1.0×10^6	2.0×10^6	8.3×10^2
^{131}I	8.0×10^5	1.0×10^6	4.1×10^2
^{67}Ga	8.0×10^7	2.0×10^8	8.3×10^4
^{201}Tl	3.0×10^8	4.0×10^8	1.6×10^5
^{32}P	8.0×10^6	1.0×10^7	4.1×10^3
^{89}Sr	6.0×10^6	1.0×10^7	4.1×10^3

Table 5.5 Surface contamination limits

Category of areas	Limit of surface contamination
1. Monitored areas (e.g. inside fume hood)	10^{-3} μCi/cm^2 (37 Bq/cm^2)
2. Laboratory areas (surveyed)	10^{-4} μCi/cm^2 (3.7 Bq/cm^2)
3. Other non-active areas	10^{-5} μCi/cm^2 (0.37 Bq/cm^2)

References

1. International Union of Pure and Applied Chemistry. Compendium of chemical terminology gold book. Version 2.3.3. 24 Feb 2014; 2014.
2. Smith DS, Stabin MG. Exposure rate constants and lead shielding values for over 1,100 radionuclides. Health Phys. 2012;102(3):271–91.
3. Madsen et al. The American Association of Physicists in medicine task group 108: PET and PET/CT shielding. Med Phys. 2006;33(1).
4. Kharrati H, Agrebi A, Karaoui M-K. Monte Carlo simulation of x-ray buildup factors of lead and its applications in shielding of diagnostic x-ray facilities. Med Phys. 2007;34:1398–404.
5. Shimizu A, Onda T, Sakamoto Y. Calculation of gamma-ray buildup factors up to depths of 100 mfp by the method of invariant embedding, (III) generation of an improved data set. J Nucl Sci Technol. 2004;41:413–24.
6. International Commission on Radiological Protection. Occupational intake of radionuclides: part 1. ICRP Publication 130. Oxford: Pergamon Press; 2015.
7. International Commission on Radiological Protection. Occupational intake of radionuclides: part 2. ICRP Publication 134. Oxford: Pergamon Press; 2016.
8. International Commission on Radiological Protection. Occupational intake of radionuclides: part 3. ICRP Publication 137. Oxford: Pergamon Press; 2017.
9. International Commission on Radiological Protection. Occupational intake of radionuclides: part 4. ICRP Publication 141. Oxford: Pergamon Press; 2019.
10. International Commission on Radiological Protection. Age-dependent doses to members of the public from intakes of radionuclides: part 1. ICRP Publication 56. Oxford: Pergamon Press; 1989.
11. Stather JW, Greenhalgh JR. The metabolism of iodine in children and adults. UK National Radiological Protection Board Report NRPER140 Chilton. 1983.

Occupational and Public Exposure to Nuclear Medicine

6

Abstract

Exposure due to nuclear medicine practice can be either external or internal. External exposure is fairly straightforward, while internal exposure is quite complex and often involves approximation and conservative estimation. This is due to many uncertainties such as biological clearance of isotopes, differential uptake by certain tissues and limited knowledge on the transport of these isotopes from one organ or system to another. These uncertainties hold well even while estimating patient dose. However, apart from optimizing patient dose, recording patient dose for radiological protection is not done as the patient dose is a part of neither occupational nor public exposure. Even when a radiation worker undergoes any diagnostic procedure involving radiation, the incurred dose is not considered part of the occupational dose.

Identification of sources of exposure, types of exposure, situations, categories of exposed individuals, typical exposure patterns in occupational setup and order of doses involved in various procedures are being discussed in this chapter.

6.1 Introduction

Most of the isotopes used in nuclear medicine practice have a short half-life in nature. This is to ensure minimal patient dose after the diagnostic procedure and to ensure the public's safety in particular. This also helped in the safe handling of radioactive waste generated during the procedure. The safe disposal criteria for this waste are strictly based on stringent dose limits considering a conservative approach. Such disposals should not lead to excess exposure to either public or occupational workers, even under extreme conditions [1].

6.2 Type of Exposures

The types of radiation exposure depend on the exposure scenarios. The exposure may be inadvertent and unavoidable and may involve lifesaving tasks, routine and deliberate with the intent of medical need. They can be classified into various scenarios.

6.2.1 Planned Exposures

In this scenario, the safety and dose limits are in-built. The exposure that can happen due to the planned operation of a radiation facility can belong to this category.

6.2.2 Emergency Exposures

This scenario is called during unintentional or malicious intents leading to unexpected radiation exposure scenarios requiring prompt interventions to avoid larger damage or life-threatening situations.

6.2.3 Existing Exposure Situations

This includes exposure due to natural background, residual radioactive material from past practices or those remains after an emergency operation is concluded.

6.3 Categories of Exposures

Depending on the exposed individual(s), we can categorize the nature of exposures. The exposed individual may be a member of the public, a patient, a comforter, a radiation worker and emergency workers such as fire, disaster management, etc.

6.3.1 Occupational Exposure

Exposure is due to one's occupation with the installed radiation device (medical facilities, accelerators, reactors, etc.) or naturally existing (mines, radon or other airborne radioactive materials due to excavation, etc.). Either external or internal exposure will amount to the recording of occupational exposure. In most cases, these exposures are within the dose limits. Doses exceeding these set limits call for constraints on the work of occupational workers. The constraints depend on the severity of exposure, which may amount to shifting radiation workers to a non-radiation-related work or simply a revised dose limit for an applicable period. Occupational exposure does not include natural background and any other exposure not involving one's occupation.

6.3.2 Public Exposure

The exposure that can involve members of the public due to any man-made radiation technol-ogy or activity, including medical procedures, in which the public is not directly involved can be recorded under this category. Public dose limits and constraints need to be applied while optimizing any radiation facility or the number of facilities near public occupation to ensure adherence to dose limits collectively by all the facilities. Public dose limits and constraints are far less than occupational dose limits.

6.3.3 Medical Exposure

Either diagnostic or therapeutic exposure to radiation for any individual/patient with the intent of medical reason is recorded in this category. No dose limits are applicable in this category as the exposure is done for the patient's benefit. Dose constraints are applied to ensure and minimize the patient's dose.

6.4 Identification of Exposed Individuals

Identification of exposed individuals based on both exposure and category of exposure.

6.4.1 Workers

Exposures related and only due to occupation, excluding medical and background exposures, are considered here. Workers will be considered public when they are off-duty or not directly involved in an occupation or undergo medical exposure.

6.4.2 Members of the Public

Individuals who are not directly involved in any of the exposure categories, most likely exposed under 'existing exposure situations', are considered. Since members of the public include sensitive populations such as children, infants and pregnant women (more precisely, foetus), this cohort is usually considered the most sensitive one, and minimal dose limits are applied. Besides, members of the public are not monitored individually for recording dose; they do not derive direct benefit though they may be indirect beneficiaries, they do not get/earn monetary benefits as in workers, and they are likely to have their occupational risks; hence, for ethical reasons, minimal dose limits are considered.

6.4.3 Patients and Comforters

Patients are the direct beneficiaries of medical procedures. The exposure may be intentional as in therapeutic ones or unintentional as in diagnostic procedures where the purpose is not to deliver the dose but to get the image/information. Patient comforters such as relatives/friends are usually considered volunteers, not the direct beneficiaries but are involved for the benefit of caregiving, which is voluntary. In most cases, the exposures are in one-time instance type, and hence, the dose limits/constraints are relaxed for them. No dose limits are applicable for patients as the medical procedures are intended to benefit while overweighing the risks involved.

Dose constraints have been prescribed for comforters and visitors of the patients. For adults, it is 5 mSv, and for the child, it is 1 mSv.

6.5 Death of Patient-Administered with Radiopharmaceutical

When a patient dies with the therapeutic amount of radiopharmaceuticals in the body, the body should only be released to family when the levels of activity in the body are below the regulatory authority's allowed limit for cremation, burial, post-mortem or embalming. If the amount of activity is higher than the required limits, the body should be kept in the morgue to allow the activity to decay and then released when the levels fall below the prescribed limits. If the situation necessitates immediate body release, the organs with accumulated activity are removed first. Furthermore, even after removing important organs containing radioactivity, the amount of activity at the moment of the body's discharge

must not exceed the limit. The radioactive organs should be disposed of according to the disposal rules.

6.6 Possibilities of Exposure in Nuclear Medicine

When a patient undergoes nuclear medicine procedure, they will carry the radioactive isotope in their body. The isotope distribution may be nearly uniform or maybe highly localized, depending on the case. Most of these isotopes emit gamma rays and, hence, can lead to exposure to individuals nearby, such as nursing staff, comforters and members in the residence and public after discharge of the patient. Besides, many wastes generated by the patient, such as medical items (swabs, syringes, etc.), urine, faeces, sweat, clothes, etc., which are finally disposed of, can also contribute dose to those in the vicinity of those handling them [2].

The estimation of the external dose is fairly simple. Even though the isotopes are distributed in the body and act as large area or complex area sources, the geometry can be approximated using appropriate phantom geometries such as cylindrical or line sources for dose estimations. For internal exposures, estimation methods are very complicated, depending on the biochemistry of the radionuclides, their route of entry, physical and biological half-life for clearance and many more factors that need to be considered.

The major component of artificial exposure (man-made) sources is the use of radiation in the medical field. More than 80% of artificial radiation dose comes to people from medical diagnostic procedures. The average background dose resulting from natural radioactivity and man-made is about 2–3 mSv. Eighty-seven percent of this is contributed by natural radiation, and the remaining 13% is from man-made sources. Therapeutic exposures are not considered as the intention is deliberate in such a scenario. This is shown in Fig. 6.1 in a pie chart.

With the availability of many artificially produced short-lived radioisotopes, medical diagnosis has become easier and more specific in many

Fig. 6.1 Percentage contribution of natural and man-made radiation to background radiation

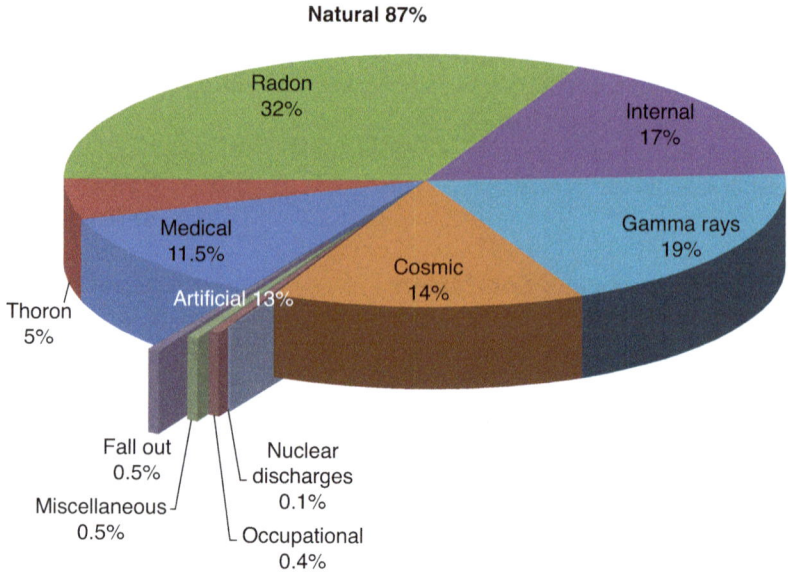

Fig. 6.2 Pathways of radioactivity by the internal route

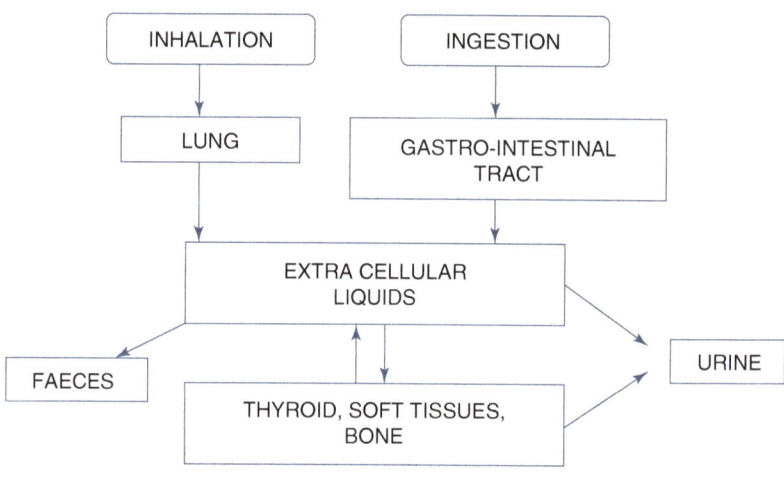

diseases. However, using these radioisotopes in unsealed form has potential radiation hazards to the persons handling the radiation sources and the people around if they are not used properly and the facilities provided are not adequate. The patients are also likely to be affected if misadministration takes place. The radioisotopes used in medicine are ^{32}P, ^{67}Ga, ^{89}Sr, ^{99m}Tc, ^{131}I, ^{201}Tl, etc. ^{32}P, ^{89}Sr, ^{131}I, ^{177}Lu, ^{186}Re, ^{188}Re and ^{153}Sm are mostly used in therapeutic applications.

Radiation hazards are classified into two categories: (a) external hazard, which is due to the sources present in the surroundings, such as in storage, during transportation, while handling, etc., and (b) internal hazard, which is due to the radioactive materials which get into the body of the person by way of ingestion, inhalation or through subcutaneous absorption (through wounds) (Fig. 6.2).

6.7   Elimination of Radionuclides from Internal Routes

Both uptake and elimination of radioisotopes depend on the chemical nature and biochemistry. Many a time, the radioisotopes used are part of radiopharmaceuticals, in which case, the kinetics of the same depending on the biochemistry of the pharmaceuticals. It is to be noted. Generally, there is not much difference in radioisotopes compared to their stable isotopes at low concentrations used for diagnostic or therapeutic purposes. Hence, most of the kinetics of these molecules follow similar biochemistry as that of ones having stable isotopes.

Clearance of radioisotopes can happen either due to biological elimination or physical decay. Physical decay plays an important role in short-lived radionuclides, while in relatively long-lived ones, biological half-life becomes important while assessing the clearance (Fig. 6.3).

6.8   Effective Half-Life

Two types of half-lives are defined to assess the activity accumulated in the human body. The physical half-life is well known and depends on the nature of the radionuclide. It does not depend on the chemical nature of the compound or radiopharmaceutical in which the isotope is incorporated. The biological half-life is quite complex to estimate/understand as it depends on many intrinsic and extrinsic factors and can be changed either by intervention or due to the patient's physiology. It can have many components such as incorporated fraction, humeral fraction, soft tissue incorporation, bone incorporation and digestive track clearance. Nonetheless, for short-lived radionuclides, the incorporated fraction may not play a major role while estimating the dose as the incorporated fraction will get cleared off quickly. Besides, the incorporated fraction is usually very low compared to the humeral or digestive track fraction.

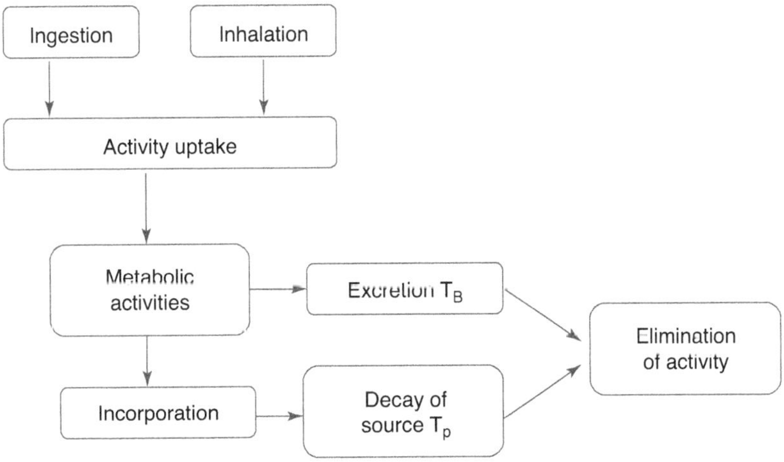

Fig. 6.3 Elimination of activity from internal systems

Table 6.1 Commonly used radionuclides in nuclear medicine

Radionuclide	Half-life	Energy (MeV)
^{99m}Tc	6.03 h	0.140γ
^{131}I	8.05 days	0.2 β (av) 0.364 γ (max)
^{18}F	109.7 min	635 keV (max)
^{68}Ga	271 days	890 keV (av)
^{177}Lu	6.647	Multiple β (497, 384 and 176 keV max) and γ (208, 113 keV)
^{3}H	12.26 year	0.0186 β
^{32}P	14.28 days	1.71 β
^{33}P	24.4 days	0.248 β
^{35}S	87.9 days	0.167 β
^{14}C	5730 year	0.156 β
^{45}Ca	165 days	0.252 β
^{147}Pm	2.62 year	0.224 β
^{36}Cl	3.08×10^{5} year	0.714 β
^{204}Tl	3.81 year	0.766 β
^{147}Pm	2.62 year	0.224 β
^{125}I	60 days	0.354 γ

The effective half-life is a combination of two half-lives, i.e. physical and biological. The physical half-life will have only one component, while the biological can have many. In most cases, only one biological half-life is considered in nuclear medicine for dose estimation or assessment of residual activity during or after the procedure (Table 6.1).

6.8.1 Estimation of Effective Dose

Effective dose can be estimated based on a contribution by external and internal exposures, and these components are additive (ICRP 60). For external dose estimation, an equivalent dose at 10 mm skin depth is considered, while for internal, all components, including ingestion, inhalation, biological incorporation, etc., need to be considered. The equation below shows three components generally used for most radionuclides for occupational exposure assessment, i.e. external, inhalation and ingestion.

$$E = H_{p}(10) + \sum_{j} e_{j\,\text{inh}}(50) I_{j\,\text{inh}} + \sum_{j} e_{j\,\text{ing}}(50) I_{j\,\text{ing}}$$

where

$H_{p}(10)$ = the personal dose resulting from external exposure

$e_{j\,\text{inh}}(50)$ = the committed effective dose coefficient for activity intake by inhalation of radionuclide j

$e_{j\,\text{ing}}(50)$ = the committed effective dose coefficient for activity intake by ingestion of radionuclide j

$I_{j\,\text{ing}}$ = the activity intake of radionuclide j by ingestion

ICRP 60 had used age-specific computational models defined by the Medical Internal Radiation Dose (MIRD) committee. The latest recommendation has adopted the computational models based on medical topographic images. The anatomy is described by Voxel (3D volume elements) (Voxel phantom).

For internal dose estimation, the dose contribution of up to 50 years is considered. For record purposes, the dose is recorded on the year of exposure in most cases. Since dose limits and constraints are applied to effective doses, the protection part is inherently addressed irrespective of the route of administration of radionuclides.

6.8.2 Dose to Extremities and Individual Organs

Hands and feet (limbs), skin areas and eye lenses are considered vital individual organs that are likely to contact either radiation or radioisotopes while working with them. These organs are not very sensitive to stochastic effects point of risk and hence possess very low tissue weighting factors. For example, the skin has a tissue weighting factor of 0.01. Assuming only the skin is exposed, the annual limit of exposure to skin corresponding to the effective dose of 20 mSv will amount to 2000 mSv of the equivalent dose. To avoid such conditions and the possibility of exposure to these organs, separate limits are imposed. These limits are usually based on the threshold of deterministic effects for chronic exposure conditions.

To measure dose to extremities, separate dosimeters are being used. For example, wrist monitors for the dose to hands, eye dosimeter is worn on the forehead, or lead goggles for eye lens dosimetry are popular. Interventional radiologists and nuclear medicine professionals are the target workers to monitor.

6.8.3 Occupational Exposure of Women

There is no difference in the bases for controlling occupational exposures of women who are not pregnant and men. The commission recommends no special dose limits for women in general. Once a pregnancy has been confirmed, the conceptus should be protected by applying a supplementary equivalent dose limit of 2 mSv to the surface of the woman's abdomen (lower trunk) for the remainder of the pregnancy, as well as limiting radionuclide intake to around 1/20 (0.05) of ALI. The dose limit of radiation exposure from all kinds of occupational exposures to the foetus for the remainder of the pregnancy period after declaration of the pregnancy is 1 mSv (ICRP 103). It is the responsibility of the woman to declare the pregnancy on the work front.

6.8.4 Apprentices and Students

No occupational exposure is permitted below the age of 18 years. The use of radiation by a student below the age of 18 should be discouraged. For students between 16 and 18 years of age, the recommended limit for effective dose is 5 mSv and the equivalent dose to the lens is 5 mSv and to the skin or the extremities, 150 mSv. These doses are about 30% of the dose limits for occupational exposures for adults. For apprentices and students above 18y, the same dose limits as occupational workers shall be applied.

6.9 Methods to Prevent or Reduce the Dose to Occupational Workers

The principle of prevention of internal dose is by reducing the effective half-life by increasing the excretion rate in case of any contamination leading to the ingestion of radionuclides. The following steps are usually taken depending on the severity of contamination.

(i) Administration of diuretic.
(ii) Prevention of uptake by administering chelating agents.
(iii) Prevention of absorption in the intestine.
(iv) Washing of wounds and body surface.
(v) Prompting vomiting when ingested.

Prevention is also possible in most scenarios by using simple protective devices routinely deployed in work areas such as nuclear medicine administration or preparation of sources. Mask, hand gloves, clean working conditions, cold traps for volatile radioactive solutions, fume hoods, portable charcoal filters with or without exhaust, lead goggles or screens or acrylic screens and many more such safety devices are being used. Remote handling devices and glove boxes are being used for handling intense sources. These devices are often provided with air filtration to avoid airborne activity.

6.9.1   Storage of Source

- When not in use, radioactive materials should be kept in an adequately shielded storage place assigned exclusively for this purpose. Radiation symbols with appropriate warning inscriptions must be displayed conspicuously outside the storage area.
- Radioactive materials should be stored in their respective adequately shielded containers.
- The storage areas should be regularly checked for possible air and/or surface contamination.
- Every radioactive source should bear an identification tag wherein details such as the nature, activity and time of activity measurement are entered.
- Appropriate records of radiation sources with details such as the nature, activity, time of measurement of the activity, persons to whom and purpose for which they were issued should be maintained.
- Bottles and vials containing radioactive liquids should be placed in vessels large enough to hold the entire contents of the bottles or vials. In addition, the inside of the vessels should be lined with sufficient absorbent material to absorb the entire contents of the bottles in case of breakage.

6.9.2   Essential Points in Planning Work

- A minimum quantity of radioisotope, which will suffice for the purpose, should be used.
- Preferences should be given to those of short $T_{1/2}$, low radiation energy and low toxicity.
- Maximum distance, compatible with effective working, should be maintained between source and worker.
- Time spent in the vicinity of the source should be minimum.
- Shielding should be used wherever necessary.
- Operations with significant quantities of unsealed sources should be conducted in fume hoods.
- Care should be taken to confine the spill to a limited area by using trays with absorbent paper. Any uncontrolled spills should be dealt with promptly.

6.9.3   Handling of Sources

- Good housekeeping procedures should be maintained in the radioisotope laboratories.
- Items such as glassware, tools and equipment meant for use with unsealed sources should be identifiable and used exclusively for this purpose.
- All manipulations involving radioactive material should be carried out in a suitable double tray with a disposable lining such as polythene. In addition, sufficient absorbent material should cover the disposable lining to soak up any possible spill in the case of solutions. The absorbent material and the lining should then be treated as radioactive waste.
- In all radioisotope laboratories, personnel should wear protective clothing such as laboratory or surgical coats, and these coats should not be used outside the radioisotope laboratory, and for this purpose, they must be identifiable. Laboratory coats and personal clothing of personnel should be kept in separate cubicles to avoid cross-contamination.
- A suitable trolley or carrier should be used to transport a consignment containing radioisotope(s). Unsafe practices such as dragging or kicking boxes should be avoided to eliminate the possibility of damage.
- Surgical gloves should be worn during work with radioactive materials to avoid contamination of hands. Gloves should be put on and removed so as not to contaminate the inner side of the gloves and the hands. For this purpose, the inside and outside of the gloves should be distinguishable.
- After use, the gloves must be washed and checked for contamination if they are meant for reuse, and in the event of persistent contamination, they should be discarded and treated as active wastes. Disposable gloves are preferred.
- Pipetting by mouth should never be done in a radioisotope laboratory. Pro-pipette or remote pipette should be employed.
- Smoking, eating or drinking should not be permitted in the areas where radioactivity is handled.

6.10   Overexposure Investigations and Follow-Up

The equivalent doses of 10 mSv and above recorded by a personnel monitoring badge in a monitoring period are considered overexposure, and the same should be intimated urgently to the concerned institution for prompt investigations. The case of overexposure has been covered in detail in Chap. 11.

6.11   Occupational Exposures and Dose Records

The results of personnel monitoring shall be preserved as dose records during the lifetime of the person concerned or at least 30 years after the cessation of work involving exposure to ionizing radiation, whichever is longer. The dose records provide information to the regulatory authority to ensure that both workers and employers comply with the stipulated regulation.

6.12   Roll of Personnel Monitoring

The personnel monitoring of the radiation workers in the nuclear medicine departments has a significant role in radiological protection. There have been numerous occasions and examples where, based on personnel exposures recorded by the personnel monitors in nuclear medicine departments, the status of working conditions and non-availability of required radiation handling devices and procedures could be anticipated and confirmed by way of subsequent radiation protection surveys. The average dose per person in the nuclear medicine department is quite low and has remained constant in the last decade.

6.13   Comparison of Occupational Nuclear Medicine Laboratories Compared to Other Medical Practices

Table 6.2 shows the growth of personnel monitoring for medical institutions in India since 1975. The growth of radiation workers in the medical profession has gone up by a factor of two between 1975 to 1990 but has shown a constant trend since then. The average equivalent dose/person has come down from 0.95 mSv in 1975 to 0.48 mSv in 2004. A similar trend is also observed in nuclear medicine departments regarding the increase in radiation workers, but the average annual dose per person has been reduced, as shown in Table 6.3. The average dose per person for the nuclear medicine department is of a higher order than radiotherapy and diagnostic radiology despite the fact that the dose rates encountered in nuclear medicine are very small compared to X-ray diagnostic radiology or radiotherapy, as shown in Table 6.4. This is because the nuclear medicine department handles unsealed sources with higher radiation hazards, whereas sealed sources and radiation-generating equipment with collimated beams and adequate shielding are handled in other medical practices.

Safety and security of sources and safe work practices incorporated in the protocols helped achieve much lower personnel doses for workers in the field of medicine in general and nuclear medicine in particular. Nuclear medicine is considered one of the safest practices in radiation applications even though open isotopes are being used and a large number of professionals, patients and the public are involved in this field.

Table 6.2 Personnel monitoring service data for medical institutions

	1975	1980	1985	1990	1995	2000	2004
No. of institutions	1165	1601	1900	2256	2199	2079	2530
No. of persons	8560	11,753	13,631	17,054	15,759	17,303	23,600
Avg. dose/person (mSv)	0.95	0.68	0.54	0.41	0.60	0.49	0.48

Table 6.3 Personnel monitoring service data for nuclear medicine laboratories

Year	No. of persons	Av. annual dose (mSv)	Av. exposed persons (mSv)
1985	554	0.94	2.10
1990	771	0.69	1.42
1995	761	0.78	1.18
1996	784	0.63	0.96
1997	706	0.76	1.12
1998	678	0.62	1.06
1999	682	0.71	1.19
2000	746	0.54	0.86
2002	1023	0.72	1.17

Table 6.4 Average effective dose (mSv) of occupational workers engaged in various medical practices in India

Year	Diagnostic radiology	Radiotherapy	Nuclear medicine
1990	0.24	0.80	0.62
1991	0.22	0.72	0.74
1993	0.21	0.61	0.62
1995	0.24	0.71	0.73
2000	0.24	0.55	0.54
2002	0.26	0.46	0.72

References

1. IAEA Safety Standards for protecting people and the environment. *Radiation Protection and Safety in Medical Uses of Ionizing Radiation*. Specific Safety Guide No. SSG-46. 2018.

2. IAEA Safety Standards for protecting people and the environment. *Occupational Radiation Protection*. General Safety Guide No. GSG-7. 2018.

Biological Bases of Radiation Protection

7

Abstract

The unregulated use of radiation technology and isotopes in the initial days of their invention has led to many radiological incidents involving untoward exposure of radiation. There was a realization on harmful effects of radiation on living systems besides its benefits. Such effects prompted the scientific community to undertake systematic studies on biological effects of radiation which further lead to the bases of radiation protection. The studies not only comprised epidemiological but also many laboratory experiments with animal models and cellular systems. Such studies not only helped to protect radiation workers in particular but also helped the public and environment.

These studies have led to better quantification and qualitative understandings and also an approach to understand mechanistic details of underlying mechanisms creating damage and their progression into complex response of life. The knowledge on risk has also helped to set limits and improve safe work practices. This chapter discusses basics of radiation biology, biological effects of radiation and their implications in radiation protection.

7.1 Introduction

All living organisms are made up of a fundamental unit of life, namely, the cell. Human beings have nearly 10^{14} cells of different types. Cell is made up of different organelles, namely, a central nucleus, cytoplasm and an envelope called the cell membrane. Cell nucleus contains the chromosomes made up of deoxyribonucleic acid (DNA) and basic proteins. DNA molecule constitutes the genetic material of the cell and is the most important target molecule for the induction of radiation damage in cells.

In the human body, we come across many different cell systems. In some organs the cells continuously divide, undergo maturation by a series of modifications (differentiation) and then give rise to functional cells. An example of such organs is the red bone marrow where the primitive type of stem cells continuously divide and give rise to matured blood cells like red blood cells, white blood cells, platelets, etc. Cells in the epidermis of the skin also divide continuously to replace the surface layer. Cells lining the intestines are also continuously replaced. In adults the cells in the testis (spermatogonia) continuously divide and differentiate giving rise to matured sperm cells. In females, oocytes divide and differentiate giving rise to matured ova. In many other organs, the cells divide very slowly, e.g.

kidney, liver, connective tissues, etc. In adults, the muscle and nerve cells are highly differentiated and do not divide at all.

7.2 Radiation Effects at Cellular Level

7.2.1 Mechanism of Damage

As mentioned earlier DNA molecule constitutes the most radiosensitive molecule in living cells. Radiation induces damage in cells either by directly ionizing the DNA molecule or indirectly by producing chemically reactive species, namely, the radiolytic products of water. These in turn react with the biological targets and inflict structural changes leading to deleterious effects.

Internal exposure to radiation occurs mostly due to inhalation or ingestion of radioisotopes. Many volatile and gaseous isotopes enter the human body by inhalation. Some of the isotopes like ^{3}H can be directly absorbed through the skin. Some of the isotopes may accumulate specifically in a particular tissue, for example, ^{90}Sr in bones and ^{131}I in the thyroid gland, whereas many others like tritium and ^{137}Cs may get distributed in the system. The isotopes that are poorly absorbed in the intestines deliver the dose mostly to the GI tract and then get excreted. The isotopes, which are highly soluble and readily absorbed into the blood stream, can irradiate many parts of the body and also deliver a significant dose to the kidney during the excretion. The hazard due to an isotope depends mainly on the effective half-life of the isotope, activity of the isotope and the energy of the radiation emitted. The hazard due to an internalized isotope can be assessed by using biokinetic models based on extensive experimentation in model animal systems. The limits for intake of radionuclides based on the level of hazard are reflected in their ALI values. Annual limit intake (ALI) of a given radionuclide is defined as the intake by inhalation, ingestion or through the skin in a year by the reference man, which will result in a committed dose equivalent to the dose limit (20 mSv). ALI is expressed in units of activity. Internal exposure is assessed by various methods such as whole-body counting (for isotopes which emit adequately energetic photons) and bioassay techniques in urine blood, other body fluids and excreta. Internal exposures by inhalation are restricted by constantly monitoring the working environment where necessary. The air concentrations of volatile radionuclides may have to be controlled by proper ventilation. Radiation workers are also monitored for external contamination by contamination monitors. The level of activity of the order of few MBq to K Bq entails only low-level radiation exposures. Their possible deleterious effects are generally restricted to an excess risk of stochastic effects.

7.2.2 Nature of Damage

Progression of damage by ionizing radiation can be classified into different categories based on nature of damage and time scale. The initial physical and chemicals steps are very fast and lead to formation of primary lesions. The final expression of biological effect takes a very long time. Expression and progression of the primary damage into biological effects involve many complex biochemical and physiological interactions.

Ionizing radiation-induced damage mainly causes breaks (single and double strand breaks) in the large DNA molecule or brings about structural changes such as base damage and cross-linking. Microscopic examination of irradiated cells undergoing division indicates various alterations in the structure of chromosomes, namely, the chromosome aberrations. Chromosome aberrations are of different types—dicentrics, deletions, transpositions and inversions. In fact, increase in the frequency of dicentric chromosomes in the human lymphocytes (WBC) is used as a biological dosimeter. An exposure over 100 mGy (10 rad) can be detected by this technique.

7.2.3 Effects at Cellular Level

Damage to DNA and chromosomes can lead to cell killing (lethal damage) or heritable alterations in the genetic material of the cell leading to mutations (non-lethal damage). Cells with heavy damage generally die during mitosis (mitotic death). This is why the expression of damage takes place much earlier in organs which have fast dividing cells (e.g. bone marrow, intestinal crypt cells, skin and reproductive system). In some of the tissues like eye lens, kidney, connective tissues, liver, etc., the expression of damage is delayed.

Non-lethal damage which can result in modification in the information content in the cells (DNA) can also be harmful. At present it is believed that the alterations in the genetic material may be the first step of radiation carcinogenesis, whereas such changes (mutations) in germ cells (sperms and ova) are the basis of induction of hereditary effects (genetic effects) which appear in future generations. It must however be emphasized that the damage at cellular level alone is not a sufficient condition for the induction of carcinogenesis and genetic effects. There are many steps where the damaged cells can be eliminated before they can cause the end effects (Fig. 7.1).

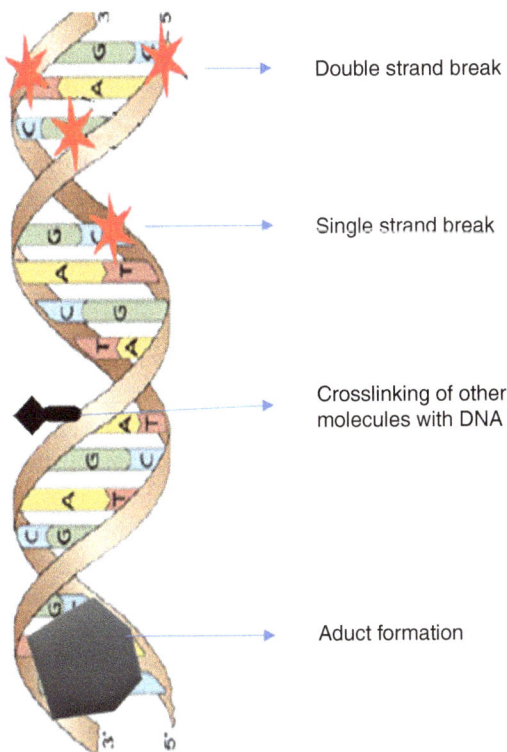

Fig. 7.1 Various types of DNA damages

7.2.4 Factors Modifying the Damage

Many physical, chemical and biological factors modify the radiation damage. Such factors are type of radiation, dose rate and dose fractionation, presence of sensitizers and protectors, type of cells, ability of recovery from damage, etc. Generally, radiations which produce dense ionizations such as neutrons and alpha particles produce more damage in cells than the sparsely ionizing radiations like X-rays, beta rays and gamma rays. Hence the same dose of alpha rays or neutrons produces more damage than X-rays or gamma rays. However due to limited penetration of alpha rays and beta rays, the hazard from this radiation is mostly due to internal irradiation, whereas the hazards from gamma rays and X-rays and neutrons are from external irradiation.

Higher dose of any radiation causes much more damage than low doses. The intensity of irradiation is also an important factor. If a large dose of radiation is delivered in a short period of time, it causes more damage than the same dose delivered over a protracted period or in many fractions. This is due to the ability of the cells to repair the radiation-induced damage. These observations hold good even for multicellular organisms. Human beings have survived radiation doses up to 15 Gy received over a period of 2–3 months. It has now been demonstrated that the extent of genetic damage is reduced by a factor of three at low dose rates.

7.3 Relative Biological Effectiveness (RBE)

As a thumb rule, densely ionizing radiation or high linear energy transfer (LET) such as charged particles and energetic neutrons has tendency to create more and complex damages in biological systems. Often such damages lead to severe effects such as death of cells or mutations. To quantify the efficacy of a radiation, relative biological effectiveness (RBE) is defined [1]. It is a measurable quantity, specific to a type, energy and biological endpoint defined. RBE can be defined as ratio of dose of reference radiation to that of test radiation under consideration to create the same biological effect.

Double strand break

Single strand break

Crosslinking of other molecules with DNA

Aduct formation

$$RBE = \frac{\text{Dose of Co} - 60 \text{ gamma radiation}}{\text{Dose of reference radiation}} \Bigg|_{\text{For a given type and level of effect}}$$

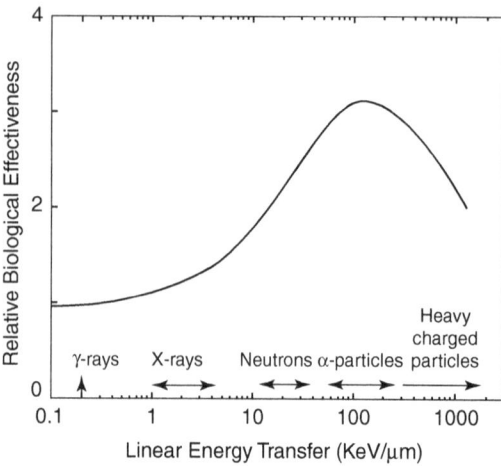

Fig. 7.2 Variation of RBE with LET

With increase in LET, RBE initially increases and shows the tendency to saturate at around 60–200 keV/μ range, which is typically the LET of α particles. Beyond this, uptrend of RBE stops and starts declining due to excessive deposition of energy and its wastage (Fig. 7.2).

7.4   Law of Bergonie and Tribondeau

Some tissues or organs are more sensitive to radiation due to the type of cells they are made up of. A thumb rule to predict sensitivity of a cell says a type of cell is radiosensitive if:

1. They have a high mitotic rate: A cell with more division rate.
2. They have a long mitotic future: A cell which has capability to divide.
3. They are of a primitive type, i.e. less differentiated.

7.5   Deterministic and Stochastic Effects

On the basis of mechanism of induction, radiation effects are broadly classified into two cate-

gories, viz. the deterministic effects and stochastic effects [2].

Radiation can result in two important effects in cells of the human body. A significant level of cell killing in any sensitive organ can result in the manifestation of clinically detectable damage. These are called deterministic effects. Of course, they cannot occur at small doses, as this does not result in significant extent of cell killing.

The second important cellular level effect of radiation is the non-lethal genetic alteration. Such alterations in genetic information in somatic cells appear to be precursor for radiation carcinogenesis, and when such changes occur in germ cells, these can result in genetic disorders in the progeny of the exposed individual. Radiation carcinogenesis and genetic effects are the two most important stochastic effects in human beings.

Deterministic effects are sure to occur at high doses and do not occur below a particular threshold dose. Furthermore, severity of this type of effect is proportional to the dose. Deterministic effects are generally associated with the killing of a large fraction of cells in a specific organ or tissue. On the other hand, stochastic effects do not have a threshold dose, and they are probabilistic in nature. The probability of incidence of such effect increases with dose rather than its severity. Deterministic effects such as radiation sickness, haematopoietic syndrome, gastrointestinal syndrome, CNS syndrome, pneumonitis, cataract, sterility, skin erythema, skin burns and damage to other individual organ systems occur after different threshold doses depending upon the sensitivity of the cells associated with the damage. Time course of manifestation of the damage depends upon the kinetics of cell division in different organs. Organs and tissues that have undifferentiated stem cells with a large fraction of actively dividing cells, such as the bone marrow, testis, small intestines (crypt epithelial cells) and basal layer of the skin, record the damage within a few days to few weeks. On the other hand, in organs such as the heart, liver, kidney, eye lens, muscle and nervous system where the cells divide very slowly or do not divide at all, the damage appears after a long latent period ranging from a few months to several years. The late effects may result due to the slow division kinetics which delays the expression of damage till the entry of these damaged cells into mitosis. Damage to stromal cells and vascular damage resulting in tissue fibrosis may be yet another important cause of delayed expression of damage.

7.6   Acute Radiation Syndrome

7.6.1   Radiation Sickness

Acute whole-body exposure to a radiation dose in the range of 1 Gy and above results in prodromal symptoms. Anorexia, nausea, fatigue, vomiting, diarrhoea, apathy, perspiration, conjunctivitis, fever, erythema, shock, agitation, ataxia and ileus are some of the prodromal manifestations associated with radiation damage. Whole-body exposure to a dose of 1 Gy induces prodromal syndromes in <10–20% of the individuals exposed, whereas an exposure of 3 Gy induces the syndromes in 100% of the individuals. These syndromes appear within a few hours of exposure and disappear within 2–3 days (at low doses). In individuals exposed to 1–2 Gy, the syndromes may appear after several hours and disappear within 2–3 days. However, exposure to higher doses may result in death of the individual and in many late effects among those who survive the initial sickness. Time course of appearance, severity and persistence of these syndromes might be a useful indicator of the dose [3].

7.6.2   Haematopoietic Tissue Damage

Haematopoietic system is composed of lymphatic tissue (lymph nodes, thymus, white pulp of spleen), which produces the lymphocytes and monocytes, and red bone marrow that produces the red blood cells, granulocytes and platelets. Mature lymphocytes are the most radiosensitive blood elements. An acute whole-body exposure to 2–3 Gy results in killing of approximately 50–75% of lymphocytes within 4–7 days. Most of the red bone marrow is distributed in flat bone and the ends of long bones. A dose of 3 Gy could result in the death of 90% stem cells. Recovery of the bone marrow has been observed after 6–8 weeks due to the few surviving stem cells repopulating the bone marrow. At any site the regeneration of the bone marrow can occur following exposure to 10 Gy by stem cell migration and repopulation. However, a local exposure to 20–30 Gy impairs the regeneration process possibly due to acute stromal injury.

A drastic reduction in the leucocytes could result in enhanced susceptibility to infection; a reduction in platelet counts results in extensive internal haemorrhage leading to further loss of blood cells. Acute reduction in red blood cells causes a reduction in oxygen supply leading to tissue hypoxia. Clinical symptoms such as chills, fever, anorexia, anaemia, breathlessness, fatigue, systemic infection, etc. manifest during 3–8 weeks post-irradiation period. A dose of 3–5 Gy can cause death among 50% of the individuals exposed within 1–2 months.

7.6.3   Gastrointestinal Tract Damage

Small intestine constitutes the most radiosensitive part of the GI tract, with the mouth, pharynx, stomach, colon, rectum and oesophagus progressively more resistant. Intestinal epithelial cell layer is replaced every 3–6 days; hence the structural and functional damage to the intestine is expressed within 3–5 days. Stem cells of the crypt of Lieberkuhn show severe functional impairment following a dose of 5 Gy. Prodromal syndromes like anorexia, nausea, vomiting, diarrhoea and abdominal swelling persist for 3–5 days, and death occurs between the first and second week due to dehydration, electrolyte imbalance, infection and circulatory collapse. Death due to GI tract damage may not be caused by a dose less than 10 Gy. Death from GI tract exposure may more likely be caused by ingestion of radioactivity. LD50 dose for internal exposure is estimated to be in the range of 35 Gy.

7.6.4   CNS Syndrome

Following whole-body exposure to doses in the range of >25 Gy, prodromal syndromes like agitation, apathy, convulsions, ataxia and coma appear within hours resulting in death within the first 3 days. Exposure to lethal doses cause severe headache within 3–4 h. Most of the CNS damage-induced deaths (18 deaths from 23 accidents) have occurred in criticality accidents involving doses in the range of 40–50 Gy. Death generally occurred within 100 hours after exposure.

Experience from fractionated radiotherapy shows that the CNS can tolerate a dose of 55 Gy delivered over a period of 5–6 weeks. However, 43 Gy delivered within 3 days can cause brain damage within 20 months. Doses greater than 70 Gy causes severe necrosis of the brain tissue.

7.7   Damage to Individual Organs

7.7.1   Skin

7.7.1.1 Early Effects
Epithelial cells of the basal layer of the epidermis, situated at an average depth of 70–120 µm, constitute most radiosensitive part of the skin. Exposure to doses above 3 Gy could result in transient erythema, which appears within a few hours and disappears within a day or two. Temporary epilation (loss of hair) can also occur following exposure to 3–5 Gy. Depilation generally occurs during 14–18 days post exposure.

Exposure to X-ray doses up to 6–8 Gy results in fixed erythema, which appears between 2 and 3 weeks after exposure. Damage to arterioles of the skin (small arteries) has been considered to be responsible for this type of erythema. Threshold dose for erythema is only 2 Gy for fast neutrons. Exposure to doses above 5 Gy of low LET radiation could also result in the falling off of the top layers of the skin, exposing the dermis (dry desquamation).

Higher doses in the range of 15–20 Gy could result in exudation of fluids from exposed dermis (wet desquamation). The threshold dose for moist desquamation is 8 Gy and ED_{50} dose is 20 Gy. Dose in the range of 15–25 Gy can cause blisters within 15–25 days. Ulceration generally occurs in 2–3 weeks. Doses above 25 Gy could result in necrosis involving degeneration of the skin tissue after 21 days. Very high doses can shorten the time course of manifestation of the different types of damage. Different degrees of damage, its dose response relationship and kinetics of evolution are described in Figs. 7.3 and 7.4. Generally the radiation with higher energy causes greater damage to the skin due to the involvement of

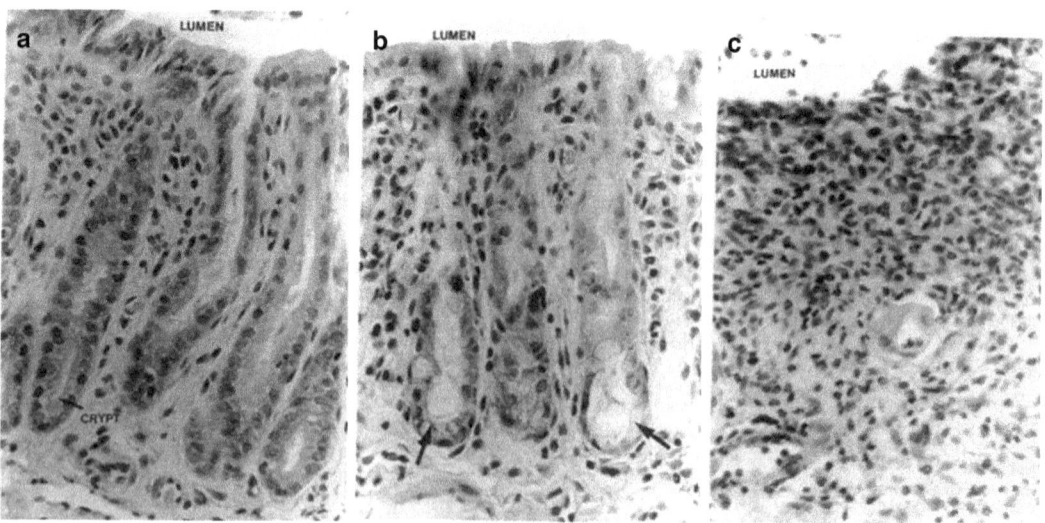

Fig. 7.3 Degradation of lumen in irradiated mice

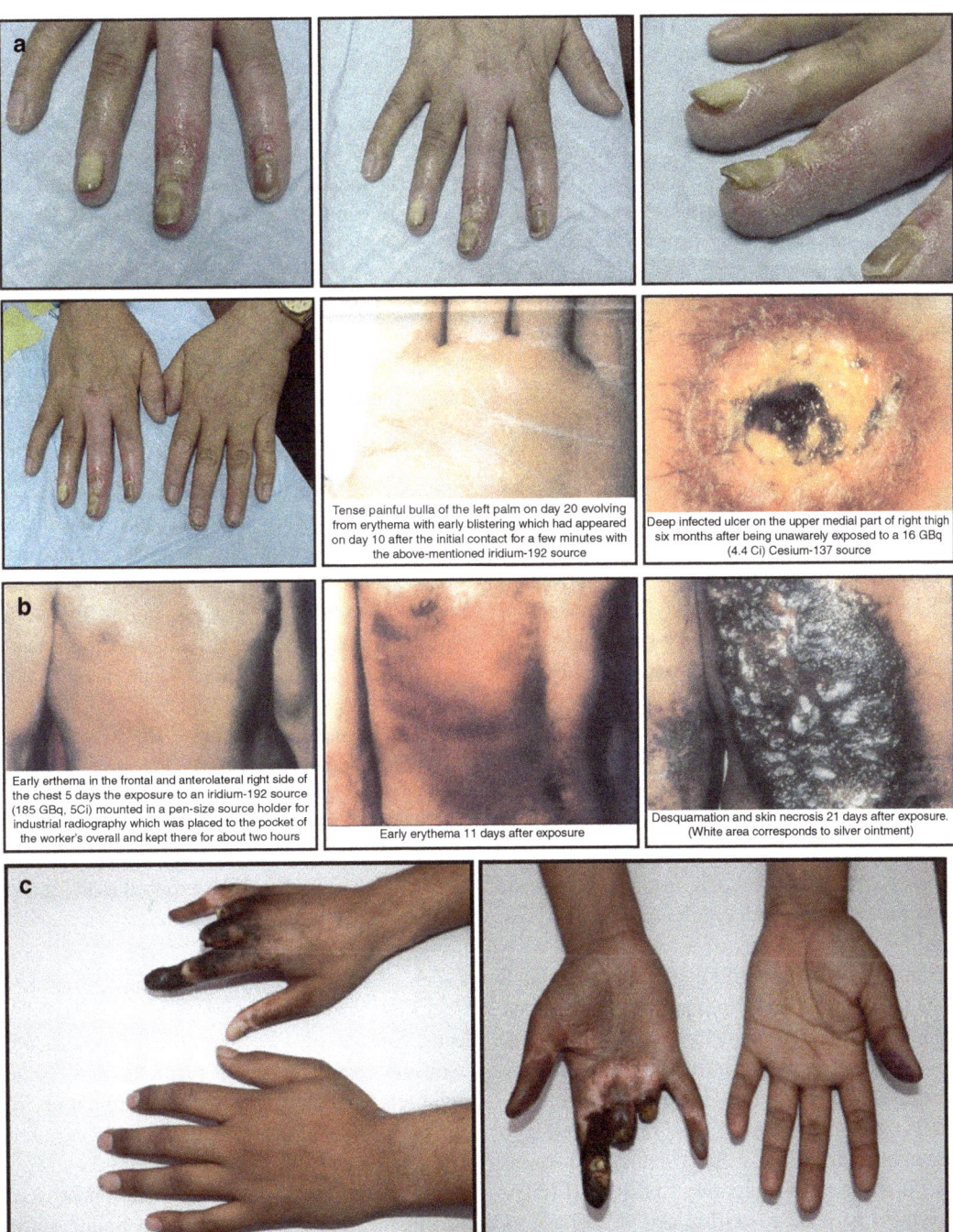

Fig. 7.4 (a) Different degrees of skin damage induced by radiation. Image originally printed in IAEA pamphlet. (b) Progression of radiation-induced skin damage with time. Image originally printed in IAEA pamphlet. (c) Severe damage to the skin and deep tissue due to highly localized exposure

Tense painful bulla of the left palm on day 20 evolving from erythema with early blistering which had appeared on day 10 after the initial contact for a few minutes with the above-mentioned iridium-192 source

Deep infected ulcer on the upper medial part of right thigh six months after being unawarely exposed to a 16 GBq (4.4 Ci) Cesium-137 source

Early erthema in the frontal and anterolateral right side of the chest 5 days the exposure to an iridium-192 source (185 GBq, 5Ci) mounted in a pen-size source holder for industrial radiography which was placed to the pocket of the worker's overall and kept there for about two hours

Early erythema 11 days after exposure

Desquamation and skin necrosis 21 days after exposure. (White area corresponds to silver ointment)

underlying tissues. When radiation is delivered in fractions over a period of 5–6 weeks, the skin can tolerate doses up to 60 Gy.

Several late effects of skin irradiation have been documented in radiotherapy patients. Such effects are contraction and atrophy of the skin, loss of elasticity, non-functioning of sweat glands, epilation resulting from total destruction of hair follicles and extensive dermal fibrosis. Such skin undergoes premature ageing and is also incapable of readily healing any injury.

7.7.1.2 Skin Late Effects

Fractionated, protracted and chronic exposure to radiation can result in skin damage but at much higher doses compared to acute exposure. Besides, after recovery from skin injury due to acute exposure, late effects can manifest after months or years. These effects are typically observed among cancer therapy patients who have undergone localized exposures. Loss of elasticity, sensation, lack of hair growth and discoloration are observed depending on the nature of exposure.

7.7.2 Gonads

7.7.2.1 Males

In males, the spermatogonia in the testis are the most radiosensitive cells of the reproductive system. Doses of the order of a few tenths of a Gy can induce significant killing of spermatogonia B cells in the early part of the spermatogenic series. Reduction in sperm number occurred in around 4 months among the Japanese fishermen exposed to radioactive fallout in Bikini in 1954 (1.7–6.9Gy). Acute exposure to 1–2 Gy can cause temporary sterility in most individuals. The lowest dose that can cause temporary sterility is 0.15 Gy. However, permanent sterility can be induced by doses of the order of 5–6 Gy. Chronic exposure to 1 mGy/day does not impair fertility; hence a yearly exposure of 0.3–0.4 Sv/year may not be associated with any somatic damage.

7.7.2.2 Female

In females the maturing oocytes are the most radiosensitive germ cells. Due to the fixed number of oocytes in females, the dose required to induce sterility in women decreases with increasing age. For example, a dose of 3 Gy can induce permanent sterility in 40-year-old women, while the same dose can induce only temporary amenorrhoea in 20-year-old females. Temporary sterility can be induced by a dose of 1 Gy and may last from a few months to 1–2 years after exposure.

7.7.3 Eye Lens

Eye lens is among the most radiosensitive tissues in the human body. Cornea and conjunctiva are moderately sensitive, while the retina and the optic nerve are radio resistant. Damage to anterior epithelial cells that normally proliferate throughout the lifetime is the cause of cataract induction. With the inhibition of the actively proliferating cells, there is an interference with the normal differentiation process. Following exposure to radiation, damaged cells and the breakdown products migrate and accumulate in the posterior pole of the lens, resulting in opthalmologically detectable opacities. These enlarge to 3–4 mm size and progress to the anterior pole. However, opacities following exposure to doses less than 1–2 Gy do not seriously impair vision. An acute dose of 7.5 Gy can induce vision-impairing cataract in all the exposed individuals.

7.7.4 Lungs

Lung consists of 40 different cell types. Acute exposure to 10 Gy could result in fatal acute radiation pneumonitis (ARP, inflammation of the lungs) in 84% of the individuals exposed and result in 50% mortality. Time, dose and volume exposed are crucial in the seriousness of the lung damage. Mortality from pneumonitis generally occurs during 1–5 months after exposure but most of them during 60–90 days. Those who survive the initial phase of acute inflammation may develop late effects such as fibrosis of alveolar walls and loss of fine vasculature. These late damages render the lung extremely susceptible to infections and pneumonias. Other tissues in the respiratory system are relatively radioresistant.

7.7.5   Endocrine System

Thyroid glands are relatively more radiosensitive compared to the rest of the endocrine glands. Generally, the endocrine glands show greater sensitivity in children as compared to adults. Many of the Marshall islanders (children) exposed to 7–14 Gy of thyroid dose exhibited growth retardation and remarkable decrease in thyroid activity. In adults, the threshold dose for thyroid injury is estimated to be 25–30 Gy for fractionated exposures. For the rest of the endocrine system, which is more resistant, the threshold dose may be in the range of 45–60 Gy. Female breast development may be impaired by a dose of 10 Gy delivered during childhood. The rest of the human system shows radiosensitivity only during childhood; in adults most of the organs show appreciable radio-resistance and exhibit threshold dose values above 30–50 Gy of low LET radiation.

7.7.6   Chronic Radiation Sickness

Chronic radiation sickness develops as a result of prolonged exposure to radiation doses in excess of 0.1 Gy per year to cumulative doses >1.5 Gy. Chronic radiation sickness may appear during 1–10 years depending upon the dose rate and the accumulated dose. This has been reported in Russian radiation workers with cumulative doses in the range of 2–10 Gy. The most important observation is the rapidly progressive leukopenia and the haematological disturbances. But progressive changes in a number of tissues in the endocrine, neurological and cardiovascular systems have also been seen.

7.8   Stochastic Effects

Radiation carcinogenesis and genetic effects of radiation are the only effects that can occur among individuals exposed within permissible limits of radiation. Both these effects occur frequently as a result of exposure to many other environmental agents such as chemicals, heat and unknown agents.

Epidemiological surveys of radiation workers over several decades do not suggest excess cancer risk among the radiation workers. As for the genetic effects of radiation, there is no evidence for the excess incidence of any serious hereditary diseases among the children born to exposed parents at Hiroshima and Nagasaki. Only 2 out of 27,000 children showed minor variation in the enzyme structure. A small increase in the incidence of untoward pregnancy outcomes and childhood cancers related to genetic effects has been observed among the progeny of survivors.

7.8.1   Carcinogenesis in Human Beings

Most of the human data on carcinogenesis is derived from people exposed to heavy doses during the bombing at Hiroshima and Nagasaki and others exposed to large doses for therapeutic reasons. Information from these sources suggest that the bone marrow, female breast, lungs and thyroid are very sensitive to radiation carcinogenesis. Cancers generally appear after a latent period of several years. Leukaemia begins to appear within 3–5 years with a peak incidence between 5 and 7 years [2]. Many solid tumours appear only after 10 years, and the incidence appears to be in excess even after 30–40 years. Children have been found to be sensitive to the induction of leukaemia (acute lymphocytic) and thyroid cancers. Female in the age group of 10–19 years have been found to be particularly sensitive to the radiation carcinogenesis of the breast. Females are two to five times more sensitive to thyroid cancer but less sensitive than males, for radiogenic leukaemia. Carcinogenesis risk decreases with age at irradiation. Children in the age group of 10–20 years may have five times the risk compared to people in an age group of 50–60 years. Females always have a higher risk compared to males (due to the breast and thyroid).

Most of the radiogenic cancer incidence data on human beings is based on exposures in the dose range of 1–10 Gy. Since there is no certainty about the initial shape of dose response curve, risk evaluation is done on the assumption of linear non-threshold type response. Risk at low doses is assessed by linear extrapolation. The International Commission of Radiological Protection (ICRP) has suggested a risk value of 400 cases of excess cancers over the entire lifetime as a result of exposure of 100,000 people occupationally exposed to an average dose of 100 mGy.

7.8.2 Genetic Effects

As a result of absence of significant human data for the induction of genetic damage, risk evaluation is based mainly on the results from model experimental system, i.e. mice. On the basis of visible genetic disorders like those caused by recessive mutations, dominant skeletal abnormalities and dominant cataract induction, genetic risk for human beings is evaluated. It is estimated that a dose of 1 Gy doubles the incidence of mutations (*doubling dose*) [4]. However, since many genetic diseases (over 90%) are multigenic and involve complex interactions, a doubling dose of radiation increases the genetic disorder only by 10–15%. Natural incidence of genetic disorders among human beings is almost 10–20% of all live born. Exposure of the germ cells to a dose of 10 mSv can increase the incidence of genetic disorders by 50–150 cases per million live born.

It may be concluded that the estimated risk arising from low-level radiation exposure is very small. Present levels of occupational exposure of radiation workers are associated with a mortality risk of 50–100 cases per million per year. Hence radiation industry can be classified as a safe occupation from the point of view of occupational mortality risk.

Table 7.1 and Fig. 7.5 summarize [4] various level of exposures and their relevance from the

Table 7.1 Significance of different levels of radiation exposure [5]

Exposure	Significance
1–2 mSv/annum	Background radiation level at sea level outdoors
0.5–5 mSv	Most diagnostic radiological examinations
1 mSv/annum	Limit for non-occupational exposure
2.5 mSv/annum	Average occupational exposure
10 mGy (whole body)	Risk of cancer mortality—About 5 per 10^4 exposed Risk of leukaemia—About 5 in 100,000 exposed
20 mSv/annum	Limit for occupational exposure of the whole body
150 mSv/annum	Exposure limit for eye lens
500 mSv/annum	Limit for exposure of the skin and extremities
100 mGy (whole body)	Detectable increase in chromosome aberrations. No detectable clinical injury or sickness
1 Gy acute (whole body)	Threshold dose for radiation sickness (5–10% of individuals exposed) This is 50 times annual dose limit to workers
1 Gy reproductive system	(a) Dose that induces the same number of mutations as spontaneous induction (doubling dose) (b) Dose which can cause temporary sterility in males (c) Dose that doubles the frequency of mutations in germ cells
2–3 Gy acute whole body	(a) Threshold for epilation (b) Threshold for induction of cataract (c) Radiation sickness in most individuals exposed (d) Early erythema induction (e) Leukopenia (f) Death of a small percentage of individuals (10–30%)
3–5 Gy acute whole body	(a) LD 50(60) for human beings (untreated) (b) Severe leucopoenia, purpura, haemorrhage infection and depilation
6 Gy (X-rays)	(a) Threshold for skin erythema (b) Permanent sterility in both males and females exposed (c) Whole-body exposure results in death of more than 50% individuals, even with the best care

Table 7.1   (continued)

Exposure	Significance
>6 Gy (gamma, partial body)	(a) Threshold dose for skin erythema (b) Causes permanent loss of hair
>10Gy (skin)	Threshold dose for dry desquamation
>20Gy (skin)	Threshold for wet desquamation and necrosis Larger doses cause skin burns within a few days
>30 Gy (skin)	Ulceration and necrosis
40–60 Gy (localized)	Total radiation dose used in fractionated radiotherapy of cancer

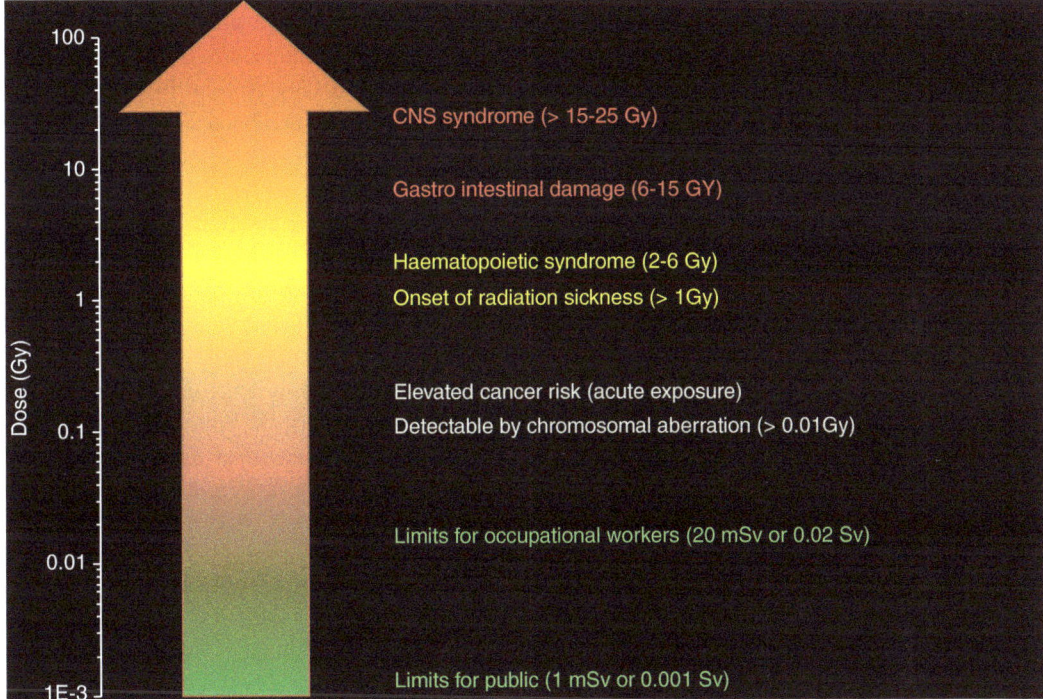

Fig. 7.5   Significance of various levels of exposure for biological effects

point of protection, medical management and biological effects. It is ensured that limits are set well below the threshold doses of serious deterministic effects and minimized stochastic risks to acceptable levels.

7.9   Summary

- High doses of radiation (several Gy) received within a short duration can cause deleterious effects in human beings. A significant level of cell killing in any organ or tissue can result in **deterministic effects**. These effects cannot occur under normal working conditions but can be seen in **accident situations** or in cancer radiotherapy involving several Gy of radiation. These include radiation sickness, serious damage to blood-forming organs, the digestive system and sensitive organs such as the reproductive system, eye lens, oral mucosa, small intestines, lungs, thyroid glands, skin, etc. Whole-body exposure to 4 Gy may result in death among 50% of those exposed, within 60 days [LD_{50} (60)].

- High-dose radiation results in increased frequency of cancers in many organs such as the bone marrow (leukaemia), lungs, stomach, colon, breast, thyroid, oesophagus, bladder, liver and skin, decades after exposure to radiation. Some of the organs are at greater risk than others (e.g. the bone marrow, lungs, stomach, colon, breast and thyroid are more sensitive than the skin, ovaries, small intestine, liver, brain, kidney, etc.).

• *Human being is more sensitive to radiation effects before birth and also in childhood, than during adulthood.* Irradiation in utero (prenatal, before birth) to doses of the order of 100 mGy entails the risk of malformation, mental retardation and increased risk of childhood cancers (during the first decade after birth). Smaller doses may not involve any significant risk. However, as a measure of precaution, ICRP has recommended a dose limit of 1 mSv to the foetus during the entire period of pregnancy.

• Low-level radiation exposures (few tens of mGy) do not result in clinical manifestation of damage. However, a whole-body dose of 100 mGy, received within a short time, results in detectable increase in the frequency of dicentric chromosomes.

• **Stochastic effects** of radiation are caused by the modification of genetic information of cells. Modification of somatic cells may cause **cancers**, whereas the mutation in germ cells (sperm/ovum) has potential to cause **genetic disorders** in the future generations. These are the two important stochastic effects in human beings.

• Low-level radiation exposure (a few mGy) for prolonged period may entail an increased risk of cancer induction in old age. But human populations exposed to low-level radiation from occupational exposure, routine diagnostic X-rays and nuclear medicine procedures or to elevated natural background radiation do not show a statistically significant increase in cancer incidence as compared to control populations. The cancer risk estimation is based on the excess risk seen in human beings exposed to large doses (1–6 Gy) of radiation (e.g. A-bomb survivors). The dose response curve is extrapolated to lower doses, assuming a linear non-threshold hypothesis.

• Children are two to three times more sensitive than adults for the induction of radiation carcinogenesis. Females are 20–30% more sensitive than males.

• Radiation-induced genetic effects have been seen in experimental animals exposed to large doses (several Gy). Among the 70,000 children born to A-bomb survivors exposed to moderate doses of $\cong$ 500 mSv, there is no increase in the incidence of genetic disorders as compared to the children of the unexposed. *Hence, there is no human evidence to the induction of serious genetic effects by radiation.*

• Low-dose rate exposures are far less hazardous as compared to acute exposures. This is due to the ability of biological systems to repair a large fraction of damage caused by low-intensity exposure (protracted or chronic exposures). The risk associated with low-dose exposures are considered to be *one-half* of that of acute exposures.

References

1. Hall EJ, Giaccia AJ. Radiobiology for radiologists. Philadelphia: Lippincott Williams & Wilkins; 2006.
2. Status of the dosimetry for the radiation effects research foundation (DS86); 2001.
3. Recommendations of the international commission on radiological protection. ICRP publication 60. Ann ICRP. 1991;21(1–3):1–201.
4. Health risks from exposure to low levels of ionizing radiation BEIR VII; 1998.
5. The 2007 recommendations of the international commission on radiological protection. ICRP publication 103. Ann ICRP. 2007;37:2–4.

Planning and Design of Nuclear Medicine Imaging Facilities

8

Abstract

Presently, diagnostics nuclear medicine is considered a very reliable, non-invasive modality, which has found its application in most medical specialties, such as oncology, endocrinology, nephrology, cardiology, neurology, haematology, urology, respiratory system, etc. The availability of short-lived radionuclides such as ^{18}F-FDG, ^{99m}Tc, etc. has revolutionized this branch of medical diagnosis. For performing nuclear medicine procedures, the main requirement is an approved nuclear medicine department with all facilities for handling radionuclides/radiopharmaceuticals. Radiation safety is an important factor to be considered in all nuclear medicine procedures; it is the responsibility of the institution to give due consideration to this aspect in the planning stage itself to protect the nuclear medicine staff, patients and nearby public from ionizing radiations emitted by the radiopharmaceuticals while using them in the healthcare programme. In this chapter, all important aspects of the planning of a diagnostic nuclear medicine laboratory are described.

8.1 Diagnostic Nuclear Medicine Facility

8.1.1 Site Selection

In case the nuclear medicine facility is located inside a hospital, its location should be away from other equipment/departments where ionizing radiations are being used, such as diagnostic X-ray equipment and radiotherapy department. The radiation from other sources will create interference with nuclear medicine imaging equipment, thereby distorting the image quality. Preferably, nuclear medicine facility should be located at the end of the hospital where movement of the general public can be restricted. As the nuclear medicine facility caters the diagnostic need of other clinical departments, the same should be well connected with other departments.

8.1.2 Layout and Area Requirement

The understanding of a diagnostic nuclear medicine facility, in general, applies to a facility having a gamma camera (planar or with SPECT/SPECT-CT installation) or PET-CT installations for imaging purposes, an in-house radiopharmacy, facilities for decontamination, storage and safe disposal of radioactive waste, dose administration

© The Author(s), under exclusive license to Springer Nature Singapore Pte Ltd. 2022
P. Tandon et al., *Radiation Safety Guide for Nuclear Medicine Professionals*,
https://doi.org/10.1007/978-981-19-4518-2_8

room, pre- and post-administration waiting area for patients and general areas for reception and records, staff sitting place, toilets, etc.

The typical layout plans are shown in Figs. 8.1, 8.2 and 8.3. The total area required for such a facility is about 200 sq. metres. The gamma camera room should have an area of about 30–40 sq. metres, depending upon the size of the equipment. For comfortable working, the radiopharmacy room should be at least 12 sq. metres. It will be advantageous to have separate rooms for extraction of ^{99m}Tc and other radiopharmacy works such as dispensing of PET radiopharmaceuticals or preparation of ^{68}Gallium or ^{177}Lutetium radiopharmaceuticals. As injectable compounds are prepared in the in-house radiopharmacy, this room must be maintained dust-free and very clean. A laminar airflow (LAF) system is preferred to be installed in this room for the preparation of injectable compounds.

Area requirements for other rooms, such as for nuclear medicine physician, nuclear medicine physicists/technologists, in vivo counting, dose administration, patient waiting, etc., may be flexible. About 10–15 sq. metres for each of these purposes is considered as reasonable for comfortable working. Nuclear medicine facility performing low-dose therapeutic procedures, such as treatment of thyrotoxicosis using ^{131}I, palliative treatment of painful bone metastasis using ^{32}P or ^{89}Sr, ^{177}Lutetium therapies and alpha therapy, which do not warrant hospitalization of patients after administration of radioactivity, may be carried out by providing additional facilities for the respective procedure such as a room where the patient after administration of activity can wait for observation. However, for handling and administration of therapeutic quantities of ^{131}I, including low-dose therapeutic procedures, a separate fume hood with charcoal filter is essential. For handling ^{32}P or ^{89}Sr for low-dose therapeutic procedures, a fume hood may not be essential, but a beta shielding device, made of Perspex sheets of about 12 mm thickness, must be provided. For ^{177}Lutetium therapies, a separate room with a drainage system directly connected to the main sewerage line is sufficient and does not require a delay and decay tank.

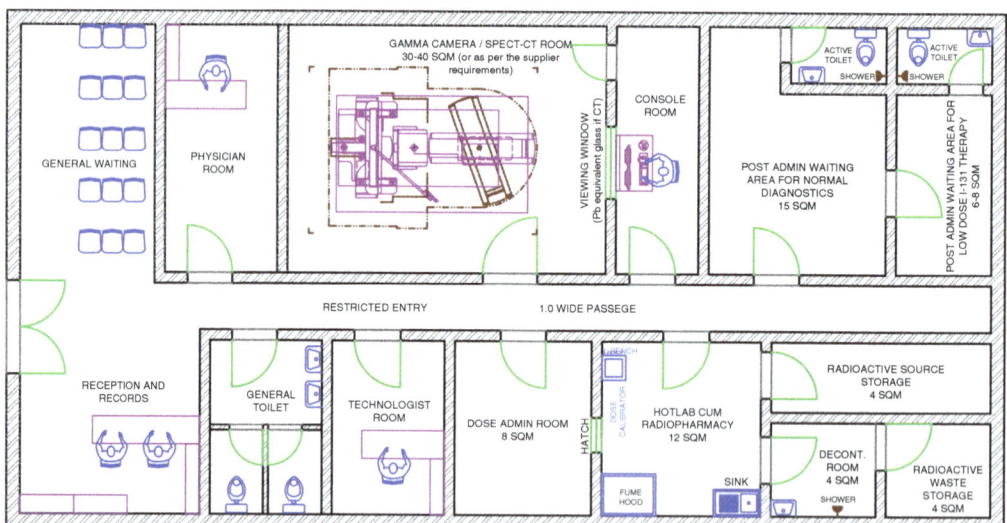

Note: All walls / Partitions of the Nuclear Medicine laboratory should be made of 9" brick or 6" concrete. Fume Hood to be installed, if required.
Decontamination Room is Optional. The active toilet may be used for personnel decontamination by providing a shower & wash basin.

Fig. 8.1 Typical layout of SPECT-CT facility

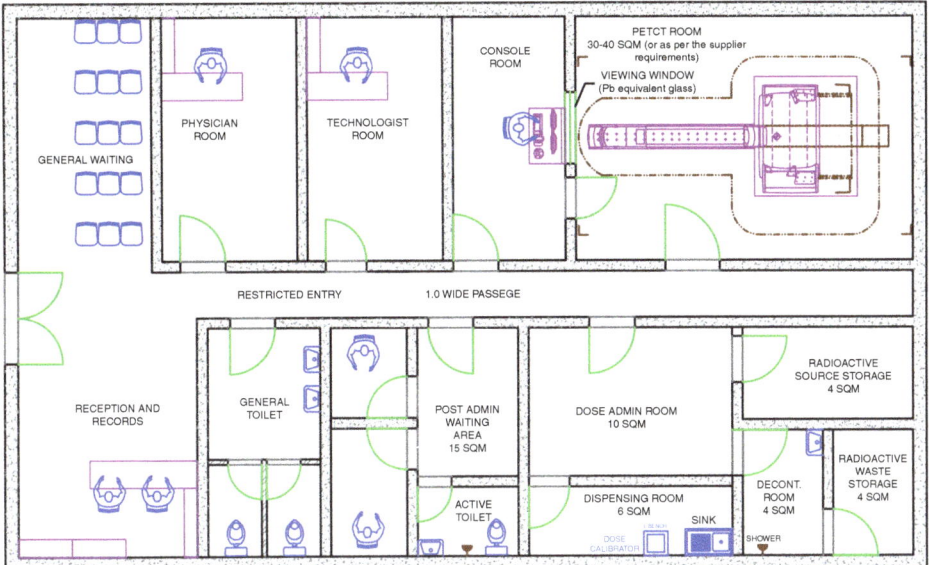

Note: All walls / Partitions of the PETCT Imaging Room and Radiopharmacy area should be made of concrete only, the thickness of which will depend on the required workload.

Decontamination Room is Optional. The active toilet may be used for personnel decontamination by providing a shower and wash basin.

Facilities using Ge-68/Ga-68 generator will require a Hot Lab of 12 Sq. m area.

Radioactive waste storage room for PETCT facility is optional. However, an alternative place has to be identified in the Hot Lab for storage of decayed Ge-68 / Ga-68 generators.

Fig. 8.2 Typical layout of PET-CT facility

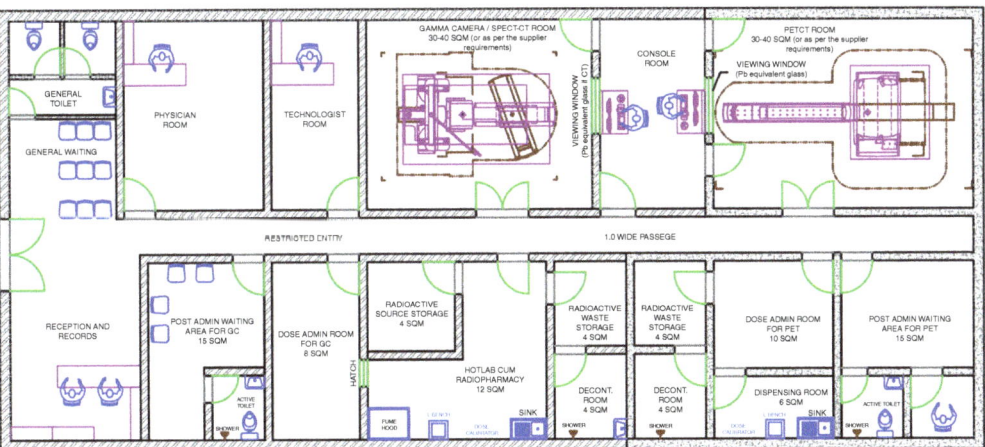

Note: All walls / Partitions of the Nuclear Medicine laboratory should be made of 9" brick or 6" concrete only but the wall of the PETCT Imaging Room and Radiopharmacy area to be made of concrete only, the thickness of which will depend on the workload.

Decontamination Room is Optional. The active toilet may be used for personnel decontamination by providing a shower and wash basin.

Facilities using Ge-68/Ga-68 generator will require a Hot Lab of 12 Sq. m area.

Radioactive waste storage room for PERCT facility is optional. However, an alternative place has to be identified in the Hot Lab for storage of decayed Ge-68 / Ga-68 generators.

Fig. 8.3 Typical layout of SPECT-CT and PET-CT facility

8.1.3 Equipment and Accessories

For measurement of activity before administration to patients, a calibrated isotope calibrator (dose calibrator as it is often called) and a portable contamination monitor for monitoring of work surfaces, body parts, etc. should be procured and maintained in the department. For a nuclear medicine facility, a properly ventilated fume hood for iodination work is an additional requirement. As the iodination work leads to a considerable amount of radioactive waste, appropriate containers for storage of liquid and solid radioactive waste must be provided in the laboratory.

8.1.4 Staff

In the nuclear medicine facility, administration of radiopharmaceuticals in humans has to be done by trained nuclear medicine physicians. Hence, in these facility, at least one nuclear medicine physician having an appropriate qualification as prescribed by the regulatory authority (RA) should be available. For the preparation of radiopharmaceuticals, administration to patients, quality assurance, dosimetry, etc., an adequate number of qualified physicists/technologists as per the qualification prescribed by RA should be appointed. It is needless to mention that an adequate number of supporting staff for helping in scintigraphy work, nursing, reception and record-keeping, etc. should also be employed. Apart from the above mandatory staff, a RSO (Radiological Safety Officer), approved by the competent Authority, should be appointed to take care of radiation safety aspects in the nuclear medicine facility.

8.1.5 General

The drainage system for the nuclear medicine facility should be a separate one, without passing through other departments in the building and connected to the main sewage line. The walls and doors in the facility should be painted with hard, washable paint to enable easy decontamination whenever the need arises. All the work surfaces should have a smooth, non-absorbent finish, such as Sunmica. All the sinks in the facility should be of smooth finish, such as stainless steel, and should have elbow-operated taps. The floor should be covered with linoleum or similar material, which is non-absorbent and has a smooth surface. The ventilation system in the facility should be such that the direction of airflow should be from the area having low activity to the area having high activity and then be driven out of the facility through a properly laid exhaust system, such as a fume hood or any other type of exhaust system.

8.2 Shielding Requirement in a Diagnostic Nuclear Medicine Facility

Since most of the diagnostic procedures in conventional nuclear medicine facility (gamma camera or SPECT-CT/ SPECT-CT) are carried out using ^{99m}Tc, which emits 140 keV gamma rays, normally, there is no necessity for additional structural shielding. However, in places where large activities are stored or radioactive waste containers, such as carboys containing large activities of spent ^{99}Mo, are stored, additional shielding using interlocking lead bricks should be provided. Similarly, where the ^{99m}Tc generator is kept, a barrier of lead bricks should be provided to bring down the radiation level in the working area (controlled area) to within the maximum permissible level (0.01 mGy/h, for 40 working hours in a week). The principle of ALARA (as low as reasonably achievable, taking the socio-economic factors into account) should be employed while planning the laboratory. The shielding data for some of the important radionuclides used in nuclear medicine has been given in the chapter in Tables 5.2 and 5.3.

8.3   Shielding Calculation for SPECT-CT and PET-CT Facilities

The SPECT-CT facility requires the normal structural shielding, i.e. 15 cm concrete or 23 cm brick, which is good enough for performing procedures using Tc-99 m-labelled radiopharmaceuticals. However, for the PET-CT facility, where F-18 (511 keV gamma)-labelled radiopharmaceuticals are commonly used, additional shielding is required based on the size of the room and occupancy all around the room.

The dose rate constant of F-18 is 5.3 R/h-mCi at 1 cm; however, due to absorption of activity in the patient's body, the dose rate constant to be considered for calculation is 3.4 R/h-mCi at 1 cm, which leads to an absorption factor of 0.36. Further, in an uptake and imaging room, where the patient with large activity spends most of its time and requires appropriate shielding, thus as an example, we have shown here the shielding calculation for these two rooms specifically.

8.3.1   Shielding Calculation for Uptake Room in the NM Facility (Fig. 8.4)

The transmission factor, 'B', is defined as the ratio of the radiation flux 'ϕ' passing through the

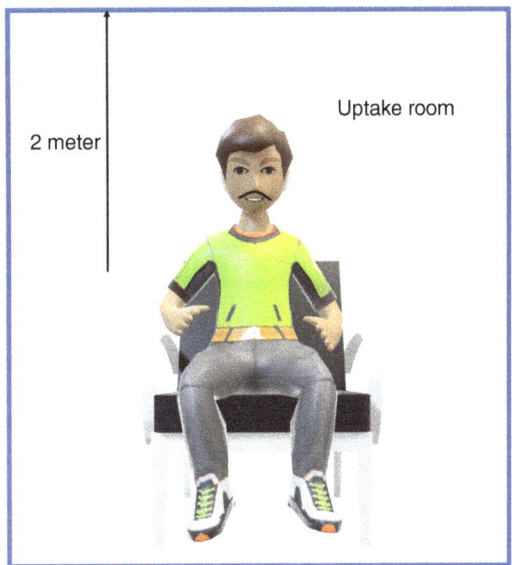

Uncontrolled area

Uptake room

2 meter

Fig. 8.4   Uptake room

medium to the flux 'ϕ_0' incident upon the surface of the medium 'B' = ϕ/ϕ_0. The transmission factor in any area depends upon the activity of the radioactive source, distance from the source, the occupancy and the permissible dose limit. The equations used for the calculation of transmission factors for uncontrolled and controlled areas as per AAPM Task Group-108 [1] are given below.

(a) *For uncontrolled areas:*

$$B = 5.89 \times d\,(\mathrm{m})^2 / \left(T \times N_\mathrm{w} \times A_0\,(\mathrm{mCi}) \times t_\mathrm{u}\,(\mathrm{h}) \times R_{\mathrm{tU}}\right) \tag{8.1}$$

(b) *For controlled areas at ALARA levels:*

$$B = 29.5 \times d\,(\mathrm{m})^2 / \left(T \times N_\mathrm{w} \times A_0\,(\mathrm{mCi}) \times t_\mathrm{u}\,(\mathrm{h}) \times R_{\mathrm{tU}}\right) \tag{8.2}$$

where
B = Transmission factor.
d = Distance from source to barrier (m).
T = Occupancy factor.
N_w = Number of patients per week.
A_0 = Administered activity (mCi).
t_u = Uptake time (h).
R_{tU} = Dose reduction factor over uptake time.

Example 8.1 *What is the transmission factor and subsequently the thickness of the wall required in the uptake room for uncontrolled areas? The patient is administered with 10 mCi of F-18 FDG, there is a workload of 40 patients per week and the uptake time for the patient is 45 min.*

$T = 1$.

$d = 2$ m.

$A_0 = 10$ mCi.

$N_w = 40$ patients per week.

$t_U = 45$ min.

$R_{tU} = 0.87$.

Here, during the calculation of 'R_{tU}', it is known that the tracers used for PET procedures are of short half-life, and thus total dose received by the patient is less than the product of initial dose rate and the time spent, as activity decays during the time the patient stays in the uptake room.

$$R_t = D(t)/(D(0) \times t)$$
$$= 1.443 \times (T_{1/2}/t) \times (1 - \exp(-0.6932 t/T_{1/2}))$$
(8.3)

For F-18, the 'R_t' factors for $t = 30, 45, 60$ and 90 min are 0.91, 0.87. 0.83 and 0.76, respectively.

Thus, 'B' is calculated by using Eq. (8.1) as follows:

$$B = 5.89 \times 2^2 / (1 \times 40 \times 10 \times 0.75 \times 0.87)$$
$$= 0.09$$

Now using Table 8.1, the thickness of concrete required is 20 cm.

Table 8.1 Transmission factor for concrete at different thickness

Thickness in (cm)	Transmission factor for concrete (2.35 g/cm³)
0	1.0000
1	0.9583
2	0.9088
3	0.8519
4	0.7889
5	0.7218
6	0.6528
7	0.5842
8	0.5180
9	0.4558
10	0.3987
12	0.3008
14	0.2243
16	0.1662
18	0.1227
20	0.0904

8.3.2 Shielding Calculation for Imaging Room in the NM Facility

Usually, the calculation carried out for the imaging room is the same as that used for the uptake room until the tomograph provides any shielding. However, as there is a delay after the administration of activity till the patient is taken for imaging procedure, the patient activity decreased by a factor 'F_U', which is calculated as follows:

$$F_U = \exp(0.693 \times t_U (\text{min})/110)$$

Before imaging, the patient will void in most of the cases; thus about 15% of the activity administered will be removed, which will decrease the dose rate by 0.85.

(a) *For uncontrolled areas:*

$$B = 5.89 \times d\,(\text{m})^2 / \left(T \times N_w \times A_0\,(\text{mCi}) \times 0.85 \times F_U \times t_I\,(\text{h}) \times R_{tI}\right) \tag{8.4}$$

(b) *For controlled areas:*

$$B = 29.5 \times d\,(\text{m})^2 / \left(T \times N_w \times A_0\,(\text{mCi}) \times 0.85 \times F_U \times t_I\,(\text{h}) \times R_{tI}\right) \tag{8.5}$$

where.

B = Transmission factor.

d = Distance from source to barrier (m).

T = Occupancy factor.

N_w = Number of patients per week.

A_0 = Administered activity (mCi).

F_u = Uptake time decay factor

t_I = Imaging time (h).

R_{tI} = Dose reduction factor over imaging time.

The decay factor for 'F_U' for F-18 for 45 min is equal to exp (−0.693 × 45/110) = 0.75.

Example 8.2 *If a patient is administered 10 mCi F-18 FDG, what will be the required transmission factor for uncontrolled area at 3 m from the patient during imaging using PET? Consider workload of 40 patients per week, uptake time 45 min and 30 min as average time for imaging.*

Thus,

$T = 1$

$d = 3$ m

$A_0 = 10$ mCi

$N_W = 40$ patients/week

$t_i = 30$ min (0.5 h)

$R_{tI} = 0.91$ for 30 min

So by Eq. (8.1), the transmission factor is calculated as follows:

$$
\begin{aligned}
B \quad &= 5.89 \times 3^2 / \left(1 \times 40 \times 10 \times 0.85 \times 0.75 \times 0.5 \times 0.91\right) \\
&= 53.01 / 116.03 \\
&= 0.46
\end{aligned}
$$

Thus, from Table 8.1, for the transmission factor for 0.46, the thickness of concrete required is 9 cm (Fig. 8.5).

8.3.3 Calculation of the Thickness of the Ceiling above the PET-CT Facility

Typically, during the calculation of ceiling thickness, the radioactivity source (i.e. patient) is assumed to be 1 m above the floor. The calculation for dose rate 0.5 m above the floor level for the room located above the source is given as follows (Fig. 8.6):

Example 8.3 *Find the required shielding for an uncontrolled area (room) above the uptake room for PET. Activity administered to the patient is 10 mCi of F-18, uptake time is 45 min, the workload is 40 patients per week and the floor-to-floor distance is 4.3 m.*

Uncontrolled area

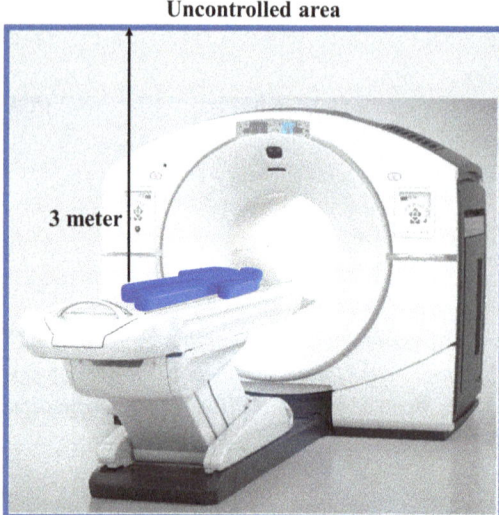

3 meter

Fig. 8.5 Imaging room

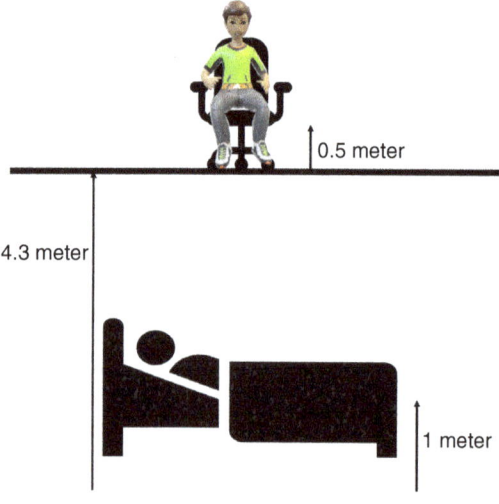

0.5 meter

4.3 meter

1 meter

Fig. 8.6 Distances for calculation of thickness of the ceiling above the PET-CT facility

N_w = 40 patients/wk.
$T = 1$
A_0 = 10 mCi
t_u = 45 min
d = 4.3 m
R_{tU} = 0.87 for 45 min by using Eq. (8.3)

Thus, 'B' is calculated by using Eq. (8.1) as follows:

$$B = 5.89 \times d^2 / (1 \times 40 \times 10 \times 0.75 \times 0.87).$$

$$
\begin{aligned}
D &= (4.3 - 1) + 0.5 \\
&= 3.8\,\text{m} \\
&= 5.89 \times 3.8^2 / (1 \times 40 \times 10 \times 0.75 \times 0.87) \\
&= 85.051 / 261 \\
&= 0.325
\end{aligned}
$$

From Table 5.3 requires, about 12 cm of the concrete thickness of the ceiling to reduce the radiation level below the permissible limit.

Since the dose rate constant is high for PET, the doses to the hands of personnel engaged in the drawing and administering of PET radiopharmaceuticals activities may be significant. An unshielded syringe containing 15 mCi (555 MBq) F-18 will give dose rate of 33 mSv/h at 5 cm distance. Use of tungsten shield for the syringe will reduce the doses to the hands by 88%; however, the additional (~0.8 kg) weight of the shield will make injection process difficult. Diving the injection duties among the staff and use of an automatic dispensing system can reduce the doses to the hands.

Similarly, at the times of patient positioning, the dose rate at 1 m is approximately 20 µSv/h assuming an administration of 10 mCi (370 MBq) of F-18 so always take precautions while with the patient during patient positioning. During construction of an uptake room, it is always ensured that there should be sufficient distance between the two patients, and while calculating the thickness of the wall between the two patients, it is considered that there should not be dose to the patient due to the activity administered to another patient who is sitting in the adjacent room as he/she is already receiving dose due to the activity administered to him/her.

The regulatory constraints should be considered while a new PET-CT facility is planned. Imaging rooms and uptake rooms for PET facility should be located away from the areas having high occupancy and any uncontrolled areas. Normally, CT associated with PET does not require any additional shielding as the half-value layer for 511 keV photon is much higher than the half-value layer for CT photons. It is always kept in mind that reasonable efforts are to be made to keep sensitive counting equipment like thyroid uptake probe and scintillation well counter away from the PET-CT facilities to avoid unambiguous reading in the counting equipment.

In conclusion, the PET-CT facility requires a different approach for shielding calculation than any other diagnostic imaging facility. This different approach is due to the fact that the patient in PET diagnosis serves as a constant source radiation source during procedure and the annihilation photon has high energy. The careful planning helps to bring out an effective design of the facility while considering the radiation safety-related aspects.

Reference

1. Madsen MT, Anderson JA, Halama JR, Kieck J, Simpkin DJ, Votaw JR, Wendt RE, Williams LA, Yester MV. AAPM report no. 108: PET and PET/CT shielding requirements. Med Phys. 2006;33(1):4–15.

Planning and Design of High-Dose Therapy Facility

Abstract

In today's era, because of the development of therapeutic radiopharmaceuticals, there is a tremendous development of nuclear medicine in the treatment of various diseases non-invasively. Therapeutic applications of radionuclides, though limited to a few types of procedures, are known to be very effective in specific cases. The use of ^{131}I for the treatment of thyroid cancer and thyrotoxicosis is a well-established practice. As per the international literature, ^{177}Lu is very effective for treating neuroendocrine tumours and castration-resistant prostate cancer (CRPC) ^{186}Re, ^{90}Y, ^{225}Ac and a few newer radioisotopes are also found to be useful in therapeutic applications.

For performing any nuclear medicine procedures, the regulatory requirement is to have an approved nuclear medicine facility with all associated rooms for handling radionuclides/radiopharmaceuticals. As we are aware that radiation safety is an important factor to be considered while performing nuclear medicine procedures, it is the responsibility of the institution to give due consideration to this aspect in the planning stage itself to protect the staff, patients and the general public from effects of ionizing radiations emitted by radiopharmaceuticals while using them in the healthcare programme. Here, all important aspects of planning a high-dose therapy facility for patients administered with large doses of ^{131}I or ^{177}Lu for therapeutic purposes are described.

9.1 Therapeutic Nuclear Medicine

9.1.1 Site Selection

While choosing a site for constructing a high-dose therapy ward, it should be borne in mind that there should not be any impact on the general public and environment from the use of radioactive material in the laboratory as being practised while planning a diagnostic nuclear medicine facility. The high-dose therapy department should be constructed at the corner of a nuclear medicine facility. The location should be such that there is no interference to other departments having the counting equipment because of the high radiation level in the high-dose therapy department. Normally, it is preferred that the nuclear medicine facility with the isolation ward is located at the end of a hospital block where the movement of the public is restricted.

9.1.2 Layout and Area Requirement

The understanding of a high-dose therapy ward, in general, applies to a department having an iso-

© The Author(s), under exclusive license to Springer Nature Singapore Pte Ltd. 2022
P. Tandon et al., *Radiation Safety Guide for Nuclear Medicine Professionals*,
https://doi.org/10.1007/978-981-19-4518-2_9

lation ward, an in-house radiopharmacy, a place for decontamination, a storage place for radioactive waste generated during procedure, an attendant room for a patient who got admitted and general areas for reception and records, staff sitting place, toilets, etc. In addition to this, a delay and decay tank is required if ^{131}I is used to treat Ca-thyroid.

Nuclear medicine facilities performing low-dose therapeutic procedures, such as treatment of thyrotoxicosis using ^{131}I and palliative treatment of painful bone metastasis using ^{32}P or ^{89}Sr, which do not warrant hospitalization of patients after administration of radioactivity, may be carried out by providing additional facilities for the respective procedure such as the room where the activity is administered and a post-administration area where the patient after administration of activity waits for observation. However, for handling and administering therapeutic quantities of ^{131}I, including low-dose therapeutic procedures, a separate fume hood if handling liquid radioiodine with a charcoal filter is essential.

Nowadays, ^{177}Lu-based radiopharmaceuticals are commonly used to treat neuroendocrine tumours and prostate cancer. In most countries, for the treatment using ^{177}Lu-based radiopharmaceuticals, there is a requirement for an isolation ward with a delay and decay tank. However, in India, this has been relaxed since 2021 after the pilot study in an institute that does not have a delay and decay tank. After application of the theoretical model and analytical equation using the collected data obtained from the pilot study, it has been concluded that ^{177}Lu radioisotope-based radiopharmaceuticals can be used in an isolation ward, where the attached toilet is connected to the main sewerage line and does not require a dedicated delay and decay tank which is required in the case of ^{131}I. A fume hood may not be essential for handling ^{32}P or ^{89}Sr for low-dose therapeutic procedures, but a beta shielding device made of Perspex sheets of about 12 mm thickness must be provided. These radioisotope-based radiopharmaceuticals are used to treat patients with OPD.

Nowadays, the use of such alpha-emitting radionuclides for organ−/tissue-specific radionu-

clide therapy (targeted alpha therapy (TAT)) is gaining wide popularity worldwide. Because of its inherent properties, adopting certain specific radiation safety measures for TAT may be required. In keeping with the ALARA principle, unlike beta/gamma sources which require suitable shielding from external radiation, shielding requirements for alpha particles are not much. However, alpha particles deposit their energy over a very short range causing greater localized dose, and therefore incorporated alpha emitters possess high biological effectiveness. Radioprotection measures shall focus on avoiding internalization from accidental intake. There is a risk of contamination or incorporation from the spread of radioactivity, bodily fluids such as blood, vomiting, urine, etc.

Nevertheless, the most relevant path of incorporation is inhalation. Contamination or incorporation is strict to be avoided. External radiation doses to radiation workers or caregivers of alpha therapy patients are not significant. Radiation protection measures shall focus on avoiding internalization from accidental intake. The dose rates from patients administered with alpha activity are low, and treatment on an outpatient basis can be permitted. Although accompanied gamma emission allows for monitoring with standard equipment, as a good practice measure, monitoring equipment to detect α particles, such as ZnS (Ag) scintillators, should be available while handling α emitters in addition to Geiger-Mueller survey meters. Further, a well-ventilated hood and double gloving may also be used for handling alpha activity to minimize radiation exposures to handling personnel.

9.1.3 Equipment and Accessories

To measure activity before administration to patients, a calibrated isotope calibrator (dose calibrator as it is often called) and a portable contamination monitor for monitoring work surfaces, body parts, etc. should be procured and maintained in the department. In addition, if liquid iodine needs to be handled, then a properly ventilated fume hood for iodination work is required.

As the iodination work leads to a considerable amount of radioactive waste, appropriate containers for storage of liquid and solid radioactive waste must be provided in the laboratory. Apart from this, a gamma area monitor must be installed outside the isolation ward. Mobile protective barrier with viewing window requires an iodine decontamination kit while handling the ^{131}I.

9.1.4   Staff

Usually, an isolation ward is a part of a nuclear medicine facility, but even if it is a standalone facility, the facility has to comply with the regulatory requirements of the country. Apart from nuclear medicine physicians, the facility should employ nuclear medicine technologist/nuclear medicine physicist and radiological safety officer (RSO) approved by the competent authority. Presently, RSO (NM) is approved for low-dose therapy, and diagnostic procedure needs to obtain a certificate for successful completion of their training programme in high-dose therapy from the recognized institute if they desire to become RSO for high-dose therapy facility. For operational requirements, radiation safety-trained staff nurses and housekeeping staff would also be needed.

9.1.5   General

In general, the drainage system for the isolation ward should be a separate one, without passing through other departments in the building and connected to the delay and decay tank if using ^{131}I. After due monitoring of the radiation level at the delay and decay tank outlet, if the radiation level is below the regulatory limit, the outlet of the delay and decay tank can be opened, which is connected to the main sewage line. The walls and doors in the department should be painted with hard, washable paint to enable easy decontamination whenever the need arises. All the work surfaces should have a smooth, non-absorbent finish, such as Sunmica, stainless steel, etc. All the sinks in the department should be of smooth finish, such as stainless steel, and should have elbow-operated taps. The floor should be covered with linoleum or similar material (vitrified tiles), which is non-absorbent and has a smooth surface. The ventilation system in the laboratory should be such that the direction of airflow should be from the low activity area to the medium activity area to the high activity area and then be driven out of the laboratory through a properly laid exhaust system, such as a fume hood or any other type of exhaust system.

9.2   Isolation Ward for Hospitalization of Patients

The activity limit above which the patients are to be hospitalized is decided by the regulatory authority of the respective country. In India, the Atomic Energy Regulatory Board (AERB), the regulatory authority is responsible for enforcing radiation safety in the country. As per the AERB Safety Code on Nuclear Medicine Facilities AERB/RF-MED/SC-2(Rev. 2) [1], normally, patients with the administered activity of more than 1110 MBq (30 mCi) of radioiodine are to be hospitalized in an isolation ward having barrier nursing and attached toilet facility.

Figure 9.1 gives a typical layout of a two-bedded isolation ward for hospitalization of patients treated with ^{131}I for thyroid cancer treatment and the attendant room. In such patients, the activity of ^{131}I administered varies from 1.5 GBq to 10 GBq. The location of the isolation ward should be carefully planned and should have minimum impact on the surroundings and the general public. It is preferred to have the isolation ward in a separate building if space permits. Otherwise, it should be located at a corner of the building, where the movement of people should be restricted, except for nursing staff and relatives of the patient who may be required to render some help, under the supervision of the RSO, as and when required. If there is a residential premise in the neighbourhood of the isolation ward, then it should have a minimum distance of 3 ft. from it. The isolation ward should be air-

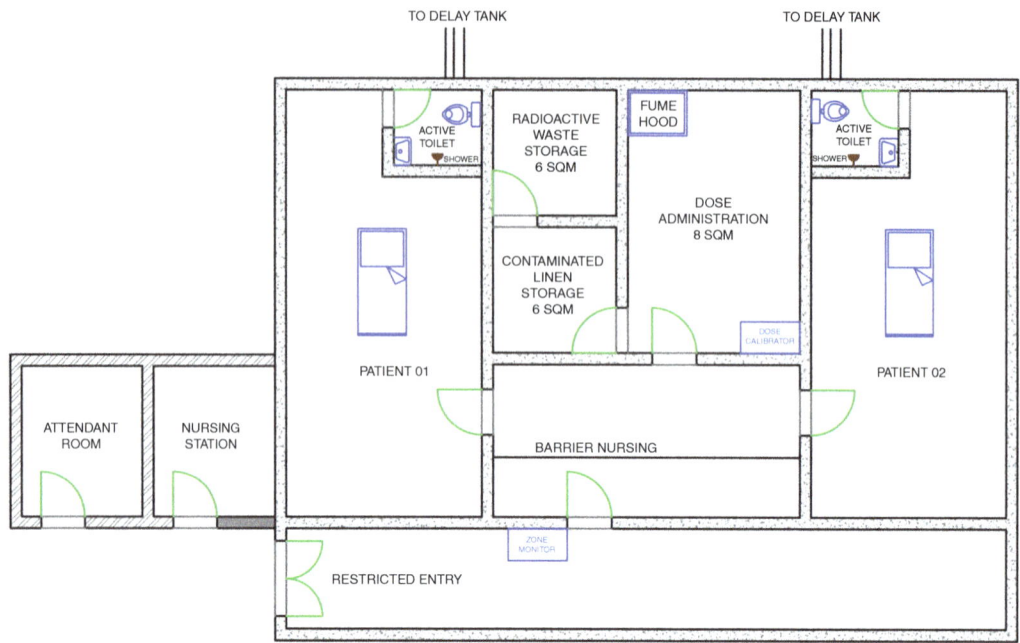

Note: All walls / Partitions of the facility should be made of concrete only, the thickness of which will depend on the area of the rooms, postion of the bed and occupancies all around. The ducting line of the fume hood should be shown in the plan. Decontamination Room is optional. The active toilet ma be used for personnel decontamination by providing a shower & wash basin. Also the plumbing and ducting line from the isolation ward to the delay tank is to be shown clearly.

Fig. 9.1 Layout of a two-bedded isolation ward

conditioned for the patient's comfort, but the AC duct should not be connected to the centralized air handling unit (AHU). Windows, if any, should be provided at the height of 2 m from the floor level to bring down the radiation level outside to a minimum as most of the administered activity is concentrated below the neck region.

The therapeutic dose of ^{131}I should be administered in an adjacent room having a fume hood for dispensing/handling and administering the dose. A properly calibrated isotope calibrator should be provided in this room to measure the activity before administration. An adequate buffer area should be provided outside the isolation ward for the movement of the nursing staff. The nurse's room should be located not too close to the isolation ward, but at the same time, it should be possible to monitor the patient's movement from the place and render any help to the patient as and when required. A gamma zone monitor has to be installed outside the isolation ward to

monitor the radiation level in the corridor and if the patient moves out of the ward. There is a sudden increase in radiation level and the beeper sounds.

9.3 Shielding Requirement in an Isolation Ward

Unlike in a diagnostic laboratory, the patients undergoing large dose therapy (activity in the range of 1.5 GBq–10 GBq of ^{131}I) have a considerably high radiation level around the patient. As indicated above, ^{177}Lu, ^{32}P, ^{89}Sr, ^{90}Y, ^{153}Sm, etc. labelled radiopharmaceuticals can be used to treat patients on an OPD basis. However, the use of ^{131}I and ^{177}Lu, the most commonly used radioisotopes for treatment purposes, requires an isolation ward for the hospitalization of patients administered with radioactivity. Hence, adequate measures must be taken to reduce the radiation

Table 9.1 Half-value thickness of ^{131}I in lead, concrete and brick

Sl. no.	Material	Half-value thickness (cm)
1	Lead	0.3
2	Concrete	3.0
3	Brick	4.5

level outside the isolation ward to be within the prescribed limit for the general public, which is 1 mSv per year. This can be achieved by providing concrete walls of sufficient thickness. Table 9.1 gives the half-value thickness of ^{131}I gamma rays in lead, concrete and brick. Using these values, the specific gamma ray constant of ^{131}I (2.2 R/mCi at 1 cm) and the occupancy and use factors, one can calculate the desired thicknesses of the isolation ward walls to achieve the permissible exposure levels outside.

Moreover, in places where large activities are stored or radioactive waste containers, such as carboys containing large activities of spent ^{131}I, are stored, additional shielding using interlocking lead bricks should be provided. Similarly, inside the fume hood where ^{131}I is kept, a barrier of lead bricks should be provided to bring down the radiation level in the working area (controlled area) to within the maximum permissible level (0.01 mGy/h for 40 working hours in a week). The principle of ALARA (as low as reasonably achievable, taking the socio-economic factors into account) should be employed while planning the laboratory.

If space is not a limiting factor, the distance factor can reduce the radiation level by providing large area rooms and having brick walls. Normally, for hospitalizing one patient, the room size should be about 3 × 6 sq. metres, with an attached toilet facility, as shown in Fig. 9.2. For a maximum administered activity of 250 mCi of ^{131}I, considering the effective decay of the activity in the patient, using a distance factor of 3 m from the patient to the wall and for a full occupancy outside the ward, the thickness of the concrete wall should be about 20 cm, to bring down the radiation level outside the isolation ward to be within the limit prescribed for the general public (calculation shown below). If the room is small, mobile bed shields of adequate thickness, made of lead, may be used.

$$\text{Specific gamma ray constant}\,(\Gamma) \quad = 2.2\,\text{R}/\text{h}\;\;\text{mCi at 1 cm}$$
$$\text{The activity}\,(A_0)\,\text{in the patient is} \quad = 250\,\text{mCi}$$
$$\text{Distance}\,(d) \qquad\qquad\qquad\qquad = 3\,\text{m}$$
$$\text{Thus, radiation level at 3 m is} \quad = 2200 \times 250 \times 1/(300)^2$$
$$= 6.11\,\text{mR}/\text{h}$$

Fig. 9.2 Isolation ward

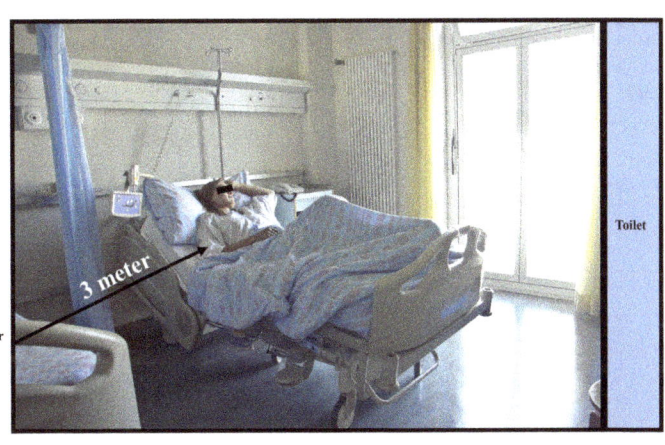

So, to bring the radiation level to 0.1 mR/h for the general public, we require 1 TVT and 3 HVT, and by substituting the HVT of concrete from Table 9.1, a concrete wall of thickness of 20 cm is required.

9.4 Delay-Decay Tank for Storage of Radioactive Waste

As the uptake of ^{131}I in Ca-thyroid patients (metastasis) is very small (less than 10%), most of the activity is discharged through urinary excretion, faeces, sweat, etc. It has been estimated that about 70–80% of the administered activity is excreted through urine within 48 h. If about 5.55 GBq (150 mCi) of ^{131}I is administered to a patient, about 4 GBq of ^{131}I is discharged through urine within 48 h. Hence, this activity should not be directly discharged into the sewage system. A suitable facility with a delay and decay tank should be provided for storage and disposal of the patients' urine and other effluent's radioactive waste.

The delay-decay tank should be located so that there is no movement of the general public near or around it. Only the authorized workers should be allowed to control the flow of effluents from the tank to the main sewerage chamber. The tank should be leak-proof, corrosion-free and have a smooth surface from the inside. The outlet of the delay tank should be at a much higher level than the inlet pipe of the sewerage chamber to ensure no backflow from the sewerage to the delay tank, as shown in Fig. 9.3.

The volume of the tanks depends on the outflow of effluents from the isolation ward into these tanks. A dual tank system is convenient for effective storage for delay and decay after measuring the concentration of activity in the effluent. During 1 month, the effluent from the patient toilet may be allowed to go to only one tank using a valve provided on the inlet pipeline to the tank. When this is nearly full, the inlet to this tank should be closed, and the effluent from the patient toilet now should be allowed into the second tank. During the collection of effluent in the second tank, the radioactivity in the first tank undergoes decay so that it can be conveniently disposed of when the discharge limit prescribed by the competent authority is achieved. In India, the Atomic Energy Regulatory Board, the Regulatory Authority of India, has prescribed that the average monthly concentration of ^{131}I in the effluent waste at the discharge point to the public sewerage system should not exceed 22.2 MBq per cubic metre (i.e. 0.6 μCi/L). The size of the delay and decay tank and the duration of holding should be such

Fig. 9.3 Delay and decay tank. A dual delay and decay tank system for collection and safe disposal of radioactive waste from the isolation ward. The capacity of the delay tank for a single patient is 3000 L each

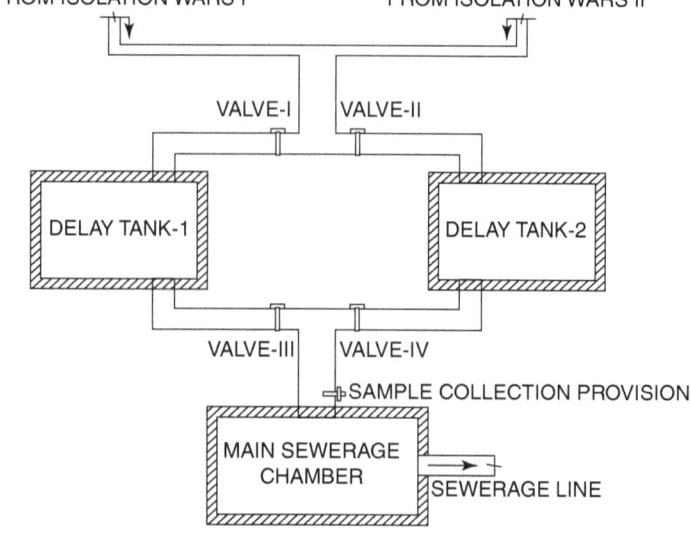

that this limit is not exceeded. For one-bedded isolation ward, a dual tank system having a volume of 3000 L each tank may be adequate for effective collection and storage before disposal.

Reference

1. AERB Safety Code on Nuclear Medicine Facilities AERB/RF-MED/SC-2 (Rev.2).

Planning and Design of Medical Cyclotron Facility

10

Abstract

Radionuclides produced by the medical cyclotron are becoming popular for molecular imaging in nuclear medicine (NM) facilities. The important clinical information provided by this modality (NM) is superior to the other imaging modalities. Several hospitals like to have medical cyclotron in their nuclear medicine facility. Since radionuclides produced by the medical cyclotron have energy higher than other radionuclides used in nuclear medicine facilities for diagnosis, the user should be aware of the regulatory and administrative requirements and aspects of radiation safety before installing of medical cyclotron in their institution. Generally, medical cyclotrons are of two types, unshielded and self-shielded. Therefore, one should know and follow the important steps for planning and design of these facilities.

10.1 Selection of Site

The first and foremost requirement for installing a medical cyclotron, if in the hospital premises, is that the site for the installation should be located within the hospital/institution premises close to the nuclear medicine laboratory and at a place where the movement of the general public is minimum. Presently, as per the regulatory requirement in India, there should be a minimum distance of 30 m from the medical cyclotron vault to any residential premises in the near vicinity of the medical cyclotron site. The layout plan should also indicate the distance of the tallest building close to the site.

While submitting the site plan, it is always advisable to first decide on the make and model of the medical cyclotron unit and go for layout planning of the facility later. It is mainly because the structural shielding requirement for cyclotron housing and adjoining rooms would depend on the energy and model of the cyclotron. Further, one should ensure that the particular make and model of the cyclotron unit has a valid no objection certificate from regulatory authority, and after that, the unit has to be type approved by them.

10.2 Approval of Layout Plan of Medical Cyclotron

Once the make and model of medical cyclotron are selected, the layout plan of the facility may be drawn as per the needs and requirements by giving details of medical cyclotron housing and all the adjoining rooms, including the occupancy above the ceiling and below the floor. From a radiological safety point of view, it is recommended that the medical cyclotron should be installed in the basement of a building. This arrangement would enhance radiological safety around the cyclotron and also would shed/reduce the cost of constructing structural shielding significantly.

At the time of submitting the layout plan of medical cyclotron for approval, one is required to provide the following details. These details enable the evaluation of the radiation shielding adequacy for the medical cyclotron housings.

1. Make and model of the cyclotron.
2. Energy(ies) of accelerated particle(s).
3. Maximum current rating(s) of the unit.
4. Operational current rating of the unit (proposed).
5. The maximum energy(ies) of neutrons and gamma radiation emitted using a particular target.
6. Neutron and gamma radiation dose rate at 1 m in (cGray/min/mA) from the target in 4π direction (i.e. in $0°$, $45°$, $90°$, $135°$, $180°$, $225°$ and $275°$), applicable for non-self-shielded cyclotron.
7. Tenth-value thickness (TVT) for neutron and gamma radiation separately in ordinary concrete of 2.35 Kg/m^3 density or the other form of concrete used for the construction of vaults of the cyclotron.
8. Tenth-value thickness (TVT) for neutrons in a high-density polystyrene plastic (if planning to use).
9. The permissible radiation level for which shielding thickness has arrived.
10. Details of safety interlock system in cyclotron unit.
11. Operating time of medical cyclotron in a week.
12. Occupancy all around the cyclotron housing.
13. Location of radioactive waste (disused activated targets, etc.) stored in the facilities.
14. Details of the radiation shield of chemistry modules and the supplier of the modules.
15. Maximum activity that would be handled at a time in the chemistry modules.
16. Details of air circulation (ventilation system in the cyclotron unit, chemistry modules and other parts of the building).
17. Height of the stack of ducts above the ceiling of the building.

10.3 Staff Requirement in Medical Cyclotron

The following staff members are the minimum requirement for smooth functioning and ensuring radiological safety in the facility.

1. Trained and qualified medical cyclotron operator.
2. Radiological safety officer (RSO).
3. Trained and qualified radiopharmacist/NM technologist.
4. Other supporting staff trained in radiation safety.

10.4 Personnel Monitoring of Staff Members

Personnel involved in the handling of radioactivity such as medical cyclotron operation, radiopharmacy, radioactivity dispensing, radiochemistry, radiopharmaceutical administration to the patient, packing of radiopharmaceuticals, etc. should be provided personnel monitoring services. There are such accredited laboratories from which these services can be availed.

10.5 Supply of Cyclotron-Produced Radionuclides to Users and Transportation

Cyclotron-produced radionuclides may be supplied to various users in the country with prior permission and approval from the regulatory authority. The radionuclides are transported from one place to another in a properly designed transport container(s), satisfying the transportation conditions of radioactive material by road or air. The design of container(s) meant for transportation of radioactive material also requires approval from the regulatory authority.

10.6 Radiation Monitoring Devices

The following are the monitoring types of equipment required.

1. G.M. type survey meter.
2. Ionization type survey meter.
3. Contamination monitor.
4. Neutron survey meter.
5. Pocket dosimeters.
6. Isotope dose calibrator.
7. Area zone monitor.
8. Air monitor.

10.7 Radiation Safety Devices

As a large quantity of radionuclides is produced during each run of the medical cyclotron, and the radiation emitted by these radionuclides is high, it is important to install radiation safety equipment for fast processing and application the product. The essential ones are:- an automatic chemistry module, automatic synthesis module, L-bench and fume hoods to handle smaller activity required for QA tests, an automatic dispensing unit for smaller dosage, lead bricks, other shielding devices made of high-Z material such as tungsten (source container, syringe shield transport container, etc.) and remote handling devices (such as cap sealer, de-capper, tweezers and long vial holder).

10.8 Transport of Individual Dosages

A suitable arrangement is required to transport an individual's smaller dosages to the dose administration room from the hot cell. The weight of transport containers of such doses is heavy, and manual transportation may be inconvenient.

10.9 Typical Model for the Medical Cyclotron

Broadly, medical cyclotron is categorized under two types, viz. self-shielded cyclotron and unshielded cyclotron. The basic difference between the two, as the name suggests, is that the first type of cyclotron does not require any additional shielding on the walls; however, on the second type, the walls should be provided with additional shielding to bring the radiation level outside the wall within the prescribed limit. Usually, the shielding thickness depends upon the size of the vault where the cyclotron is installed and the occupancy all around.

Typical models and the arrangement of rooms for the installation of self-shielded and unshielded medical cyclotron are given in Figs. 10.1 and 10.2.

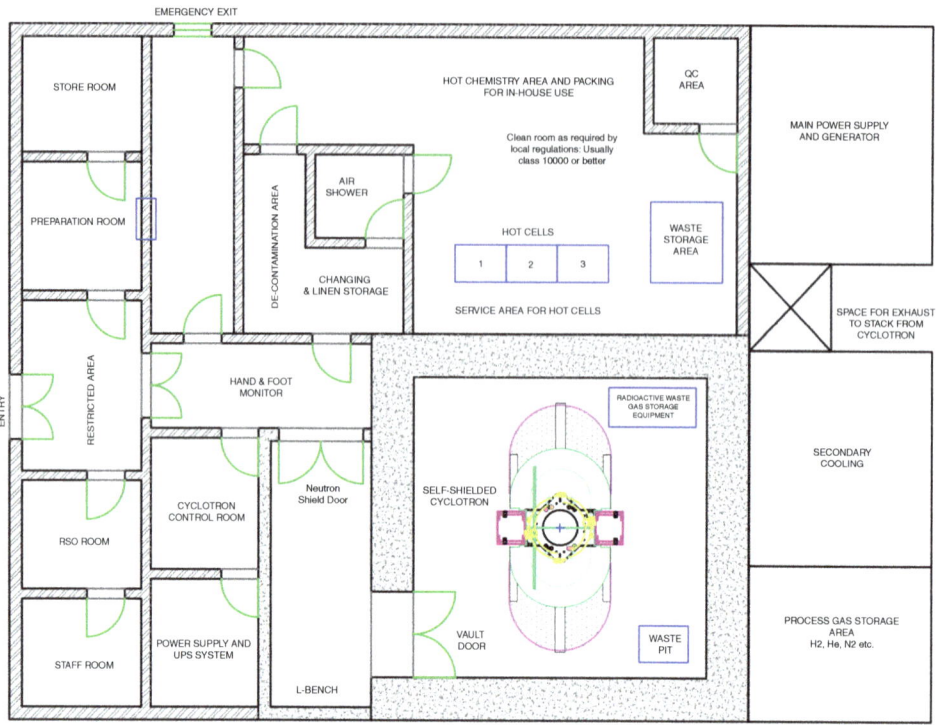

Fig. 10.1 Typical layout plan for the unshielded medical cyclotron

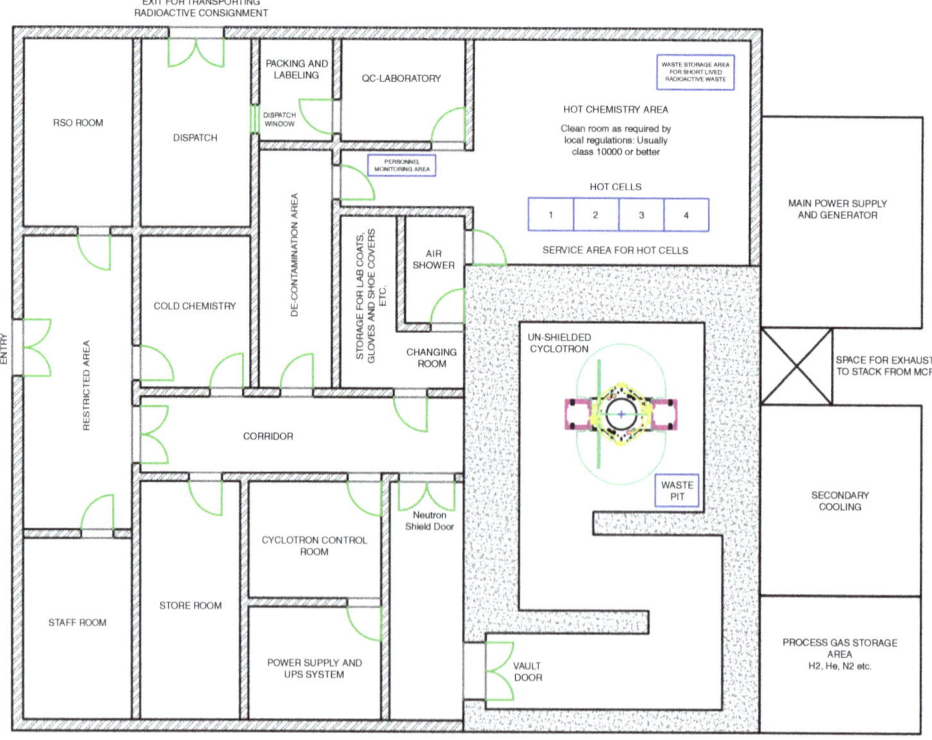

Fig. 10.2 Typical layout plan for the non-self-shielded medical cyclotron

10.10 Shielding Calculation for Medical Cyclotron

10.10.1 Shielding Calculations for Unshielded Medical Cyclotron

For calculation of transmission factor 'B_n' in rem-cm^2, the following equation is used:

$$B_n = 2.8 \times 10^{-7} \left(H_m \times d^2 / \phi_0 \times T \right) \quad (10.1)$$

where H_m = Dose limit (0.1 mR/h for general public and 1 mR/h for radiation worker)

d = Distance in metre.

T = Occupancy Factor

ϕ_0 = Neutron fluence rate at 1 m (cm$^{-2} \times$ s^{-1})

The quantity ϕ_0 is provided by the vendor, and after substituting the other parameters, the transmission factor is calculated by using Eq. (10.1) given above. In case the ϕ_0 is not given by the vendor, then in such case, there are graphs published by NCRP in Report No. 51 (Radiation Protection Design Guidelines for 0.1–100 MeV Particle Accelerator Facilities).

Example 10.1 *Calculate the concrete shielding barrier thickness for the fast neutron from a 1 cm diameter, 10 μA and 8 MeV proton beam incident on a thick carbon target. Assume 5 m distance between the reference point outside the barrier and the target. The space outside the barrier is considered a controlled area with a full occupancy factor.*

Where the neutron fluence rate $\phi_0 = 1.6 \times 10^7$ m^{2}cm^{-2} s^{-1}:

$$H_m = 40 \, \text{mR} / \text{week}$$
$$= 1 \, \text{mR} / \text{h}$$
$$T = 1$$

So by using Eq. (10.1).

$$B_n = 2.8 \times 10^{-7} \times 1 \times 5^2 / 1.6 \times 10^{-7} \times 1$$
$$= 0.43 \times 10^{-2} \, \text{rem cm}^2$$

Now, Appendix F-9, given below as Fig. 10.3 from NCRP Report No. 51, can be referred to find the thickness of concrete shielding. From curve 'G' derived for 8 MeV proton beam, a thickness of 320 gmcm^{-2} is obtained. For ordinary concrete of density 2.35, a linear thickness of 136 cm (54 in.) is required.

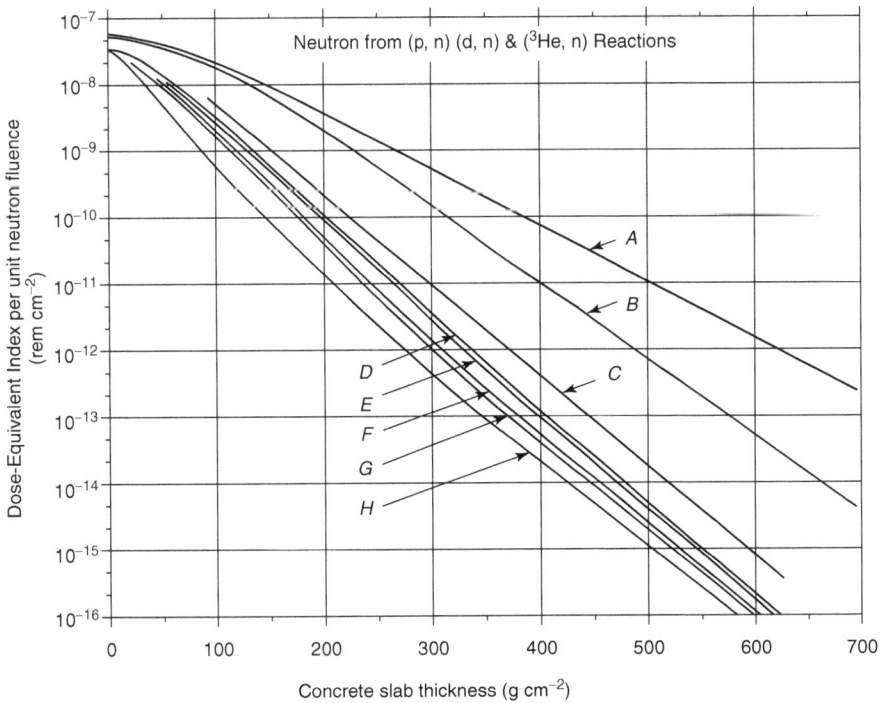

Fig. 10.3 Dose equivalent index transmission through the concrete of neutron from ion-induced reaction. Image originally printed in NCRP Report. No. 51 [1], Appendix F-9. (Reprinted with permission)

10.10.2 Shielding Calculations for Self-Shielded Medical Cyclotron

Here, the box model is used for carrying out the shielding thickness as shown in Fig. 10.4.

The equation used for calculating the dose rate is as follows:

$$H_R = \left(r_0 / R\right)^2 \left(H_n \times 10^{-X/\text{TVTn}} + Hg \times 10^{-X/\text{TVTg}}\right)$$

(10.2)

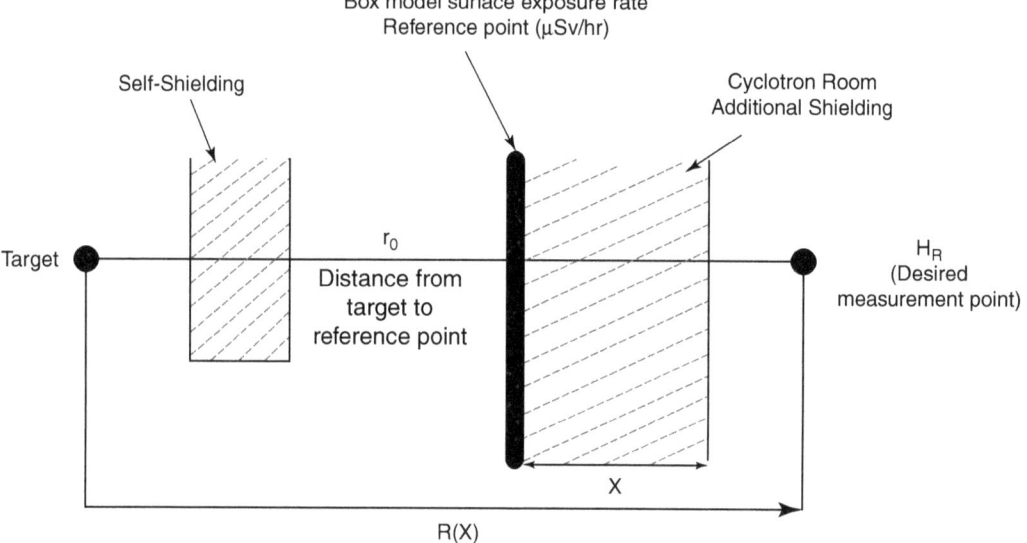

Fig. 10.4 Box model surface exposure rate reference point (μSv/h)

where H_R = Dose rate due to both neutron and gamma.

r_0 = Point of reference to target distance r_0(cm).

R = Point of measure to target distance R(cm).

X = Thickness of shielding material (cm).

H_n = Neutron dose rate at reference point (μSv/h).

Hg = Gamma dose rate at reference point (μSv/h).

For calculation few assumptions are taken:

Point of reference to target distance r_0(cm) = 280.
Point of measure to target distance R(cm) = 325.
Thickness of shielding material (cm) = 50.
Neutron dose rate at reference point (μSv/h) = 5.
Gamma dose rate at reference point (μSv/h) = 24.
Neutron TVT_n (cm) = 43.
Gamma TVT_g (cm) = 38.

Now, substituting these values in the above Eq. (10.2), we have the following equation:

$$H_R = (280/325)^2 \left(6 \times 10^{-50/43} + 20 \times 10^{-50/38}\right)$$
$$= 0.74\left(6 \times 0.0687 + 20 \times 0.0483\right)$$
$$= 1.01\,\mu Sv/h$$
$$= 0.1mR/h$$

This is within the prescribed limit in the controlled area (for the public); it shows that the shielding thickness of 50 cm concrete is good enough to bring the radiation level within the prescribed limit if the distance from the reference and the distance for the measurement is 280 cm and 325 cm, respectively. In the above problem, we have taken neutrons, and the gamma dose rate is 6 and 20, respectively. Further, TVT for neutron and gamma is 43 cm and 38 cm, respectively.

Reference

1. National Council on Radiation Protection and Measurements, Washington, DC (USA). Radiation protection design guidelines for 01 to 100 MeV particle accelerator facilities (NCRP--51). United States;1977.

Personnel Monitoring and Radiation Protection Survey in Nuclear Medicine

Abstract

Radiation sources are potentially hazardous; hence proper precautions need to be taken while handling. Monitoring personnel and the workplace plays a significant role in radiation safety. Persons handling radiation sources, those in the vicinity and members of the public are the ones who are likely to be affected. Radiation hazards can be both external and internal. External hazards are due to radiation sources outside the body. In contrast, internal hazards are due to radiation sources inside the body, entering through the air, water, food, wound, etc. To get maximum benefits of radioisotopes, the hazards are to be minimized. The external hazards can be controlled by a judicious combination of basic principles of radiation protection, i.e. time, distance and shielding. The internal hazards can be minimized by minimizing the release of activity in the air and water and avoiding the presence of activity on work surfaces. These can be achieved by adequately handling radioactivity, radioactive waste, cleanliness, good housekeeping and good work practices. This chapter explains various limits of personnel monitoring, including radiation workers in planned and emergency exposures, dose limits for medical exposures, exposure of a pregnant radiation worker, handling of overexposure cases and workplace monitoring.

11.1 Introduction

The purpose of personnel monitoring or individual monitoring is to estimate the mean equivalent dose due to occupational exposure. Occupational exposure, as defined by ICRP-103 (2007) [1], is "all exposure incurred by workers in the course of their work, except for, (a) excluded exposures and exposures from exempt activities involving radiation or exempt sources; (b) any medical exposure; and (c) the normal local natural background radiation. Here, excluded exposure means the exposures beyond the control of legislation, whereas exempted exposure means unwarranted controls or need not be regulated. Examples of excluded exposure are potassium-40 (^{40}K) incorporated into the human body & exposure to cosmic rays at ground level. The exempted exposures are such exposure that arises from trivial doses which give whole-body exposure of the order of 10–100 µSv/year, also called "tens of microsieverts per year" (ICRP 104) [2]. It is important to note that the term 'occupational exposure' relates to the exposure from a source or practice but not to where a worker is working.

A radiation protection survey is carried out to evaluate radiation exposure levels at locations occupied by radiation workers and members of the public to ensure that the exposure levels are as low as reasonably achievable (ALARA). There are three types of radiation protection surveys: pre-commissioning, routine and emergency surveys. The pre-commissioning survey is carried out to ensure that the department is constructed as per the approved plan and the availability of safe handling devices, trained personnel, etc. The routine survey ensures that the conditions are not changed and ensures the absence of contamination, etc. The emergency survey is to tackle any accidental spillage of radioisotope or release of activity into the atmosphere.

11.2 Objectives of Personnel Monitoring

The monitoring of radiation workers ensures that the exposures are kept as low as reasonably achievable and the authorized limits conforming to ICRP dose limitation systems are not exceeded. In India, the maximum permissible annual equivalent dose for radiation workers used to be 50 mSv until 1990. After that, the regulatory body 'Atomic Energy Regulatory Board (AERB)' gradually reduced it to 40 mSv in 1991, 35 mSv in 1992 and 30 mSv in 1993, keeping in view the recommendations of ICRP-60. Since 1994, the cumulative effective dose constraint for 5 years is 100 mSv, i.e. average of 20 mSv per year for the 5-year block (calculated on a 5-year sliding scale, i.e. current year and previous 4 years), and shall not exceed 30 mSv in any single year. The same dose limit is continued to date as the recommendations of ICRP-103 (2007) have not changed from ICRP-60 (1991) recommendations as far as planned exposure situations for radiation workers are concerned.

11.3 Benefits of Personnel Monitoring

Benefits that occur from a personnel monitoring program are:

- The adequacy of supervision, training and engineering standards must be demonstrated.
- Using data from people and groups, evaluate and implement a radiation protection pro-

gramme. Epidemiological research, cost-benefit analyses and medico-legal purposes may benefit from such information.

- The motivation of workers to reduce their exposure because of the information they provide.
- In case of any significant exposure, prompt action and follow-up can be taken.

As a rule, no person can work in the radiation field without using a proper personnel monitoring device unless exempted and authorized by the regulatory authority. Hence, the operational management should decide to provide individual monitoring but should be subjected to review by the regulatory authority. The decision should be influenced by three primary technical criteria [3]:

1. The expected level of dose or intake about the applicable limits.
2. The likely variation in the dose and intake.
3. The complexity of the measurement and interpretation procedure comprises the monitoring program.

Personnel monitoring exemption is also based on the minimum measurable limit, amount of radioactivity present at any time in the department and the capability of measuring personnel monitoring devices. In general, nuclear medicine facilities handling activity of more than a few MBq must be covered by personnel monitoring.

11.4 Devices

The personnel monitoring devices commonly used are thermoluminescent dosimeter (TLD) badge and pocket dosimeters. The most used personal dosimeter is TLD; however, film badge dosimeters were used in earlier days. The TLDS have proven superiority, accuracy, cost-effectiveness, simplicity in the readout, automation and commercial availability of various systems [4]. The working mechanism of TLDs and pocket dosimeters has been explained in Chap. 15.

11.5 Dose Limits in Planned Exposure Situations for Radiation Workers

Based on recommendations from ICRP 103 (2007) [1] and ICRP statement on tissue reactions issued in April 2011 [5], the International Atomic Energy Agency (IAEA) (Specific Safety Guide—46) [6] published dose limits for planned exposure situation for radiation workers. The summary of those dose limits is as follows:

For radiation workers above 18 years of age:

- The whole-body effective dose of 20 mSv per year averaged over 5 consecutive years, i.e. 100 mSv in 5 years with not more than 50 mSv in any single year. In India, the maximum permissible whole-body effective dose in a year is 30 mSv, whereas the total effective dose of 100 mSv averaged over 5 years, i.e. 20 mSv per year, is the same as per ICRP recommendations.
- The equivalent dose for the eye lens is 20 mSv per year, averaged over 5 consecutive years with 100 mSv in 5 years and not exceeding 50 mSv in any single year. The maximum equivalent dose to the eye lens in any single year in India is 150 mSv. As per the meeting held in Niyamak Bhavan, Bhabha Atomic Research Centre, Mumbai, on 18 March 2016 [7], it is concluded that the revised limits prescribed by the International Commission on Radiological Protection (ICRP) can be implemented in the coming time. However, the implementation is yet to be seen in official documents.
- The equivalent dose for extremities, i.e. hands and feet or the skin, is 500 mSv in a year.
- Further restrictions apply to female radiation worker who has notified pregnancy or breast-feeding. The embryo or the foetus or the breastfed infant has assumed the same level of protection as required for the public members.

For trainees/apprentices 16–18 years of age:
The trainees/apprentices of age 16–18 years are required to work under the supervision of trained radiation workers. Following dose limits are recommended for this category:

- The whole-body effective dose is 6 mSv in a year.
- The equivalent dose for the eye lens is 20 mSv in a year. In India, it is 50 mSv as of now.
- The equivalent dose to the extremities or the skin is 150 mSv in a year.

Individuals below the age of 16 years shall not be allowed to work with radiation.
For a member of public exposure:

- The whole-body effective dose is 1 mSv in a year. The same 5-year average with a total of 5 mSv is allowed here also in exceptional cases.
- The equivalent dose for the eye lens is 15 mSv in a year.
- The equivalent dose to the skin is 50 mSv in a year.

11.6 Dose Limits in Emergency Exposure Situations for Radiation Workers

The International Atomic Energy Agency (IAEA), in its document General Safety Requirements (GSR Part 7) 2015 [8] and General Safety Guide (GSG-7) 2018 [9], publishes dose limits in emergency exposure situations for radiation workers. The same are as follows:

- The employer is responsible for ensuring that emergency workers are not exposed to more than 50 millisieverts (mSv) (annual dose limit for a single year), except:
  - In case of life-saving situations or to prevent serious injuries.
  - In case of prevention of severe deterministic effects.
  - In case of prevention of catastrophic conditions that can affect members of the public and the environment.
  - In case of aversion to large collective doses.

- To avert a large collective dose, the dose limit is 100 mSv (GSR Part 7, Table I.1) [8].
- *In life-saving conditions*, the dose limit is 500 mSv from all the sources, such as external, internal or skin contamination. This limit can exceed only when the expected benefit to others outweighs the emergency worker's health risks. The second condition is that the emergency worker should volunteer to take action during an emergency to understand the consequences of health risks on their own and accept the same.

11.7 Dose Limit for Medical Exposures

The exposure of patients, comforters, carers and volunteers of biomedical research comes under the 'medical exposure' category (IAEA, SSG-46) [6]. The medical exposure is primarily for the patients undergoing radiological procedures, but carers, comforters and volunteers also fall under the same category. The meaning of 'patient' is the one who is under the radiological procedure. The other patients, including those waiting for the radiological procedure, are considered members of the public. Again, the relatives who are waiting in the waiting room and the patient or the casual visitor who came to meet a patient-administered high-dose therapy are also considered members of the public.

The term 'carer and comforter' is defined as: 'The person willingly and voluntarily helping (other than in his/her profession) in the care & support of the patient or comforts the patient during the radiological procedure for the medical diagnosis or treatment is called carer and comforter (GSR Part 3) [10]'.

Medical exposure is different from occupational and public exposures as the patient is deliberately, directly and knowingly exposed to radiation, weighing risks versus benefits. A limit cannot be applied in such cases of medical exposure as limiting this may limit the benefit for the patient.

The ICRP recommends three principles of radiation protection, viz. justification, optimisation and dose limit. The only two, i.e. justification and optimisation, are applicable in the case of medical exposures. Justification does the role of gatekeeper and determines whether or not the medical exposure takes place. If yes, what can be the minimum radiation exposure sufficient to achieve the target, i.e. optimisation.

11.8 Dose Record of Occupational Exposures

The dose record of each occupational worker shall be maintained for a minimum of 30 years post-cessation of the work or they attained the age of 75 years with no less than 30 years after cessation of work. This record shall be maintained irrespective of whether the worker is alive or not (IAEA GSR Part 3) [10].

11.9 Personal Monitoring During Pregnancy

Protection of unborn children is the same as the members of the public. Soon after the pregnant worker informs the employer about her pregnancy, they shall ensure that employment conditions are as such where the *equivalent dose to the unborn child does not exceed 1 mSv* for the remainder period of pregnancy [10]. The pregnant worker should not be employed in work with a significant risk of radionuclides intake or bodily contamination. However, the limitation *does not mean that pregnant workers should avoid working with radiation* but that the employer should review the working conditions for normal and potential exposure situations. Also, any discrimination in routine work should not be observed. Pregnant women might be restricted from *spending much time* in radiopharmacy or working with radioiodine liquids (Para 4.146, SSG-46) [6].

The foetal radiation dose should be monitored using an appropriately positioned additional dosimeter. Personal electronic pocket dosimeters are very useful in assessing such radiation doses. The embryo or foetal dose is unlikely to reach 25% of the external exposure readings recorded by the personal dosimeter (Para 4.149, SSG-46) [6]. However, this value depends on the penetrating power of incident radiation, i.e. the energy of the photon in use. The foetal doses can be overestimated by more than 10 times in the case of diagnostic radiology professionals; however, it can go about 100 times if the dosimeter is worn wrongly outside the lead apron (Para 3.135, SSG-46) [6]. Since higher photon energies are used in nuclear medicine and radiotherapy, lead aprons are generally not worn. Still, the foetal dose is not likely to exceed 25% of dosimeter readings (Para 144, ICRP 84) [11].

In the interpretation of the dose limit of 1 mSv after the declaration of pregnancy, one should *not* create unnecessary discrimination against pregnant women. Each year, tens of thousands of pregnant women are exposed to ionizing radiation (Para 1, ICRP 84) [11]. The anxiety or panic creation is due to a lack of knowledge, and termination of pregnancy is unnecessary. For most patients, medical exposure is appropriate, and the risk to the foetus is minimal [11].

11.9.1 Radiation-Induced Malformations

The malformations induced by the radiation have a threshold dose of 100–200 mGy and are related to central nervous system problems. Foetal doses of 100 mGy are not reached even with 3 pelvic C.T. scans or 20 conventional diagnostic X-ray examinations (Para 25, ICRP 84) [11]. The fluoroscopically guided interventional procedures may reach these doses during pelvic procedures or radiotherapy. The central nervous system is sensitive during 8–25 weeks of conception. Some reduction in intelligence can be seen if the foetal doses are over 100 mGy. In the case of foetal doses around 1000 mGy, severe mental retardation or microcephaly can be seen [11] (Table 11.1).

Table 11.1 Probability of bearing healthy children as a function of radiation dose. Table originally published in ICRP publication 84, Pregnancy and Medical Radiation, Mar 2000 [11]. Reprinted with permission

Dose to conceptus (mGy), above natural background	The probability that child will have no malformation, %	The probability that a child will not develop cancer (age 0–19 years), %
0	97	99.7
0.5	97	99.7
1.0	97	99.7
2.5	97	99.7
5	97	99.7
10	97	99.6
50	97	99.4
100	Close to 97	99.1

11.10 Overexposure Investigation and Follow-Up

The equivalent doses of over 10 mSv in a monitoring period/30 mSv in a calendar year/100 mSv in a block of 5 years recorded by the TLD badge are considered overexposure. The same is intimated immediately to the concerned institution for prompt investigations and the regulatory body. The cases of overexposures are viewed seriously by the regulatory authority. The heads of the institution/radiation safety officer (RSO) investigate the reason(s) for the overexposure and submit the overexposure report (OER) with their findings along with an individual statement to the regulatory authority for receiving advice regarding remedial and follow-up measures.

The timeline allotted to respond with OER is 21 days in India, but this can vary depending from country to country. In case no response is submitted by the institution to the regulatory authority, appropriate regulatory action(s) as the deemed fit is initiated against the institution.

The report of excessive exposure from any institution reflects the safety status of the institution. Such incidences are considered the failure of responsibility by the employer/licensee and, more particularly, the RSO, who is responsible for ensuring radiation safety in the institution. Detailed root cause analysis and identification of additional measures undertaken by the institution to prevent such exposure in the future are submitted to the regulatory authority. In many cases, the reported overexposures are non-genuine and related to improper TLD storage after the radiation work. If TLDs are appropriately stored in a transparent glass-doored box with earmarked place for each TLD, checked and verified daily, as shown in Fig. 11.1, such cases of overexposures can be avoided.

In case the TLD is left inadvertently in the radiation area and is known soon after, the same should be reported by the licensee/RSO immediately to the laboratory from whom the personnel monitoring service (PMS) is being availed before its routine processing by marking a copy to the regulatory authority. The immediate notification can be an email with the request to despatch the

new TLD for the remaining period. The reading of such dose records is neither considered excessive exposure nor recorded in individual lifetime dose records.

The following check-off list on 'proper use of TLD badges' may be implemented in radiation facilities to prevent any occurrence of excessive exposure due to non-genuine reasons:

- Ensure that the TLD cards are received periodically from the laboratory without undue delay.
- If there is a delay in receiving new TLD badges for more than 10 days for a monitoring period, the laboratory should be informed about the same.
- The arrangement is to store radiation workers' control TLD and other TLDs in a radiation-free area.
- During the periodic change of TLDs, RSO should ensure that TLD cards are loaded in the cassettes (card holders) before handing over TLD badges to the workers.
- All workers are to be briefed about the purpose, use and storage of the TLD badges by the RSO at the initial stage of handing over the TLD and at regular intervals.
- All the workers are to be informed that the TLD badge is not used for protection but rather to measure the dose received by the individual to ensure safety.
- All the workers must be informed that before entering the radiation area, they must wear a TLD badge at the chest level (and wrist badge, if applicable).
- All the workers must be informed that if they use a lead apron, they have to use TLD *below the lead apron.*
- All the workers must be informed that if the TLD badge is left inadvertently in the radiation area, the RSO/licensee should be intimated promptly.
- All the workers must be informed that TLDs are to be stored in the radiation-free area after working in the radiation area.
- As the safety supervisor, RSO should periodically monitor the workers using TLD badges while working in the radiation area.

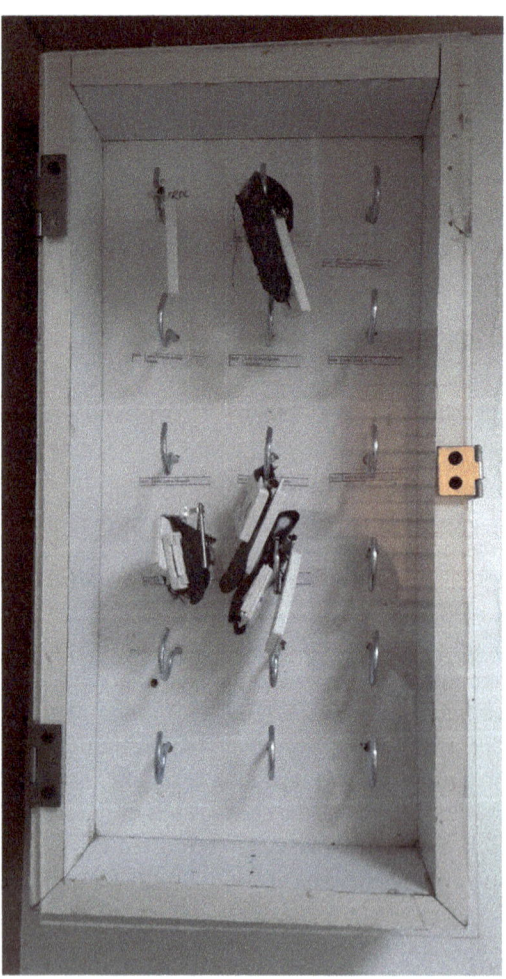

Fig. 11.1 TLD box

- Staff should be identified to ensure whether the workers store the TLD badges in designated radiation-free areas after work.
- All the TLD cards received are sent back to the PMS lab for processing without leaving behind any TLD cards.
- Dose records of the workers are to be informed to them periodically.
- Maintain the dose reports properly in the institution.

In nuclear medicine, there is one more challenge of the non-genuine case, as unsealed liquid radioactive materials are used. Any splash of radioactivity during injection may contaminate the TLD, especially the wrist TLD, which can give rise to overexposure which is not a genuine case of overexposure. To prevent such exposures, wrist TLDs should be covered under hand gloves.

For all the reported matters of overexposure, a committee consisting of experts is constituted by the regulatory authority to analyse the circumstances under which the overexposure has occurred considering the following data: (1) investigation report received from the RSO and the statement of the individual, (2) pattern of the T.L. readouts and (3) chromosome aberration test report (for exposures > 100 mSv only) to recommend the necessary steps to be taken to minimize the future exposures including restriction on radiation work for the person exposed.

The chromosomal aberration test report or bio-dosimetry is carried out only in cases where exposure is more than 100 mSv in a single monitoring period. The decision to carry out bio-dosimetry can be taken by the committee set up by the regulatory authority. In addition to the bio-dosimetry, a complete blood count (CBC) is also carried out to look for infection. CBC has no role in radiation monitoring unless the dose is not more than 500 mSv. Since, in infectious cases, bio-dosimetry may become a little complicated, CBC is carried out. This can be helpful at times for record-keeping also.

Once the investigations are over, the institution is asked to submit an action plan to prevent further overexposure cases.

11.11 Situations Not Warranting Personnel Monitoring

Personnel monitoring may not be necessary in many cases as the dose received by the workers is likely to be within the dose limits set for the general public. Such workers can be those people who work in the vicinity of radioactive sources or the radiation-generating equipment and include:

- Security personnel of radiation facility where the doses are within the dose limits for a member of the public.
- The reception staff of the radiation facility do not enter the controlled areas and less frequently enter the supervised areas.
- The workers who handle exempt quantities of radioactivity.
- The workers work in areas where the estimated dose is below the detectable limit of thermoluminescent dosimeters (TLDs).

11.12 Survey of Nuclear Medicine Facility

The radiation survey of workplaces in the nuclear medicine department is an essential requirement from the radiation safety point of view. This monitoring includes taking appropriate measures on people in the workplace and the environment and interpreting the results to limit worker's exposure and environmental contamination. Furthermore, monitoring is critical for the safety of sensitive and costly imaging equipment such as the PET-CT machine, the gamma cameras, the SPECT-CTs and other imaging devices that may become inoperable owing to radioactive contamination.

External radiation monitoring, surface contamination monitoring and air contamination monitoring are types of workplace monitoring. It should be done whenever radiation sources are produced, processed, handled, stored, transported or disposed of. The following expertise is required to conduct the monitoring:

- *Operational monitoring*: This is undertaken by a radiation worker as a standard procedure, especially when operations pose a higher risk.
- *Routine monitoring*: The radiation safety officer (RSO) conducts routine monitoring to ensure safe working conditions. This should be carried out at appropriate regular intervals but not on a set schedule.

The monitoring is meant to confirm the adequacy of steps taken to control internal and external hazards and look for any deterioration in radiation safety standards. The record of monitoring should be maintained preferably for 5 years from the time of the survey undertaken (para 7.272 of IAEA GSG-7) [9].

There are two aspects of radiation monitoring:

1. *External*: When a radiation survey of surroundings is carried out, it is called external monitoring.
2. *Internal*: When radioactive material is monitored for deposition of radioactive material inside the body, it is called internal monitoring.

External monitoring is again subdivided into (a) *personnel monitoring* and (b) *area and environmental monitoring*. Personnel monitoring can be performed using personnel monitoring badges, whereas area and environmental monitoring can be performed with survey meters. The personnel monitoring equipment has been covered in Chap. 15.

11.13 Area and Environmental Monitoring

A survey of the facility involves area and environmental monitoring. The radiation survey is to be carried out to test the adequacy of built-in safety features and to ensure that the radiation levels are within the satisfactory level. It should be done regularly throughout the operation to identify the working radiation levels and control

accumulated doses. An ideal area and environmental survey monitor should be able to detect all types of penetrating radiation, portable and simple to operate and display exposure rate/dose rate, preferably in different units. Because it is impossible to construct a single instrument that can meet all these requirements, different instruments have been developed to measure different radiation types.

In a nuclear medicine facility, the most used radiation monitoring equipment is Geiger-Muller (G.M.) survey meter, ionization chamber type survey meter (gun monitor) and contamination monitor for area survey and contamination check. The areas where one can expect high radiation fields, such as the radiopharmacy room and radionuclide therapy room gun monitor, should be used to measure high radiation levels. On the other hand, a G.M. survey meter can be used to measure low radiation fields.

The recorded survey data should:

- Demonstrate adherence to rules and regulations.
- Examine the working environment to see whether there have been any significant changes.
- Include information about radiation surveys, such as the date and time, location, dose rate, air-borne activity concentration, surface contamination levels, instruments used, name of surveyor and any other relevant information.
- Describe any relevant actions that were taken.

11.14 External Contamination Monitoring

A small amount of radioactivity can cause a significant internal radiation hazard, even though they are a minor external hazard. Therefore, the contamination monitors must be more sensitive than radiation survey meters [12]. This necessity for increased sensitivity prompts contamination monitors to have built-in amplification systems such as G.M. counters or scintillation detectors [12].

11.14.1 Surface Monitoring

There are two methods of surface contamination monitoring:

1. Direct Surface Contamination Monitoring.

    This is the most basic and practical method of contamination monitoring, and it is used to determine the presence of contamination on surfaces such as benchtops, floors, clothing and hands. The most practical instrument is a contamination monitor with an audio alarm feature, which emits a loud audio alarm when contamination is detected.
2. Swipe Sampling (Indirect Surface Contamination Monitoring).

    Swipe sampling is an indirect way of determining the level of contamination on a surface. They have been used to detect very low contamination levels or monitor contamination in areas with high radiation. Mostly, a filter paper is wiped over a defined surface area (minimum 100 cm²) and counted in a system with known counting efficiency. If the surface is contaminated with a pure beta emitter of low energy (³H, ¹⁴C), this is the only method using a liquid scintillation counter.

11.14.2 Air Monitoring

Air sample monitoring is carried out in areas where air-borne contamination is likely to occur. Particulate air-borne activity is measured using a suction pump and a filter paper. The air suction volume is known, and the same is filtered through filter paper. This filter paper is then counted in a high-efficiency detection system. Activated charcoal filter paper is used for areas where Iodine-131 is used. The sample holder is kept at working height, and activity in the air is calculated with the following formula:

$$\text{Activity in the air} = \text{cps} \times \frac{100}{\text{efficiency}} \times \frac{1}{\left(\text{volume in m}^3\right)} \qquad (11.1)$$

Example 11.1 An air sample is collected in 60 min with 10 L/min flow rate. The collected sample is counted for 60 s and shows 1156 counts. Standard source of I-131 of activity $1.4 \times 10^{-3} \mu\text{Ci}$ gives 550 counts in 60 s. Determine the activity of the sample.

Solution *Equipment efficiency calculation:*

$$\text{The activity of I-131 in standard source} = 1.4 \times 10^{-3} \, \mu Ci$$
$$= 1.4 \times 10^{-3} \times 3.7 \times 10^4 \, \text{Bq}$$
$$= 51.8 \, \text{Bq}$$

$$\text{Measured counts per second}\left(\text{cps}\right)\text{for the standard source} = \frac{550}{60} = 9.17 \text{counts / s}$$

$$\therefore \text{The efficiency of the counting device} = \frac{\text{cps}}{\text{dps}} \times 100$$
$$= \frac{9.17}{51.88} \times 100$$
$$\text{Efficiency} = 17.67\%$$

Determination of activity in the sample:

$$\text{The volume of the air sucked} = 10 \times 60 \; \text{L} = 600 \; \text{L}$$

$$\text{The volume of air sucked in m}^3 = 600 \times 10^{-3} = 0.6\text{m}^3 \left(\because 1\text{L} = 10^{-3}\,\text{m}^3 \right)$$

$$\text{The formula for activity calculation} = \text{cps} \times \frac{100}{\text{efficiency}} \times \frac{1}{\left(\text{volume in m}^3 \right)}$$

Putting value from the question,

$$\begin{aligned}
\text{Activity in the sample} &= \frac{1156}{60} \times \frac{100}{17.67} \times \frac{1}{0.6} \text{Bq} / \text{m}^3 \\
&= \frac{115,600}{636.12} \\
&= 182 \,\text{Bq} / \text{m}^3
\end{aligned}$$

The derived air concentration for I-131 is 410 Bq/m³ (Chap. 5, Table 5.4).

11.15 Monitoring and Surveillance Procedures in Nuclear Medicine

During handling of any unsealed source of radioactivity, there is always a possibility of spillage occurring, thus resulting in contamination of the personnel or working areas. Usually, if spillage has occurred with the worker's knowledge, it can be easily contained and decontaminated. However, when personnel are not aware of the spillage, there is a high risk of contamination spreading from one place to another. Therefore, it is essential to carry out radiation monitoring of various radioactive laboratories to check and control the spread of contamination.

In the diagnostic nuclear medicine facility, areas that are highly susceptible to contamination are the radiopharmacy room and the injection room. If dynamic studies are being performed, the scintigraphy rooms are also likely to be contaminated. G.M. survey meters may not always be suitable for monitoring in these facilities. The best technique to detect contamination is to take swipe samples from different surfaces in such situations. For example, the surfaces which are more likely to be contaminated in the radiopharmacy room are the floor near the fume hood, dispensing table, radioactive sink, radioactive dustbin, etc. Due to the handling of ^{99m}Tc, there is no risk of air-borne contamination occurring unless one is performing aerosol studies. If these studies are undertaken, it is essential to take air samples in the area to ensure that there is no air-borne radioactivity.

In a therapy department, the patients administered with therapeutic doses pose a high risk of external and internal radiation hazards to the nursing staff, relatives, etc. A routine survey of the radiation field in the area and the patients should be carried out to ensure that the radiation levels are below permissible limits. The dispensing room and isolation rooms/wards should be checked for surface contamination by taking swipe samples. The personnel handling therapeutic doses should be monitored for any possible contamination.

Regular and proper monitoring or surveillance procedure will ensure that radiation field and contamination levels in and around a nuclear medicine facility are kept below permissible limits.

11.16 Conclusion

The personnel monitoring of the radiation workers and workplace monitoring in the nuclear medicine facilities plays a significant role in radiological protection. There had been numerous occasions in practice, and examples were based on personnel exposures recorded by the personnel monitors in nuclear medicine facilities. The status of working conditions and non-availability of required radiation handling devices and procedures could be anticipated and confirmed through subsequent radiation protection surveys.

References

1. ICRP. The 2007 recommendations of the International Commission on Radiological Protection. ICRP publication 103. Annals ICRP 37 (2-4). Amsterdam: Elsevier; 2007.
2. ICRP. Scope of radiological protection control measures. ICRP publication 104. Annals ICRP 37 (5). Amsterdam: Elsevier; 2007.
3. ICRP. 1990 recommendations of the International Commission on Radiological Protection. ICRP Publication 60. Annals ICRP 21 (1-3). Oxford: Pergamon Press; 1991.
4. Pradhan A. "Requirement of thermoluminescence dosimeters for personnel monitoring" in National Conference on Luminescence and its Applications, October 13–15, 1997
5. Boal MPTJ. Dose limits to the lens of the eye: International Basic Safety Standards and related guidance. International Commission on Radiation Protection (ICRP)—2013 PRO; 2013.
6. International Atomic Energy Agency. Radiation protection and safety in medical uses of ionizing radiation. IAEA Safety Standards Series no SSG-46. Vienna: IAEA; 2018.
7. Theme meeting on 'New Dose Limits For Eye Lens-Measurement, Monitoring And Regulation' held at Niyamak Bhavan. Atomic Energy Regulatory Board, Bhabha Atomic Research Centre, 18 March 2016. [Online]. https://www.aerb.gov.in/images/PDF/image/thememetmar2016.pdf. Accessed 31 Mar 2022.
8. IAEA. IAEA safety standards series no. GSR Part 7, preparedness and response for a nuclear or radiological emergency. General safety requirements. Vienna: International Atomic Energy Agency; 2015.
9. IAEA. Occupational radiation protection, general safety guide, IAEA safety standards series no. GSG-7. Vienna: International Atomic Energy Agency and International Labour Office; 2018.
10. IAEA. IAEA safety standards series no. GSR part 3, radiation protection and safety of radiation sources: international basic safety standards. General safety requirements. Vienna: International Atomic Energy Agency; 2014.
11. ICRP. Pregnancy and medical radiation. ICRP publication 84. Ann ICRP. 2000;30(1)
12. Alan Martin SAH. An introduction to radiation protection. Cham: Springer; 1996. ISBN: 978-1-4899-4543-3.

Radiation Safety Considerations in Nuclear Medicine

Abstract

In the last few decades, significant developments have been seen in nuclear medicine's diagnostic and therapeutic field. Thus, we do not have only ^{131}I but many other radionuclides such as ^{177}Lu, ^{153}Sm, ^{32}P, ^{90}Y, etc., based on radiopharmaceuticals in use. Apart from this, ^{99m}Tc-based radiopharmaceuticals are used for many studies like myocardial perfusion imaging studies etc. Therefore, it becomes necessary to plan the nuclear medicine facility by taking due care. It becomes essential to handle these radionuclides by taking all safety precautions. The primary concern is the risk of irradiation and contamination due to the unsealed radionuclide underuse. Therefore, the occupational worker handling these sources has to take care of the general public, patients and the environment to be protected. This chapter covers the radiation safety issues arising from the handling of diagnostic and therapeutic radionuclides in nuclear medicine.

12.1 ^{99m}Tc Products

^{99m}Tc radionuclide has favourable characteristics from a dosimetric point of view also and is thus commonly used in carrying out diagnostic procedures by administering doses of few mCi (MBq) activity. Although the half-life is 6 h, total amount of activity handled is significant in a day. Thus, during dispensing, formulation, administering to the patients and imaging, appropriate care must be taken because of contamination and exposure to nearby personnel.

12.2 Cyclotron Products

12.2.1 SPECT Product

^{111}In-based radiopharmaceuticals are mainly helpful in delayed imaging needs, involving slow pharmacokinetics and better compatibility as a label for biological carrier molecules. Amongst the more critical ^{111}In products, special mention can be made of the ^{111}In-labelled peptide and ^{111}In-labelled octreotide for imaging tumours of neuroendocrine origin.

^{123}I-metaiodobenzylguanidine (^{123}I-MIBG) is a very good marker for myocardial imaging. Still, due to its logistics problems, accessibility to pure ^{123}I has precluded broader utilisation of these products, despite its logistics problems and their importance in clinical practice.

12.2.2 PET Products

1. PET Products in Neurology: As glucose is the primary energy source for our brain, ^{18}FDG

becomes the choice of radiopharmaceuticals for calculating cerebral blood flow. Similar lines include 15O- and 11C-based radiopharmaceuticals for neurological disorders.

2. In Cardiology: ^{18}FDG is the choice of radiopharmaceutical for performing the viability studies of the heart.

3. In Oncology: ^{18}FDG plays an essential role in tumour staging during PET imaging.

12.3   Radionuclide Therapy (RNT)

In RNT, it is a daunting task to give the maximum dose to the cancerous cells by sparing the normal cells. The choice of radionuclide depends on physical characteristics and the ease of production and logistics. In the Indian context, therapeutic radionuclides of high relevance and practicability are ^{90}Y, ^{153}Sm, ^{166}Ho, ^{186}Re and $^{186\ +\ 188}$Re apart from the commonly used ^{131}I and ^{32}P.

12.3.1 ^{131}I-MIBG

This radiopharmaceutical is used to treat neuroblastoma, pheochromocytoma and medullary thyroid carcinoma. Due to the renal route of excretion, essential precautions must be taken during the safe discharge of ^{131}I.

12.3.2 Products for Radiation Synovectomy

Radiation synovectomy or radio-synoviorthesis is an alternative to conventional surgery where the inflammation of joints is treated using the locally instilled formation made of the beta-emitting radionuclide. Typically, 2.27 MeV of ^{90}Y for large joints and 0.3 MeV of ^{169}Er for small joints like finger joints are used.

12.3.3 Product of Radio-immunotherapy (RIT) and Radio-Peptide Therapy (RPT)

RPT and RIT use radiolabeled peptides and monoclonal antibodies for the targeted lesions. ^{131}I-labelled B1-anti-C20 monoclonal antibody used for lymphoma was the first product that got approval.

12.3.4 Products for Loco-Regional Delivery for RNT

There are products for treating hepatic cancers using ^{131}I-Lipiodol by intrahepatic arterial injection. $^{188/186}$Re-Sulphide-Lipiodol   and ^{166}Ho-Chitosan are used for such therapy.

12.3.5 Products for Endovascular Radionuclide Therapy (EVRT)

To prevent the neo-intimal proliferation of arterial walls after balloon angioplasty, ^{32}P-coated stent or ^{192}Ir source and ^{32}P/^{89}Sr-coated source attached to the catheter is used. Alternative to the above approaches, ^{188}Re, ^{166}Ho and ^{90}Y radionuclides are better options.

12.4   Other Products and Techniques

Radioisotope-guided surgery (RIGS) is a new, promising, radiotracer approach for surgical oncology.^{188}Re-DMSA is being explored to treat medullary carcinoma of the thyroid with minimal excision of tissue of lesion. The treatment can be carried out using ^{99m}Tc-sestamibi or

Table 12.1 Radionuclides for therapy

Radionuclide	Half-life (days)	Emission	Energy β_{max}/β_{av} (MeV)	Principle gamma energy (MeV)	Maximum range in tissue
^{131}I	8.0	β^-, γ	0.61/0.20	0.364	2.4 mm
^{32}P	14.3	β^-	1.71/0.69	–	8.7 mm
^{89}Sr	50.5	β^-	1.50/0.58	–	8.0 mm
^{90}Y	2.67	β^-	2.28/0.93	–	12.0 mm
^{153}Sm	1.95	β^-, γ	0.81/0.22	0.103	3.0 mm
^{186}Re	3.77	β^-, γ	1.08/0.35	0.137	5.0 mm

^{99m}Tc-tetrofosmin. We can precisely locate the lesion in breast cancer. A fine collimated hand-held radiation detector probe can be used during surgery to identify the exact region of radioactivity, i.e. the tumour lesion, because the non-lesion tissues do not accumulate the tracer—Table 12.1 shows the most commonly used radionuclides for therapy in nuclear medicine.

12.5 ^{131}I Administration

The nuclear medicine facility approved by the regulatory authority must comply with the staff requirement mentioned in their safety code, i.e. a qualified nuclear medicine physician, nuclear medicine technologist/physicist and a radiological safety officer (RSO). We are well aware that ^{131}I is the most commonly used unsealed radionuclide for treating thyroid diseases like thyroid cancer and hyperthyroidism and is associated with issues with its therapeutic applications. Since ^{131}I is highly volatile and used in liquid form, it requires proper handling procedures. During its handling, as used in large quantities, it can cause a risk of internal contamination in addition to the external radiation hazard it poses, and therefore, stringent safety measures need to be taken to avoid this internal hazard because 1 μCi (37 KBq) of ^{131}I gives an absorbed dose nearly 1.3 cGy (1.3 rad) for a thyroid mass of 20 g for a 30% thyroid uptake. The radiation worker involved in the administration of the activity should wear disposable rubber gloves and protective clothing.

^{131}I is used in both liquid and capsule form; however, handling ^{131}I in capsule form possesses fewer hazards than handling in solution form. Before planning administration to the patient, activity should be ascertained by using a calibrated isotope calibrator. The three cardinal/principles of radiation protection, i.e. time, distance and shielding, should be effectively followed to reduce personnel exposure. The liquid ^{131}I should be handled in fume hood to avoid contamination during radioiodine administration. Before the patient drinks, a small amount of water is put into the vial. If the patient coughs or activity spills, the contamination can be minimized because the droplets are of diluted activity.

The radioactivity of the oral cavity is washed by giving water in a disposable paper cup. It is always advisable to place absorbent papers on the edge of the fume hood and on the floor to restrict the spread of contamination before treatment. Moreover, in the treatment of neural crest tumours, the i.v. line should be ascertained before injecting ^{131}I-MIBG into the dextrose saline bag.

It is always necessary that the radionuclide is transported in the isolation room before starting treatment by taking all radiation safety precautions. Once the treatment is completed, all personnel involved should be monitored for contamination. It is the responsibility of the RSO to maintain a record of activity administered and exposure rate readings over the stomach and at 1 m from the patient after administration and before the discharge of the patient.

12.6   Radiation Safety Precaution During Pre- and Post-therapy

As per the protocol, to enhance absorption of ^{131}I from the stomach, the patient is advised to fast for 2 h before and after the administration. This minimizes the radiation dose to the stomach, but if the patient develops nausea, it reduces the volume of the vomitus. Also, it is advised to remove dentures and removable bridges before administering the ^{131}I solution. This would help avoid contamination to the surface and personnel in case of removal after administration.

The radiation safety procedures have become more stringent and elaborate for bed-ridden patients. In such cases, the patient has to be catheterised for at least 24–48 h, so the patient needs time to adjust the catheter and complications can be avoided. This reduces the radiation exposure to the occupational worker attending to the patient.

Further, the patient is put on a liquid diet 2 days before therapy to reduce bowel activities during isolation, which may be a source of contamination and exposure. In the case of female patients, the details of menstrual history should be obtained to rule out pregnancy, as pregnancy is a contraindication for radioiodine therapy.

The patient needs to be admitted to the isolation ward approved by the regulatory authority till the activity level reduces below the limit prescribed by the regulatory authority. The two main reasons for admitting the patient to the isolation ward are: (a) the patient is a source of radiation exposure and (b) to avoid the possible risk of contamination from bodily fluids such as urine, saliva and perspiration, especially in the first 24–48 h. Staff must make the patient and their relatives understand the term isolation. Ideally, it is informed to the patient that cooperation is essential to minimize radiation exposure to occupational workers and visitors.

12.7   Radiation Protection for the Nursing Staff

The nursing staff, also as occupational workers involved in patient care, should be familiar with radiation safety procedures during treatment.

Being an occupational worker, personnel monitoring badges be provided. Nursing care such as medication, food, etc. should always be given across a shoe barrier, and direct contact with the patient is avoided.

Further, blood and urine sampling for laboratory tests should be avoided to prevent contamination in the nuclear medicine facility. However, the same is taken in consultation with RSO only in critical cases. The nursing staff should have the telephone numbers of the nuclear medicine physician and RSO on hand in the event of an unfortunate incident involving a medical emergency.

12.8   Radiation Protection for the Visitors

The isolation ward is not a general ward. Therefore, the visitors have to follow all the instructions, i.e. the duration of the visit and restrictions on distance from the patient, and abstain from physical contact. Small children and women with pregnancy should not be allowed for visiting the patient.

12.9   Patient Monitoring and Discharge Criteria for Isolation Ward Patients

For patient monitoring (exposure rate), the ionization chamber type of radiation survey meter is used. These measurements are taken over the surface of the neck, metastatic site, stomach and thigh at a 1-m distance from the patient to know whether there is a desired concentration of ^{131}I in tissues. RSO must record these measurements, enabling them to determine the discharge date.

Every country has a limit at which the patient can be discharged. In India, patients are released only if their total ^{131}I activity is less than or equivalent to 30 mCi (1.1 GBq), as per the regulatory authority's limit. This limit corresponds to an exposure rate of about 50–60 μSv/h (5–6 mR/h) at a distance of 1 m.

At the time of patient discharge, some basic instructions/guidelines like the time they should be away from the spouse, children and pregnant women at home and travelling in public transportation should be given. Keeping the patient in the isolation ward during treatment minimizes the radiation risk to their family members/relatives and the general public.

12.10 Optimisation of Radiation Dose to Non-target Tissues

As indicated earlier, the radiation dose to the stomach can be minimized if the administration of ^{131}I is carried out on an empty stomach. It is observed that sialadenitis is one of the acute complications resulting from therapy because the salivary gland concentrates radioiodine; patients are instructed to chew a lemon or any other substance that stimulates salivation during the first 2 days of treatment. Similarly, the patient is asked to drink plenty of fluids and voids as frequently as possible to minimize the radiation dose to the kidneys, bladder and gonads because radioiodine is primarily excreted through urine.

Because ^{131}I concentrates on the lactating breast, nursing mothers should refrain from breastfeeding their children (at least 4 weeks). The mother should be encouraged to suck out the milk with a breast pump during the first 2 days after therapy.

12.11 Handling Emergency Situations

The nuclear medicine facility should have made an emergency/accidents programme for handling any untoward incident. The most common incident that may occur during the use of ^{131}I is misadministration, spillage and death of the patient.

However, a proper pre-planning to handle a situation like the patient's death is required. Every country has its regulations governing the disposal of cadavers containing radioactivity. In India, the guidelines for the safe disposal of cadavers are provided by the Atomic Energy Regulatory Board (AERB), the regulatory authority. These guidelines vary depending upon procedures like burial, cremation, autopsy or embalming to be carried out. In any of the mentioned procedures, prior authorisation and specific safety precautions are to be followed by RSO after taking due approval from regulatory authority.

RSO must recommend dose reduction methods to the personnel involved in washing, preparing and transporting the cadaver. Plastic sheets and disposable gloves are used to control the spread of contamination. The autopsy and embalming procedure are not being performed/encouraged. However, if required, the same is performed under the supervision of RSO only.

The nuclear medicine facility should have a decontamination kit to deal with spillage contamination. During any of the following situations, misadministration is considered to have occurred:

1. Administration of a radionuclide to the wrong patient.
2. Administration of a radionuclide by the wrong route.
3. Administration of a radionuclide other than the one intended and.
4. Administration of an amount far in excess of that desired.

To avoid such situations, always ensure the identity of the radionuclide, the patient, the route of administration and the isotope calibrator's performance.

12.12 Conclusions

Due to the advancement of technology and availability of therapeutic radionuclides, the future of nuclear medicine is very bright. Thus, awareness of radiation safety design, monitoring and practices is greater in nuclear medicine facilities. ALARA (*As Low As Reasonably Achievable*) can be accomplished by developing and implementing a good radiation safety programme.

Radiation Safety Consideration in Medical Cyclotron

13

Abstract

The medical cyclotron (MC) is the most widely used particle accelerator for producing medically important radionuclides. In this chapter, the various safety-related issues during the operation of MC are narrated. In today's era, the construction of MC is usually taken by the vendor on a turnkey basis, whether the installation is done at nuclear medicine facility or the production and distribution centre at commercial premises. During the maintenance of MC or modification of equipment, the various components that get activated due to the production of gamma rays and neutrons may pose a radiation hazard to the maintenance engineer. Further, the management of radioactive waste generated during the operation of MC should also be given due care to minimize radiation exposure not only to occupational workers but also to the general public and the environment. It is always a challenge for the MC management to adhere to the requirement imposed by the regulatory authority of the country and follow all the radiation safety guidelines for the safe operation of MC.

13.1 Radiation Surveillance Programme in Medical Cyclotron Facility

The need for a radiation surveillance programme in an MC facility is to ensure that the radiation field and contamination levels in the workplaces and external environment around the facility during operation are maintained within the permissible levels. Therefore, every facility should have a well-organized system of radiation protection existing to aid and assure that proper operating procedures are followed and the radiation exposures to the occupational workers and general public are below the limits prescribed by the regulatory authority.

For an adequate and satisfactory regulation of radiation safety in a MC facility, the responsibility for safe operation must rest with one or a few designated, qualified and trained individuals. The local radiation safety committee of the institution should periodically review the safety procedures, practices and protocols for further optimization of radiation exposures. Before starting the irradiation or production process, a procedure shall be in place to ensure that the high radiation and exclusion areas inside the MC vault are first cleared of any personnel, ideally referred to as search and clear operation. The RSO should make a personal on-the-spot survey of these areas inside the MC vault to ensure no person's occupancy during the irradiation process. The MC installation in the approved premises should be made so that there should not be any access inside the cyclotron vault during its operation. The same should be verified by checking the functioning of all the safety interlocks at the defined periodicity. However, there may be occasions during some troubleshooting or problems wherein the interlock systems may be required to be bypassed. Though planned and made available in the facility, such bypass systems can be permitted only after carefully monitoring the situation. Only authorized and designated individuals under the supervision of the RSO shall be responsible for removing all bypasses before commencing regular production. The interlock system should be undertaken only when essential and as far as possible should be kept to the minimum. In addition to the interlock systems, emergency scram switches or buttons should also be made available. Any person accidentally trapped or caught inside the MC vault during irradiation can have easy and ready access to these scram switches in such events. These switches should do the intended functions as desired during an emergency.

The radiation safety problems are much less in a self-shielded MC where the radiation levels are quite low in the vault, even during the bombardment. Though the interlock switches are advisable at the entrance door, the lights which can indicate the situation, like the functioning of the magnet, radio-frequency power and the beam, should also be provided at the control console or entry door.

The radiation exposure to the staff operating the MC generally results from the following work practices;

1. Routine search and clear operation.
2. Checking and replacing target vials.
3. Changing filters.
4. Troubleshooting and.
5. Preventative maintenance.

The radiation exposure during routine procedures like (1), (2) and (3) mentioned above is generally very minimal. The average exposure is in the range of 1–3 μSv per procedure. At performing procedures (2) and (3), the staff should wear protective clothing and gloves. The filters and target vials replaced should be stored behind the lead shielding as these items are associated with some amount of trapped or induced radioactivity.

During routine preventative maintenance procedures in the MC facility, there is a potential for significant external irradiation and contamination of the personnel involved in the maintenance. The various jobs that are typically undertaken during preventive maintenance procedures are:

1. Removal and cleaning of the target assembly.
2. Removal and replacement of Havar foil.
3. Opening and cleaning of vacuum tank.
4. Inspection of carbon carrousels.
5. Inspection and cleaning of collimators.
6. Inspection and cleaning of ion sources and.
7. Maintenance of LTF (liquid transfer function) panel.

In the MC operation, the radiochemistry laboratory is usually in conjunction with the production facility. Before any automated transfer or delivery of the radioactivity from the MC to the radiochemistry hot cell, the radiochemists need to confirm the three important steps. Firstly, whether all the necessary reagents, liquid nitrogen traps, separation and purification chromatography columns, etc., required for the synthesis of ^{18}F-FDG are loaded in the hot cell. Secondly, the door of the hot cell receiving the radioactivity is closed, and the radiochemist is aware of the transfer of radioactivity about to be made. Any accidental transfer with the hot cell door open would result in phenomenally high radiation exposure to the laboratory personnel that may even exceed the dose limits of occupational workers as specified by the regulatory authority. Thirdly, whether negative pressure is maintained inside the hot cell before the radioactivity is transferred, and whether arrangements exist to sustain this negative pressure throughout the synthesis process. The absence of negative pressure would result in the airborne radioactivity of ^{18}F along with traces of ^{13}N activity escaping out of the hot cell, thus causing contamination of the personnel (internal hazard) and the work environment and the environment external to the facility.

In normal routine conditions, the radiation exposure contributed to the radiochemist primarily is from the following procedures:

1. Synthesis.
2. Retrieval of the product vial from the dispensing module.
3. Transfer of product vial into the transport package or pneumatic duct.
4. Quality control.

The average cumulative exposure received by the radiochemist during procedures (1), (2) and (3) indicated above generally for ^{18}F-FDG preparation is usually in the range of 4–10 μSv per synthesis. Likewise, the exposure during quality control procedures has been about 1–3 mSv per procedure.

13.2 Pregnant Occupational Worker

Moreover, in properly controlled and regulated conditions, pregnant female radiation workers involved in the routine operation of MC are unlikely to receive a dose of more than 2 mSv during their pregnancy period. However, handling components having high induced radioactivity during preventive maintenance tasks may cause significant exposure, and thus these practices should be avoided during pregnancy. In radiochemical synthesis procedures, the dose is likely to exceed 2 mSv, and the pregnant staff, therefore, should undertake these practices only after consultation with the RSO. They may undertake quality control procedures as they are associated with the least risk of exposure and contamination, but that is also to be performed with due care by taking all radiation safety precautions.

13.3 Management of Radioactive Waste

Generally, the various components and the surrounding materials of the MC are subject to activation. A cocktail of radioisotopes forms.

However, few of them have a long half-life and have radiation safety concerns during their disposal. Few of the activation products having long half-life are ^{49}V (330 d), ^{53}Mn (3.74 My), ^{54}Mn (312.3 d), ^{55}Fe (2.73 y), ^{57}Co (271.7 d), ^{59}Ni (76000 y), ^{63}Ni (100.1 y), ^{65}Zn (244.3 d), ^{179}Ta (1.82 y), ^{181}W (121.2 d), ^{108m}Ag (418 y), ^{110m}Ag (249.9 d), etc. Usually, a lead pot or a storage area sealed with a lead sheet is provided in the MC vault to keep these radioactive components. Special handling procedures and care should be undertaken while dealing with these activated components.

It is always advisable that negative pressure is maintained inside the MC vault so that the radioactivity generated can be exhausted. However, it is noted that the air containing ^{16}N and ^{41}Ar exhausted from the MC does not pose any radiation hazard to the general public. Generally, with hospital-based MC, the presence of the air exhausted into the atmosphere is believed to be of negligible hazard to the public.

The radioactive waste in the radiochemistry laboratory comprises the ^{18}F separation cartridge, ^{18}F-FDG separation column, rinse water, gloves, filters, tissue papers, etc. The recovered ^{18}O water might contain traces of ^{110m}Ag radioactivity and should be handled with caution. Since most radioactive waste is associated with ^{18}F (110 min) activity, these should be allowed to decay for ten half-lives and later disposed of as general waste.

13.4 Radiological Surveillance

In radiological surveillance, the facility must carry out area monitoring, contamination monitoring and personnel monitoring.

13.4.1 Area Monitoring

Installed gamma, neutron and hand-held survey meters are used to carry out area monitoring of the MC facility.

1. The places that are likely to have radiation levels higher than the permissible levels require gamma and neutron monitors.

2. The console of the MC should display the readings obtained by the radiation monitors installed at a fixed location.
3. There should be a provision to set the alarm based on the locations of the monitors.
4. The gamma and neutron survey meters should be available at MC to measure the ambient radiation.
5. All the radiation survey instruments available with the facility should be calibrated at a defined periodicity.
6. It is desirable to have an air sampler to measure airborne radioactivity.

13.4.2 Personnel Monitoring

1. Occupational workers are covered under personnel monitoring services.
2. The facility should also keep the calibrated digital pockets dosimeters, called direct reading dosimeters, to carry the instantaneous exposure measurement.
3. The service engineer should be provided with a direct reading dosimeter during the maintenance of MC.
4. The facility should also arrange to provide the ring or headband consisting of a thermoluminescent dosimeter (TLD) to estimate extremity doses.

13.4.3 Contamination Monitoring

1. The facility should install hand and foot monitors outside the radiopharmacy lab to monitor the contamination of body parts of occupational work after completing the work.
2. During contamination of work surface, monitoring is done either by detecting directly or taking the swipe sample of the contamination for counting.

3. The MC should periodically monitor the contamination of the activation product generated either in the vault, radiochemistry or quality control laboratory.
4. The personnel and equipment decontamination kit should be available.
5. In case of contamination, the decontamination procedures should be carried out promptly.

13.5   Log Book Keeping

Radiological safety officer should maintain the logbook of the following;

1. No. of hours of operation of the facility.
2. Radiation survey of the MC facility.
3. Personnel monitoring records, including direct reading dose records.
4. Incident/accident occurrence record.
5. Record of radioactive waste disposal.
6. Breakdown/maintenance record, etc.

13.6   Decommissioning

The issue of radiation safety for decommissioning MC becomes much more of a concern in a non-self-shielded MC as the volume of radioactive waste generated is much larger than in a self-shielded MC. Therefore, the facility should have a plan for managing the disposal of radioactive waste to be generated at the time of decommissioning the MC. Decommissioning of MC should be carried out only after due approval of the regulatory authority.

Further Reading

1. Atomic Energy Regulatory Board (AERB) safety guide for medical cyclotron facilities, no. AERB/RF-RS/SG-3, October 2016.

Radiation Safety Considerations During Radiopharmaceutical Preparation

14

Abstract

Radiopharmaceutical preparation and handling involve working with open source of radioactivity, which can significantly expose staff, patients, their attendants and the environment if the radiation safety aspect has not been adhered to correctly. There are various factors such as the design of the nuclear medicine department, trained workforce and equipment functioning that contribute towards safe radiation safety practice. The trained staff should have in-depth knowledge about the subject, including radiopharmaceutical preparation; quality control; knowledge of its equipment; safety protocols; standard operating procedures; knowledge of doses to patients, including paediatric and obese patients; knowledge about troubleshooting; management of radiation hazards; radioactive waste management; and record-keeping of all. The chapter summarizes key information needed to fulfil the requirement described above.

14.1 Introduction

The radiopharmacy in the nuclear medicine department is the place where radiopharmaceuticals are prepared, stored and dispensed. The radi-ation safety of patients, staff and the environment depend on many factors, including:

- Design of the nuclear medicine department and the work practice.
- Trained workforce.
- The equipment working conditions.

The more often equipment used in radiopharmacy is the dose calibrator. Good knowledge about it is an essential requirement as part of radiation safety. However, other equipment such as the biosafety cabinet, laminar flow, survey meter, contamination monitor, shielding accessories, etc. are also crucial for radiation safety.

14.2 The Dose Calibrators

The dose calibrator (Fig. 14.1) is a gas-filled detector and works in the ionization chamber region of the voltage-response curve (shown in Fig. 14.2, discussed in Sect. 14.2.6). It gives a unique response to every radioisotope but does not distinguish the two in the case of a mixed sample as spectrometers do.

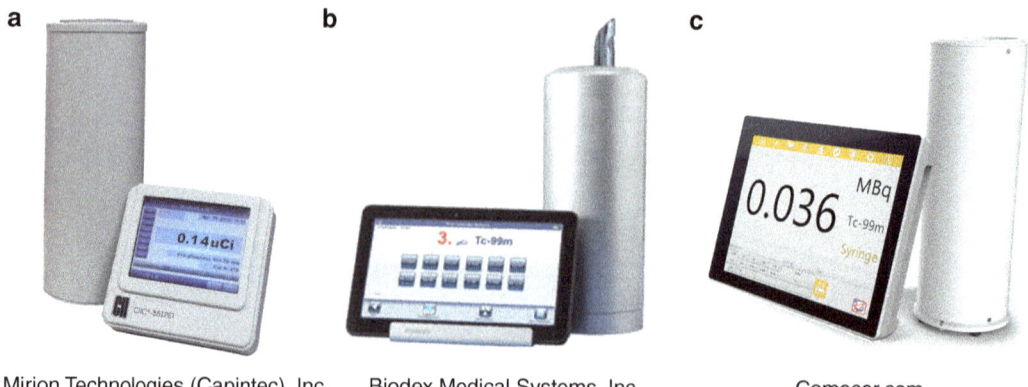

a   **b**   **c**

Mirion Technologies (Capintec), Inc.   Biodex Medical Systems, Inc.   Comecer.com

Fig. 14.1 Dose calibrators from different vendors. (**a**) and (**b**) copyright: Mirion Technologies (Capintec), Inc., printed with permission. (**c**) copyright—Comecer.com, printed with permission

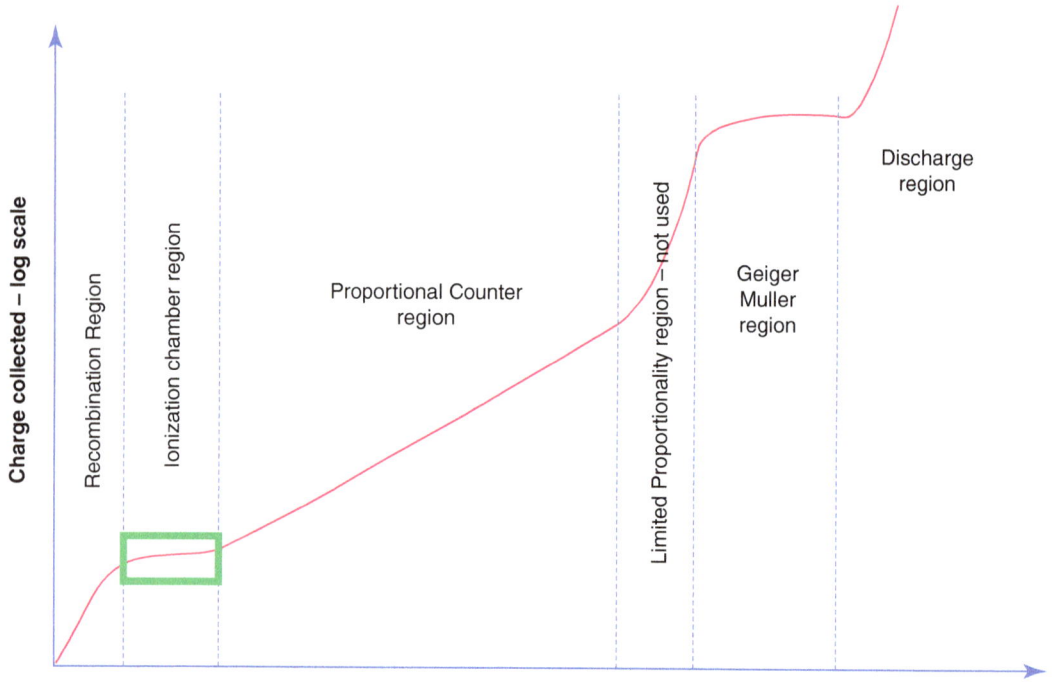

Fig. 14.2 Voltage-response curve

14.2.1 Various Names of Dose Calibrators

Different organizations have used different terminologies. However, the common term is 'dose calibrator'. Table 14.1 summarizes the name of organizations and their corresponding names used for the dose calibrator. The American Association of Physicists in Medicine (AAPM) [1] uses the term 'radionuclide calibrator' as 'dose' term is broadly used for 'energy absorbed per unit mass' of irradiated material.

Table 14.1 Terminologies used for dose calibrators

Terminology	Used by
Dose calibrator	• Common usage
	• US Pharmacopeia (USP)
	• The American National Standards Institute (ANSI)
Radionuclide calibrators	• International Electrotechnical Commission (IEC)
	• National Physical Laboratory (NPL, UK)
Radionuclide dose calibrator	• Food and Drug Administration (FDA)
Radionuclide activity calibrator	• National Institute of Standards and Technology (NIST)
Radionuclide calibrator	• The American Association of Physicists in Medicine (AAPM)

14.2.2 Physical Characteristics of Dose Calibrator

Dose calibrators have a well-type ionization chamber, cylindrical and usually constructed of aluminium, filled with argon gas under pressure. The pressurized chamber can have 1.4–20 atmospheric pressure depending on the model. It is hermetically sealed and designed to simulate ideal 4π geometry (Fig. 14.3). The ionization chamber is connected to a microprocessor-controlled electrometer within the chamber assembly, providing calibrated measurements for a range of common radionuclides.

The chamber is shielded with lead from the outside to reduce the effect of local environmental radiation and the exposure to the operator. Depending on the radioisotope used in the vicinity and for measurement, the operator must add more shielding to reduce the background effect in the chamber and protect the operator from the radiation safety point of view, especially in the case of positron-emitting radionuclides (Fig. 14.4). A plastic liner (Fig. 14.5) is placed

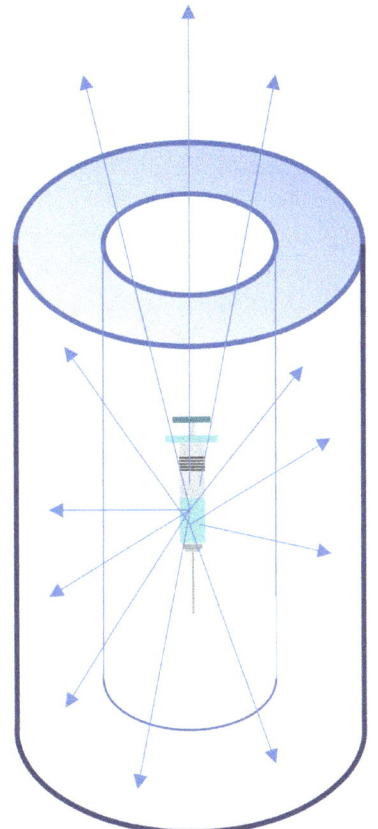

Fig. 14.3 The chamber design

within the well to protect the chamber from contamination during measurement.

The dose calibrator function is based on several parameters such as the amount of activity, the energy of the photons and the nature of emissions such as particulate or non-particulate. The reproducibility of the sample placement within the well is crucial. The same is ensured using a source holder (Fig. 14.5). Due to a combination of geometry and electrical influences, the response of most chambers decreases near the top and bottom of the chambers. The chamber volumes are in the range of several thousand cm^3 (e.g. typical Capintec CRC 55t-PET chamber volume is 9434 cc).

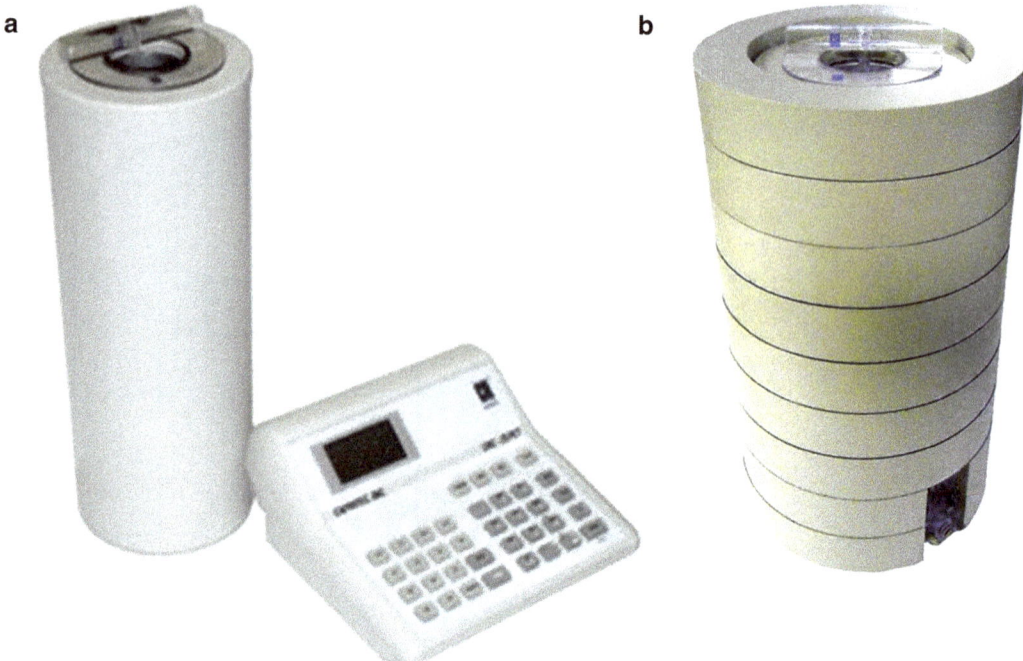

Fig. 14.4 (**a**) Dose calibrator. (**b**) Shielding of well chamber for PET radionuclides. Image copyright, Mirion Technologies (Capintec), Inc. Printed with permission

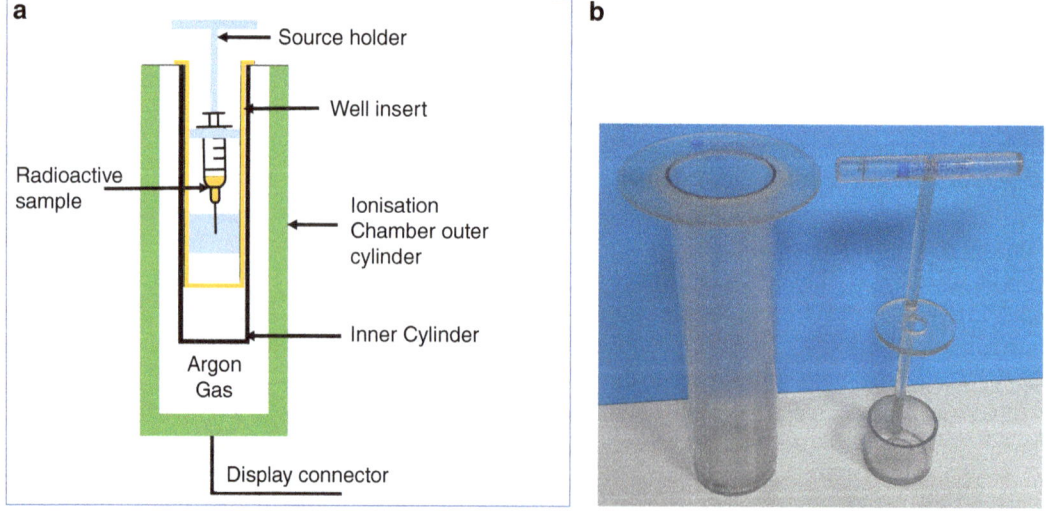

Cross-sectional diagram of well-chamber   Plastic Liner or Well insert & Source Holder

Fig. 14.5 (**a**) Cross-sectional diagram of Ionization chamber. (**b**) Dose calibrator accessories

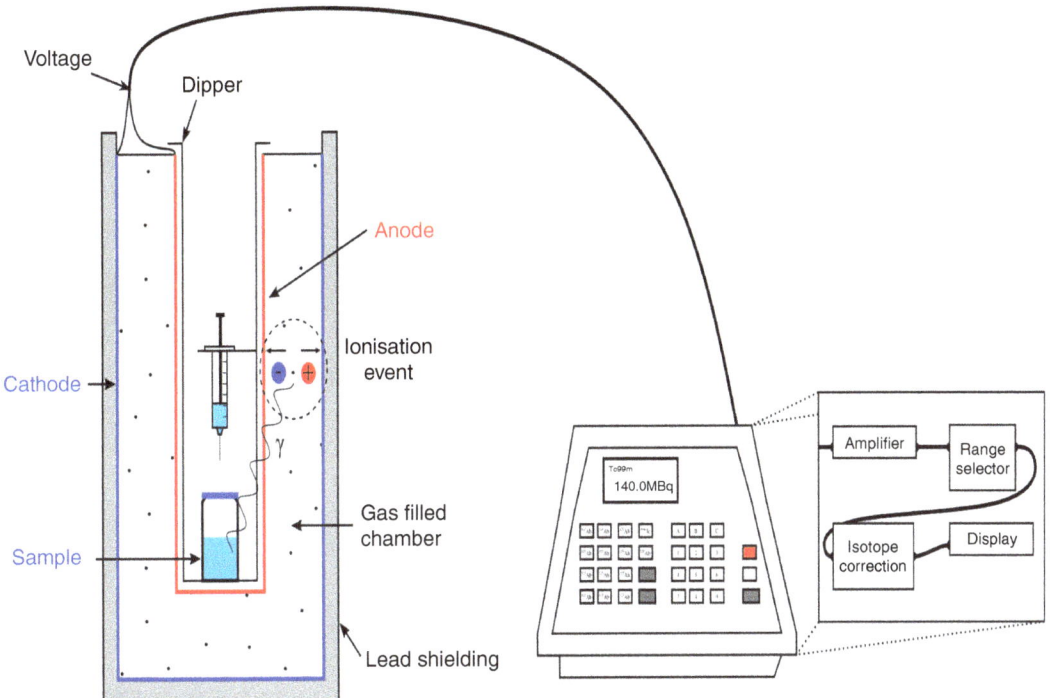

Fig. 14.6 Working diagram of dose calibrators. Image copyright, Frank Brewster, printed under creative commons license attribution-share alike 4.0 international

14.2.3 Working Mechanism of Dose Calibrators

The dose calibrator works on the following mechanism (Fig. 14.6):

- Photons interact with gas molecules and leave behind ion pairs.
- Positive ions drift to the cathode, whereas negative ions move to the anode.
- A high voltage is supplied within the dose calibrator chamber, and a battery acts as a capacitor.
- The battery keeps the voltage on the cathode and anode constant and functions as a backup.
- Two equal radioactive materials generate different amounts of current due to differences in their energy and emission, e.g. particulate or photon.

14.2.4 Choice of Gas in Dose Calibrators

Gases with an affinity for electrons, such as oxygen, water vapour and halogens, should not be utilized in an ionization chamber of dose calibrators. Other gases, such as helium, neon, argon, hydrogen, nitrogen, carbon dioxide and methane, seldom combine with electrons to generate negative ions and can be employed as filler gases [2]. Table 14.2 summarizes the properties of various potential gases.

Noble gases may be the preferred choice compared to other gases because the full valence electron shells make them highly stable and unlikely to form chemical bonds in most physical conditions. Table 14.3 shows the comparison of the properties of noble gases. Radon is radioactive and has emissions, xenon is relatively rare and expensive, and

Table 14.2 Properties of potential gases for gas-filled detectors

Element/ compound	Atomic number (Z)/ effective atomic number (Z_{eff})	Density (kg/m³)	Average energy required to produces ion pair (eV/ion pair) [3]	First ionization potential (eV) [4]	Remarks
Argon (Ar)	18	1.7572	26.4	15.7	Noble gas is the third-most abundant gas on earth and cheap
Helium (He)	2	0.1761	41.3	24.5	Noble gas, very low density and Z
Hydrogen (H_2)	1	0.0887	36.5	15.6	Very low Z and density
Nitrogen (N_2)	7	1.2323	34.8	34.8	Low Z and relatively higher energy for ion pair production
Air	7.6	1.2740	33.8	–	Low Z and relatively higher energy for ion pair production
Oxygen (O_2)	8	1.4076	30.8	12.5	Low Z and affinity for electrons
Methane (CH_4)	5.05	0.7057	27.3	14.5	Flammable

Table 14.3 Properties of noble gases

Property	Helium	Neon	Argon	Krypton	Xenon	Radon
Atomic number	2	10	18	36	54	86
Electronic configuration	$1s^2$	[He] $2s^2\,2p^6$	[Ne] $3s^2\,3p^6$	[Ar] $3d^{10}\,4s^2\,4p^6$	[Kr] $4d^{10}\,5s^2\,5p^6$	[Xe] $4f^{14}\,5d^{10}\,6s^2\,6p^6$
Density (g/L) at 0°, 1 atm pressure	0.17847	0.899	1.784	3.75	5.881	9.73
First ionization potential (eV) (approx)	24.8	21.4	15.7	14	12	11

krypton is not very abundant in our planet's atmosphere. At a given temperature, heavier atoms move slowly and cannot be used in gas-filled detectors. Helium and neon have very low Z and density and hence possess low stopping power.

Argon is the cheapest among other noble gases (Wikipedia, 2004 data). It is relatively denser with higher Z and just need 15.7 eV of energy for the first ionization potential (better than helium and neon). All these qualities make argon a choice of gas for the dose calibrators.

14.2.5 Current Conversion

1 μCi (3.7×10^4 Bq) of ^{60}Co produces a current in the order of 10^{-13} amperes in a typical 10^4 cm³

active volume chamber. The current produced per unit activity (MBq) for common radionuclides ranges from tens of femtoamperes (fA) to microampere. For high-energy beta emitters, it is up to tens of picoamperes (pA), whereas the high-energy, high-yield photon emitters and high-activity assays can involve microampere (μA) currents (AAPM report 181 [1]).

The response of the chamber differs for 1 Bq of ^{99m}Tc (140 keV) and 1 Bq of ^{131}I (364 keV). The unique current depends on a specific gamma constant (discussed in Chap. 5). Higher activities generate more photons and generate more current. For the dose calibrator to display one millicurie (mCi) for both isotopes, a conversion factor called 'calibration factor' or 'calibration coefficients' is applied.

14.2.5.1 Calibration Factors

Calibration factors convert the measured ionization chamber current to a nominal activity. The reciprocal calibration factor represents the efficiency of the ionization chamber for the radionuclide [5]. Over 240 calibration factors for more than 80 radionuclides are provided in Capintec dose calibrator manuals and similarly by other manufacturers in their respective manuals. They are to be entered the same way as guided in manuals (Figs. 14.7 and 14.8).

Dose calibrators used in dedicated positron emission tomography (PET) facilities are designed with reduced gas pressure to detect high radioactivities. Table 14.4 gives details of various dose calibrator models with chamber pressure and measurement range.

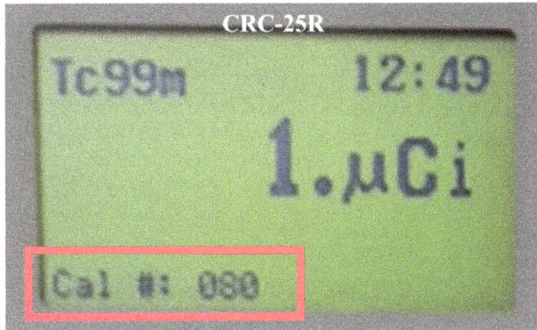 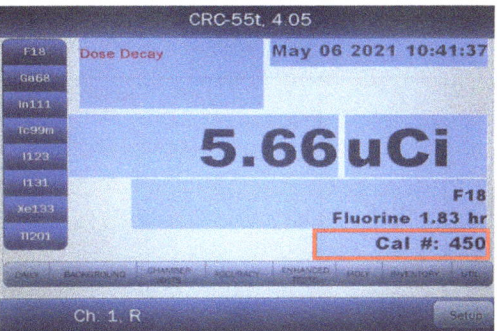

Fig. 14.7 Calibration factor seen in the display of Capintec dose calibrators

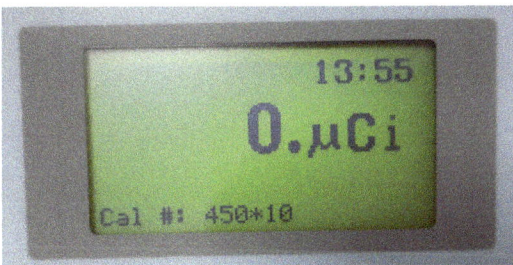

Lutetium - 177

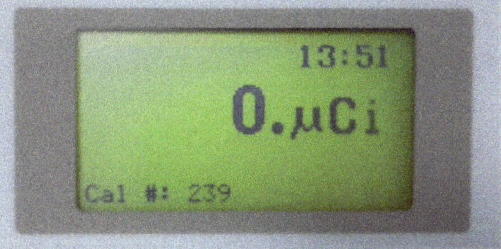

Samarium - 153

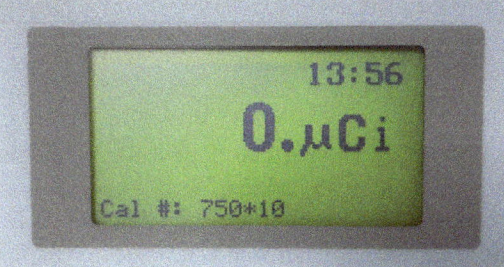

Phosphorus – 32
For Estimation use only

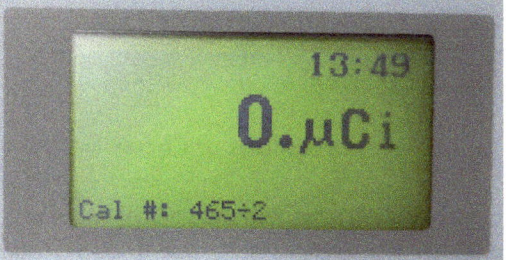

Yttrium - 88

Fig. 14.8 Various calibration factors used in Capintec dose calibrator CRC-25R

Table 14.4 Chamber pressure and measurement range

Make	Model	Chamber pressure (as per manual)	Measurement range[a]
Capintec	CRC-15R/25R	20 bar ultra-pure Argon	Up to 6 Ci
	CRC-25PET	20 bar ultra-pure Argon	Up to 20 Ci
	CRC-55tR	12 bar ultra-pure Argon	Up to 6 Ci
	CRC-55tPET	5 bar ultra-pure Argon	Up to 20 Ci
Biodex	Atomlab 400	2.48 bar Argon	Up to 40 Ci
	Atomlab 500	2.36 bar Argon	Up to 100 Ci
Comecer	VIK-202	14 bar abs. Argon	Up to 6 Ci
	VIK-203	1.4 bar abs. Argon	Up to 20 Ci

[a]Range is different for different radioisotopes. Please see the manual for details

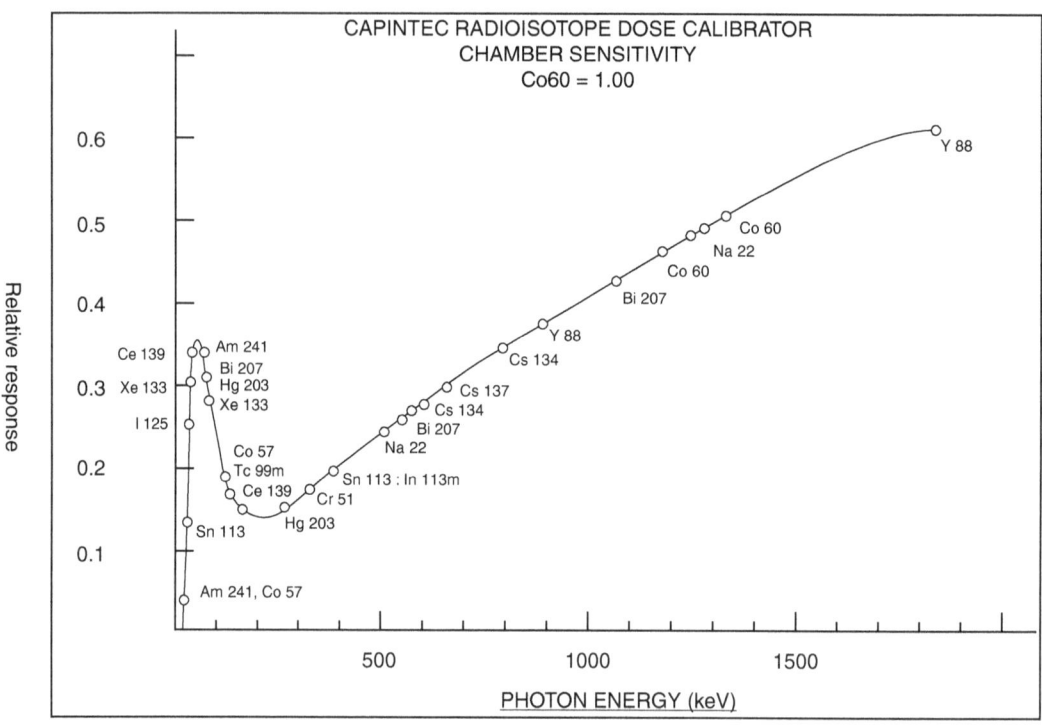

Fig. 14.9 Dose calibrator energy-response curve. Image copyright, Mirion Technologies (Capintec), Inc. Printed with permission

14.2.6 Energy-Response Curve

Figures 14.9 and 14.10 show the energy-response curve of dose calibrators from Capintec and Biodex, respectively. The graphs are plotted with limited calibration coefficients and measured using standard reference sources. Chamber response is calculated, normalized to a reference radionuclide and plotted for the energy-response curve [6]. Using this curve, calibration coefficients for other radionuclides can be calculated. For aluminium-walled chambers, photons below 13 keV energies are stopped. The actual cutoff depends upon the source volume, container wall material and thickness, thickness of the source holder, plastic liner and chamber wall [1].

The ionization current increases rapidly and then abruptly decreases in the energy-response curve, yielding a peak at ~50 keV. The peak results from the competing effects of 'photoelectric (PE) absorption versus Compton scattering'. Above 200 keV, Compton scatter dominates, and sensitivity increases linearly. Let us check the photoelectric absorption and Compton scattering with the energy change.

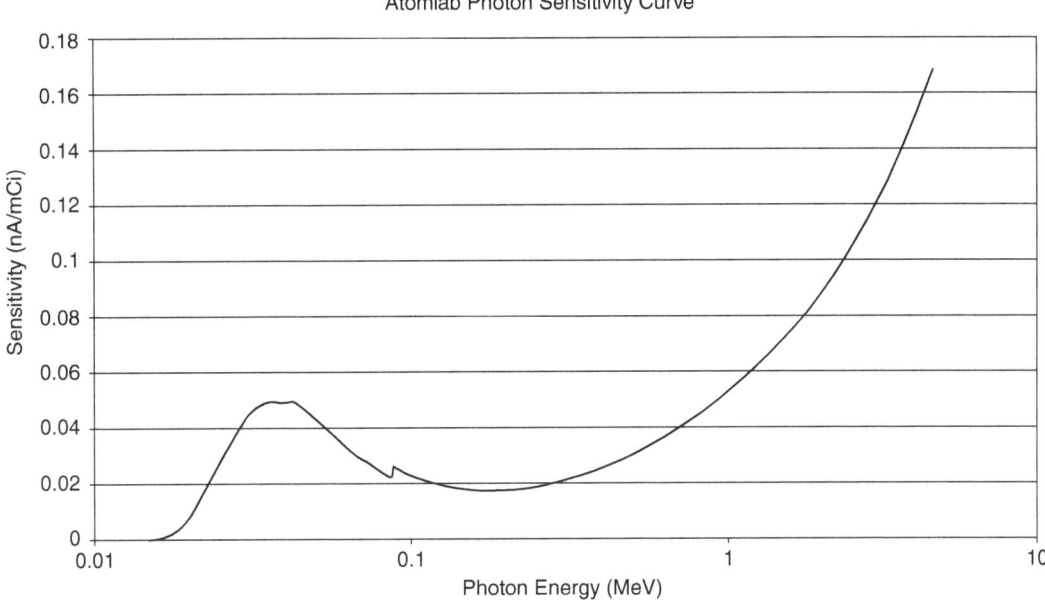

Fig. 14.10 Biodex Atomlab™ 500 dose calibrator's energy-response curve. Image copyright, Mirion Technologies (Capintec), Inc. Printed with permission

Table 14.5 Energy of photons vs probability of photoelectric effect

Energy of photon	Photoelectric effect ($Z^{3\ or\ 4}/E^3$)	
25 keV	$(18)^3/(25)^3$	0.37
50 keV	$(18)^4/(50)^3$	0.83
75 keV	$(18)^4/(75)^3$	0.25
100 keV	$(18)^4/(100)^3$	0.10

The values of probability observed in the table are in congruence with the observed graph

14.2.6.1 Theoretical Dependence on Energy of Photons for Competitive Photoelectric Effect and Compton Scattering

- The probability of the photoelectric effect is $\sim Z^n/E^3$. The exponent n varies between 3 and 4 depending upon the gamma-ray energy.
- The Compton process is a prominent interaction ranging from 100 keV to 10 MeV and is almost independent of atomic number Z.

14.2.6.2 Photoelectric Effect Probability Dependence on Energy of a Photon in Argon Gas

- Probability of PE effect $\sim Z^{3\ or\ 4}/E^3$.
- Z of Argon = 18 (Table 14.5).

14.2.6.3 Compton Scattering Probability

The energy of the scattered photon hv' and the Compton electron E_e are calculated using the formula:

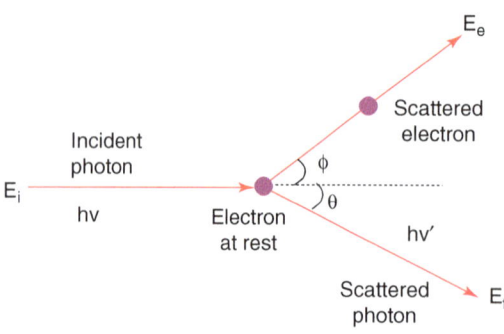

$$hv' = hv \, \frac{1}{1+\alpha\left(1-\cos\theta\right)}$$

$$E_e = hv \, \frac{\alpha\left(1-\cos\theta\right)}{1+\alpha\left(1-\cos\theta\right)}$$

$$\text{Where}\,\alpha = \frac{hv}{m_0 c^2}$$

Fig. 14.11 Compton scattering

Here, hv is the incident photon's energy, while $m_0 c^2$ is the electron rest mass-energy, i.e. 0.511 MeV (Fig. 14.11). For maximum energy transfer to the recoil electron, the scattered photon will be scattered straight back, i.e. at an angle of 180° or $\theta = 180°$, whereas the angle of the recoil electron will be at 0°, i.e. $\varphi = 0°$, and moves straight forward. With $\theta = 180°$, $\cos\theta = -1$, and the above formulae reduce to:

$$E_{e(\max)} = hv \, \frac{2\alpha}{1+2\alpha} \qquad hv'_{\min} = hv \, \frac{1}{1+2\alpha} \qquad \text{where}\,\alpha = \frac{hv}{m_0 c^2}$$

Table 14.6 demonstrates how the amount of energy transmitted to electrons varies with photon energy and shows a strong linear relationship between energy growth and the probability of Compton scattering.

14.2.7 Major Sources of Error in Measurements [4]

1. Calibration factor.
2. Electronics.
3. Statistical variations.
4. Ion recombination.
5. Effects of background.
6. Size and shape of source container and effects of volume.
7. Effects on the source position.
8. Source adsorption on the container surface.

14.2.7.1　Calibration Factor
For most radioisotopes, a decent approximation for radiopharmaceutical in a plastic syringe or glass syringe is around 5 g of radioactive solution in a standard source ampoule constructed of about 0.6-mm-thick borosilicate glass as standard radioactive sources [6]. Since this standard source measurement has uncertainty in the range of 1–3% (for ^{99m}Tc and ^{131}I), all the calibration factors derived have the same uncertainty. The calibration factor for various containers and/or capacities may differ dramatically from the established calibration factors.

14.2.7.2　Electronics
Electrometer with a dynamic range of 10^8 measures current output from tens of femtoamperes to microamperes. Discontinuities may occur depending on the linearity properties of each range. Figure 14.12 shows variation in chamber response due to electronics.

14.2.7.3　Statistical Variations
The radioactive decay process is probabilistic, and it is always associated with statistical uncertainty and so are the radioactive measurements.

Table 14.6 Relationship of photon energy with energy transfer to electron

Incident photon energy			
50 keV	140 keV	511 keV	1 MeV
$\alpha = \dfrac{50 \text{ keV}}{511 \text{ keV}} = 0.1$	$\alpha = \dfrac{140 \text{ keV}}{511 \text{ keV}} = 0.27$	$\alpha = \dfrac{511 \text{ keV}}{511 \text{ keV}} = 1$	$\alpha = \dfrac{1000 \text{ keV}}{511 \text{ keV}} = 1.95$
$E_{e(\text{max})} = 50 \text{ keV} \times \dfrac{2 \times 0.1}{1+(2 \times 0.1)}$ $\approx 8 \text{keV}$	$E_{e(\text{max})} = 140 \text{ keV} \times \dfrac{2 \times 0.27}{1+(2 \times 0.27)}$ $\approx 50 \text{ keV}$	$E_{e(\text{max})} = 511 \text{ keV} \times \dfrac{2 \times 1}{1+(2 \times 1)}$ $\approx 341 \text{ keV}$	$E_{e(\text{max})} = 1000 \text{ keV} \times \dfrac{2 \times 1.95}{1+(2 \times 1.95)}$ $\approx 796 \text{ keV}$
$h\nu'_{\min} = 50 \text{ keV} \times \dfrac{1}{1+(2 \times 0.1)}$ $\approx 42 \text{ keV}$	$h\nu'_{\min} = 140 \text{ keV} \times \dfrac{1}{1+(2 \times 0.27)}$ $\approx 90 \text{keV}$	$h\nu'_{\min} = 511 \text{ keV} \times \dfrac{1}{1+(2 \times 1)}$ $\approx 170 \text{ keV}$	$h\nu'_{\min} = 1000 \text{ keV} \times \dfrac{1}{1+(2 \times 1.95)}$ $\approx 204 \text{ keV}$
Energy transferred to the Compton electron = 16%	*Energy transferred to the Compton electron = 35.7%*	*Energy transferred to the Compton electron = 38.5%*	*Energy transferred to the Compton electron = 79.6%*

Fig. 14.12 Variation in chamber response due to electronics. *Image copyright: National Physical Laboratory, reprinted with permission*

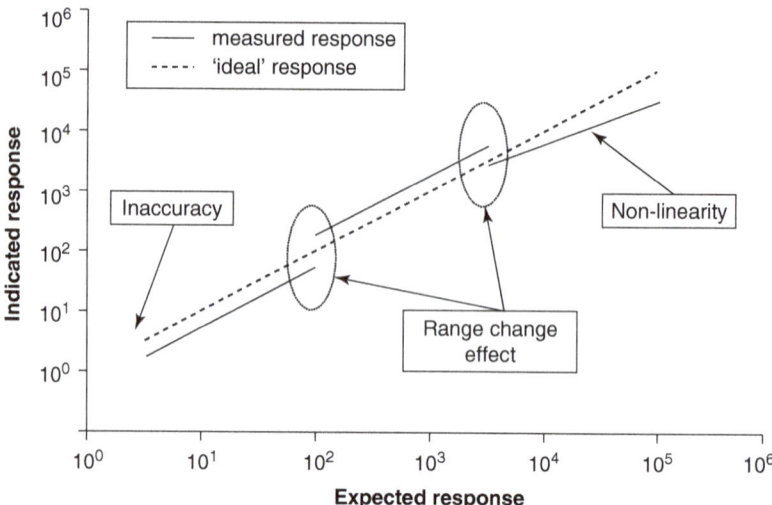

14.2.7.4 Ion Recombination

The likelihood of recombination increases as source activity increases. This can become severe at high source activities, resulting in a drop in the measured current.

14.2.7.5 Effects of Background

The dose calibrator still records 'activity due to background radiation' even when the source holder is empty. An increase in background radiation may interfere with measurements. If needed, corrections should be applied. Background is determined at least each morning and recorded.

14.2.7.6 Size and Shape of Source Container and Effects of Volume

The source container's composition and thickness have an impact on the measured activity (Figs. 14.13 and 14.14). It is noticeable with low-energy photon emitters and pure beta emitters.

14.2.7.7 Effects on the Source Position

The source holder holds the source in a fixed position. Changes in position caused by a few millimetres of movement are usually minor. But this can be significant if the holder moves upward or downward.

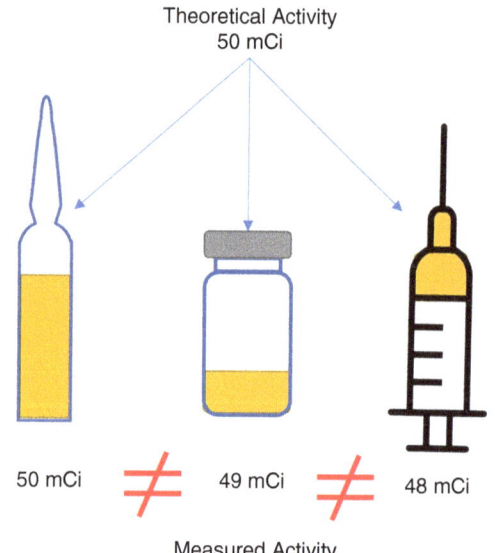

Fig. 14.13 Variations due to container

14.2.7.8 Source Adsorption on the Container Surface

It has been noticed that some radiopharmaceuticals adsorb to the container's surface. Adsorbed activity can account for a large portion of total activity. When a facility uses syringes from a different manufacturer, the likelihood of activity adsorption should be considered.

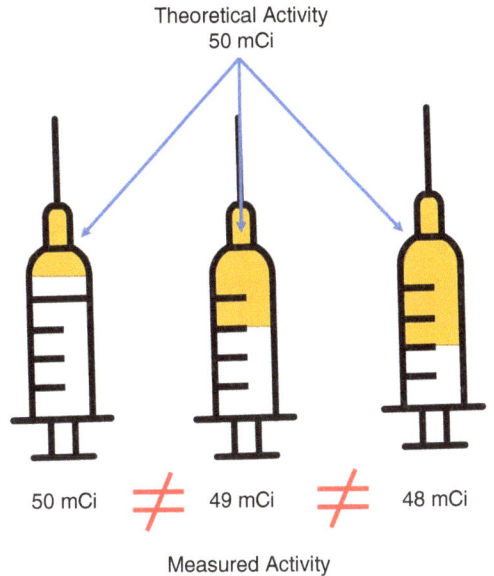

Theoretical Activity
50 mCi

50 mCi ≠ 49 mCi ≠ 48 mCi

Measured Activity

Fig. 14.14 Variations due to volume

14.2.8 Measuring Pure Beta Emitters

Ionization chambers have low detection effectiveness for beta radiation. The responses are almost entirely from bremsstrahlung radiation. To establish an exact estimate, a significant amount of activity is required. Generally, beta emitters employed in therapies require significant activities; hence, the activity estimate is possible, but efficiency can be variable. Clinically useful radioisotopes such as ^{90}Y, ^{89}Sr and ^{32}P can be measured accurately [7]. Data from five different manufacturers showed that all systems had [7]:

- A good calibration for ^{32}P.
- A reduction in efficiency of approximately 10–20% for ^{89}Sr.
- A wide divergence in efficiency for ^{90}Y: The results vary between 64 and 144% of the true value.

^{153}Sm and ^{186}Re measurements: ^{153}Sm (103 keV, 28% abundance) and ^{186}Re (137 keV, 9.5% abundance) are beta gamma-emitting radionuclides. In these cases, the gamma contribution is the primary determinant of ionization chamber efficiency, and the accuracy is usually within ±10%.

14.2.9 Effects of Contaminants

Dose calibrators do not differentiate energies like spectrometers. Even <1% of contaminants can affect the measured activity [7]. The presence of high-energy contaminants has an adverse effect on image quality.

14.2.10 Dose Calibrators Acceptance and Routine Testing

Many factors influence the dose calibrator's accuracy and performance, and quality control parameters must be reviewed regularly. The results are to be checked for consistency. The performance tests must be performed at the time of acceptance and at a regular interval, as specified in the following paragraphs:

14.2.10.1 Accuracy and Constancy
- Accuracy is verified by comparing activity measurements to a known calibrated reference source.
- Constancy, also called reproducibility, is determined by performing many measurements of the same source. Long-lived sealed ^{57}Co and ^{137}Cs sources can be utilized daily to confirm this. A reading that differs from what is expected could indicate a malfunctioning calibrator.
- *Frequency: Accuracy - Annually; Constancy - Daily.*

14.2.10.2 Linearity
It is a measurement of linear response from high to low range of activity for the same source. There are various methods for assessment of the same and are as follows:

- Source-decaying method.
- Multiple dilution method.
- Graded attenuator method.
- *Frequency: Linearity - Annually.*

14.2.10.3   Geometry Response

The activity measurement may vary in the following conditions:

- The source's location within the well.
- The vial or syringe's composition.
- The liquid volume in the vial or syringe.
- Correction factors can be calculated for various quantities or containers.
- *Frequency: At calibrator acceptance and then for any change in sample geometry.*

14.2.11   Materials Needed for Quality Control of Dose Calibrators

1. Sealed low-, medium- and high-energy gamma radiation sources calibrated to less than 5% overall uncertainty. The examples are ^{57}Co ($\approx$1 mCi or 37 MBq), ^{137}Cs ($\approx$100 µCi; 3.7 MBq) and ^{60}Co ($\approx$50 µCi; 1.85 MBq).
2. Unsealed radionuclides such as ^{99m}Tc, ^{18}F and ^{131}I.
3. Vials, syringes, source holder and tongs to handle radioactivity.

14.2.12   Test of Accuracy and Constancy

The test reveals random errors or accuracy loss owing to background variations, scatter (due to increased shielding), changes in chamber gas pressure and gradual drift, among other things.

14.2.12.1   Procedure

1. Take a note of the background activity.
2. Record the background-subtracted reading after measuring the activity (A_i).
3. Calculate the mean, standard deviation (SD) and percentage of variation (PV) using the formula below.

$$\%PV = \frac{SD}{Mean} \times 100$$

% PV value should be less than 5%.

4. A quality control chart that plots the percent PV readings shows the trend of instrument performance.

5. To determine accuracy, calculate the percentage difference between the mean measured activity 'A' and the activity of the source after decay correction on the measurement day, A_i, for each source, i.e.

$$Accuracy = \frac{(A - A_i) \times 100}{A}$$

6. Repeat accuracy tests for various energies, such as ^{57}Co, ^{137}Cs and ^{60}Co.
7. Accuracy for each source should be within 5%.

14.2.13   Measurement of Linearity of Dose Calibration

1. Elute ^{99m}Tc from the ^{99}Mo-^{99m}Tc generator.
2. Fill a sample vial with 100 mCi or more with ^{99m}Tc activity (the initial activity should be equal to or more than the highest activity for which the instrument is utilized).
3. Measure and record the activity every 1 hour up to 36 hour. Subtract the background activity for the net measurement.
4. Draw the line of decay pattern on a semi-log graph paper, and fit the line with the least square method.
5. If the instrument is linear, these points should be in a straight line.
6. The variation in values corresponding to the straight line should be within 10%.

Note: It's critical to ensure that the ^{99m}Tc solution is devoid of impurities and that there is no ^{99}Mo breakthrough.

14.3   Radiopharmaceutical Dispensing to Patients

The procedural guidelines published by the Society of Nuclear Medicine and Molecular Imaging (SNMMI) and the European Association of Nuclear Medicine (EANM) provide reference dosages for various radiopharmaceuticals. The EANM and North American Consensus Guidelines have prepared dosage cards for high-quality images at low radiation doses [8, 9]. In its book 'Nuclear Medicine Physics: A Handbook for Teachers and Students' [4], the IAEA has used a scaling factor for dosage calculation of heavyweight patients for a few radiopharmaceuticals. Table 14.7 shows the dosage of various radiopharmaceuticals for an adult of 70 ± 20 kg weight (SSG-46, 2018, para 4.205).

14.3.1 Dosage Calculation for Heavyweight Patients

For commonly used radiopharmaceuticals, a simple scaling factor as a power function of body weight has been calculated to deliver a constant effective dosage as described in the International Commission on Radiological Protection (ICRP)

Table 14.7 Radiopharmaceutical doses for adults (weight = 70 ± 20 kg)

Radiopharmaceuticals	Recommended dosage for adults in MBq (mCi)	References
^{18}F-FDG whole-body PET-CT	350–630[a] MBq (9.5–17 mCi)[b] Formula = 14 (MBq min/bed/kg) × patient weight (kg)/ emission acquisition duration per bed position (min/bed)	[10]
^{18}F-FDG brain PET-CT	125–250 MBq (typically 150 MBq) 3.4–6.8 (typically 4 mCi)	[11]
[18]F-NaF bone scan	185–370 MBq (5–10 mCi)	[12]
[18]F-Fluciclovine	370 MBq (10 mCi)	[13]
^{18}F-PSMA	4.0 ± 0.4 MBq/kg	[14]
^{18}F-DCFPyL	310.8–327.1 MBq (8.40–8.84 mCi)	[15]
^{18}F-DOPA	3–4 MBq/kg	[16]
^{68}Ga-PSMA	1.8–2.2 MBq (0.049–0.060 mCi) per kg bodyweight	[17]
^{68}Ga-DOTA	100–200 MBq (2.7–5.4 mCi)	[18]
^{68}Ga-DOTA-exendin-4	74–111 MBq (2–3 mCi)	[19]
^{99m}Tc-MDP	500–1110 MBq (~13–30 mCi)	[20]
^{99m}Tc-DTPA, ^{99m}Tc-MAG3, ^{99m}Tc-EC	90–200 MBq (2.4–5.4 mCi)	[21]
^{99m}Tc-DMSA	111 MBq (3 mCi)	[21]
[99m]Tc-Mebrofenin	111–185 MBq (3–5 mCi)	[22]
[99m]Tc-Aerosol for lung ventilation	In nebulizer, ^{99m}Tc-DTPA or sulphur colloid with 900–1300 MBq (25–35 mCi) activity. The patient receives 20–40 MBq (0.5–1.0 mCi) to the lungs	[23]
	[133]Xe: 200–750 MBq (5–20 mCi)	
	[81m]Kr: Continuous inhalation of ≈40–400 MBq (1–10 mCi)	
[99m]Tc MAA for perfusion	40–150 MBq (1–4 mCi) aiming 200,000–700,000 particles. Slow injection over three to five respiratory cycles in supine position.	[23]
[99m]Tc-Pertechnetate (thyroid imaging)	74–370 MBq (2.0–10.0 mCi)	[24]
[131]I—Thyroid scan (with whole body) in the case of Ca thyroid	37–185 MBq (1.0–5.0 mCi)	[24]
[123]I—Sodium iodide for thyroid scintigraphy	7.5–25 MBq (0.2–0.6 mCi)	[25]
[123]I—Thyroid scan (with whole body) in the case of Ca thyroid	37–185 MBq (2.0–5.0 mCi)	[24]
[99m]Tc-Pertechnetate (Meckel's diverticulum)	296–444 MBq (8–12 mCi)	[26]
^{99m}Tc-HMPAO or 99mTc-ECD (brain)	555–1110 MBq (15–30 mCi)	[27]
^{99m}Tc-HMPAO (infection)	185–370 MBq (5–10 mCi)	[28]
[99m]Tc-Sulphor colloid (liver/spleen)	• 111–222 MBq (3–6 mCi) for planar • Up to 370 MBq (10 mCi) for single-photon emission computed tomography (SPECT)	[29]
[99m]Tc-Labeled red blood cells (RBCs)	555–1110 MBq (15–30 mCi)	[30]
[99m]Tc-Sulphor colloid (gastric reflux)	18.5–74 MBq (0.5–2 mCi)	[31]

(continued)

Table 14.7 (continued)

Radiopharmaceuticals	Recommended dosage for adults in MBq (mCi)	References
^{99m}Tc-SestaMIBI for parathyroid	740–1110 MBq (20–30 mCi)	[32]
Myocardial perfusion imaging: ^{99m}Tc-SestaMIBI/Tetrofosmin (1-day protocol rest-stress or stress-rest)	Up to total 1480 MBq (40 mCi)	[33]
Myocardial perfusion imaging: ^{99m}Tc-SestaMIBI/Tetrofosmin (2-day protocol for stress and rest)	1110 MBq (30 mCi) for each injection	[34]
Myocardial perfusion imaging: ^{201}Tl-chloride	74–148 MBq (2–4 mCi)	[34]
Myocardial perfusion imaging: ^{82}Rb	1100–1850 MBq (30–50 mCi)	[34]
Myocardial perfusion imaging: ^{13}N-ammonia	370–740 MBq (10–20 mCi)	[34]
^{67}Ga-Citrate infection imaging	150–220 MBq (4–6 mCi)	[35]
^{123}I-mIBG	400 MBq (10.8 mCi)	[36]
^{131}I-mIBG	40–80 MBq (1.2–2.2 mCi)	[36]

[a]For patients weighing more than 90 kg and with L(Y)SO crystals, FDG activities higher than 530 MBq are not recommended; instead, an increase in bed position time is recommended to improve the image quality
[b]Dosage calculation based on 2 min/bed position. It is to be reduced to half for bed overlap of >30%

Table 14.8 The power factor 'a' for common radiopharmaceuticals. Adapted from the IAEA book 'Nuclear Medicine Physics A Handbook for Teachers and Students', reprinted with permission [4]

Radiopharmaceutical	a value	Radiopharmaceutical	a value
^{18}F-FDG	0.782	^{123}I or ^{131}I iodide	1.11
^{99m}Tc-DTPA	0.801	^{99m}Tc-DMSA	0.706
^{99m}Tc-IDA	0.840	^{99m}Tc-Tetrofosmin	0.834
^{99m}Tc-MAG3	0.520	^{99m}Tc-Red cells	0.859
^{99m}Tc-HMPAO	0.849	^{99m}Tc-White cells	0.869
^{99m}Tc-MAA	0.842	^{99m}Tc-Sestamibi	0.871
^{99m}Tc-Phosphonate	0.763	^{67}Ga-Citrate	0.931

publications 53, 80 and 106. An equation $(W/70)^a$ is used, where W denotes the person's weight and a denotes the radiopharmaceutical's power factor. A power factor 'a' value for common radiopharmaceuticals is shown in Table 14.8. This method is useful but should be used with caution [4].

Example 14.1 Calculate ^{99m}Tc-DTPA dosage for 110 kg patient using power factor a.

Solution From the formula $(W/70)^a$,
  Scaling factor $= (110/70)^{0.801} = (1.571)^{0.801} = 1.44$.
  That is, 1.44 times or 144% of the average dose of ^{99m}Tc-DTPA.

14.3.2 Paediatric Dosage Calculations

For paediatric dosage calculation, traditionally, various formulas (summarized below) have been used in the past (Table 14.9). However, the EANM dosage card and North America Consensus guidelines provide guidelines for radiopharmaceutical dosage of various nuclear medicine procedures starting from 3 kg to 68 kg in paediatric patients, and the same can be applicable for low weight adults. EANM has developed a mobile application named EANM PedDose and an online paediatric dose calculator on their website. SNMMI website also calculates

Table 14.9 Traditional mathematical formulas used for paediatric dose calculation

$$\text{Webster formula} = \frac{\text{Age} + 1}{\text{Age} + 7} \times \text{Adult dose} \quad (\text{Age in years})$$

$$\text{Young's formula} = \frac{\text{Age}}{\text{Age} + 12} \times \text{Adult dose} \quad (\text{Age in years})$$

$$\text{Clark's formula} = \frac{(\text{Child's weight in pounds})(\text{Adult dose})}{150 \, \text{pounds}}$$

various radiopharmaceuticals dosages, including showcasing comparative differences between SNMMI and EANM recommendations.

EANM tracer-dependent dosage cards (published in 2016, currently under review) have two sets of tables to calculate the dosing of radiopharmaceuticals. In Fig. 14.15, the left-hand side table has weights ranging from 3 kg to 68 kg. It contains three classes of radiopharmaceuticals: *class A* contains tracers for renal studies, and *class B* contains all other tracers except those used for thyroid studies and ^{89}Sr therapy, which belong to *class C*. The right-hand side table contains the radiopharmaceuticals, their corresponding 'class', 'baseline activity' for calculating total injected activity and the 'minimum recom-

mended activity'. The 'minimum recommended activity' for each radiopharmaceutical is recommended for the lowest body weight of 3 kg and should guarantee good image quality at the lowest acceptable radiation burden [4].

Example 14.2 Calculate injected activity for a patient of age 14 kg undergoing ^{18}F-FDG whole-body PET-CT.

Solution *Identify the class and baseline activity from the right-hand side table.*

For ^{18}F-FDG, class is B, and baseline activity is 25.9 MBq.

Identify weight factor from the left-hand side table:

$$\text{The weight factor for } 14 \, \text{kg weight} = 3.57$$
$$\therefore \text{Amount of } ^{18}\text{F} - \text{FDG activity to be injected} = 3.57 \times 25.9$$
$$= 92.5 \, \text{MBq}$$

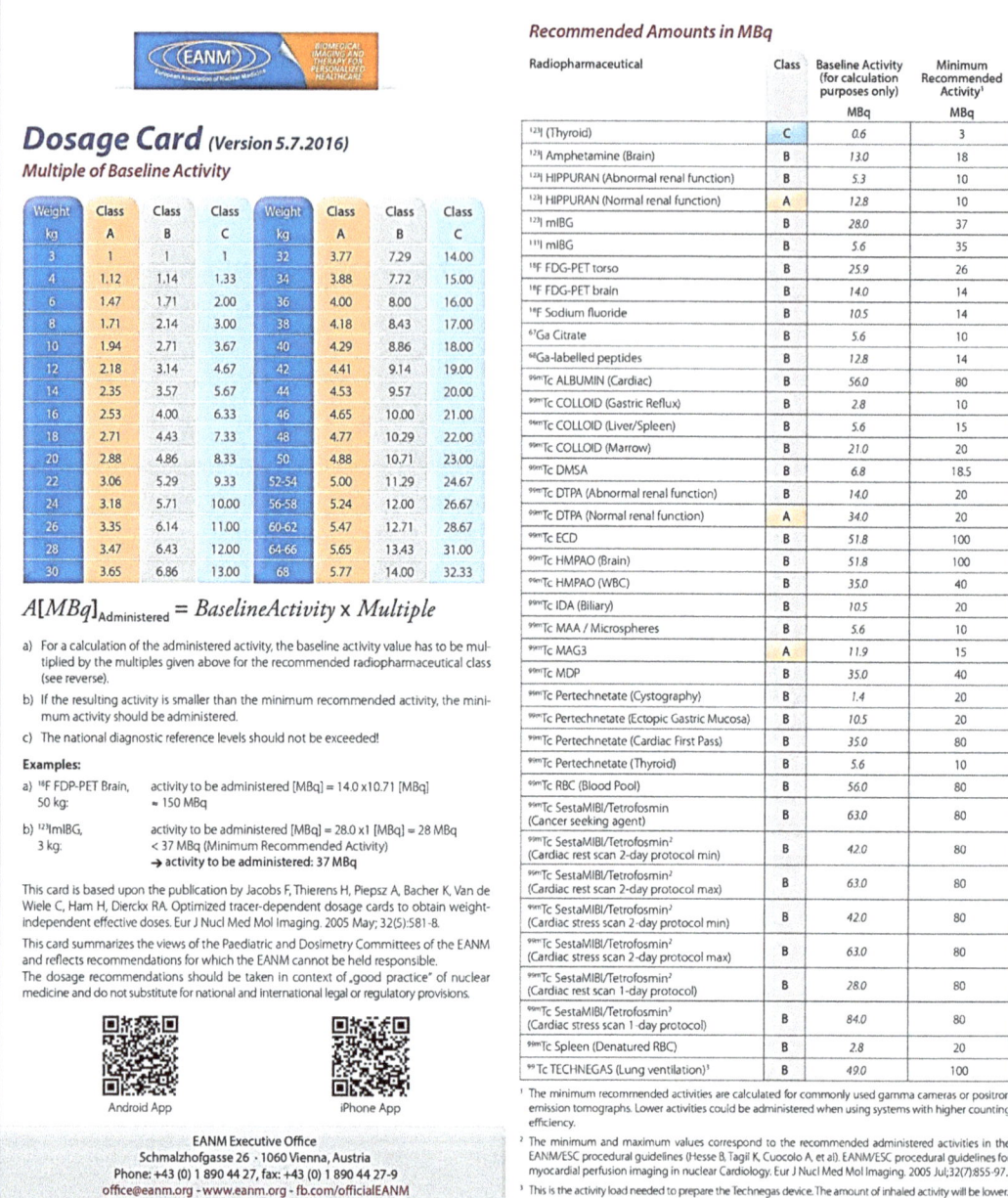

Fig. 14.15 EANM paediatric radiopharmaceutical dosage card (2016). Copyright EANM reprinted with permission. Further unauthorized use is prohibited

14.4   Medical Events (Formerly Misadministration)

The terminology misadministration was changed to Medical Events in 2002 [37]. As per the new definition (the *United States Nuclear Regulatory Commission, Code of Federal Regulations-10 CFR-35.3045*) [38], medical events are:

A dose, that differs from the prescribed dose

1. For diagnostic: 50% or more.
2. For therapeutic: More than 20%.

Or dose that would have resulted from the prescribed dosage by more than 50 mSv (5 rem) effective dose equivalent, 500 mSv (50 rem) to an organ or tissue or 500 mSv (50 rem) shallow dose equivalent to the skin from any of the following:

1. Administration of *wrong radiopharmaceuticals.*
2. Administration through the *wrong route.*
3. Administration of radiopharmaceuticals to the *wrong patient.*
4. Administration of the *wrong dosage.*

The code further says the radiopharmaceutical dose administration to a patient not intended to receive a dosage, e.g. patient's attendant or a foetus in a pregnant woman, who is administered without confirming the pregnancy or such other cases, would also be a medical event.

14.4.1 Medical Event Reporting

When a medical event is detected, requirements for reporting the occurrence and patient management must be followed. All medical incidents must be reported to the radiation safety officer, the regulatory authority, the referring physician, and the patient or person affected. The complete records must be available for 10 years of duration. The reporting procedure is as follows (Extract of USNRC Regulations 10 CFR Part 35.3045):

1. The licensee shall notify the regulatory authority no later than the *next calendar day* after discovering the medical event. A detailed written report containing the following information is to be submitted to the regulatory authority within 15 days:
   (a) The name of the licensee.
   (b) The name of the prescribing physician.
   (c) A brief description of the event.
   (d) The reason why it occurred.
   (e) The effect, if any, on the patients/individual who has received the administration.
   (f) The actions, if any, planned to prevent such recurrence.

   (g) Certification that the licensee notified the individual (or the individual's responsible relative or guardian). If not, then why.
2. The report may *not* contain the individual's name or any such information that could reveal the individual's identity.
3. Within 24 hours, the licensee shall notify the event to the referring physician and the individual subjected to the medical event, unless the referring physician personally informs that they will inform the individual based on the medical judgment stating the subject would be harmful. The licensee shall not inform the individual without consulting the referring physician. If the referring physician cannot be reached within 24 hours, the licensee shall notify the individual as soon as possible after that. But because of any delay in notification, the licensee shall not delay any appropriate medical care to the individual, including any necessary remedial care needed because of the medical event.

14.5 Control of Radiation Hazards in Radiopharmacy

14.5.1 Safety Aspects

All radiopharmaceutical applications such as planning, procurement, preparation, use of radiopharmaceuticals and radioactive waste disposal require proper care at every stage.

14.5.2 Essential Points in Planning Work

- A minimum quantity of radioisotope should be used, which suffices for the need.
- Maximum distance with effective working should be maintained between source and worker.
- Plan the work so that minimum time is spent near the source.
- Appropriate shielding should be used wherever it can be.

- Fume hoods should be used whenever needed, especially while handling volatile radioisotopes.
- In case of spillage, care should be taken to confine it with absorbent sheets. Any uncontrolled spills should be dealt with appropriately.

14.5.3 Handling of Sources

- Good housekeeping procedures should be maintained in radiopharmacy.
- Items meant for use with unsealed sources should be used only for this purpose.
- The absorbent sheets are used on work surfaces involving liquid radioactive material to contain or absorb any spills.
- In radiopharmacy, personnel should wear full sleeve protective clothing, e.g. laboratory coats.
- Surgical gloves should always be worn during work with radioactive materials to avoid contamination of the hands. Gloves should be put on and removed not to contaminate the hands.
- After use, the hands must be washed and checked for contamination.
- Pipetting by mouth should never be done in radiopharmacies.
- Smoking, eating or drinking should not be permitted in the areas where radioactivity is handled.

14.5.4 Storage of Source

- When the source is not in use, it should be stored in a properly shielded location dedicated solely to this function. Outside the storage area, radiation symbols must be prominently visible.
- The storage areas should be regularly monitored for possible air and/or surface contamination.
- All sources should bear an identification tag containing details such as the nature, activity and measurement time.

- Appropriate records of each radiation source should be maintained.
- Radioactive liquids in bottles and vials should be placed in large vessels to accommodate the full contents. Furthermore, the inside of the vessels should be lined with enough absorbent material to absorb the entire contents of the bottles in the event of any breakage.

14.5.5 Monitoring

Radiation levels in and around the areas where radioisotopes are handled should be routinely measured. All personnel working with radioisotopes are to wear personal monitoring devices such as TLD. In nuclear medicine, wearing wrist TLD is also mandatory while handling radiation. Some people also prefer to wear digital pocket dosimeters while actively handling radioisotopes.

14.5.5.1 Wipe Tests and Daily Surveys

Radiopharmacy areas should be checked at the end of each working day to check for surface contamination. Surveys should be conducted first with a survey meter to ensure no radioactive sources are present. After that, a contamination monitor with a probe should be used to examine all surfaces for contamination. The beta emitters with low energy are difficult to detect with an external probe and should be tested using a wipe test. A $100\,cm^2$ area should be wiped and checked, preferably in a well counter or pancake probe. The acceptable limit for contamination is $10^{-5}\,\mu Sv/cm^2$ and 22 dpm/cm^2 for gamma and beta and for alpha $10^{-6}\,\mu Sv/cm^2$ and 2.4 dpm/cm^2. In general, a dry wipe removes one-tenth of the contamination, whereas wet wipes remove one-fifth.

14.6 Decontamination of Working Area and Equipment

In the case of a short-lived contaminant, the material can be isolated and stored for decay. Wet methods should preferably attempt decontamination of surfaces. In serious contamination where the contamination is well fixed, entry into this area should be strictly forbidden until the radiation level reduces to below the permissible limit.

14.6.1 Personnel Decontamination

14.6.1.1 Internal Contamination

This is likely to occur through ingestion, inhalation, wounds and skin penetration.

Immediate action should be taken to remove the contaminant from the mouth, GI tract or respiratory tract by inducing vomiting or expectorants. Open wounds, cuts, etc. should be quickly cleansed. Bleeding should be encouraged. If required, the same may be performed under medical and radiation safety officer supervision.

14.6.2 External Contamination

External contamination can cause skin injury by local irradiation, penetration of the skin or transfer into the body by ingestion or inhalation. Remedial action should be to remove the contamination as early as possible.

The contaminated area should be washed with lukewarm water and not abrasive or alkaline soap for decontaminating body surfaces. A soft brush may be used. While removing contamination from the face, care should be made to protect the eyes and lips. If the above method fails, mild detergents may be used. Washing should be stopped if any abrasion is observed on the skin. Acids, alkalies and organic solvents should be avoided in the washing.

14.6.3 Surface Decontamination

All surfaces can be washed initially with water and soap or detergents using the brush. This can be repeated. Glass surfaces can be decontaminated using chromic acid or trisodium phosphate. Metal surfaces can be decontaminated using solvents such as inorganic acids, acid mixtures, dilute nitric acid or a 10% sodium citrate solution.

14.7 Radioactive Waste Disposal

Patient excretions, vomitus, contaminated clothes, syringes, needles, vials, gloves, unused/disused radioactive sources and other radioactive waste created in nuclear medicine might be solid, liquid or in the form of gas. At the point of generation itself, the waste is to be separated and stored in an appropriate shielding with radioactive marking. The three primary methods of disposing of radioactive waste are as follows:

1. Delay and decay.
2. Dilute and disperse.
3. Concentrate and contain.

However, only the first two methods are used in nuclear medicine. In the delay and decay method, the radioisotope is stored for ten half-lives before disposal (a half-life of fewer than 100 days is allowed for the delay and decay method).

The vapourized radioactivity produced during radioiodine therapy is expelled through the fume hood containing an activated charcoal filter system. Once radioactive waste decays to the background level, it should be disposed of as per hospital–/institution-approved policy in their respective categories.

For the dilute and disperse method, the radioactive effluents must be soluble or dispersible in water. The concentration and total quantity of liquid radioactive wastes disposed of must not exceed the limits set by the regulatory authorities. The effluents originating from the high-dose isolation room should be collected in purpose-built delay tanks, which are sufficiently shielded, fenced and leakproof.

Cleaning staff should be adequately trained to handle radioactively contaminated bins and warned about the potential hazards associated with this with detailed instructions on avoiding it. Cleaning personnel should be made aware of contaminated bins and must be alerted to the potential hazards and carefully instructed about the proper precautions.

Old, sealed sources such as ^{137}Cs, ^{57}Co, ^{153}Gd, ^{68}Ge, etc. should be kept in a secure store and disposed of as per the method prescribed by the regulatory authority.

14.8 Record-Keeping in Radiopharmacy [4]

Record-keeping can be subdivided into four parts: quality control records, records of receipt of radiopharmaceuticals, radiopharmaceutical preparation and dispensing and radioactive waste management. Each section should contain the following minimum documents:

14.8.1 Quality Control Records

- Acceptance testing record of dose calibrator.
- Daily QA record of dose calibrator.
- Radiopharmaceutical QC record includes:

  - Time of elution.
  - The volume of eluate.
  - Radionuclide activity.
  - ^{99}Mo breakthrough record.

14.8.2 Records of Radioactive Materials (RAM) Received

- Name of the radionuclide, activity, chemical form, supplier, batch number and purchase date.
- If the package containing RAM is suspected of being damaged upon delivery, it should be monitored for leakage with a wipe test. A survey meter should be used to check for unusually high external radiation levels.
- The supplier should be informed if the package is damaged or appears to be damaged. The record of the event is to be maintained.

14.8.3 Radiopharmaceutical Preparation and Dispensing Records

Each preparation's records should include the following information:

- Name of radiopharmaceutical.
- Batch number of the cold kit.
- The manufacturing date of the cold kit.
- Final product batch number.
- Result of radiochemical purity.
- Date of expiry.

A record must be maintained for each patient with the dose dispensed along with the following information:

- The patient's name.
- The radiopharmaceutical's name.
- Measured radiopharmaceutical with timing details.
- Date and time of measurement and administration.
- Route of administration.

14.8.4 Radioactive Waste Disposal Records

The packages containing radioactive waste, including sharps and glass containers, should be stored in a radioactive waste storage area with markings as follows:

- The name of radioisotope(s).
- The dose rate at the surface and 01-m distance.
- The date of storage.
- The expected date of release.

The release date can be ten half-lives of the longest half-life radioisotope in the package. On the day of disposal, check and record the dose rate, the date of disposal and the name of the person who disposed of the waste. Only authorized personnel can dispose of the radioactive waste.

References

1. AAPM Report 181. The selection, use, calibration, and quality assurance of radionuclide calibrators used in nuclear medicine. 2012. ISBN: 978-1-936366-18-7.
2. Hendee WR, Russell Ritenour E. Medical imaging physics. 4th ed. London: Wiley; 2002. ISBN: 978-0-471-38226-3.
3. ICRU report 31. 1979. Average energy required to produce an ion pair.
4. Smart RC. Physics in the radiopharmacy. Chapter 9. In: Nuclear medicine physics: a handbook for teachers and students. Vienna: International Atomic Energy Agency; 2014. ISBN: 978-92-0-143810-2.
5. Cherry SR, et al. Physics in nuclear medicine. Chapter 7. 2012. ISBN: 978-1-4160-5198-5.
6. CRC-55t, owner's manual. 2010. p. 257/288.
7. Candelaria G, Irwin D. The science of measurement: a primer on radioactivity dose calibrators. Vol 15, Lesson 4. 2010.
8. Lassmann M, Ted Treves S. EANM paediatric dosage card, pediatric radiopharmaceutical administration: harmonization of the 2007 EANM pediatric dosage card (version 1.5.2008) and the 2010 North American consensus guidelines. Eur J Nucl Med Mol Imaging. 2014;41(8):1636.
9. Ted Treves S, et al. 2016 update of the North American consensus guidelines for pediatric administered radiopharmaceutical activities. J Nucl Med. 2016;57(12):15N–8N.
10. Ronald B, et al. FDG PET/CT: EANM procedure guidelines for tumour imaging: version 2.0. Eur J Nucl Med Mol Imaging. 2014;42(2):328–54. https://doi.org/10.1007/s00259-014-2961-x.
11. Andrea V, et al. EANM procedure guidelines for PET brain imaging using [18F] FDG, version 2. Eur J Nucl Med Mol Imaging. 2009. https://doi.org/10.1007/s00259-009-1264-0.
12. Segall G, et al. Sodium 18F-Fluoride PET-CT bone scans. J Nucl Med. 2010;51(11):1813–20. https://doi.org/10.2967/jnumed.110.082263.
13. Nanni C, et al. [18F] fluciclovine PET/CT: joint EANM and SNMMI procedure guideline for prostate cancer imaging—version 1.0. Eur J Nucl Med Mol Imaging. 47(3):579–91. https://doi.org/10.1007/s00259-019-04614-y.
14. Piron S, et al. Optimization of PET protocol and inter-rater reliability of 18F-PSMA-11 imaging of prostate cancer. EJNMMI Res. 2020;10:14.
15. Zsolt S, et al. Initial evaluation of [18F]DCFPyL for prostate-specific membrane antigen (PSMA)-targeted PET imaging of prostate cancer. Mol Imaging Biol. 2015;17(4):565–74. https://doi.org/10.1007/s11307-015-0850-8.
16. Santhanam P, et al. Role of 18F-FDOPA PET/CT imaging in endocrinology. Clin Endocrinol. 2014;81:789–98.
17. Fendler WP, et al. 68Ga-PSMA PET/CT: joint EANM and SNMMI procedure guideline for prostate cancer imaging: version 1.0. Eur J Nucl Med Mol Imaging. 2017;44(6):1014–24.
18. Irene V, et al. Procedure guidelines for PET/CT tumour imaging with 68Ga-DOTA-conjugated peptides: 68Ga-DOTA-TOC, 68Ga-DOTA-NOC, 68Ga-DOTA-TATE. Eur J Nucl Med Mol Imaging. 2010;37:2004–10. https://doi.org/10.1007/s00259-010-1512-3.
19. Pallavi UN, et al. Molecular imaging to the surgeon's rescue: Gallium-68 DOTA-exendin-4 positron emission tomography-computed tomography in preoperative localization of insulinomas. Indian J Nucl Med. 2019;34:14–8.
20. Bartel Twyla B, et al. SNMMI procedure standard for bone scintigraphy 4.0. J Nucl Med Technol. 2018;46(4):398–404.
21. Donald BM, et al. The SNMMI and EANM practice guidelines for renal scintigraphy in adults. Eur J Nucl Med Mol Imaging. 2018;45(12):2218–28.
22. Mark T, et al. SNM practice guideline for hepatobiliary scintigraphy 4.0. J Nucl Med Technol. 2010;38(4):210–8.
23. Bennett S, et al. SNM practice guideline for parathyroid scintigraphy 4.0. J Nucl Med Technol. 2012;40(2):111–8.
24. ACR–SNM–SPR practice guideline for the performance of thyroid scintigraphy and uptake measurements. Revised 2009 (Res. 17).
25. Balon Helena R, et al. Society of Nuclear Medicine procedure guideline for thyroid scintigraphy. Version 3.0. 10 Sept 2006.
26. Spottswood Stephanie E, et al. SNMMI and EANM practice guideline for Meckel diverticulum scintigraphy 2.0. J Nucl Med Technol. 2014;42(3):163–9.
27. Juni Jack E, et al. Procedure guideline for brain perfusion SPECT using 99mTc radiopharmaceuticals 3.0. J Nucl Med Technol. 2009;37(3):191–5. https://doi.org/10.2967/jnmt.109.067850.
28. Palestro Christopher J, et al. Society of Nuclear Medicine procedure guideline for 99mTc-exametazime (HMPAO)-labeled leukocyte scintigraphy for suspected infection/inflammation, version 3.0, approved 2 June 2004.
29. ACR–SNM–SPR practice guideline for the performance of liver and spleen scintigraphy. Revised 2010 (Res. 27).
30. Dam Hung Q, et al. The SNMMI procedure standard/EANM practice guideline for gastrointestinal bleeding scintigraphy 2.0. J Nucl Med Technol. 2014;42(4):308–17. https://doi.org/10.2967/jnmt.114.147959.
31. ACR–SNM–SPR practice guideline for the performance of gastrointestinal scintigraphy. Revised 2010 (Res. 29).
32. Greenspan Bennett S, et al. SNM practice guideline for parathyroid scintigraphy 4.0. J Nucl Med

Technol. 2012;40(2):111–8. https://doi.org/10.2967/jnmt.112.105122.

33. ACR–SNM–SPR practice guideline for the performance of cardiac scintigraphy. Revised 2009 (Res. 14).

34. William SH, et al. Procedure guideline for myocardial perfusion imaging 3.3. J Nucl Med Technol. 2008;36(3):155–61. https://doi.org/10.2967/jnmt.108.056465.

35. Palestro Christopher J, et al. Society of Nuclear Medicine procedure guideline for gallium scintigra-phy in inflammation. Version 3.0, approved 2 June 2004.

36. Emilio B, et al. 131I/123I-metaiodobenzylguanidine (mIBG) scintigraphy: procedure guidelines for tumour imaging. Eur J Nucl Med Mol Imaging. 2010;37:2436–46. https://doi.org/10.1007/s00259-010-1545-7.

37. Ziessman HA, et al. Nuclear medicine: the requisite. 4th ed. Amsterdam: Elsevier. ISBN: 978-0-323-08299-0.

38. UNNRC 10 CFR 35.3045 report and notification of a medical event.

Working Mechanism of Radiation Detectors Used in Nuclear Medicine

15

Abstract

We use various radiation detectors in our day-to-day routine. They all work intending to detect the radiation but work on different techniques. The basic principle of radiation detection is discussed in this chapter which includes the interaction of radiation with matter, the qualities and characteristics of a good radiation detector, the types of various radiation detectors and their working mechanism. The voltage-response curve of gas-filled detectors and detectors based on ionization chamber, proportional counters and Geiger-Muller counters is discussed. The scintillation and semiconductor detectors used in nuclear medicine have also been explained. The mechanism and types of thermoluminescent dosimeters is discussed at the end.

15.1 Introduction

Radiation detection is of primary need in nuclear medicine facilities as it uses radioactivity for diagnosis and treatment. Radiation detection is needed for patients, staff and area monitoring. Various detectors are used in imaging and localizing the internal distribution of radioactivity within the patient. Examples of such detectors are the gamma camera, PET-CT, SPECT-CT, PET-MRI, thyroid probes, gamma probes for sentinel node detection, organ-specific (cardiac, breast, etc.) gamma cameras, liquid scintillation counters for beta detection of blood and urine samples, etc. There are several other uses, such as dose measurement, personnel monitoring and area monitoring, for which various other radiation detectors are used. Examples of such detectors are dose calibrators, radiation survey meters (Geiger-Muller counter, gun monitors), contamination monitors (well counters, GM counters), personal monitoring devices (thermoluminescent dosimeters, analogue and digital pocket dosimeters, hand and feet monitors, whole-body counters), area monitors (gamma area monitors), etc. This chapter limits our discussion to covering broad mechanisms of working principles of all these detectors.

15.2 Interaction of Radiation with Matter

The radiation used in nuclear medicine is ionizing. Examples are alpha, beta, gamma, positron, characteristics X-rays, bremsstrahlung radiation, auger electrons, etc. These may be divided into particulate and non-particulate or electromagnetic radiation. The radiation detection process uses the properties of the energy transfer mechanism to the matter, and the same is different for particulate and non-particulate.

While interacting with matter, particulate radiation transfers its energy and causes excitation and ionization of the atoms. In another type of interaction, due to columbic forces of the positively charged nucleus, the velocity of incident electrons is decelerated, and in response, bremsstrahlung radiation is produced (Fig. 15.1). The non-particulate radiations cause multiple interactions depending upon the photon's energy and is a complex process. The interactions in nuclear medicine are the photoelectric effect and Compton scattering (Fig. 15.2) which is used for detection. These effects have been discussed in detail in Chap. 14.

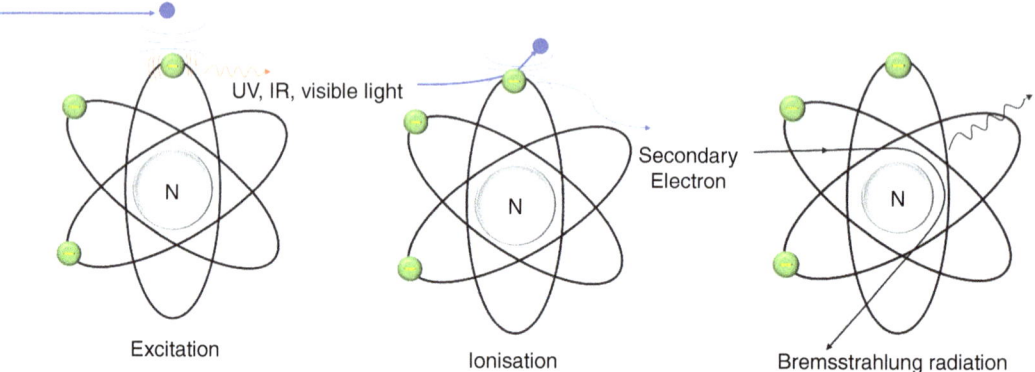

Fig. 15.1 Charged particle interaction with matter

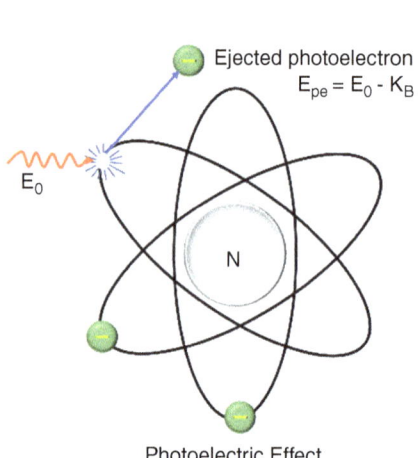

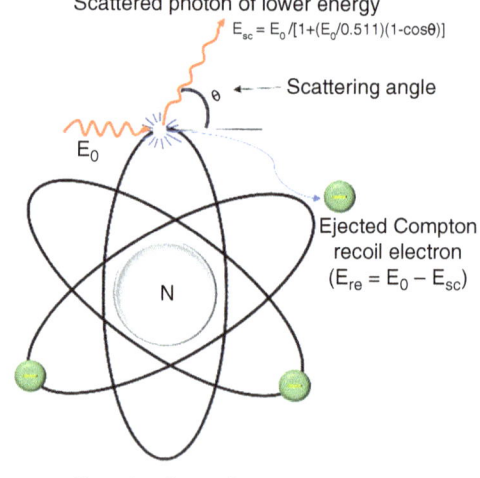

Fig. 15.2 Interaction of photons with matter

15.3   Definition of Radiation Detector

A sensor that produces signals upon interaction with radiation can be processed to quantify. Some detectors can characterize the incident radiation too.

15.4   Characteristics of Good Radiation Detection System

- Energy resolution—This is the ability to resolve and discriminates different radiation energies efficiently.
- Counting efficiency—Ability to count radiation $\left(\dfrac{cps}{dps} \times 100 \right)$.
- Inherent dead time—Ability to count without being paralyzed.

15.5   Types of Detectors

- Gas-filled detectors.
  - Ionization chamber detectors.
  - Proportional counters.
  - Geiger-Muller (GM) tube detectors.
- Solid-state (semiconductor) detectors.
- Scintillation detectors.
  - Inorganic scintillation detectors.
  - Organic scintillation detectors.

15.6   Radiation Detection Mechanism

15.6.1   Gas-Filled Detectors

Figure 15.3 depicts a cross-sectional diagram of a standard gas-filled detector. The primary modes of interaction in gas-filled detectors are the excitation and ionization of gas molecules and the particle or photon track. On ionization, the neutral gas molecule produces a positive ion and free electron, i.e. a negative ion, referred to as an ion pair. This ion pair can generate an electrical pulse that can be detected using a potential difference. The ions can be made either by direct interaction or through secondary processes. Still, irrespective of the involved technique, the total number of formed ion pairs plays a role in detecting and measuring radiation.

The first ionization potential (energy required to remove the outermost electron) required for creating an ion pair in most of the gases that can be filled in an ionization chamber is 10–25 eV. There are other processes by which the incident radiation may lose energy within the gas without creating ion pairs. Therefore, a minimum of 25–35 eV per ion pair (refer to Chap. 14, Table 14.2) is required by the incident radiation to produce an ion pair. The completely stopped 1 MeV particle can create about 30,000 ion pairs [1] (Fig. 15.4).

Fig. 15.3 Typical design of the gas-filled detector

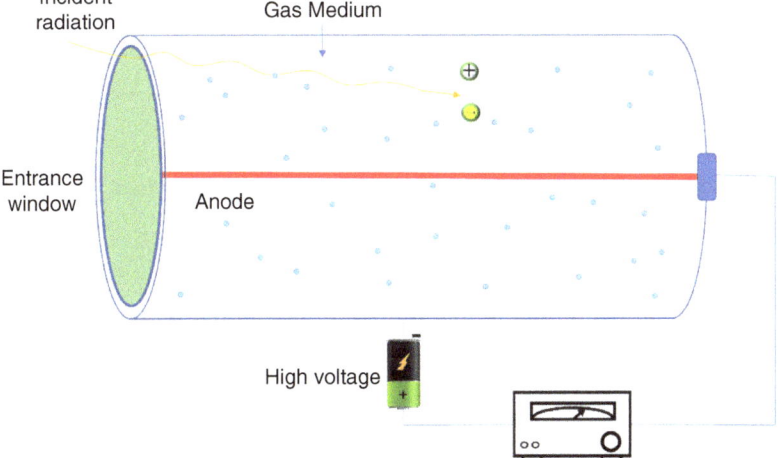

Incident radiation · Gas Medium · Entrance window · Anode · High voltage

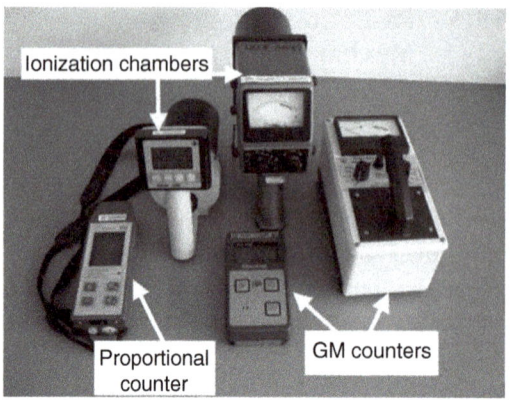

Fig. 15.4 Examples of gas-filled detectors. (Image copyright IAEA. Reproduced with permission) [2]

15.6.1.1 Voltage-Response Curve

For gas-filled detectors, a graph can be plotted on semi-logarithmic paper with an *x*-axis demonstrating applied voltage on a linear scale and the *y*-axis showing the amount of charge collected on a logarithmic scale. This graph is called a voltage-response curve and shows the variation of ion pair charge collection with respect to the applied voltage (Fig. 15.5).

When charged particle or photon travels within the gas-filled chamber, it creates ion pairs along its path. For the optimum efficiency of the detector, the voltage difference between positive and negative electrodes within the chamber

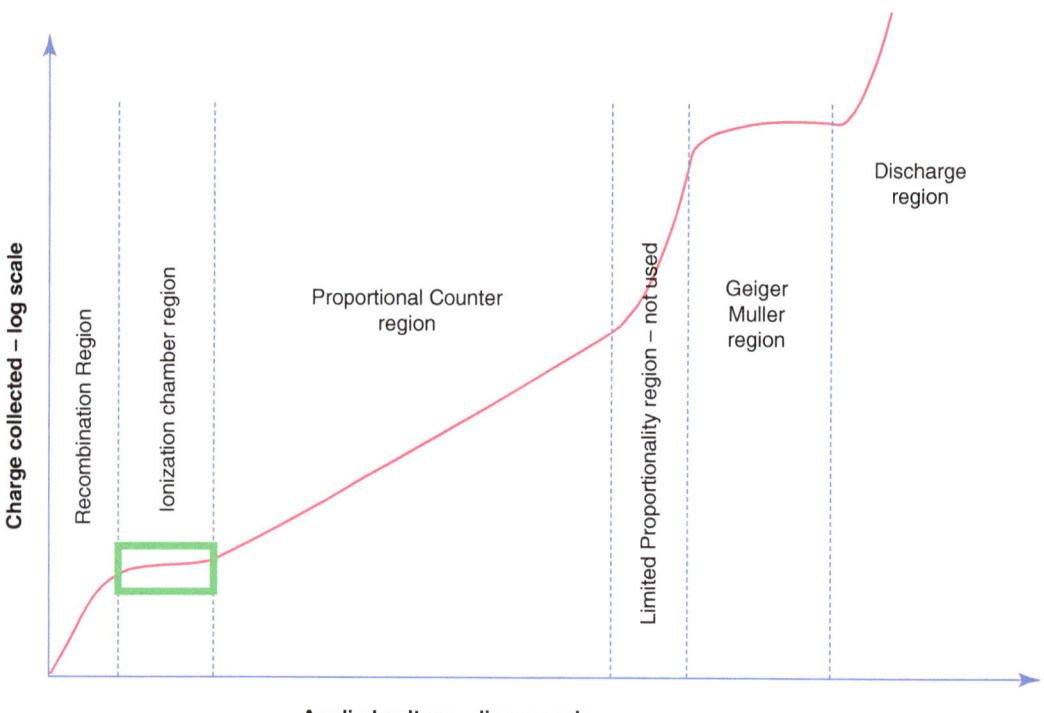

Fig. 15.5 Voltage-response curve: variation of ion pair charge with applied voltage

should be sufficient to ensure the collection of all ion pairs created by the incident radiation. To avoid recombination, the charge collection should be fast enough. In case the voltage is too low, some of the ions recombine with each other and do not contribute to the final electrical output. This area of the graph is called as recombination region. Based on the patterns of recombination, it is categorized into two types:

- Initial or columnar recombination.
- Volume recombination.

The initial or columnar recombination is the one that takes place along the track of the incident radiation, whereas the volume recombination is due to the different tracks encountering each other on their way to the collecting electrodes. The volume recombination is dependent on dose rate as more density of ions increases the probability of recombination. The high-Linear Energy Transfer (LET) tracks created by particles such as those formed by α-particles, initial recombination is more likely [3].

Ionization Chamber Region
As the voltage increases and at sufficient voltages, all ion pairs are collected at the electrodes, and no further recombination occurs. The ion pairs produced and collected become independent of the voltage level, but they still depend on the amount of deposited energy in the gas volume. This part of the graph is called the saturation region, and detectors working in this graph segment are called ionization chamber-based radiation detectors. Examples of such detectors are dose calibrators, pocket dosimeters, gun monitors, etc.

Proportionality Region
When the applied voltage increases, the negative ions, i.e. electrons, are accelerated towards the central positive electrode, causing secondary ionization within the gas volume. The energy dissipated by the incident particle is proportional to the number of ion pairs collected. Hence this increases the number of ion pairs collected. With

an increased applied voltage, the proportionality or the gas amplification rises to as high as 10^6 [4]. Since the size of pulses produced by a particle is related to the number of primary ions at a given voltage, operating the detector in this region allows one to distinguish between various energies and radiations. An example of such detectors is proportional counters used for monitoring neutrons in cyclotrons and reactors.

Region of Limited Proportionality
The free electrons are swiftly collected as the voltage rises, but the positive ions move considerably slowly due to their greater mass. This produces a cloud of positive ions, which, if concentrated enough, can act as a space charge, altering the structure of the electric field and causing nonlinearities in the output pulse, making it unsuitable for radiation detection. The pulse amplitude grows with an increasing number of ion pairs created at the beginning. This graph region is called limited proportionality and is not linear to be used as a detector.

Geiger-Muller (GM) Region
The space charge formed by the positive ions becomes dominating when the applied voltage is sufficiently high. Due to a cloud of positive ions, magnitude of the electric field in the area of the anode wire begins to diminish. The gas amplification is terminated because gas multiplications require a sufficient electric field. The termination develops at nearly the same total charge regardless of the formed primary ion pairs. As a result, each output pulse shows the same amplitude and does not reflect any property of incident radiation. The Geiger-Muller (GM) detectors operate in this region and provide a more significant pulse height. The gas amplification in GM detectors is 10^6 to 10^8 times, and the detectors working in this region are susceptible [4].

Region of Continuous Discharge
With further increase in voltage, the positive ions strike the cathode harder, resulting in cathode surface-emitting UV radiation, as well as electrons and more ionization, is triggered.

A continuous discharge starts to occur, and sparking ensues. This area is never utilized for radiation detection and is always avoided for radiation detection.

15.6.2 Ionization Chamber Detectors

Ion chambers detectors are simple gas-filled detectors. They operate mainly by collecting direct ionizations created within the gas by the incident radiation by applying an electric field. Depending upon the application, different geometries such as parallel plate, cylindrical or spherical are used for the electrodes in gas-filled detectors. Air is commonly filled with gas in ionization chamber-based detectors. However, denser inert gas (argon) is filled in dose calibrators to increase ionization density. Air has the property to form negative ions readily and is used in gamma-ray exposure measuring instruments. The filled gas pressure can be at 1 atmosphere; however, higher gas pressures can also increase the sensitivity. A supporting insulator (guard ring) is used between the two electrodes with all designs. Examples of insulating materials can be high-resistivity synthetic plastics or inorganic materials such as ceramics.

These detectors operate in the saturation region, and saturation voltage (Vs) typically begins at Vs ≈ 50–300 V [4].

15.6.2.1 Pocket Dosimeters

Pocket dosimeters, used for personal monitoring, work on the principle of charge integration. It contains a small gas-filled ionization chamber, which is kept in a charged condition before being used. When it interacts with the radiation, the voltage drops, and this fall in voltage shows the magnitude of radiation interacted. A scale showing deflection in voltage is calibrated to show the reading in the desired unit.

Consider the capacitor is charged to a specific voltage V and exposed to the radiation field where electrodes gather ΔQ electrical charge. Therefore, the voltage difference (ΔV) across the capacitor is calculated as follows:

$$\Delta V = \frac{\Delta Q}{C} \qquad (15.1)$$

The pocket dosimeter contains an integrated quartz fibre electroscope which can be read against the light. When it is fully charged, the initial reading on the electroscope remains at zero. These instruments can provide accurate readings within a few percent of the variation when utilized within the specified range.

The total accumulated dose can be monitored regularly by noting the degree of chamber discharge on the electroscope. The leakage current, which unavoidably flows over the insulator surface over time, limits the ultimate sensitivity of such devices (Fig. 15.6).

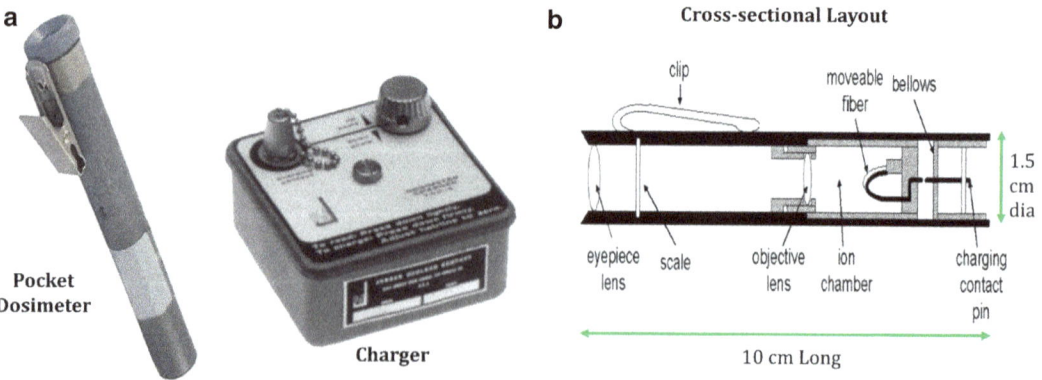

Fig. 15.6 Analogue pocket dosimeter. (**a**) copyright Elsevier. Reprinted with permission [4]. (**b**) copyright ORAU, originally published on https://www.orau.org. Reprinted with permission [5]

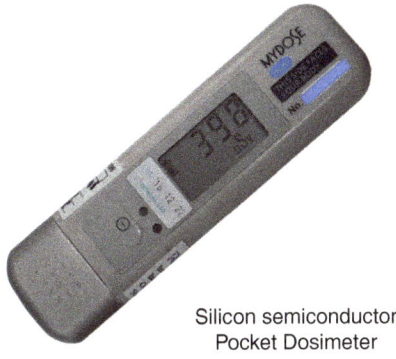
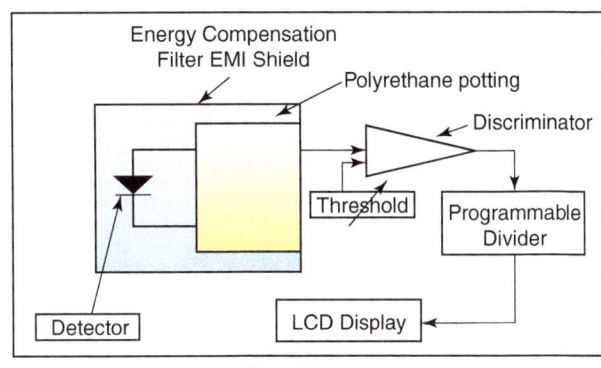

Fig. 15.7 Digital pocket dosimeter

15.6.2.2 Digital Pocket Dosimeter

The digital pocket dosimeters are semiconductor-based detectors. It is discussed here since the analogue pocket dosimeter is discussed in the previous section; however, in the latter part of the chapter, semiconductor detectors have been discussed in detail. Digital pocket dosimeters are available in different models with slight variations in their manufacturing and capabilities (one model is shown in Fig. 15.7). It is made up of PN junction silicon (Si) semiconductor detector and is suitable for X- and gamma radiation of >40 keV. The reverse bias of a silicon rectifier diode works at approximately 4 V voltage. In the presence of ionizing radiation, a charge-sensitive amplifier, with the help of a low-power CMOS IC, amplifies the pulses from the detector. All the circuit components combinedly form the detector. An energy compensation filter is included in the amplifier, which also serves as electromagnetic shielding and is housed with an anti-vibration polyurethane substance. The amplifier's pulses are supplied into a discriminator with a threshold voltage set to turn off the noise pulses. The output pulses of the discriminator are routed into a programmable divider circuit. This is calibrated such that the dosimeter shows one count equals one micro Sv (μSv) of ^{137}Cs gamma. The counts are accumulated in a digit-counter-LCD-display module. It uses coin type lithium batteries, and some models may indicate the 'Battery Low' symbol.

15.6.2.3 Gun Monitor

Gun monitors measure high exposure rates, such as the radiation level of high-dose therapy patients in nuclear medicine. When measuring such exposures, the ionization induced by all secondary electrons generated from the different beta, gamma or X-ray ranges must be considered. It must be followed for several metres. One needs to follow it for several metres. Direct measurements of such radiation are impractical, and the principle of compensation is used [6].

A gun monitor is a pistol shape ionization chamber-based radiation detector. The ion chamber is constructed of plastic material (air equivalent material) with an effective volume of $\approx$600 cc (may vary from model to model) of air. All interior surfaces of the chamber are coated with a conductive graphite surface. The central electrode runs the entire length of the chamber and is utilized for ion collection. A battery-powered electrometer circuit is used to measure the saturated ion current. The instrument has three different ranges of operation: 0–50, 0–500 and 0–5000 mR/h [7]. Figure 15.8a shows a commercially available gun monitor, whereas Fig. 15.8b shows the cross-sectional layout. Both figures are from different manufacturers, and little difference in dimensions from Fig. 15.8b to Fig. 15.8a is possible.

These instruments provide relatively accurate measurements of gamma-ray exposure from the higher flux while avoiding considerable attenuation in the chamber's walls or entrance window.

a

b

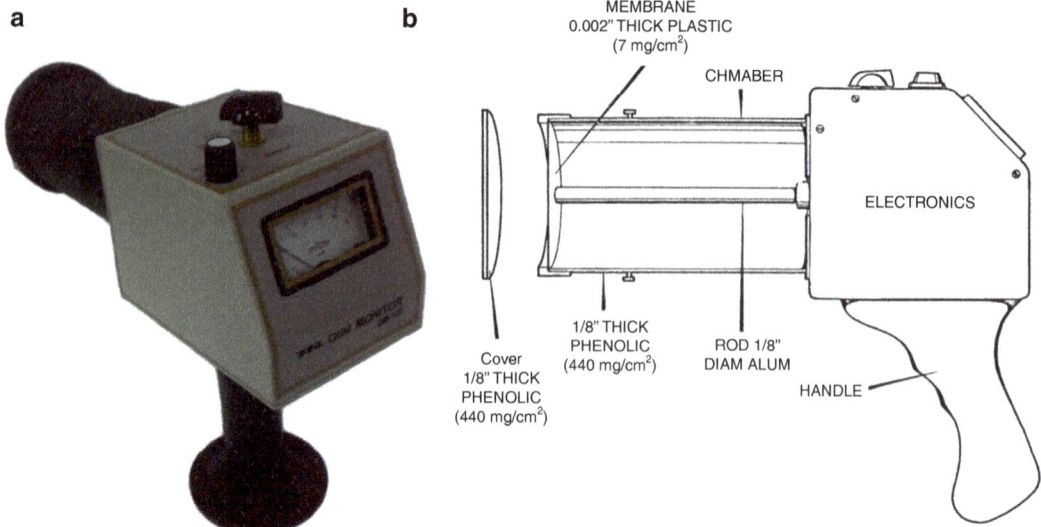

Fig. 15.8 Gun monitor and its cross-sectional layout. (**a**) Courtesy: Pla Electro Appliance Pvt. Ltd., Mumbai, India, printed with permission [8]. (**b**) Image originally printed in W.P. Howel and R.L. Kathren. Calibration and field use of ionization chamber survey instrument. BNWL-SA-2096. 1969 [7]

15.6.2.4 Dose Calibrators

Dose calibrators used in radiopharmacies are also ionization chamber-based radiation detectors and discussed in detail in Chap. 14.

15.6.3 Proportional Counters

- Gas-filled detectors that operate in the proportional region or ascending portion of the voltage-response curve are called proportional counters (Fig. 15.5).
- The detection of neutrons and the spectroscopy of low-energy X-rays are two important uses of proportional counters.
- In nuclear medicine, proportional counters do not have a direct role, but they are used during radioisotope production, such as in cyclotrons and reactors for neutron monitoring (Fig. 15.9: tritium detector used in cyclotron as area neutron monitor).
- The geometry of these detectors is preferred as cylindrical or spherical because charge multiplication is more easily achieved. These chambers contain very thin wire at the centre with positive voltage to act as the anode. The wall of the chamber is normally grounded.

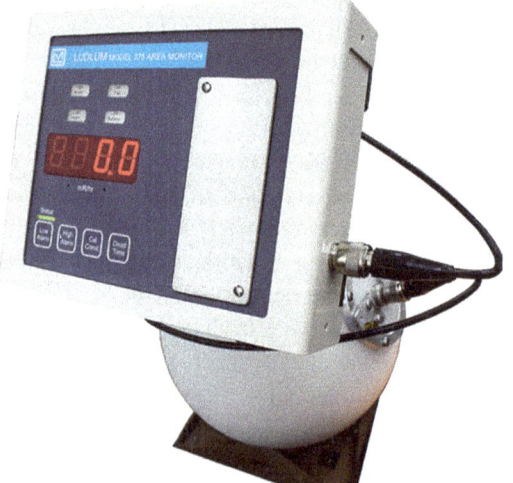

Fig. 15.9 Two atmospheric pressure tritium (^{3}H) neutron area monitor with a thermalizing polyethene sphere diameter of 22.9 cm. Image courtesy: https://ludlums.com, printed with permission [9]

- The electrostatic field inside the chamber is radial, and field intensity increases rapidly as the central wire approaches.
- As primary ionization occurs inside the chamber, the ion pairs move towards the cathode and anode. With increased field intensity towards the centre, the electrons move rapidly.

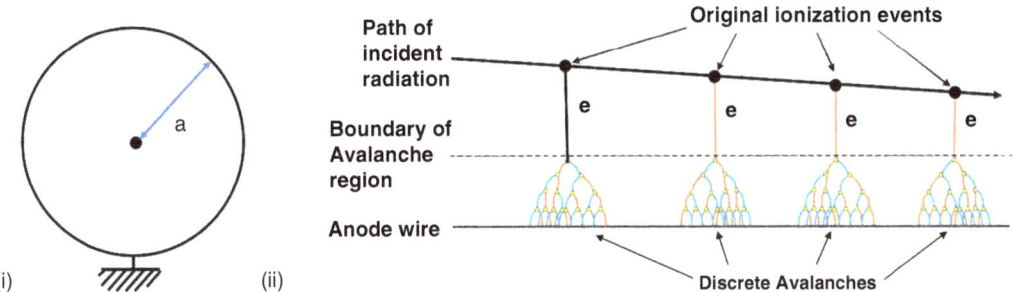

Fig. 15.10 (i) Cross-sectional diagram of the proportional counter; (ii) gas amplification near central anode wire. Image courtesy: CC BY-SA 3.0 with necessary change

- On the way, they collide with other gas molecules and produces secondary electron. These electrons interact further with other gas molecules, producing more electrons (Fig. 15.10). This phenomenon is called gas multiplication or charge multiplication or avalanche. The gas multiplication in proportional counters is in the range of 10^4.

- Since the radius of the central wire is very thin (25–100 μm or 1/1000th of an inch diameter made of tungsten or platinum) compared to the radius of the cylindrical detector, an extremely strong electric field is produced in a fraction of the chamber's volume. This volume is so small that the probability of primary ionization from incident radiation is negligible.

- In addition to the secondary electrons produced by collisions, electrons are also produced by photoelectric interactions and bombardment of the cathode surface by positive ions.

- The number of electrons captured at the anode and the resulting pulse height is proportional to the initial ionization's ion pairs. This property enables proportional counters to distinguish between different forms of radiation and analyse the energy spectrum using pulse height analysis.

- Proportional counters have a faster response than GM counters and can respond to higher radiation intensity and exposure rates than GM counters.

- The voltage applied to proportional counters ranges between 800 and 2000 V [10].

- The filled gas depends on the function of the detector to perform. Commonly used gas is P-10 and is a mixture of 90% argon with 10% methane. Krypton or xenon can be used for high-efficiency application of γ-ray detection. For thermal neutron detection, BF_3 or ^3He gases and fast neutron spectroscopy low-Z gases such as hydrogen, methane, helium, etc. are used. A tissue-equivalent mixture of 64.4% methane, 32.4% carbon dioxide and 3.2% nitrogen is recommended for dosimetry purposes [11].

15.6.4 Geiger–Müller (GM) Counters

- Hans Geiger invented the GM counter in 1908; however, Walther Müller collaborated with Geiger to develop it further in 1928. Sidney H. Liebson invented the modern halogen-filled GM tube in 1947 [12].

- It is an extensively used radiation detector developed in various designs, simple, rugged, economical and insensitive to environmental conditions and has a substantial output signal.

- They are built and operated similarly to proportional counters, but the electric field near the central wire is very high.

- The GM detectors are filled with noble gases such as helium, argon or neon with a 5–10% mixing of halogens (e.g. chlorine or bromine) or polyatomic organic vapour (e.g. butane or ethanol). The halogens are commonly filled with neon, argon or krypton, whereas organic vapours are filled in helium.

- The halogen gases ionize faster than organic vapour allowing operations at a much lower voltage, i.e. 400–600 V instead of 900–1200 V. They have several other benefits, such as longer plateau length and longer self-life of the detector. The organic vapours slowly get destroyed, whereas halogens recombine over time and make it an unlimited lifetime [12].
- The gases filled inside the GM tube have low pressure, i.e. about 0.1 atmospheric pressure.
- When radiation interacts with gaseous molecules, ion pairs are formed. The electrons are attracted by the strong electric field surrounding the central wire and produce an avalanche of ion pairs with several other excitations of atoms and molecules within the medium. This avalanche soon spreads throughout the detector volume. During this period, the anode wire constantly collects electrons. On the other hand, slow-moving positive ions stay in the detector volume near the anode and form a sheath around the centre wire. As a result, the electric field at the anode is reduced, and no more pulses may be formed until the positive ion sheath reaches a threshold distance. The time to collect the electrons in a typical detector is 1 μs; however, it takes 1 ms for positive ions. The time in which no more pulses are formed is called 'dead time'. Typically, the dead time of GM counters is around 100–300 μs, and during this time, no event is detected (Fig. 15.11).
- When positive ions finally strike the cathode, they eject further electrons. Since, by that time, the field has been restored to its original high value, a new avalanche starts, and the process described is repeated.
- The gas multiplication in GM counters is of the order of 10^{10} [4]. This allows considerable simplification for associated electronics and eliminates the need for external amplification.
- The repetitive discharge is prevented with the help of quenching gases, which eventually donate electrons to neutralized positive ion clouds. The quenching gas also absorbs ultraviolet (UV) photons formed during the avalanche and prevents any avalanche trigger. This type of quenching is called self-quenching.
- There is another type of quenching called external quenching. Here, the operating voltage of the detector is reduced once the discharge starts, and this is done until the ions reach the cathode to a value for which the charge multiplication factor is negligible.
- Generally, ion chamber detectors are three to four times and proportional counters are seven

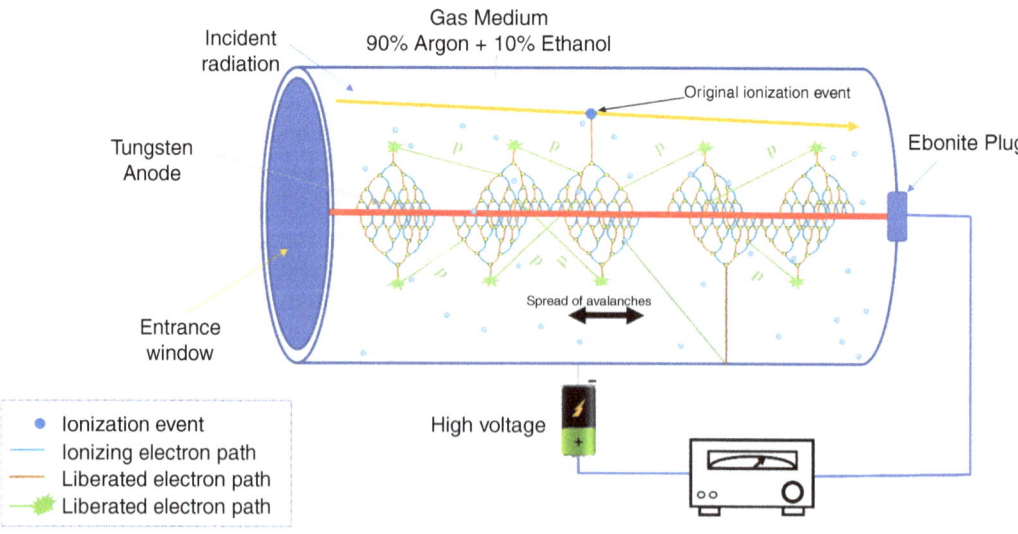

Fig. 15.11 GM counter. Image courtesy: CC BY-SA 3.0 with necessary change

to eight times more expensive than GM tube counters.

- The drawback of GM detectors is that:
  - Their signals are independent of the particle type and its energy.
  - They are unsuitable for use in high count rate areas because of relatively long dead time.

15.6.5 Scintillation Detectors

- Sir William Crookes, a British scientist, was the first person to observe scintillation with alpha particles impinging on a ZnS screen [13].
- Scintillation is the process of emitting light photons or producing sparks. Some of the material from solids, liquids or gases emits light photons when radiation deposits its energy in it. With the help of electronics, these light photons can be converted into electric pulses and be used for radiation detection, spectroscopy and measurement.
- The scintillators detectors are of three types:
  - Inorganic scintillators.
  - Organic scintillators.
  - Gaseous scintillators.

- The choice among these detectors is based on a trade-off between the following qualities:
  - Radiation kinetic energy should be converted to visible light with high scintillation efficiency and proportional to deposited energy over a wide range by the detector material.
  - The decay time of the produced scintillation should be short enough that fast signals can be processed and dead time is minimum.
  - The detector should be transparent to the wavelength of its emissions and can be grown in large sizes.
- Inorganic crystals (e.g. sodium iodide, caesium iodide, BGO, LSO), organic liquids and plastics are the most widely used scintillator detectors. The inorganic materials offer good light output and linearity, but their response time is slow. On the other hand, organic scintillators are often faster but produce less light. In gamma-ray spectroscopy, high-Z and high-density containing inorganic crystals are utilized. On the other hand, organics are preferable for beta spectroscopy and fast neutron detection (as they contain hydrogen) (Fig. 15.12).

NaI (Tl) (Sodium Iodide Thallium activated)
Image Courtesy: Saint-Gobain

CsI (Tl) (Cesium Iodide Thallium activated)
Image Courtesy: www.amcrys.com

BGO ($Bi_4Ge_3O_{12}$ - Bismuth Germanate)
Image Courtesy: Saint-Gobain

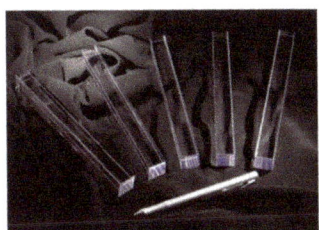

LYSO ($Lu_{1.8}Y_2SiO_5$:Ce - Lutetium-yttrium oxyorthosilicate cerium activated)
Image Courtesy: Saint-Gobain

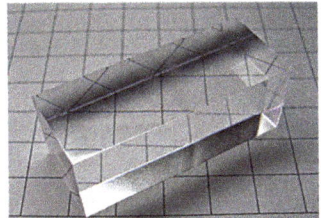

YSO(Y_2SiO_5:Ce - Yttrium Oxyorthosilicate cerium activated)
Image Courtesy: www.amcrys.com

Plastic scintillators for n-ϒ discrimination
Image Courtesy: www.amcrys.com

Fig. 15.12 Scintillator crystals. Images are copyrighted with Saint-Gobain and Amcrys, reprinted with permission

In routine clinical practice of nuclear medicine, scintillator detectors are used in gamma cameras, positron emission tomography (PET) machines, thyroid probes, organ-specific (cardiac, breast, etc.) gamma cameras, gamma probes for sentinel node detection, well counters and as CT detectors in PET-CT machine.

The energy levels of the material's crystalline lattice determine the scintillation mechanism in inorganic materials. Electrons revolve around the nucleus at various energy shells in an atom, and the outermost electrons are called valence electrons. When valence electrons receive sufficient energy, they eject from their orbit and become free in the crystal lattice. These free electrons always have greater energy than valance electrons and help conduction. The energy range of valence electrons is called a valence band, whereas the energy range of conduction electrons is called a conduction band. In a pure crystal, no electrons are found between these two bands, and this region is called a forbidden band or 'forbidden gap'.

If we diagrammatically represent these energy bands, the lower band is shown as the valence band, whereas the upper band is the conduction band. The gap between these two bands is forbidden band (Fig. 15.13).

The electrons in the valence band are raised into the conduction band when the crystal absorbs energy from radiation, creating a hole in the valence band. On de-excitation, they return to the valence band, and in the process, they emit light photons. Generally, in a pure crystal, the energy gap between the valence band and conduction band is high, and the emission of a photon is an inefficient process. The resulting photons also do not lie within the visible range.

Small amounts of an impurity, called the activator, are commonly added to enhance the probability of visible photon emission. Such deliberately added impurities create special sites in the lattice, which modify the normal energy band structure. The conduction and valence band electrons of such activators have energy less than that of the pure crystal's full forbidden gap, and transitions can now give photons of visible range. This serves as the basis of the scintillation process, where the activator forbidden gap determines the emission spectrum.

When a charged particle passes through the detecting medium, it creates several electron-hole pairs and gives electrons enough energy to excite the conduction band. The positive holes quickly drift within the crystal. Since the ionization energy of the impurity is less than that of a typical lattice site, the drifted holes ionize the activator site by taking electrons and creating a hole inactivation centre.

Meanwhile, the excited electron in the conduction band is free to migrate across the crystal until it comes across an ionized activator. The electron can now enter the activator site, forming a neutral configuration with its own set of excited energy states. De-excitation occurs swiftly, with

Fig. 15.13 Energy level diagram of activated crystalline scintillator

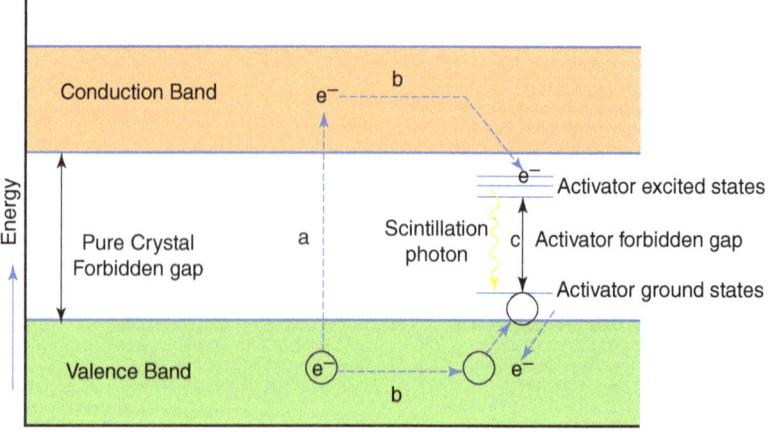

a high probability of an emission of a corresponding photon. Activators are carefully chosen to produce emissions within the visible energy range. The lifetimes of such excited states are typically 30–500 ns [14]. Tables 15.1 and 15.2 provide various characteristics of scintillation crystals suitable for gamma and PET cameras, respectively (Fig. 15.14).

Table 15.1 Properties of common crystals used in gamma camera

Principle properties	NaI (Tl)	CsI (Tl)	CsI (Na)
Effective atomic number	50	54	54
Density (g/cm^3)	3.67	4.51	4.51
Hygroscopic	Yes	Slightly	Yes
Wavelength of emission max (nm)	415	550	420
Decay time (ns)	250	1000	630
Light yield (photon/keV)	38	54	41
Comments	General purpose, good energy resolution	High Z, rugged, good match to photodiodes	High Z, rugged

Table 15.2 Properties of common crystals used in PET detectors

Principle properties	BGO $Bi_4Ge_3O_{12}$	LSO $Lu_2SiO_5(Ce)$	LYSO $Lu_{1.8}Y_{0.2}SiO_5(Ce)$
Effective atomic number	73	66	66
Density (g/cm^3)	7.1	7.4	7.1
Hygroscopic	No	No	No
Attenuation length for 511 keV (cm)	1.0	1.15	1.2
Energy resolution (%) @662 keV	12	9	8
Wavelength of emission max (nm)	480	420	420
Decay time (ns)	300	40	36
Light yield (photon/keV)	9	26	33
Comments	High Z, compact detector, low after glow	Bright, high Z, fast, dense background from [176]Lu activity	Bright, high Z, fast, dense background from [176]Lu activity, better energy resolution and decay time

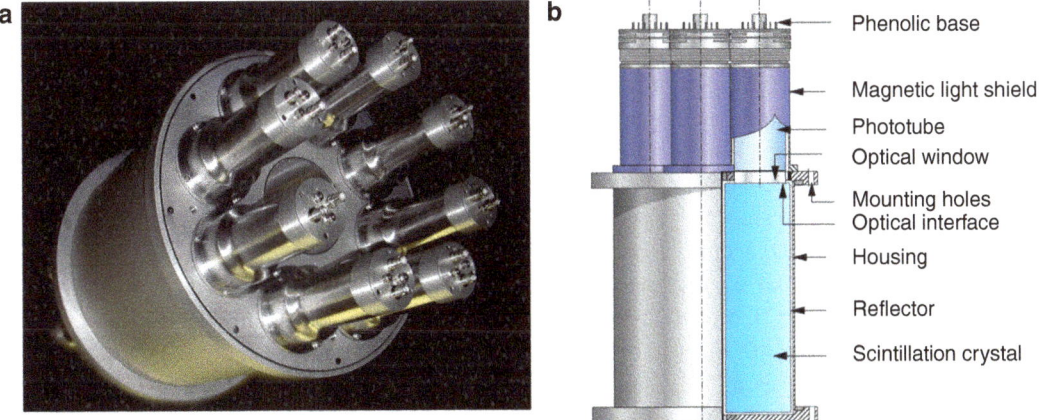

Fig. 15.14 NaI (Tl) detector assembly with demountable photomultiplier tubes. (Image courtesy: www.crystals.saint-gobain.com)

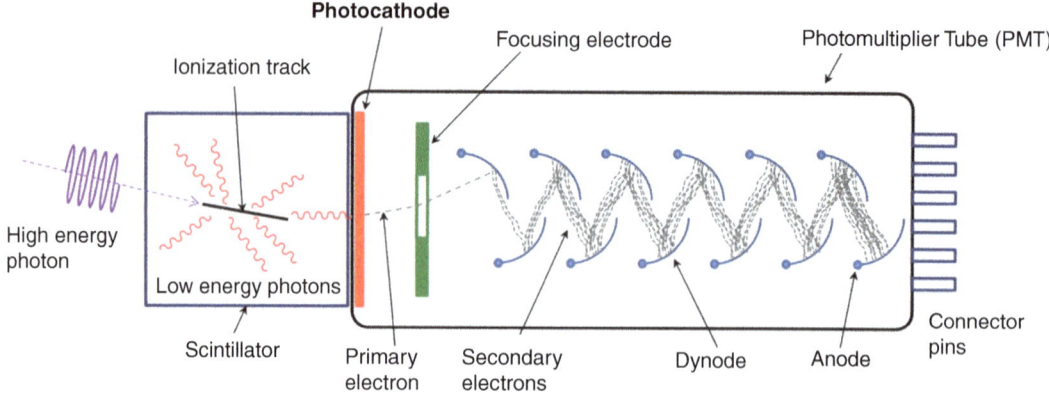

Fig. 15.15 Scintillation detectors with PM tube assembly. (Image courtesy: Creative Commons, used under license 3.0 Author: Qwerty123uiop)

The scintillator detectors are always coupled with photomultiplier (PM) tubes or photodiodes. This converts scintillation light into electric pulses and amplifies these pulses sufficiently to be read by other electronics. A transparent optical 'coupling grease' is placed between the crystal and the PM tube to reduce internal reflections. The PM tube converts each pulse of light into an electron using a photocathode material and amplifies it to give a voltage pulse that can be read and processed. The amount of radioactivity is determined by the number of these pulses recorded over time. Still, the specific energy of the radiation is determined by the size of each created pulse (Fig. 15.15).

15.6.6 Computed Tomography (CT) Detectors

CT scintillation detection differs from the scintillation detection mechanism in nuclear medicine. In CT, it always operates in current mode, whereas nuclear medicine ones operate in pulse mode. When event rates are very high, current mode can provide a stable current, but pulse mode preserves individual events' energy and temporal information. In CT, the radiation flux is very high (in the order of 10^9 quanta/sec/mm^2 at the detector level), whereas in SPECT and PET (1–100 quanta/sec/mm^2 for SPECT and a typical PET application), it is about 100 quanta/sec/mm^2 [15]. Therefore, individual counting in CT is not feasible, whereas low count rates ease the electronics in SPECT and PET.

Inorganic crystals or transparent ceramic scintillators have been used in X-ray/computed tomography equipment. The detectors are chosen for their good light output, absence of long afterglow and resistance to radiation damage over prolonged use. Scintillators such as caesium iodide thallium-activated CsI (Tl) crystals, self-activated cadmium tungstate ($CdWO_4$), gadolinium oxysulfide (Gd_2O_2S), yttria and gadolinia with Eu-activated (Y, Gd)$_2$O$_3$:Eu or $Y_{1.34}Gd_{0.60}Eu_{0.06}O_3$, etc. have been used. General Electric (GE) has used (YGd)$_2$O$_3$ in more than 8000 computed tomography (CT) scanners since 1988 [16]. They have recently used gemstone detectors which are complex rare-earth-based oxides [17]. Siemens has used proprietary detectors such as Ultra-Fast Ceramics (UFC) and Stellar detectors [18]. Table 15.3 represents the properties of various detectors used for medical applications [19].

CT detectors can be in millions of numbers arranged in a pixelated form in different arrays of channels. Like in nuclear medicine, radiation (X-rays) falls on the detectors, and scintillation light is produced, converted into electrical signals by photodiodes. After that, the pixel electronics are connected to the additional circuitry, which many detector channels share. A general representation of the CT detection system is shown in Fig. 15.16, whereas a graphical representation is shown in Fig. 15.17.

Table 15.3 Important characteristics of X-ray scintillators used in imaging. Copyright Elsevier [19]. Reprinted with permission

Material	Density (g/cm³)	Thickness to stop 99% of 140 keV X-rays (mm)	Emission (nm)	Light yield (ph/MeV)/ temperature coefficient (%/°C)	Primary decay time (μs)	Afterglow (% at 3 ms)
CsI (Tl)	4.52	6.1	550	54,000/0.02	1	0.5
CdWO₄ (CWO)	7.9	2.6	495	28,000/−0.3	2, 15	0.05
Gd₂O₃:Eu⁺³	7.55	2.6	610	–	–	–
Y₁.₃₄Gd₀.₆O₃:Eu, Pr, Tb (YGO)	5.9	6.1	610	42,000/0.04	1000	5
Gd₂O₂S: Pr, Ce, F (GOS)	7.34	2.9	520	50,000/−0.6	2.4	<0.1
Gd₂O₂S: Tb (Ce) (GOS)	7.34	2.9	550	50,000/−0.6	600	0.6
La₂HfO₇:Ti	7.9	2.8	475	13,000/−	10	–
Gd₃Ga₅O₁₂:Cr, Ce	7.09	4.5	730	39,000/−	150	<0.1

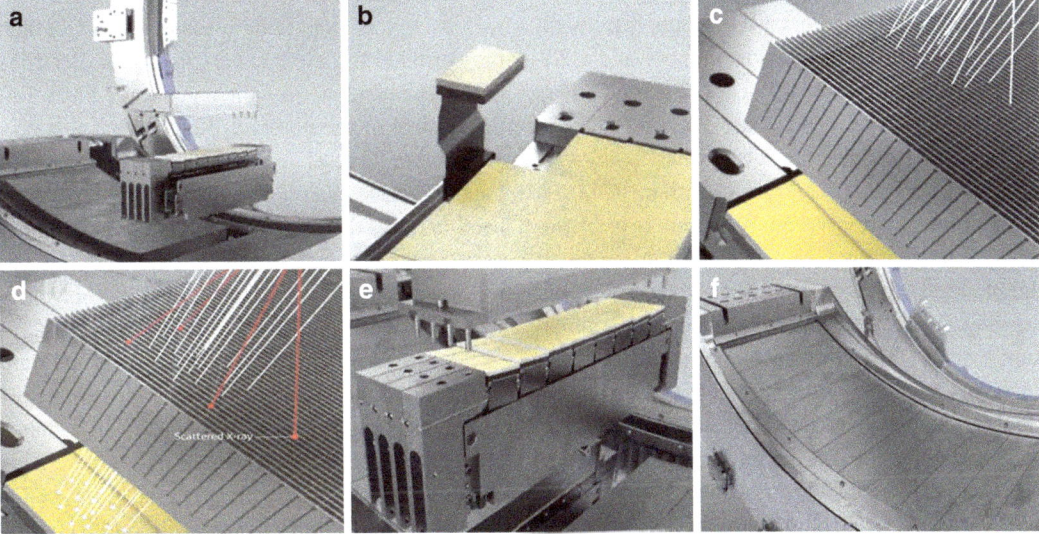

Fig. 15.16 Array of CT detectors. (**a**) Detector block, (**b**) detector module, (**c**) 3D collimator to prevent scatter, (**d**) collimated X-rays interact with the detector, (**e**) focally aligned detectors, (**f**) covered detector seen from outside. Image courtesy: GE Healthcare, printed with permission

In CT imaging, 1 mm spatial resolution is often required at the detector; nevertheless, quick gantry rotation and many projections necessitate a high temporal resolution of 100 s. Gadolinium oxysulfide has a light yield of 35–60 photons/keV, whereas cadmium tungstate has a light yield of 20 photons/keV [15]. The scintillator crystals are cut into small pieces of roughly 2–5 mm³ and put next to each other with reflective material to prevent lateral scintillation light crosstalk between them [15]. The detector elements are positioned in a 1-m-long arc of a circle, dubbed the 'detector banana' by some. A scintillator crystal and an attached photodiode make up each detector pixel. After that, the photodiode is connected to the electronics channel consisting of an amplifier and an analogue-to-digital converter (ADC) configured for the CT detector operation.

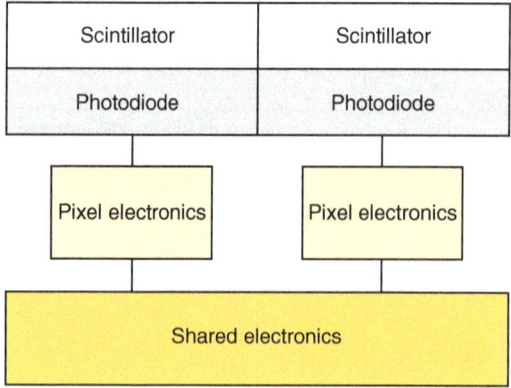

Fig. 15.17 General representation of CT detection

15.6.7 Semiconductor Detectors

Semiconductor detectors used in nuclear medicine are electronic pocket dosimeters and CZT-based gamma cameras. Pocket dosimeters are discussed in the early part of the chapter. Semiconductor detectors, also called solid-state detectors, operate like ion chamber detectors with more sensitivity and better energy resolution. Semiconductor detectors can also be of small size with faster timing characteristics and varying effective thickness to match the application's needs. Two drawbacks are small size and high susceptibility to performance degradation due to radiation-induced damage.

Silicon and germanium have been used extensively as semiconductor detectors. Silicon has been used for charged particle spectroscopy whereas germanium for gamma-ray spectroscopy. Several semiconductor compounds have been developed for radiation detection with slow developments. Examples include cadmium telluride (CdTe), cadmium zinc telluride ($Cd_{0.8}Zn_{0.2}Te$), mercuric iodide (HgI_2), gallium arsenide (GaAs), lead iodide (PbI_2), thallium bromide (TlBr), etc. The properties of some of these detectors are outlined in Table 15.4.

15.6.7.1 Mechanism of Detection

Si and Ge contain four electrons in their outer orbit; these electrons are called valence electrons. In a pure state of crystal (at $0°K$ and in the absence of any radiation), these four electrons are shared covalently with neighbouring four atoms leaving no free electron available for conduction. On receipt of energy by any means such as heat or radiation, the valency electron gets sufficient energy to leave its state and become free, which can drift throughout the crystal. This creates a positively charged atom called a 'hole'. The 'electron-hole pair' is called 'charge carriers' and is equivalent to an ion pair analogue of gases in a solid state. When an external electric field is applied, these charge carriers can be collected opposite each other. The material's conductivity is influenced by the motion of both of these charges. The energy band gap shown in Fig. 15.12 is typically about 1–3 eVs (<4 eV) for semiconductors whereas 5 eV or more for insulators.

At a temperature of more than $0°K$, some thermal energy is imparted to the electrons, which may provide sufficient energy to elevate electrons across the bandgap into the conduction band. But the control of the number of charge carriers created due to thermal excitation is feasible. A small concentration of impurity called a 'dopant' is used to tailor the properties of semiconductors.

When a pentavalent impurity (a few parts per million or less) such as arsenic (As), phosphorous (Pi), antimony (Sb), etc. is added to the crystal lattice, few normal silicon atoms are substituted. Because five valence electrons surround the impurity atom, one electron remains after all covalent bonds have been formed. This extra electron is loosely linked to the initial impurity site and dislodges with very little energy, resulting in a conduction electron with no hole. These impurities are known as donor impurities, and these semiconductors are referred to as n-type semiconductors (Fig. 15.18).

Similarly, when a trivalent impurity such as lithium (Li), aluminium (Al), boron (B), indium (In), etc. is added as an impurity, they have one less valence electron than the surrounding silicon atoms. As a result, one covalent bond remains unfilled. This vacancy makes a hole. This type of semiconductor is called p-type. In p- and n-type semiconductors, n- represents the concentration of electrons in the conduction band, whereas p- shows the concentration of holes in the valence band.

Table 15.4 Properties of semiconductor detectors. Table adapted from Knoll Glenn F, 'Radiation Detection and measurements' fourth edition (2010) [20]. Copyright John Wiley & Sons, Inc. Reproduced with permission of the Licensor through PLSclear

Material	Z	Density (g/cm³)	Bandgap (eV)	Ionization energy (eV/e-h pair)
Si	14	2.33	1.12	3.61
Ge	32	5.33	0.72	2.98
CdTe	48/52	6.06	1.52	4.43
HgI$_2$	80/53	6.4	2.13	4.3
CZT (Cd$_{0.8}$Zn$_{0.2}$Te)	48/30/52	6	1.64	5.0

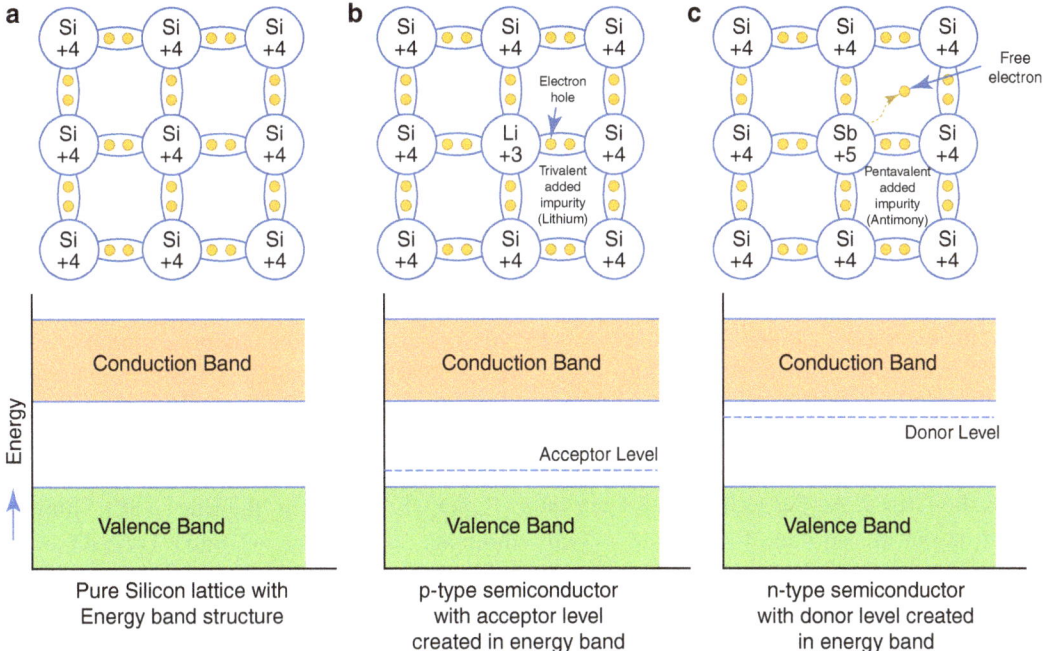

Fig. 15.18 Crystal lattice and energy band structure of intrinsic and extrinsic semiconductors

When radiation interacts with the semiconductor, energy deposition causes an equal amount of holes and electrons to form. The interaction probability for gamma rays is the same in a p-type or n-type semiconductor of equal thickness, and the range of charged particles also is the same in both types. The ionization energy of semiconductor detectors is low, giving them an advantage. The energy needed to create an electron-hole pair for silicon or germanium is around 3 eV, whereas for common gas-filled detectors, it is around 30 eV. Therefore, the number of charge carriers created 10 times higher for the same amount of energy deposited in the detector. As the overall number of charge carriers

increases, the statistical variation in carriers in each pulse is reduced.

When semiconductors are used as radiation detectors, an electric field is maintained throughout the active volume. Electrostatic forces are provided to both charge carriers, causing them to drift in opposite directions. The travel of electrons or holes generates a current that will last until the charge carriers are collected at the active volume's end.

15.6.7.2 Properties of CZT (Cd$_{1-x}$Zn$_x$Te) Detectors

CZT detectors are prepared by blending zinc telluride (ZnTe) into cadmium telluride (CdTe). This creates some favourable properties of detectors, including significantly higher resistivity. The formulation $Cd_{1-x}Zn_xTe$ contains x, which is a blending fraction and may vary from 0.04 to 0.20. The bandgap energies of the resultant semiconductors range from 1.53 eV to 1.64 eV at ambient temperature [21]. The intrinsic free carrier concentration is reduced when the bandgap widens.

Over a range of room temperatures, CZT-based gamma-ray spectrometers have significantly improved size and energy resolution. The average size of commercially available planar CdZnTe detectors ranges from 5 mm × 5 mm × 1 mm to 10 mm × 10 mm × 2 mm [21]. However, the spectroscopic performance is improved by reducing the detector thickness. These detectors are best for low-energy gamma rays but show poorly resolved gamma-ray peaks and strong 'tailing' effects for high-energy gamma rays.

The cost of organ-specific (cardiac) semiconductor-based gamma cameras is about 1.5–2 times that of general-purpose (GP) scintillation-based gamma cameras. In contrast, GP semiconductor-based gamma cameras are almost five times more expensive than GP scintillation-based. As the technology of crystal growth advances, we may expect some cuts in prices. The drawback of current semiconductor-based gamma cameras is that they can only be used for low-energy radioisotopes such as [99m]Technium and [201]Thallium. [131]Iodine scans cannot be performed with semiconductor-based gamma cameras. But their energy resolutions areas such as both isotopes [99m]Technium and [201]Thallium can be imaged simultaneously (Fig. 15.19).

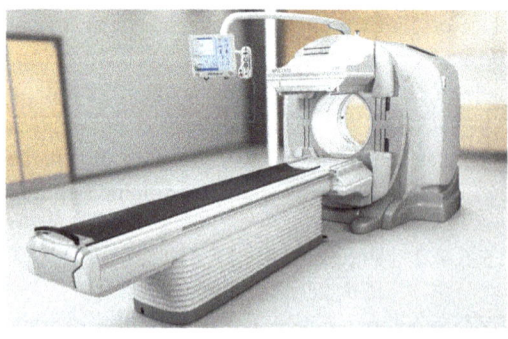

Fig. 15.19 CZT-based gamma camera (NM CT 870 CZT). Courtesy: GE Healthcare, printed with permission

15.6.8 Thermoluminescent Dosimeters (TLDs)

The thermoluminescent dosimeters (TLDs) are used for personal monitoring purposes and work on an entirely different principle than other detectors discussed above. They are passive detectors and provide reading when they are stimulated with heat. They are made up of insulating material doped with an impurity. The energy deposited by radiation is trapped at defect sites and released on heating in the form of visible light. This light can be converted into electrical signals and amplified using photomultiplier tubes. The amount of light emitted is proportional to the radiation absorbed.

As per the United Nations Scientific Committee on the Effects of Atomic Radiation (UNSCEAR) 2021 Report, about 11.4 million workers globally were monitored from 2010 to 2014, and the medical sector dominated with about 80% of the total [22]. Different countries use different materials for TLDs in the personal monitoring programme. Several materials such as calcium sulphate dysprosium activated ($CaSO_4$:Dy); lithium fluoride (LiF); lithium bromate manganese activated ($Li_2B_4O_7$:Mn); calcium fluoride dysprosium activated (CaF_2:Dy); calcium sulphate dysprosium and manganese activated ($CaSO_4$:Dy, Mn); calcium fluoride manganese activated (CaF_2:Mn); lithium fluoride manganese, copper and phosphorus activated (LiF: Mg, Cu, P); etc. have been tried to use as TLD material. India, Australia and Brazil use calcium sulphate dysprosium ($CaSO_4$:Dy)-activated TLDs because of their availability and better sensitivity. Lithium fluoride (LiF)-based TLDs are used in many parts of the world. The United States and many parts of Europe use LiF: Mg, Ti (commercially called TLD-100) and LiF: Mg, Cu, P (commercially called TLD-100H, TLD-600H, TLD-700H); Japan uses $Li_2B_4O_7$:Cu & $CaSO_4$:Tm; China also uses LiF: Mg, Cu, P (GR-200); and Russia uses LiF: Mn, Ti (DTG-4). Some parts of the USA use optically stimulated luminescence (OSL). Table 15.5 shows the properties of various TLDs used [23].

Table 15.5 Properties of some thermoluminescent materials. Table originally published in *Defect and Diffusion Forum Vol. 347 (2014) pp 179-227*. Reprinted with kind permission of Trans Tech Publications [23]

TLD type	Effective atomic number Z_{eff}	TL main peak (°C)	Emission maximum (nm)	Relative sensitivity[a]	Fading at 25 °C, when stored in the dark	Useful dose range
LiF: Mg, Ti	8.14	200	400	1	5%/year	20 µGy to 10 Gy
LiF: Mg, Cu, P	8.14	210	368	40	5%/year	0.2 µGy to 10Gy
LiF: Mg, Cu, Si	8.14	240	384	55	Negligible	1 µGy to 20 Gy
$Li_2B_4O_7$:Mn	7.3	220	605	0.40	4% monthly	0.1 mGy to 3 Gy
$Li_2B_4O_7$:Cu	7.3	205	368	8	10%/2 months	10 µGy to 10^3 Gy
MgB_4O_7:Dy/ Tm	8.4	190	490	67	4%/month	5 µGy to 50 Gy
BeO	7.1	190	330	≈1.00	8%/2 month	0.1 mGy to 0.5 Gy
Mg_2SiO_4:Tb	11	200	380–400	40–53	Very slight	10 µGy to 1 Gy
$CaSO_4$:Dy	15.3	220	480–570	30–40	1%/2 monthly	2 µGy to 10 Gy
$CaSO_4$:Tm	15.3	220	452	30–40	1–2%/2 month	2 µGy to 10 Gy
CaF_2:Mn	16.3	260	500	5	16%/2 week	10 µGy to 10 Gy
CaF_2 (natural)	16.3	260	380	23	Insignificant	10 µGy to 50 Gy
CaF_2:Dy	16.3	215	480–570	15	8%/2 month	10 µGy to 10 Gy
Al_2O_3:C	10.2	190	420	60	5%/year	0.1 µGy to 10 Gy

[a]As compared to LiF: Mg, Ti

TLD-100 (Thermo Fisher Scientific, formerly Harshaw Chemical Company) contains a natural abundance of ^6Li (7.5%) and ^7Li (92.5%) and responds not only to photons and electrons but also to neutrons, particularly thermal neutrons [24]. TLD 600 is sensitive to thermal neutron and gamma radiation, while TLD 700 is only sensitive to gamma radiation. TLD 600 is enriched with ^6Li, containing 95.62%. Already the TLD 700 is enriched with ^7Li, 99.99% [25]. The ^6Li has a high cross-section for neutrons than ^7Li, and therefore, the TLD 600 is more sensitive to neutrons than TLD 700 [25]. TLD-100 is used in most radiation clinics; however, it severely over-responds and does not produce a useful measure at higher energies [24].

15.6.8.1 CaSO₄:Dy TLDs

A simple and excellent method for preparing TLD-grade $CaSO_4$:Dy phosphor was reported by Yamashita et al. in 1968 [25]. $CaSO_4$:Dy (in the ratio of 50,000:1 or having 5×10^{-4} atom fractions of Dy per $CaSO_4$ molecule) is prepared by dissolving high purity calcium sulphate ($CaSO_4$) and dysprosium oxide (Dy_2O_3) in concentrated sulphuric acid (H_2SO_4) and then crystallizing out the product by distilling off the H_2SO_4 under reduced pressure. The crystallized product is annealed at the desired temperature and quenched to achieve the high sensitivity of the phosphor powder. Since this crystalline product cannot be prepared in pellet form, it is mixed with Teflon. A homogeneous mixture having $CaSO_4$:Dy phosphor powder (grain size <75 pm) and Teflon plastic resin (7A grade) in a weight ratio of 1:3 is formed to make discs weighing 280 mg each with a size of 0.8 mm thickness and 1.35 cm diameter. They are first cold-pressed and then given a 400 °C for 1-h heat treatment to be strong and flexible. These discs are mounted with mechanical clips on a nickel-coated aluminium plate having three circular holes of 12 mm diameter. In India, the commercial production is done by M/s Renentech Laboratories Pvt. Ltd., Mumbai, with 600 g/batch [23]. The Bhabha Atomic Research Centre (BARC) designed TLDs consisting of three well-defined areas:

- *Top portion*: A circular metal filter having a diameter of 13 mm on the front cover and 12 mm on the back cover with 1 mm aluminium and 1 mm copper filter, sandwiched together (the copper filter is closer to the TLD disc).
- *Middle portion*: A rectangular plastic filter size 30 mm × 20 mm on the front cover and a circular filter on the back cover having a diameter of 25 mm.
- *Lower portion*: A circular open-window region having 15 mm diameter in the front cover and 13 mm diameter in the back cover.

These TLDs can monitor beta, gamma and X-radiations and cover a wide range of doses from 0.1 mSv to 10 Sv [26]. They have high sensitivity, low fading, indigenous production, and many other useful characteristics. The readout time for these TLDs is 100 s/badge (Figs. 15.20, 15.21, 15.22, and 15.23).

Mechanism of Detection

The scintillation detectors emit a light photon when exposed to the radiation in the form of prompt luminescence. The electrons and holes are created by radiation, recombine at activator sites and release light photons, which are detected. Though electrons and holes are created in TLDs, too, these are trapped deep inside the energy gap between valence and conduction bands at luminescence and trapping centres. When thermally or optically stimulated, they release light photons in proportion to the radiation absorbed. This property allows the integration of the radiation absorbed dose in a certain time interval (Fig. 15.24).

TLDs are prepared with materials that have high concentrations of trapping centres (TCs) and luminescence centres (LCs) with energy levels within the bandgap. During exposure, the material produces no actual signal but simply builds up a population of trapped charges whose number is related to the energy deposited by the incident radiation. Since no external voltage is applied to TLDs, the electrons migrate randomly in the conduction band. Some are captured at

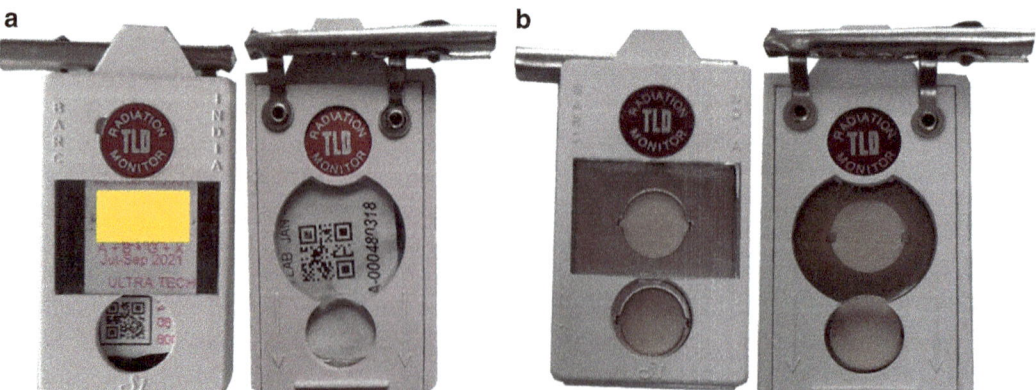

Fig. 15.20 (**a**) Thermoluminescent dosimeters from front and back. (**b**) The dosimeter inside TLDs' cover interacts with radiation in such positions. This image is for demonstration purposes only, and no one is advised to take the TLD out of the plastic cover

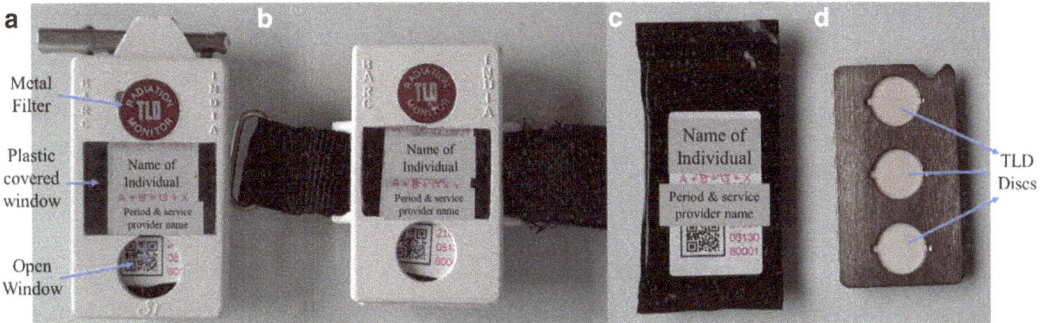

Fig. 15.21 Thermoluminescent dosimeters. (**a**) Chest TLD with three windows. (**b**) Wrist TLD. (**c**) TLDs received from the vendor in plastic cover with the name of individuals. (**d**) TLD discs mounted on 1 mm aluminium plate

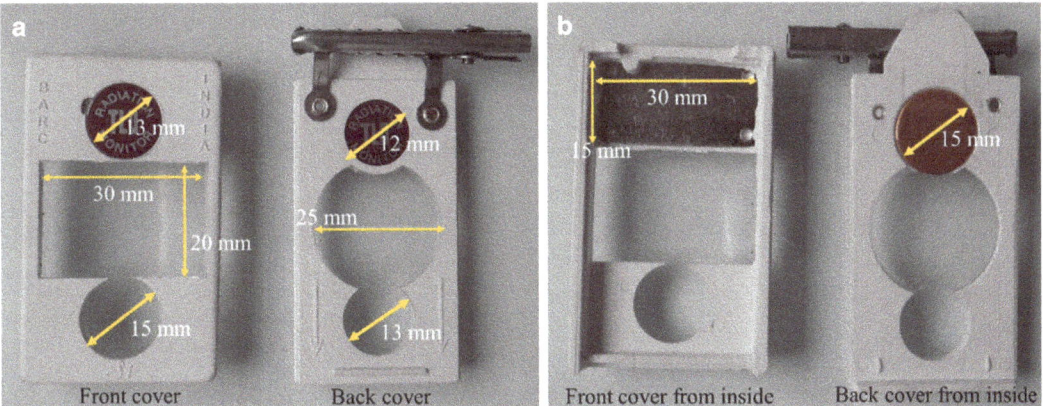

Fig. 15.22 Thermoluminescent dosimeters cover dimensions from front and back. (**a**) From outside. (**b**) From inside

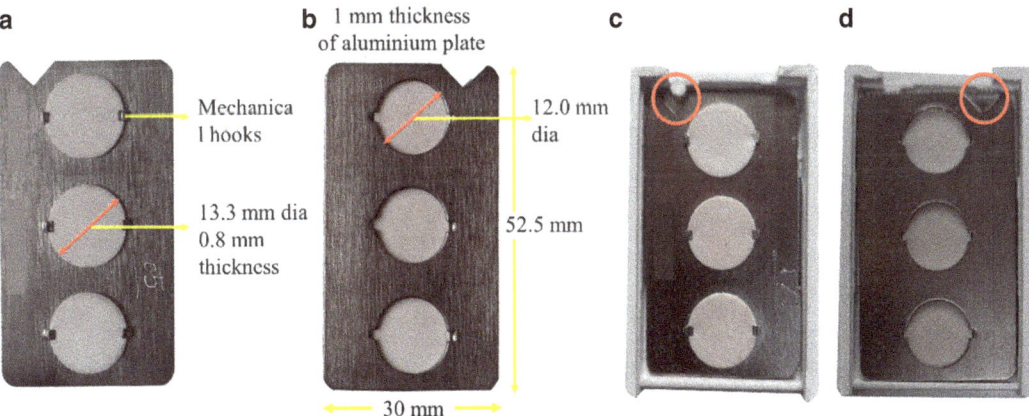

Fig. 15.23 (**a**) and (**b**): Thermoluminescent pellets mounted on an aluminium plate have mechanical clips with various dimensions. (**c**) Correctly placed TLD inside the TLD cover corresponding to V notch provided in plastic casing. (**d**) Incorrectly placed TLD in TLD cover, a complete mismatch

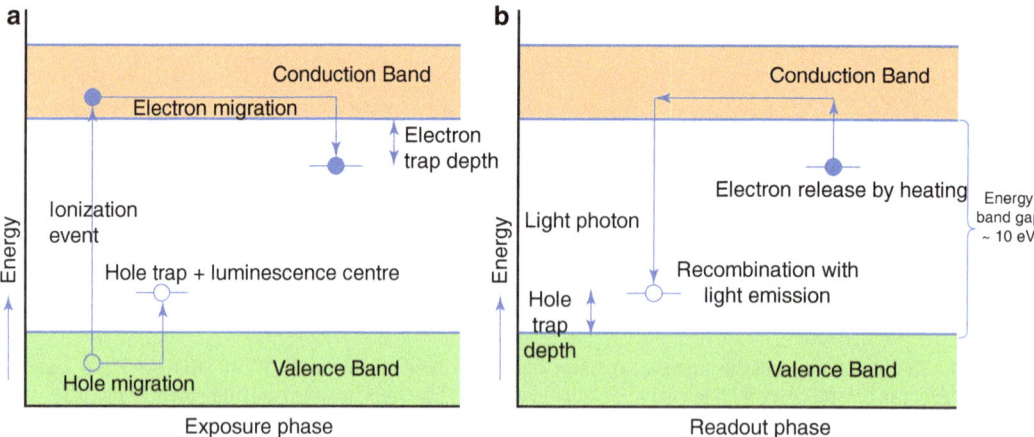

Fig. 15.24 TLDs' energy band

various trapping centres with energy levels in the bandgap. The trapped charges are released through thermal or optical stimulation and generate photons at luminescence centres in a separate readout period. Their number reflects the deposited energy integrated over the exposure period.

The readout of TLDs is generally done at the temperature range from 175 to 250 °C (TL peak of CaSO$_4$:Dy Teflon Disc occurs at about 240 °C) in a specially designed TL reader, which is commercially available. The readout at 300–400 °C (thermal oven) or post-readout annealing depletes all TCs and LCs, and the exposure record of the sample is thus 'erased'. Thus, TLD materials can be recycled, and a single sample may be reused many times without a significant change in insensitivity [27].

15.6.8.2   Disadvantages of CaSO$_4$:Dy TLDs

- The main drawback occurs in β-dosimetry.
- The dosimeter disc is thick, and the TL response is very low for β-rays with E_{max} less than 0.8 MeV.
- It does not measure neutrons.

References

1. Knoll G. Ionization chambers. In: Radiation detection and measurement. 4th ed. New York: Wiley; 2010. p. 159/857.
2. Rajan JIG. Radiation monitoring instruments. In: Radiation oncology physics: a handbook for teachers and students. Vienna: International Atomic Energy Agency; 2005. p. 101–22.
3. Attix FH. Introduction to radiological physics and radiation dosimetry. Weinheim: Wiley-VCH Verlag GmbH & Co. KGaA; 2004. p. 332.
4. Cherry SR, Sorenson JA, Phelps ME. Radiation detectors, Chapter 7, para A.2. In: Physics in nuclear medicine. London: Saunders; 2012. p. 91. ISBN: 9781416051985.
5. Frame P. https://www.orau.org. Oak Ridge: Oak Ridge Associated Universities [Online]. https://www.orau.org/health-physics-museum/collection/dosimeters/pocket/index.html. Accessed 20 Sept 2021.
6. Knoll G. Ionization chambers. In: Radiation detection and measurement. 4th ed. New York: Wiley; 2010. p. 169/857, Para IVA.
7. Kathren RL, Howell WP. Calibration and field use of ionisation chamber survey instrument. BNWL-SA-2096, 1969.
8. Gun monitor. PLA Electro Appliances Pvt. Ltd. [Online]. http://plaelectro.com/survey/med/gm125.pdf. Accessed 10 Aug 2021.
9. Model 375-31H, area monitor with neutron detector. https://ludlums.com/. [Online]. https://ludlums.

com/products/all-products/product/model-375-31h. Accessed 25 Jan 2022.

10. Tsoulfanidis N, Landsberger S. Measurement and detection of radiation. 4th ed. West Palm Beach, FL: CRC Press; 2015.

11. Knoll G. Scintillation detector principles, Chapter 8. In: Radiation detection and measurement. 4th ed. New York: Wiley; 2010. p. 263/857.

12. Wikipedia. The Wikimedia Foundation, Inc. [Online]. https://en.wikipedia.org/wiki/Geiger_counter. Accessed 5 Apr 2022.

13. den Hollander W, Kolar ZI. A centennial of spinthariscope and scintillation counting. Appl Radiat Isot. 2004;61(2-3):261–6.

14. Knoll G. Scintillation detector principles. In: Radiation detection and measurement. New York: Wiley; 2010. p. 263/857.

15. Overdick M. Detectors for X-ray imaging and computed tomography. Chapter 4. In: Advances in health care technology. New York: Springer; 2006.

16. Duclos SJ, Greskovich CD, Lyons RJ, Vartuli JS, Hoffman DM, Riedner RJ, Lynch MJ. Development of the HiLightTM scintillator for computed tomography medical imaging. Nucl Instrum Methods Phys Res A. 2003;505(1-2):68–71.

17. Chandra DLN. Gemstone detector: dual energy imaging via fast kVp switching. Med Radiol. 2010:35–41.

18. Stefan Ulzheimer JF. The Stellar detector, first fully integrated detector. Erlangen: Siemens Healthcare GmbH; 2016.

19. Lecoq P. Development of new scintillators for medical applications. Nucl Inst Methods Phys Res A. 2016;809:130–9.

20. Knoll G. Other solid-state detectors, chapter 13, para IIIB. In: Radiation detection and measurement. New York: Wiley; 2010. p. 519/857.

21. Knoll G. Other solid-state detectors, chapter 13, para IID. In: Radiation detection and measurement. New York: Wiley; 2010. p. 522–4. /857.

22. United Nations. Report of the United Nations Scientific committee on the effects of atomic radiation. Sixty-seventh and sixty-eighth sessions. New York: United Nations; 2021.

23. Bhatt BC, Kulkarni MS. Thermoluminescent phosphors for radiation dosimetry. Defect Diffus Forum. 2014;347:179–227.

24. Kry SF, Price M, Followill D, Mourtada F, Salehpour M. The use of LiF (TLD-100) as an out-of-field dosimeter. J Appl Clin Med Phys. 2007;8:169–75.

25. Cavalieri TA, Castro VA, Siqueira PTD. Differences in TLD 600 and TLD 700 glow curves derived from distict mixed gamma/neutron field irradiations. International Nuclear Atlantic Conference, volumes. November 24–29, 2013.

26. Huda W. Thermoluminescent phosphors for radiation dosimetry. Defect Diffus Forum. 2014. Trans Tech Publications, Switzerland.

27. Knoll G. Miscellaneous detector type. In: Radiation detection and measurement. New York: Wiley; 2010. p. 778–81.

Quality Control of Planar Gamma Camera and Single-Photon Emission Computed Tomography

16

Abstract

Quality control of different instruments of nuclear medicine is very important for radiation safety and is an integral portion of scheduled work. There are distinctive events and frequencies to carry out different quality control tests. A comprehensive set of acceptance tests are carried out at the time of receipt of any instrument, which shapes the premise of the successful quality control program and serves as a reference for future estimations. Based on the reference information, subsequent testing schedules such as weekly, month-to-month, quarterly and yearly can be planned. Flood field uniformity is performed and assessed on a day-to-day basis, whereas the spatial resolution is tried once a week. Evaluating parameters such as multiple window spatial registration (MWSR), collimator uniformity, system sensitivity and camera 'dead time' is done less frequently than other parameters. For single-photon emission computed tomography frameworks to perform optimally, further testing must be performed.

The centre of rotation calibration and detector registration confirmation are carried out to avoid spatial resolution fluctuations. Non-uniformities are to be removed, and noticeable collimator faults are corrected using a flood uniformity correction map based on high-count images. When employing heart and brain software, the size of the pixels is to be adjusted to eliminate attenuation and distortion correction, which would otherwise occur. Patients' movements and partial views should be examined for unsatisfactory clinical trials after completion. Finally, daily functioning checks are performed on the instrument utilized. All of these tests should be well documented, and if the findings show that anything is inadmissible, the proper remedial action should be performed immediately. It is important to note that these quality control tactics do not discourage the use of routine preventive support approaches, which must be performed regularly. Such a program's effectiveness depends on the participants' ability to recognize and accept their problems. Apart from clearly defined roles and duties, compliance with test schedules, suitable methods and processes for tracking and reporting test findings is also necessary.

16.1 Introduction

The dose of patient radiation and image quality are primary concerns in conducting nuclear medicine procedures. Understanding the image patterns and quantitative outcomes is essential for interpreting nuclear medicine image data. As previously stated, this is also true for both quality control and clinical research. Such insight is gained via observation of many instances and in-depth knowledge of the underlying principles and how to attain the best possible outcomes.

Recognizing quality degradation and artefacts in clinical images is essential as regular system performance checks with quality controls. Due to differences in the patient group, progressive degradation in image quality information in clinical images may be difficult to identify.

Recognizing quality deterioration and artefacts in clinical pictures is equally important as doing routine system quality control tests. Progressive picture quality decline in clinical imaging may be difficult to detect due to disparities in patient groups.

The National Electrical Manufacturers Association's methodologies and procedures are generally regarded for providing a description, accurate measurement and camera performance parameters which are reported and utilized to gather and analyse quality control tests.

Uniformity, resolution, collimation and hard copy equipment are all aspects that influence the final quality. Other elements, such as count rate capabilities, also play a role in certain investigations. The centre of rotation of the system, hole alignment, rotational stability of the detector head and the integrity of the reconstruction algorithms are all new characteristics that might impact clinical results with the inclusion of tomographic imaging. A reasonable amount of time may be spent executing system QC daily. Consequently, the QC program's main purpose should be to keep track of crucial parameters.
(a) System performance changes.
(b) Daily   uniformity   tests   (extrinsic   or intrinsic).

16.2   Acceptance and Reference Tests

After installation and before the equipment is used in a clinical setting, it must complete comprehensive acceptability testing. The goal is to ensure that the equipment meets its standards and serves its clinical goal. Each instrument comes with a set of base specifications, which conform to standard test procedures by NEMA (National Electrical Manufacturers Association) and IEC (International Electrotechnical Commission) [1–8]. With the supplier's support and the necessary software phantoms, basic performance data can be created. Additional testing is usually also required to follow country-specific requirements.

16.3   Routine Quality Control

After an instrument is approved for use in a clinic, it should be checked regularly using basic QCs that are sensitive to performance changes. Testing should be carried out by suitably qualified staff. All test findings should be documented, and any deviations from the reference value should be noted and necessary action taken. QC testing is a crucial element of the routine, and it should be given appropriate equipment and staff time.

16.4   Action Thresholds, Follow-Up, Record Keeping, Review, and Monitoring

The findings of the tests should be recorded in a logbook. Because the test values must be compared to calculate the action criteria for that test, a fast evaluation of the QC findings is necessary. The manufacturer's suggestions and expert guidance should be considered when determining action levels. The action taken in response to the issues highlighted and proposed remedies must be documented. If a similar situation arises, such records may be useful in finding the trouble.

16.5   Daily Operating Care and Maintenance for a Scintillation Camera

1. Create a logbook to keep track of daily temperature checks.
2. Keep the room humidity as low as possible, preferably below 50%, or as recommended by

the equipment manufacturer. Keep the environment dust-free. Humidity, particularly at higher temperatures, owing to fungal development, which causes current leakage, changes the operating properties of electronic components. This impact is amplified considerably because of dust.

3. Avoid blowing air directly onto the instruments from the air-conditioning blower.
4. For a stable electric supply, use a UPS (uninterruptible power supply). When not in use, fluctuations in the power supply may reduce the gain of the photomultiplier and monitor intensity to a bare minimum.
5. At night, turn off the monitor.
6. Do not leave the crystal unattended. Keep a collimator attached to the detector at all times.
7. Avoid storing radioactive materials on top of the collimator. Before connecting it to the detector, inspect it for contamination.
8. Examine the installation of the collimator and the detector head and any damage to the collimator.
9. Every emergency button should be tested to confirm that all gantry motion is stopped when each button is pressed. Each time the collimators are changed, the touchpads or touch a safety device must be checked.

16.6 Preparatory Steps

Several steps should be taken before beginning acceptance testing or annual testing. Arrange radionuclides in the hot lab, fill the phantoms, check the gamma camera and SPECT systems for operation, and position the source and/or phantoms for testing.

16.7 Preventive Maintenance and Calibrations

A maintenance team should do preventive maintenance before QC to ensure that mechanical movements and collimator exchangers are working properly, photomultiplier tubes (PMTs) are properly tuned, and energy and linearity adjust-

ments are within specifications. Before QC, an engineer or physicist/technologist should perform calibrations.

16.8 Radionuclides for Testing

Determine how much of each nuclide requires for the necessary tests. ^{99m}Tc, ^{57}Co, ^{123}I, ^{67}Ga, ^{75}Se and ^{131}I are the most often utilized radionuclides. The number of actions conducted by the sources depends on what is to be completed.

16.9 Test Equipment and Manuals

The availability of phantoms and manufacturer operator's manuals for the equipment should be ensured. A ^{57}Co sheet source or ^{99m}Tc fillable flood phantom, a quadrant bar phantom, COR Phantom (if it is available), and a SPECT phantom should be available in the department. After doing the QC test for decays, there should be room to store test equipment, and one should also be prepared for any possible contamination during testing.

16.10 Physical Inspection

16.10.1 Physical Condition

Pay attention to the following features in the camera system:

1. Dents, scratches, falling covers, paint damage and other mechanical faults are all examples of mechanical problems.
2. The state of the air filters in the gamma camera equipment.
3. No noise or abnormal movements in the detector or table.
4. Determine and document the status of all collimators. Check that the collimators install and unmount easily.
5. Inspect the status of all computers and display systems attached to the camera system, and note any operational issues. Check the radial detector movement, gantry rotation

and table movement by doing the homing position.

6. Check the gantry rotational limit:
   (a) Check the lowest gantry rotational angle.
   (b) Check the highest gantry rotational angle.
   (c) Determine the gantry rotational limit:
       • Determine the smallest gantry rotating angle.
       • Determine the maximum gantry rotating angle.
       • Examine the mechanical movement of the gantry by checking for regular speed consistency and examining for vibrations and mechanical sounds.
       • Verify that the gantry stops properly after the rotation. Repeat the operation in the opposite rotational direction.
7. Examine the radial limit by measuring the maximum and minimum detector radii.
8. Examine the maximum and minimum heights of the table.
9. Examine pallet electrical locks.
10. Examine body contouring.
11. Determine the pallet's overall length.
12. Check the detector's angular location by holding the bubble level over the head. Repeat for various detection systems. Check for physical vs software detector rotation radius.
13. Compare the actual detector rotation radius to the software detector rotation radius.
14. Examine the rotation speeds in both the clockwise and anticlockwise directions. Take track of the time as you acquire the tomography study in an anticlockwise orientation. Repeat the tomography study in a clockwise orientation, noting the time elapsed. The elapsed time should be the same in both investigations.
15. Inspect the wires to ensure they are not twisted or damaged when the system spins.
16. Inspect the power lines and wiring connections.
17. Examine the bed cable.
18. Examine the connections for the hand controls.
19. Examine the ECG cable and connections.

16.10.2 Safety Interlocks

Assess the operational status of security and interlock devices during acceptance

• Check emergency stop switches.
• Check patient safety devices for touch sensitivity.

16.10.3 Camera Detector Shielding

• The manufacturer shall assess the adequacy of the shielding of the detector. Use a 74 MBq ^{99m}Tc source in a lead cylinder for this purpose.
• Figure 16.1 and Table 16.1 show how to place the source in six different places around the detector head and record the count for 100 s at each site.
• Remove the source and count the background, B, simultaneously, which we used to measure the source at different places.
• Using the following formula, compute the percentage leakage at each location:

$$\%\text{Leakage} = \left(\text{max counts at each position} - \text{background}\right) / \text{max counts in front of detector} - \text{background}.$$

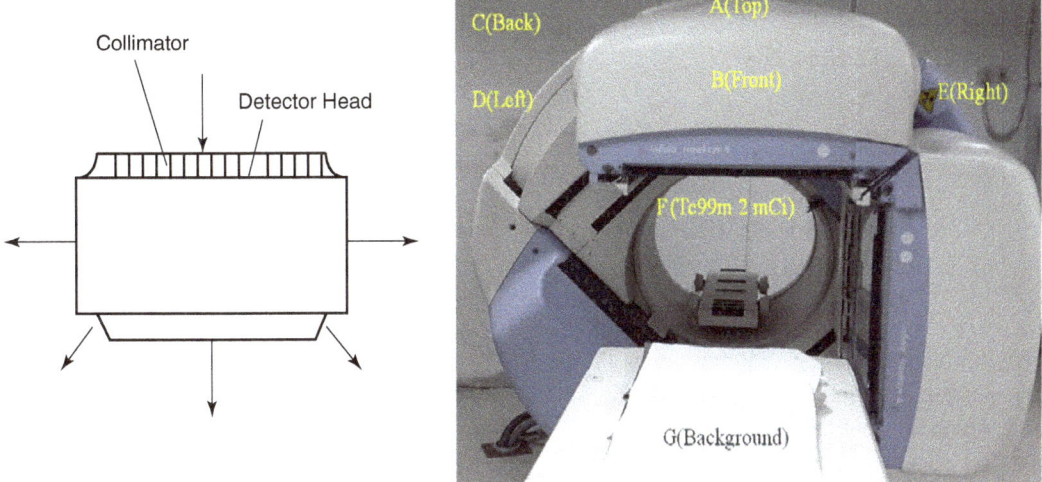

Fig. 16.1 Test of detector head shielding leakage. Six sites around detector head shielding to position point source to test for shielding leakage

Table 16.1 Test of detector head shielding leakage

Point of measurement	Detector 1 counts	Detector 2 counts
A (top)	2102	1978
B (front)	2019	2019
C (Back)	2048	2052
D (left)	2139 (maximum count)	2066
E (right)	2113	2072 (maximum count)
F (Tc99m)	44,000	42,000
G (background)	1953	1969

Example:

$$\text{Percentage of head leaking}\left(\text{Detector head 1}\right) = \frac{\left[2139\left(\max\right) - 1953\left(\text{BG}\right)\right]}{44000} = 0.42\%$$

$$\text{Percentage of head leaking}\left(\text{Detector head 2}\right) = \frac{\left[2072\left(\max\right) - 1969\left(\text{BG}\right)\right]}{42000} = 0.245\%$$

Examine the detector shielding visually for any signs of damage. If the damage is detected, a leakage survey should be done.

16.11 Computer Monitor Inspection: Monitors Used for Image Processing and Image Interpretation

The performance of the displays used by physicians for nuclear medicine interpretation and the screens used on computer stations directly connected with the gamma camera system should be checked regularly. Check the displays for dust, dead pixels and other artefacts. A physicist may evaluate annual testing. The following items should be considered in these assessments:

- Measure the monitor's maximum and lowest brightness in units of cd/m^2 in the presence of ambient light. In its Quality Control of Task Group 18, the American Association of Physicists in Medicine gives testing methods for medical displays.
- TG18 patterns include reflection, geometric distortion, brightness, spatial and angular luminance dependencies, resolution, noise, glare, chromaticity and artefacts.

Geometric distortions are measured using the TG18-QC linear test pattern, which should provide distortion coefficients of less than 2%/50% for primary/secondary displays, respectively. Check that the 5% and 95% patches are visible in the 0% and 100% brightness squares, as shown in Fig. 16.2. Find the smallest bar size in each test pattern's four

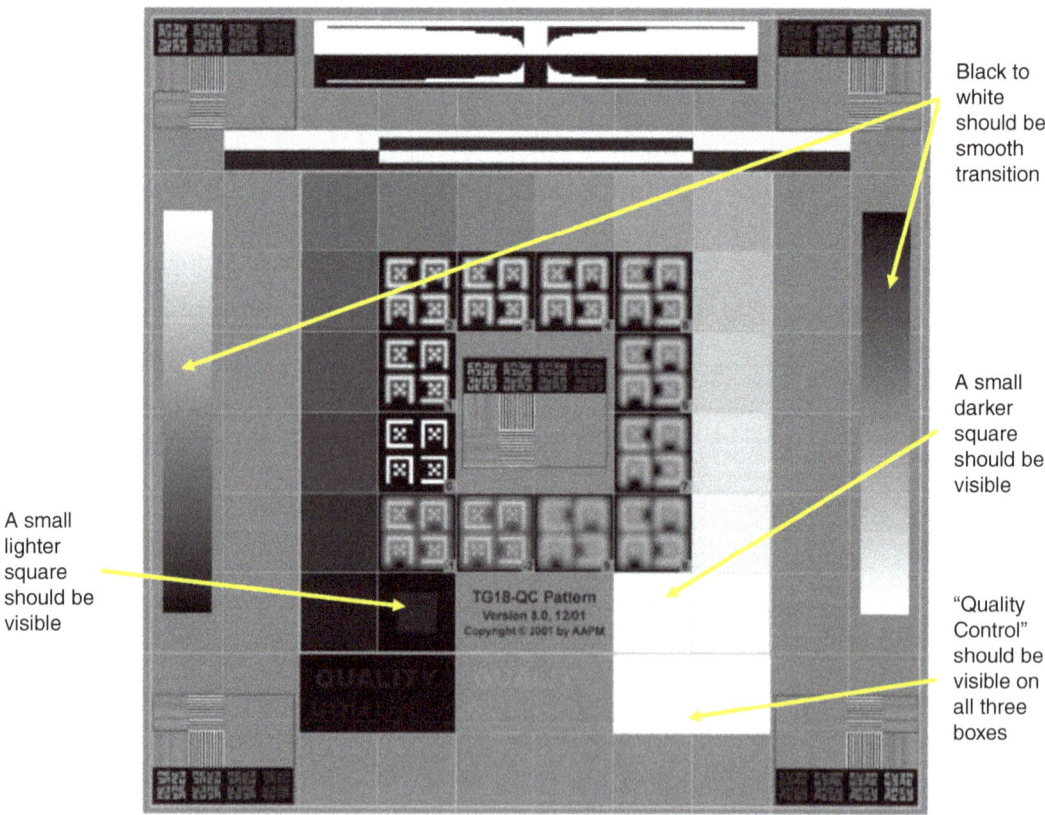

Fig. 16.2 Assessing the AAPM TG-18 QC test image. (Image copyright, The American Association of Physicists in Medicine, AAPM TG18 report, 2005 [9]. Reprinted with permission)

corners to determine spatial resolution. Identify any distortion of exhibited items over the full test pattern to determine geometric linearity. Identify how many 'Quality Control' letters are shown if the TG18QC test pattern is applied.

Report the following:

- The identity and position of nuclear monitors.
- Examining the lighting conditions in the room, particularly brightness. Any specular reflection seen on the screen should be reported.
- Workstation type, display type and grayscale calibration of the monitor.
- Specify the maximum and minimum luminance, and luminance, for each monitor.
- Test pattern, visibility of test patch at 5% and 95%, spatial resolution and geometric linearity are all parameters to consider. Indicate how many 'Quality Control' letters are shown if the TG18QC test pattern is applied.
- The ACR Technical Standard9 specifies that the depth of brighter areas, i.e. maximum luminance be a minimum of 350 cd/m² for diagnostic monitors. In contrast, monitors used for different purposes should have at least 250 cd/m². The depth of darker vicinity, i.e. minimal luminance, needs to be a minimum of 1.0 cd/m² for diagnostic monitors, whereas monitors used for other functions need to have at least 0.8 cd/m². It is also sug-

gested that the intermediate grey values be specified using the DICOM grayscale display characteristic (GSDF).

On the other hand, nuclear medicine differs from the other modalities in that its spatial resolution, signal-to-noise ratio and contrast-noise ratio are all much lower. Given the increased noise in nuclear medicine images, it is fair to assert that small intensity differences in other modalities are all that is required. Furthermore, a significant portion of nuclear medicine diagnostic data is transformed from intensity levels to colour ranges before being shown with a limited number of visicolors. According to the task group, nuclear medicine displays should have a maximum brightness of 120 cd/m², a minimum luminance of 2 cd/m² and a luminance non-uniformity of less than 20%. Compared to radiography reading rooms with view boxes, rooms with softcopy monitors may have brighter ambient lighting of 20–40 lux [10, 11].

16.12 Acceptance Reference Tests

When acquiring equipment, the most important stage is to do acceptance testing. Instruments that fail to fulfil the criteria should be rejected. Because patient safety is more important, acceptance testing should be undertaken before using the device in clinical settings. If tests fail to meet requirements, proper action should be taken. A set of acceptance tests is shown in Table 16.2.

Table 16.2 List of acceptance tests of SPECT/CT

Physical inspection	Cables, collimator and safety locks
Shielding leak test on a gamma detector	Need quantitative assessment
Uniformity of the extrinsic flood field	Apply to all multihole collimators
Uniformity of the intrinsic flood field	^{99m}Tc and other clinical radionuclides
Intrinsic uniformity in the off-peak	To get baseline images for each detector
Linearity and intrinsic spatial resolution	Slit phantom or quadrant bar phantom for ^{99m}Tc
Spatial resolution (extrinsic)	For entire low-energy collimators with a line source, ^{99m}Tc or ^{57}Co is used, as well as a quadrant bar phantom image with a low-energy collimator utilizing a ^{99m}Tc or ^{57}Co flood source

(continued)

Table 16.2 (continued)

Planar extrinsic sensitivity	At least one low-energy collimator should be ^{99m}Tc
Energy resolution	^{99m}Tc
Performance of the intrinsic count rate	At a distance, assess the peak count rate. Optionally, you may use the two-source technique using ^{99m}Tc to determine the dead time
Spatial registration with several windows (MWSR)	Optional
The spatial resolution of SPECT	^{99m}Tc is used as a line source for the measurement
SPECT gantry tilt and Centre of rotation	Analyse ^{99m}Tc projection images using a point or line source
SPECT phantom	Phantom with ^{99m}Tc and low-energy collimator and CT attenuation correction for hybrid SPECT/CT
Spatial registration of hybrid SPECT/CT	The manufacturer's recommended procedure
CT dosage in hybrid SPECT/CT	CTDI dose measurement for SPECT/CT techniques that are routinely utilized
The image quality of CT in hybrid SPECT/CT	ACR CT phantom is used to assess the picture quality of CT scans

16.13   Routine Tests

An instrument's acceptance and reference tests are required for the first and comprehensive examination of the instrument. Suppose a baseline of performance has been established for the system. In that case, frequent, elaborate and thorough quality control schedules are no longer necessary, given that the system is performing as we anticipate. Certain routine tests must be done regularly to guarantee that equipment continues to work properly over time. When choosing the most suitable regular tests, consider the instrument, how it uses in the clinic and the resources available at the institution. To verify that the equipment is operating as planned, do the following checks regularly:

Check the Integrity of the Detector Head and Collimator Mounting

(a) Verify the peak and window parameters of the analyser for the radionuclide in use.
(b) The homogeneity of the system.
(c) The operation of the camera scope and hard copy device or printer.
(d) The backdrop in the gamma camera room.
(e) The temperature of the room.

The lists of routine testing of the gamma camera/SPECT are shown in Table 16.3.

Table 16.3   Routine testing of gamma camera/SPECT

Energy peaking	The radionuclide that is used should be peaked every day before use
The uniformity of the extrinsic flood field	Low-energy collimators are used every day, while other collimators are used every 3 months
The uniformity of the intrinsic flood field	You may use it daily as an alternative to daily extrinsic flood field uniformity if you don't have a ^{57}Co flood source than use ^{99m}Tc-fillable phantom
Spatial resolution (extrinsic)	Once a week, use a quadrant bar phantom and a ^{57}Co flood sheet source or ^{99m}Tc-fillable phantom
Spatial resolution (intrinsic)	Weekly intrinsic spatial resolution may be an alternative to weekly extrinsic spatial resolution
Centre of rotation	Monthly or weekly
Jaszczak SPECT phantom	Every 3 months, use a ^{99m}Tc source and a low-energy collimator
	A ^{99m}Tc phantom with a low-energy collimator and CT attenuation correction is used to provide hybrid SPECT/CT phantom imaging
Spatial registration of hybrid SPECT/CT	Manufacturer recommended protocol

16.14 Periodical Tests

Routine tests should regularly be done on an instrument to verify that it operates at peak performance. Periodical tests are:

(a) Tests that have already been conducted as benchmarks and are now repeated on a weekly, monthly, quarterly and annual basis.

Daily or operational tests must be performed each day before the radionuclide may be utilized in a clinical setting, such as a uniformity test and an energy window for the radionuclide to be used. To remedy the failure, certain tests that are not routinely performed in quality control may need to be redone during preventative maintenance or after corrective maintenance. As a result, it's vital that these tests are conducted consistently using the same protocols and that the results are always compared to the reference data. Table 16.4 lists the periodic tests performed with a gamma camera/SPECT system.

Table 16.4 Annual testing of gamma camera/SPECT

Physical inspection	Cables, a collimator and safety latches are all included
Gamma detector shielding leak test	Need evaluation on a qualitative level
External flood field uniformity	Test all multiple hole collimators for ^{99m}Tc and other radionuclides used in clinical practice
Intrinsic uniformity (flood field)	^{99m}Tc and other radionuclides used clinically
Intrinsic uniformity (off-peak)	To obtain benchmark images for each detector
Intrinsic (spatial resolution and linearity)	Slit phantom or quadrant bar phantom using ^{99m}Tc
Spatial resolution (extrinsic)	All low-energy collimators with line source and quadrant bar phantom with low-energy collimator using either ^{99m}Tc or ^{57}Co
Planar sensitivity (extrinsic)	^{99m}Tc for at least one low-energy collimator
Energy resolution	By using source ^{99m}Tc

Table 16.4 (continued)

Intrinsic count rate performance	The peak count rate should be measured from a distance. Optionally, the dead time may be measured using the two-source approach, ^{99m}Tc
Multiple window spatial registration	Optional
SPECT spatial resolution	^{99m}Tc was used as a line source for the measurements
SPECT centre of rotation and gantry tilt	^{99m}Tc projections with a point or line source are used to evaluate the results
SPECT phantom	With phantom, ^{99m}Tc and a low-energy collimator are used The ^{99m}Tc phantom equipped with low-energy collimator and CT attenuation correction is used for hybrid SPECT/CT imaging
SPECT/CT spatial registration	The manufacturer's recommended mode of operation
CT dose in hybrid SPECT/CT	Measurement of CTDI dosage for routinely used SPECT/CT techniques is provided
CT image quality in hybrid SPECT/CT	ACR CT phantom is used to assess the quality of CT images

16.15 Gamma Camera Planar Tests

Planar tests are used to assess the overall performance of the gamma camera detectors and collimators that make up the system's hardware components. When using a multi-detector system, the planar checks should be completed for every detector and collimator combination, depending on the clinical applications specified for the camera system. The exams are divided into two categories: intrinsic and extrinsic. Intrinsic checks are done without using a collimator installed and are intended to assess the detector's performance solely. Extrinsic checks are those that use both the collimator and the detector together. This is referred to as the useful field of view (UFOV) and the detector's central field of view (CFOV). The UFOV is the complete region usable for imaging in gamma imaging. Its size is specified by the manufacturer and is typically 95% of entire FOV. CFOV, on the other hand, is 75% of the entire FOV.

16.15.1   Flood Field Uniformity

The flood field uniformity of a scintillation camera refers to the camera's capability to form a homogenous pattern once exposed to an activity with a uniform distribution. Once it involves flood field uniformity, a detector with no collimator is brought up as 'intrinsic uniformity', whereas a detector with a collimator is 'extrinsic uniformity'. For clinical imaging, the uniformity of the extrinsic flood field, measured with a sheet source or fillable flood phantom, may be the only useful calibration method. The alignment of the collimator holes over the rectangular crystal array is likely to generate aliasing artefacts, which is eliminated during the extrinsic calibration process. Once the alignment is complete, individual crystal energy response and sensitivity are assessed, the dead crystals are found, crystals and the photomultiplier tube (PMT) are assessed for replacement, and intrinsic uniformity testing is performed. Furthermore, an uncorrected flood correction intrinsic flood field uniformity test and an asymmetric energy window on both the upper and lower sides of the photopeak window may be utilized to assess the location and number of dead crystals and the existence of the hydration effect. Non-uniformity of response may be caused by several interrelated issues, including drift in photomultiplier tube (PMT) gain, broken crystal, incorrect calibration energy and optical coupling (failure of one or more than one PMT), spatial non-linearity, crystal degradation and a high-count rate.

16.15.1.1   Gamma Camera Detector Setup and Source Placement

Energy Window Width and Peak
Table 16.5 lists radionuclides used in nuclear medicine and their energy windows and levels. The energy level is the centre of the energy win-

Table 16.5 Energy and energy windows for commonly used radionuclides

Radionuclide	Gamma energy #1		Gamma energy #2	
	Level (keV)	Window (%)	Level (keV)	Window (%)
^{201}Tl	70	20	167	20
^{57}Co	122	20		
^{99m}Tc	140	15/20		
^{123}I	159	20		
^{131}I	364	20		

dow. Check whether the energy peak is aligned with the centre of the proper window. Rather, the energy level should be adjusted to ensure that it is distributed equally throughout the system. For the majority of cameras, this is a standard procedure. Keep track of the energy level used and which windows were used. Drift in photomultiplier tube (PMT) gain, cracked crystal, improper calibration energy peak, optical coupling, failure of one or more PMT, spatial non-linearity, crystal deterioration and high-count rate are just a few of the interconnected issues that can lead to non-uniformity in measurement results (Fig. 16.3).

16.15.1.2   Test of Intrinsic Flood Field Uniformity
Using a constant radioactive source, the word uniformity refers to the fluctuations in intensity that may be seen in images captured using that source. It is possible to categorize the quantification of homogeneity into two categories:

First, the integral uniformity of the count intensity over the UFOV and CFOV is the highest fluctuation possible.

Second, differential uniformity is defined as the greatest rate of change in count density across a specific distance, in this case, 5 pixels each way in both the X and Y directions in the UFOV and CFOV. The result is that a low rate of change or variation is associated with a high level of homogeneity.

Purpose of Test
A symmetric energy window over the photopeak was utilized to test the intrinsic response of a scintillation camera to a uniform flux of input photons throughout the field of view.

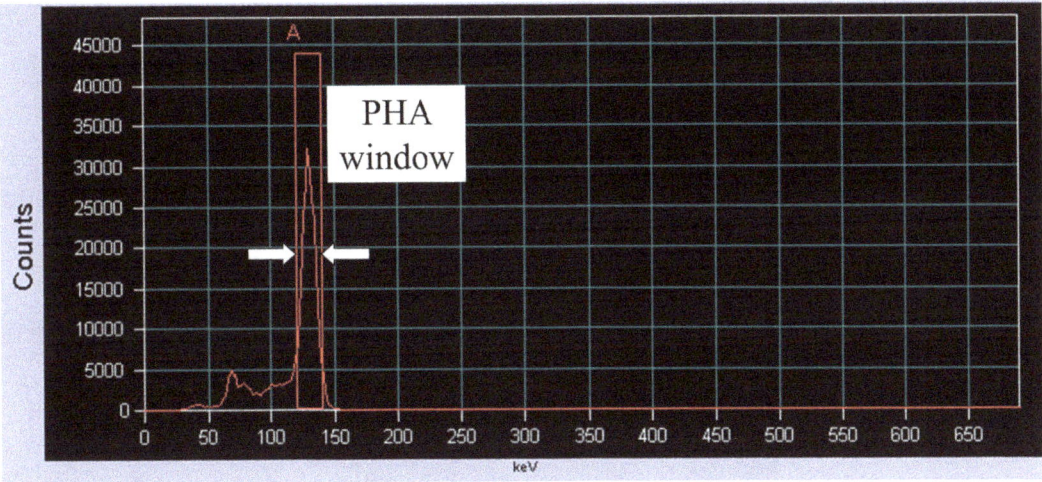

Energy (keV)

Fig. 16.3 The energy spectrum from a gamma camera was measured using a point source of $^{99m}TcO_4$ (140 keV) in the air with a 15% energy window

Materials

1. In the case of the GE camera, a point source consisting of 25 MBq (0.675 mCi) of ^{99m}Tc in 0.4 mL or less is required; however, in the case of the Siemens camera, only 1.5 MBq (0.040 mCi) is required.
2. In the case of the Ge camera, the count rate should not be larger than 40,000 count/s while using the manufacturer's preset PHA window.
3. In the case of Siemens camera, the dead time should be shorter than 10%.
4. The mounting of a point source as a source.
5. Use lead masks or decoys.

Procedure

1. Take the collimator out of the detector head and set it aside. Align the head with the crystal housing, and place the lead mask or decoy in the centre of the crystal housing.
2. Place the source at least five times the maximum UFOV away from the detector's central axis or as advised by the manufacturer.
3. Position the PHA window in the centre of the photopeak.
4. Take an image with a preset count of roughly 3×10^7 or according to the manufacturer's instructions.
5. A 64 × 64 matrix size may be utilized for any camera with a field of view (UFOV) less than 400 mm. When the UFOV is greater than 400 mm, a 256 × 256 matrix should be utilized to maximize the resolution.
6. Remove the source and lead masks from the area. Following the completion of the test, attach the collimator to the detector head (Fig. 16.4).

Analysis

In any consecutive pixels falling within the UFOV and the CFOV, compute the highest (max) and minimum (min) count in the X and Y directions for each row or column of pixels in the X and Y directions. Integral uniformity may be calculated as follows:

$$Integral\ Uniformity = \frac{(max - min)}{(max + min)} \times 100$$

This is accomplished by studying the first set of five pixels in both the X and Y directions, then moving to one pixel and assessing the following five pixels in both directions. Figure out the highest and lowest counts in each row and column group.

The differential uniformity may be calculated as follows:

$$Differential\ Uniformity = \frac{(high - low)}{(high + low)} \times 100$$

Fig. 16.4 Setting for intrinsic uniformity testing. A point source containing 25 MBq of ^{99m}Tc in 0.4 mL was filled and positioned 5 UFOV from the detector. (Image credit: Springer Nature, Journal of Nuclear Cardiology, 2018 [12]. Reprinted with permission)

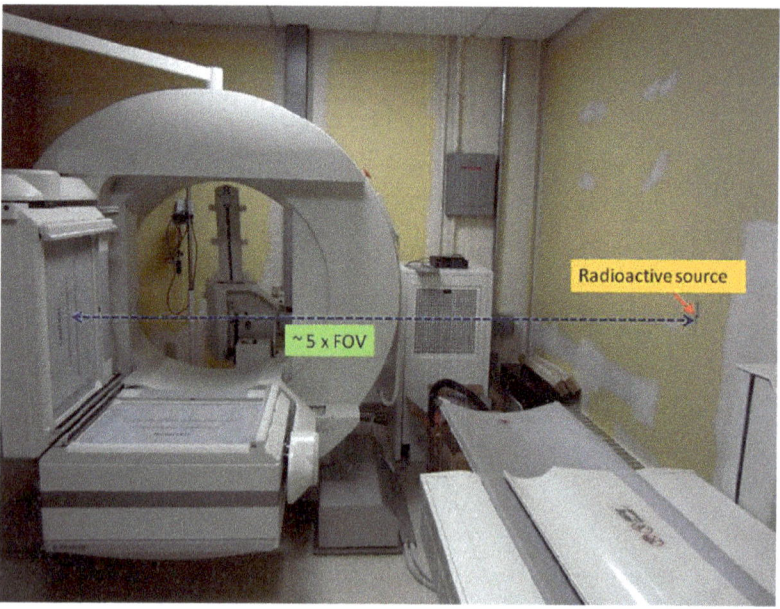

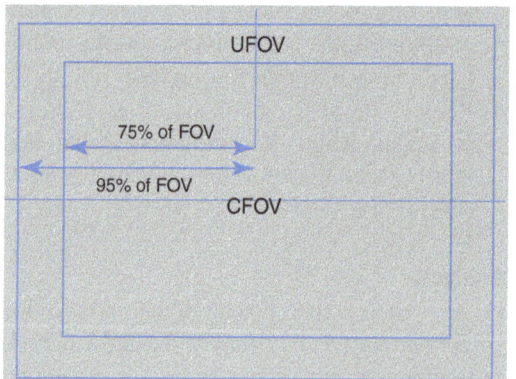

Fig. 16.5 Integral uniformity in whole UFOV and CFOV

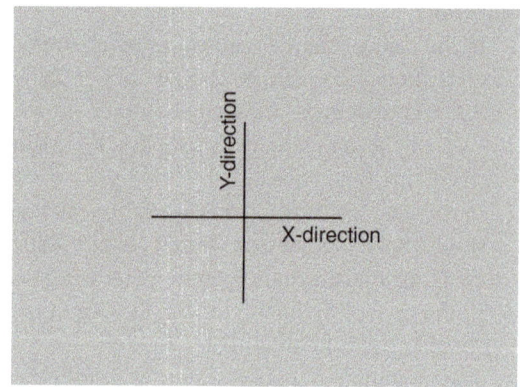

Fig. 16.6 Differential uniformity calculation in X and Y directions

High and low pixel counts are those that provide the largest maximum count difference between the two images (Figs. 16.5 and 16.6).

Image Analysis

Check the flood images for any anomalies or non-uniformities visually. Patterns, such as the PMT array, should not be shown. Notify if there are any areas of non-uniformity. A quantitative study of flood data is required to calculate the IU and DU in the UFOV and CFOV.

Report

Any non-uniformities in the ^{99m}Tc or any radionuclides analysed should be reported.

For ^{99m}Tc and all other radionuclides tested, report the IU and DU in UFOV and CFOV.

Performance Specifications

- For five million count floods over the UFOV, the IU and DU are less than 5% and 3% of the manufacturer reference value, respectively.

- The IU and DU for 30 M count floods in the UFOV should be less than 3% and 1% of the overall IU and DU, respectively.
- It may be necessary to do a fresh flood calibration to address the non-uniformities. Any non-uniformity faults that are not correctable should be asked for service and repair.

16.15.1.3   Test of System Flood Field Uniformity

Purpose of Test

Using a scintillation camera, each multihole collimator was subjected to a system flood field response test.

Materials

It is also possible to use a ^{57}Co flood source with activity comparable to ^{99m}Tc, containing 205–410 MBq (5.5–11 mCi) in solution.

Procedure

1. Connect the detector head to the collimator to be evaluated. Adjust the head to a vertically upward position.
2. Set the flood phantom or source to around 10 cm above the collimator's face.
3. Position the PHA window in the centre of the photopeak for the radionuclide to be utilized.

4. Acquire an image with a preset level of 5×10^6 counts and a matrix size of 256 × 256 pixels. If data for flood field uniformity correction has been obtained intrinsically, it should be left enabled in such instances.
5. After finishing the test, remove the flood phantom from the room.
6. Repeat the above steps for every multihole collimator that was utilized in the department.

16.15.1.4   Data Analysis

Make a visual comparison of the images, paying close attention to any increasing non-uniformities at the smaller PHA window. Using the following formulae, you may get the quantitative uniformity values stated in the intrinsic uniformity part of this chapter.

$$\text{Integral Uniformity} = \frac{(\text{max} - \text{min})}{(\text{max} + \text{min})} \times 100$$

$$\text{Differential Uniformity} = \frac{(\text{high} - \text{low})}{(\text{high} + \text{low})} \times 100$$

Report

- Note any unusual non-uniformities, such as a collimator fault or a visible PMT pattern, in the data for each collimator.
- Provide the UFOV and CFOV's IU and DU in the report (Fig. 16.7).

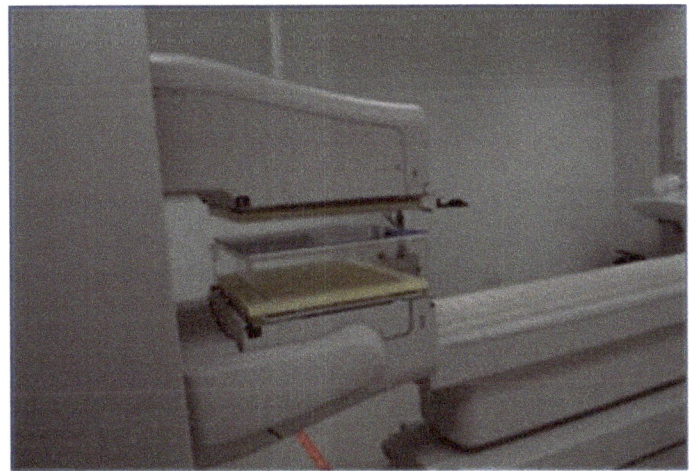

Flood Source of ^{99m}Tc or ^{57}Co

Collimator

Detector

Fig. 16.7 Test setting for extrinsic uniformity. A flood source filled with 205–410 MBq of ^{99m}Tc or ^{57}Co

Performance Specifications

- For five million count floods over the UFOV, the IU and DU are less than 5% and 3% of the manufacturer reference value, respectively.
- The IU and DU for 30 M count floods in the UFOV should be less than 3% and 1% of the overall IU and DU, respectively.
- If there is significant non-uniformity, it is required to locate the problem on the collimator. It may be necessary to remove the collimator from service and replace it with a collimator that is acceptable to the user, depending on the extent of the damage.

16.15.1.5   Intrinsic Off-Peak Flood Field Uniformity

This test examines the gamma camera detector for PMT imbalance or decoupling and crystal-related concerns like hydration. Intrinsic flood images of a ^{99m}Tc point were obtained with photo peak energies of 126 ($\pm$10% energy window) and 154 keV ($\pm$10% energy window) for both images taken simultaneously. It is visually examined for an odd pattern of non-uniformities in these off-peak flood photos. In most cases, a modest PMT pattern is normal and does not warrant worry.

Frequency

- Test each detector for baseline images as part of the acceptance process.
- No routine quality control.

Image Acquisition

Set up the detector and the point source at the specified locations to check for intrinsic homogeneity. At least five million count flood pictures in a 256 × 256 matrix with photo peaks set to 126 (Fig. 16.8a) and 154 keV (Fig. 16.8b), respectively, with a 20% energy window, should be acquired for each detector. ^{57}Co (120 keV) and ^{123}I (160 keV) radioisotope energy windows may be used instead of 126 and 154 keV acquisition asymmetric energies in camera systems when it is impossible to establish precise acquisition asymmetric energies.

Image Analysis

Take a look at the off-peak flood images to see if there are any irregularities. Expectedly, all PMTs will be recognizable. Tube imbalance or decoupling may be present if a single or a few PMTs are different from the others. Crystal hydration may be present if there are many hot patches or a 'measles-like' pattern.

Report

Report any crystal hydration flaws (Fig. 16.9a, b) or abnormal PMT patterns. Whether previous images are available, examine the patterns to see if they have changed.

Performance Specifications

If a strange PMT pattern is discovered, an engineer should be called in to fix the gamma camera detector. The camera detector should be serviced and the crystal replaced if there are any hydration issues.

16.15.2   Spatial Linearity

Gamma cameras include a performance parameter that measures how much the picture is distorted by spatial perspective. NEMA's 90° bar phantom or line phantom may measure spatial linearity by looking at the image of straight bars or lines on the phantom. Spatial distortion and non-homogeneity in the flooded field are closely connected. There will be non-uniformity if a region has significant non-linearity (Fig. 16.10a, b).

16.15.3   Spatial Resolution

Suppose the distance between two points is maintained to a minimum, where two-point sources may be treated separately. Intrinsic resolution is influenced by gamma-ray energy, crystal thickness, light guides, PHA window width and count rate. The collimator, as well as the distance between the source and the collimator, impacts system resolution.

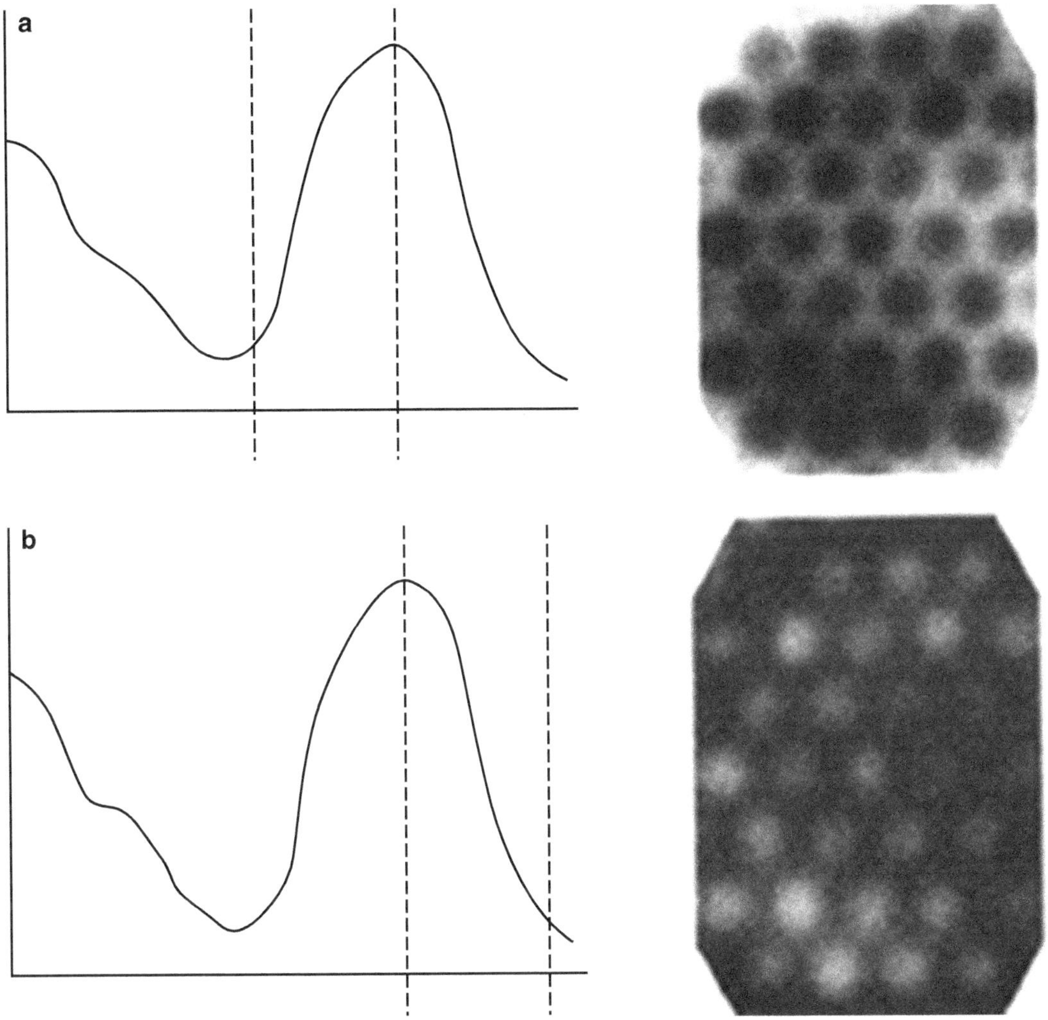

a

b

Fig. 16.8 (**a**) Peak shift at the higher side to show up PMTs hot. (**b**) Peak shift at lower side to show up PMTs cold

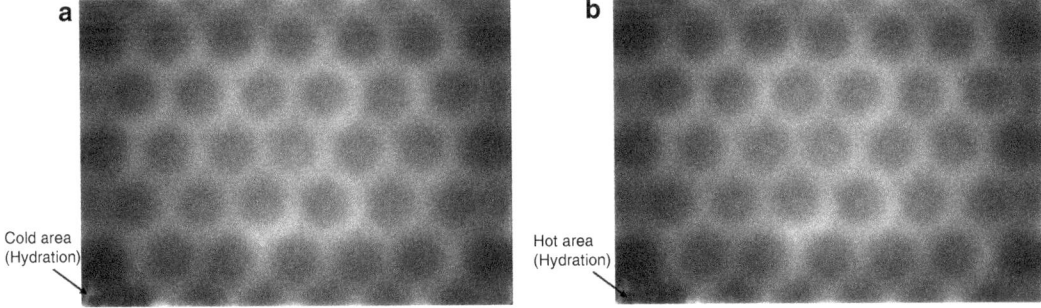

a

b

Cold area
(Hydration)

Hot area
(Hydration)

Fig. 16.9 (**a**) Measles artefacts due to crystal hydration (126 keV). (**b**) Measles artefacts due to crystal hydration (154 keV)

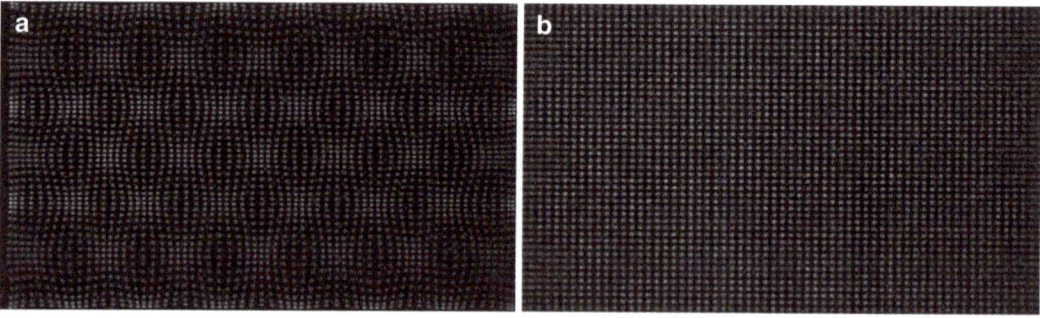

Fig. 16.10 (a) Spatial linearity image using orthogonal phantom before linearity correction. (b) Spatial linearity image using orthogonal phantom after linearity correction

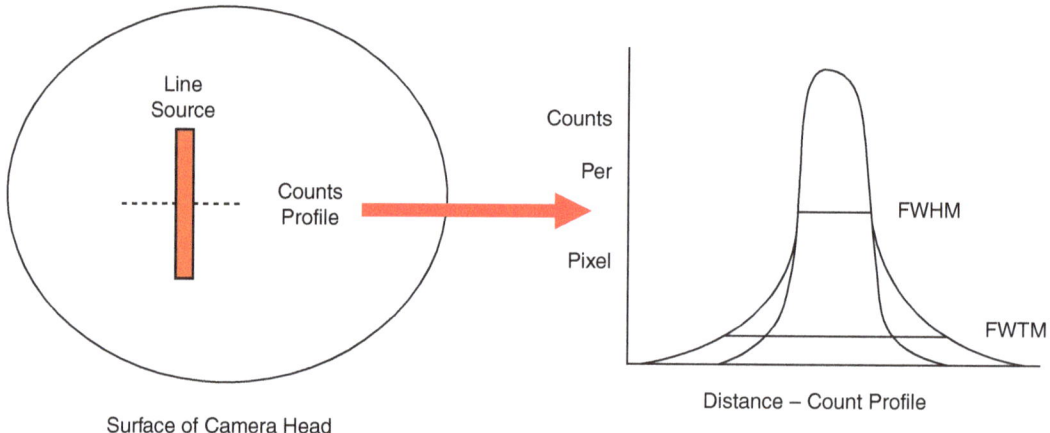

Fig. 16.11 The line spread function of a line source's full width at half maximum (FWHM)

The spatial resolution is determined by projecting an orthogonal count density profile over a line source image and measuring the width of the LSF. The full width at half maximum of the LSF at half its height is used to specify the spatial resolution (FWHM). Additionally, as shown in Fig. 16.11, a measurement at full width at tenth maximum (FWTM) may also be included. The measurements are given in mm. Image contrast losses due to collimator septal penetration and scattering are quantified using the FWTM.

Using bar phantoms to determine spatial resolution is a popular practice. As shown in Fig. 16.12, the most common bar phantom is four panels of parallel bars of decreasing size and spacing that cover the detector field of view (UFOV). This bar phantom's suitable bar width is half of the estimated intrinsic spatial resolution (the smallest commercially available bars are 2 mm). The goal of the phantom bar test is to identify the image's lowest visible quadrant of bars. LSF's FWHM is 1.75 times the size of the smallest visible bar.

Because spatial resolution and linearity measures rely on imaging a line source, they may be derived from the same image.

16.15.4   Intrinsic Tests of Spatial Resolution and Spatial Linearity

These experiments assess the gamma camera detector's spatial resolution and linearity without using a collimator, allowing for more precise observations.

Tests using a phantom (slit) such as the NU 1-2012, or a bar phantom, from which the spatial

Smallest resolver bar

Fig. 16.12 Quadrant bar phantom

resolution is computed, are required. Two separate patterns (horizontal and vertical) are often utilized to test spatial resolution in X and Y. The manufacturer should give these to you. It will be the right size to go with UFOV.

The quadrant bar phantom is enough for routine examinations and yearly assessments. When there is no slit phantom available, you may use the quadrant bar phantom for acceptance testing.

16.15.4.1 Frequency

- To assess the spatial resolution and spatial linearity of ^{99m}Tc, acceptance testing will be conducted. Slit or quadrant bar phantoms may be used. The quadrant bar phantom images are used for future reference. If available, the NEMA software from the manufacturer may be utilized for the analysis. Testing additional radionuclides often utilized in clinical practice with this system is optional.
- Get and analyse images of the slit or quadrant bar phantom with ^{99m}Tc yearly. It is possible to use this system to test other radionuclides often used in clinical practice.
- As part of normal quality assurance operations, an intrinsic slit or quadrant bar phantom may be

acquired weekly instead of an extrinsic bar phantom. The image is just examined visually.

16.15.4.2 Testing Procedure

Slit Phantom Measurement
Before conducting measurements, take off the collimator and put the slit phantom. The slit phantom must be in close contact with the detector crystal during the measurement. It is necessary to use caution to avoid damaging the crystal. The slit phantom should be attached to the detector using the methods provided by the manufacturer in the phantom design. Under some circumstances, a 2 mm film of acrylic or an equivalent substance may be placed between the slit phantom and the detector crystal. The detector areas not covered by the slit phantom will need additional lead shielding (see Fig. 16.13).

When using a slit phantom, the slit width and lead sheet thickness must comply with NU 1–2012's standards. A point source is at least 5 UFOV distant from the centre of the detector. The point source's activity should be between 250 and 450 MBq (5.5 and 12 mCi) to do this measurement or according to the manufacturer's advice.

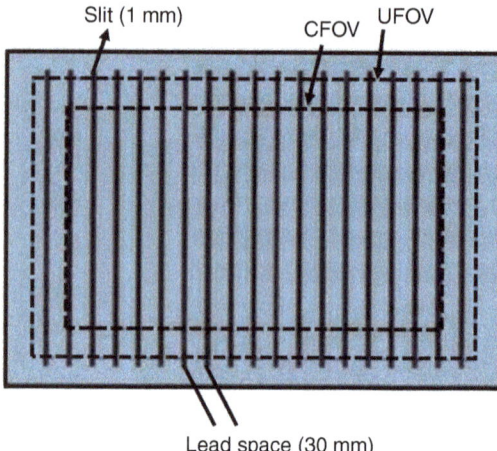

Fig. 16.13 Slit phantom for spatial linearity and resolution. (Image copyright: the IAEA, YouTube tutorial on quality control tests for SPECT systems [13]. Printed with permission)

Image Acquisition

The camera zoom factor and computer image matrix should be chosen to achieve a pixel size in the z-direction of the slit that is less than 0.2 full width half maximum (FWHM). The photopeak of ^{99m}Tc, i.e. 140 keV, should be in the centre of the 20% energy window. Acquire an image while maintaining a count rate of 10,000 counts per second. At the highest locations of the slits in the picture, a minimum of 250–300 counts per pixel should be acquired. You should obtain 16,384 K counts if your matrix size is 256 × 256. Both the X- and Y-axes should be used during the acquisition.

Analysis for Spatial Resolution

LSFs may be created on a computer workstation by placing a wide profile (about 3 cm) orthogonally across each slit. If the pixel size is more than 0.2 FWHM, the LSF peak may be identified by fitting the three biggest peak counts with a parabolic fit. The peak of the line spread function cannot be determined if the pixel size is bigger than 0.2 FWHM. Use linear interpolation to get the FWHM positions on either side of the line spread function peak.

To verify the system pixel calibration by measuring no of pixels in the 30 mm distance between the slits. We compute the average full width half maximum (FWHM) as shown in Fig. 16.14a, b since the spatial resolution does not remain constant over the field of view.

Analysis for Spatial Linearity

Check for non-linearity in the slit pictures visually. A straight slit is what you want to see in the images. Non-linearity is the most common result of a linearity correction software failure. A loss of tube balance might be indicated by a bending of the line image around a PMT, as indicated in Fig. 16.15a, b. Reporting results may be done in the following manner:

- No detectable non-linearity.
- It's hardly perceptible (non-linearity), and it's less than a millimetre.
- Substantial non-linearity, with a greater than a millimetre.

The pixel size may be used to calculate displacements.

NU 1-2012 requires differential and absolute linearity. A perfect fit line must be used to fit the count density for the image's slits. The discrepancy between true peak locations and a perfect fit line is known as relative linearity. The largest peak displacement from the best-fit line in mm is defined as perfect linearity (Fig. 16.16a, b). Report any observable non-linearity without analytical software.

16.15.5   Intrinsic Resolution Using Bar Phantom

(a) Removing the collimator and turning the detector upside down.

(b) Then, on top of the detector, place a bar phantom, with the bars being properly aligned with the detector's longitudinal and transversal axes (Figure 16.17a).

(c) As indicated in Fig. 16.17b, a source of about 600 Ci of ^{99m}Tc should be placed at a distance of five times the crystal diameter or follows the instruction of the manufacturer manual.

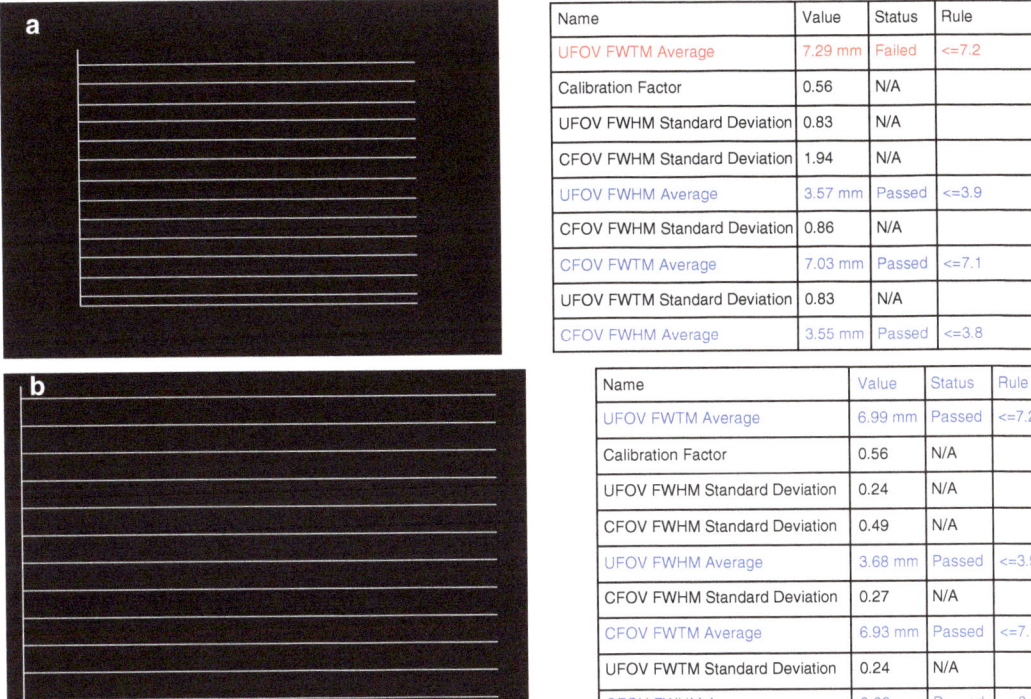

Name	Value	Status	Rule
UFOV FWTM Average	7.29 mm	Failed	<=7.2
Calibration Factor	0.56	N/A	
UFOV FWHM Standard Deviation	0.83	N/A	
CFOV FWHM Standard Deviation	1.94	N/A	
UFOV FWHM Average	3.57 mm	Passed	<=3.9
CFOV FWHM Standard Deviation	0.86	N/A	
CFOV FWTM Average	7.03 mm	Passed	<=7.1
UFOV FWTM Standard Deviation	0.83	N/A	
CFOV FWHM Average	3.55 mm	Passed	<=3.8

Name	Value	Status	Rule
UFOV FWTM Average	6.99 mm	Passed	<=7.2
Calibration Factor	0.56	N/A	
UFOV FWHM Standard Deviation	0.24	N/A	
CFOV FWHM Standard Deviation	0.49	N/A	
UFOV FWHM Average	3.68 mm	Passed	<=3.9
CFOV FWHM Standard Deviation	0.27	N/A	
CFOV FWTM Average	6.93 mm	Passed	<=7.1
UFOV FWTM Standard Deviation	0.24	N/A	
CFOV FWHM Average	3.66 mm	Passed	<=3.8

Fig. 16.14 (**a**) Resolution before calibration. (**b**) Resolution after calibration

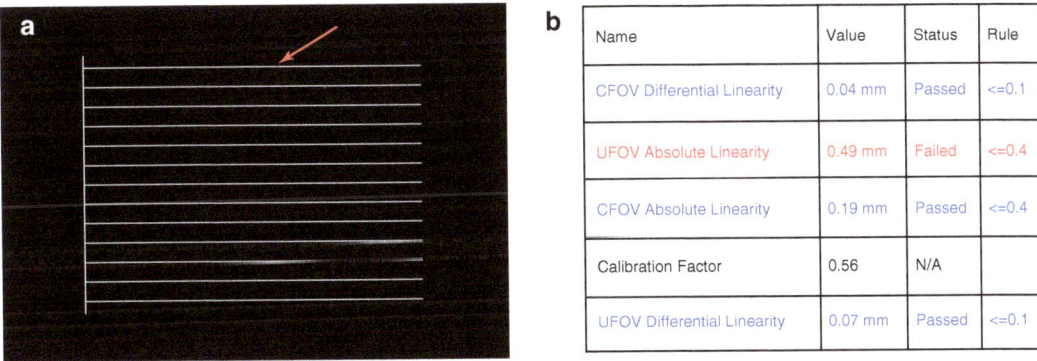

Name	Value	Status	Rule
CFOV Differential Linearity	0.04 mm	Passed	<=0.1
UFOV Absolute Linearity	0.49 mm	Failed	<=0.4
CFOV Absolute Linearity	0.19 mm	Passed	<=0.4
Calibration Factor	0.56	N/A	
UFOV Differential Linearity	0.07 mm	Passed	<=0.1

Fig. 16.15 (**a**) Non-linearity. (**b**) Non-linearity in UFOV

Name	Value	Status	Rule
CFOV Differential Linearity	0.03 mm	Passed	<=0.1
UFOV Absolute Linearity	0.14 mm	Passed	<=0.4
CFOV Absolute Linearity	0.14 mm	Passed	<=0.4
Calibration Factor	0.56	N/A	
UFOV Differential Linearity	0.03 mm	Passed	<=0.1

Fig. 16.16 (**a**) After calibration, the linearity is shown. (**b**) Demonstrating linearity in the UFOV after calibration. (Image copyright, the International Atomic Energy Agency and a YouTube video instruction on quality control checks for SPECT systems [13]. Permission was obtained to print this material)

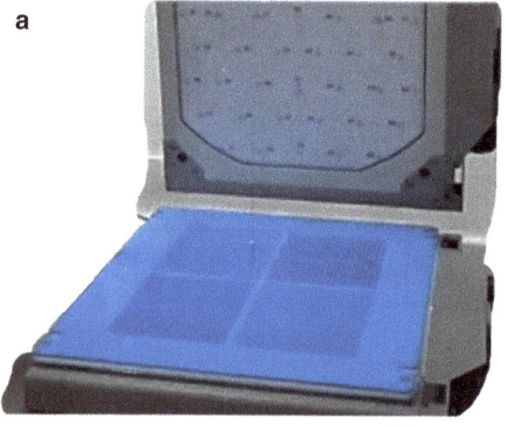

Fig. 16.17 Setup of the intrinsic uniformity test

(d) Create a 20% window and use the gamma camera to peak for ^{99m}Tc in the image.
(e) Collecting two million counts in a matrix of size 256 × 256 is required.
(f) Rotate the bar phantom by 90 degrees, 180 degrees and 270 degrees, take images, and report the results by keeping the same counts and matrix size.

16.15.5.1 Quantification of Spatial Resolution

(a) Determine the width of the bar in the 90° bar quadrant phantom that the camera can detect (let's call it 'B') as shown in Fig. 16.18.

(b) Resolution: FWHM = 1.75 B = 3 × 1.75 = 5.2 mm.
(c) Averaging the FWHM values in X and Y.

16.15.6 Extrinsic Spatial Resolution Using Bar Phantom

This test evaluates the gamma camera's overall spatial resolution with a fitted collimator. A line source in the air, 10 cm from the collimator's face, is used to take the measurement. Source-to-collimator distances other than those listed here may be examined. Each detector of gamma camera system should be checked.

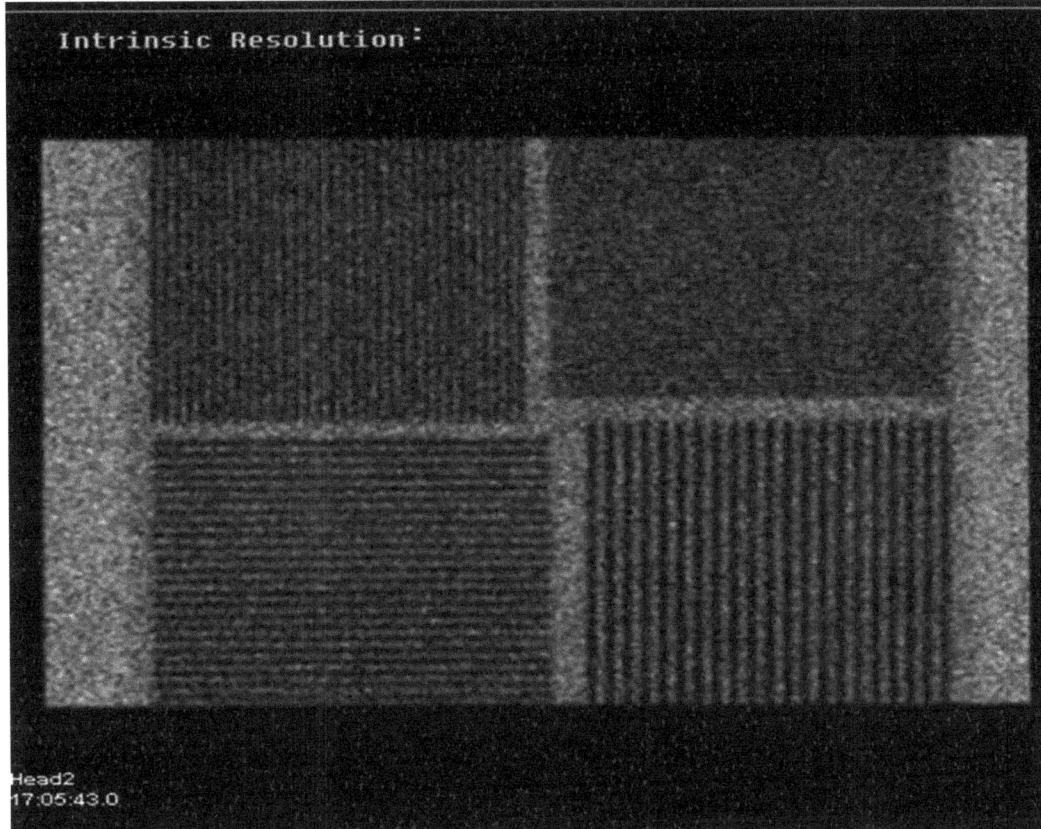

Fig. 16.18 Quadrant bar phantom image for intrinsic resolution calculation

A rectangular bar phantom with a ^{57}Co sheet source, or a ^{99m}Tc-loaded flood phantom, may be used to measure the resolution instead of a line source for low-energy collimators. In medium- and high-energy collimators, the collimator size of the hole may be bigger than the breadth of the bars, resulting in Moiré patterns that obscure the results.

16.15.6.1   Frequency

- Acceptance testing: All low-energy collimators that employ a ^{99m}Tc line source should be tested for spatial resolution. Using other collimators and radionuclides to measure extrinsic spatial resolution may be optional. A ^{57}Co sheet source or ^{99m}Tc-filled flood phantom, one of the low-energy all-purpose collimators and the quadrant bar phantom can be used to take the picture for spatial resolution. For sub-

sequent measurements, the image should be utilized as a baseline.
- When an intrinsic bar phantom cannot be recorded, a ^{57}Co sheet source or a 99mTc-filled flood phantom may be used to acquire a quadrant bar phantom for extrinsic resolution.
- Subsequent quality control: Take a quadrant bar picture using a ^{57}Co sheet or a ^{99m}Tc-filled flood phantom source, and compare it to the reference image.

16.15.6.2   Procedure for Checking

Image Acquisition

Install the collimator that is suitable for the radionuclide to be scanned. The sources (line) of activity ranging from 2 to 3 mCi (74–111 MBq) are positioned perpendicular to the measuring axis and parallel to and at 10 cm from the collimator's

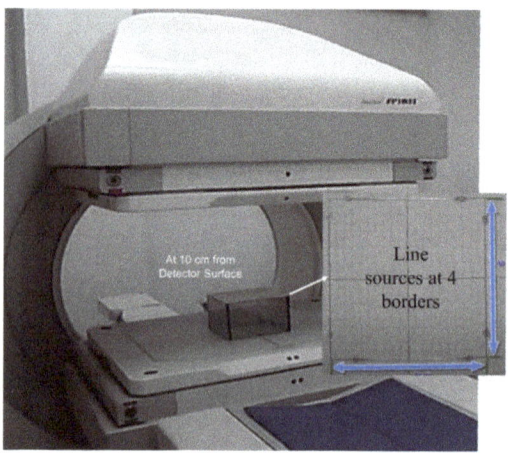

Fig. 16.19 Extrinsic resolution at 10 cm using line sources. (Image copyright, the IAEA, YouTube tutorial on quality control tests for SPECT systems [13]. Printed with permission)

Fig. 16.20 Extrinsic resolution using bar phantom

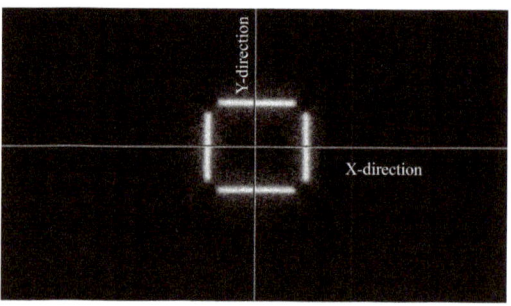

Fig. 16.21 X- and Y-axis profile of a line. Directions for calculating the FWHM were derived from online sources

surface. Take separate images along with the X- and Y-axes. The matrix should be 128 by 128 pixels in size, with one zoom. The accumulation will continue until the total number of counts exceeds five million, as shown in Fig. 16.19.

Only a LEAP should be utilized while imaging bar phantoms. The bar phantom should be placed about 10 cm away from the collimator face, and the ^{57}Co sheet source or ^{99m}Tc fillable phantom should be placed on top of it. Select the finest and highest resolution image matrix presently available with the system. Each pixel width should be smaller than one-half the width of the narrowest bars. Rotate the phantom 90 or 180 degrees between each image capture to receive at least two images per detector. As seen in Fig. 16.20, the total number of counts obtained for each image should be at least five million counts.

Spatial Resolution Analysis

To get LSFs on a computer workstation, place a wide profile (30 mm) across each line source image (X and Y directions). The peak of the LSF can be found by fitting the line spread function peak's three biggest count values to a parabolic form. This way, the LSF peak can be found. The half-maximum locations shown in Figs. 16.21,

16.22 and 16.23 can be found using linear interpolation to determine where the LSF peak is on each side.

Identify the least visible bar size in bar phantom images. For a quadrant to be deemed visible, at least half of the width of the bars must be visible in that quadrant (Fig. 16.24).

FWHM = 1.75 × B,   B = 3 mm (from Fig. 16.24) = 5.25 mm

Spatial Linearity Analysis

No pattern distortions in the line source or bar pattern. There should be a straight pattern. Bars bending around a PMT may indicate tube imbalance. The findings obtained are as follows:

- Visible and less than 1 mm.
- Visible non-linearity.
- Greater than 1 mm.

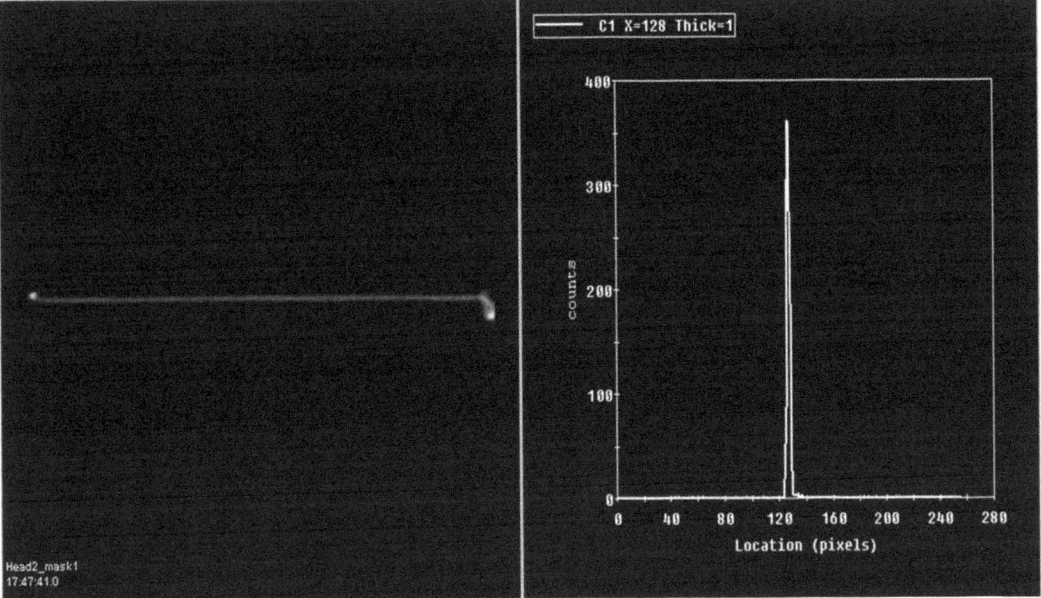

Fig. 16.22 Line profile X direction to calculate FWHM

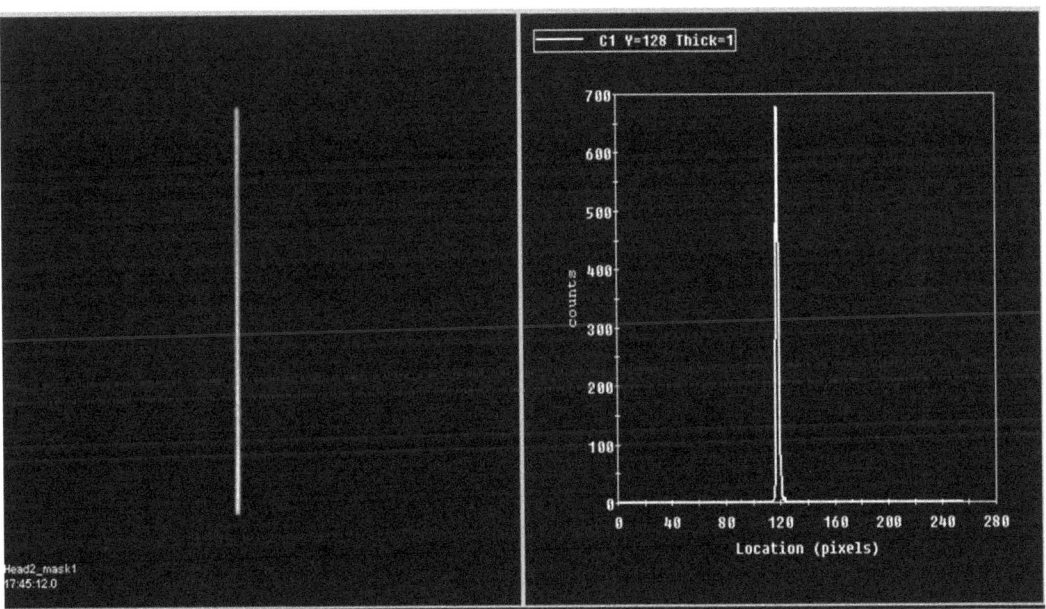

Fig. 16.23 Line profile Y direction to calculate FWHM

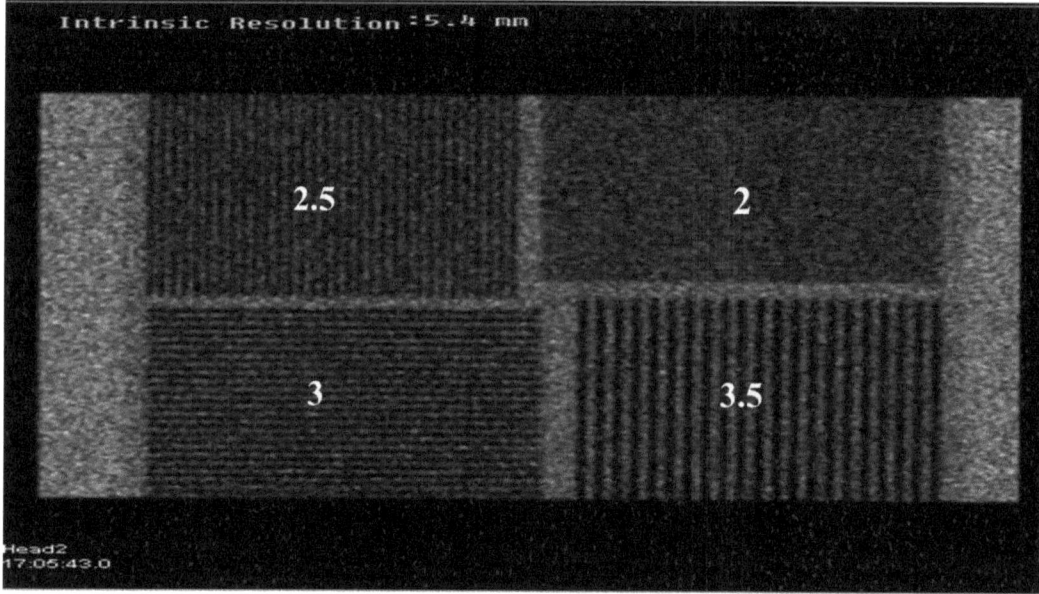

Fig. 16.24 Extrinsic resolution using bar phantom

Since spatial linearity is not represented in millimetres, the number of displacements must be determined using the pixel size.

Report

- For ^{99m}Tc and all other radionuclides and collimator combinations examined, provide FWHM values.
- Indicate the lowest bar size ever recorded for ^{57}Co or ^{99m}Tc.
- Indicate the non-linearity noticed through visual examination of the line sources and bar phantom.
- When utilizing a high-resolution, low-energy parallel-hole collimator at 100 mm, the system's spatial resolution for ^{99m}Tc should be 8 mm.

16.15.7   Energy Resolution

The energy resolution reflects its inherent capacity to discriminate between gamma rays of various energies. The advantages of increased energy resolution include the following:

- Fewer scattered events.
- Enhances the image's quality.

The energy resolution is defined as the photopeak's FWHM divided by the peak energy. The intrinsic energy resolution specified by the manufacturer for ^{99m}Tc is typically 9–11 percent. Viewing the photopeak and modifying the energy window may be approximated visually. The following approach may be used to assess the gamma camera's resolution of energy:

The NaI (Tl) crystal requires 3 eV to create a single scintillation. Scintillation has a conversion efficiency of 11%. It generates 50% scintillation photons as a result of the conversion efficiency. Ten percent scintillation photons yield a photoelectron. According to this computation, 1 keV of radiation energy results in the production of 38 photons.

$\therefore$ the number of photoelectrons produced by

$$140 \text{ keV} = \frac{140 \times 1000 \, eV}{3 \, eV} \times 0.11 \times 0.50 \times 0.10.$$

$$= 266 \text{ electrons}$$

So energy resolution:

$$\frac{\delta E}{E} = 2.35 \frac{\sigma_E}{E} = \frac{2.35\sqrt{266}}{266}$$

The high-energy resolution is necessary to reduce the scatter and differentiate many photo peaks. The number of photons created in the scintillator and the number of photoelectrons generated in the PMT detector are proportionate. To get reduced scatter, it is necessary to have gradual fluctuations and a long decay period.

16.15.7.1 Procedure for Checking

Image Acquisition
1. Turn the collimator upside down and remove it from the detector's head.
2. Place the lead mask or decoy in the middle of the crystal housing.
3. Make sure that the source of ^{99m}Tc activity is set to 600 µCi and is five times UFOV away from the detector.
4. You can see this in Fig. 16.25. The manufacturer's default PHA window should be set to 20% photopeak.
5. As shown in Fig. 16.26, it is possible to estimate the energy resolution visually by viewing the photopeak and modifying the energy window. It is possible to compute it mathematically by utilizing the formulas presented in Figs. 16.27 and 16.28.

16.15.8 Extrinsic Planar Sensitivity

The sensitivity of a gamma camera is measured in cps/kBq (cps/µCi) for radioactive sources located inside the detector's UFOV. Several factors determine the sensitivity of a system:

- The imager's geometries.
- The detector's detection power and thickness.
- The energy of radioisotopes.
- The energy resolution and window of the imager.
- The imager's source distribution.

The sensitivity may be assessed by counting radionuclides in a Petri dish with minimal attenuation and scatter. Each detector of gamma camera systems should be analysed and compared individually.

The measurement for ^{99m}Tc should be done using a low-energy parallel-hole collimator. Other clinical radionuclide and collimator combinations may be assessed. Each collimator is designed for specific radionuclides.

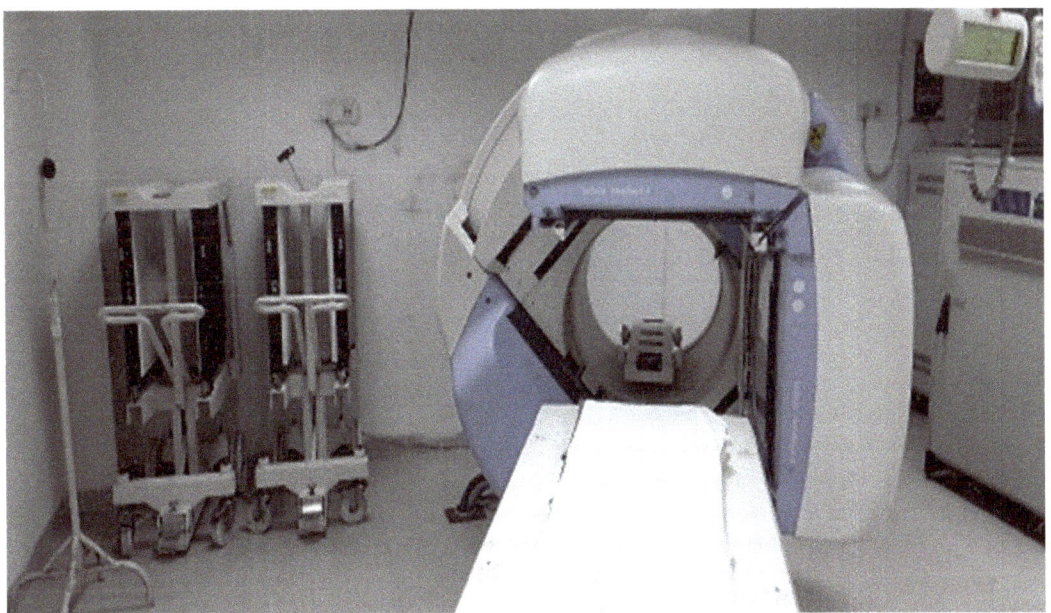

Fig. 16.25 Estimation of energy resolution for ^{99m}Tc

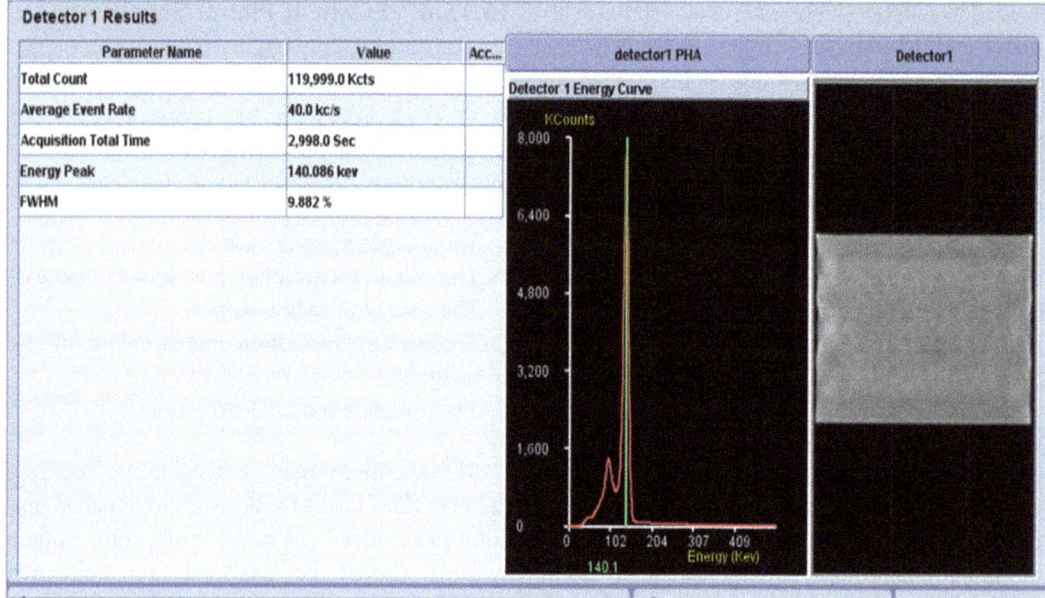

Fig. 16.26 Estimation of energy resolution visually by observing photopeak and adjusting energy window

Fig. 16.27 Energy Spectrum Acquisition and estimation of energy resolution

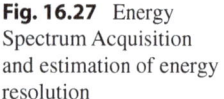

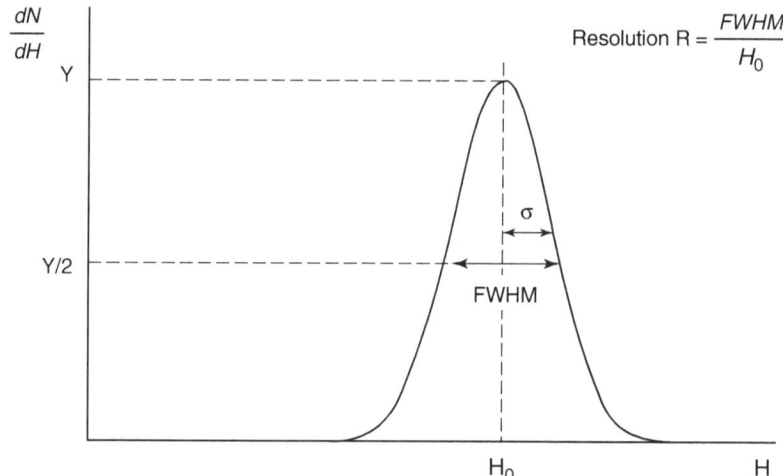

$$\text{Resolution R} = \frac{FWHM}{H_0}$$

16.15.8.1   Frequency

- At least one low-energy collimator must have its ^{99m}Tc sensitivity tested before considered 'acceptable'. Additional radionuclide-collimator combinations might be investigated.
- No routine quality control.

16.15.8.2   Procedure of Checking

Image Acquisition

Set up the sensitivity source directly above the UFOV gamma camera detector. Getting the exact distance isn't critical; nevertheless, you should be able to count on the same distance each time;

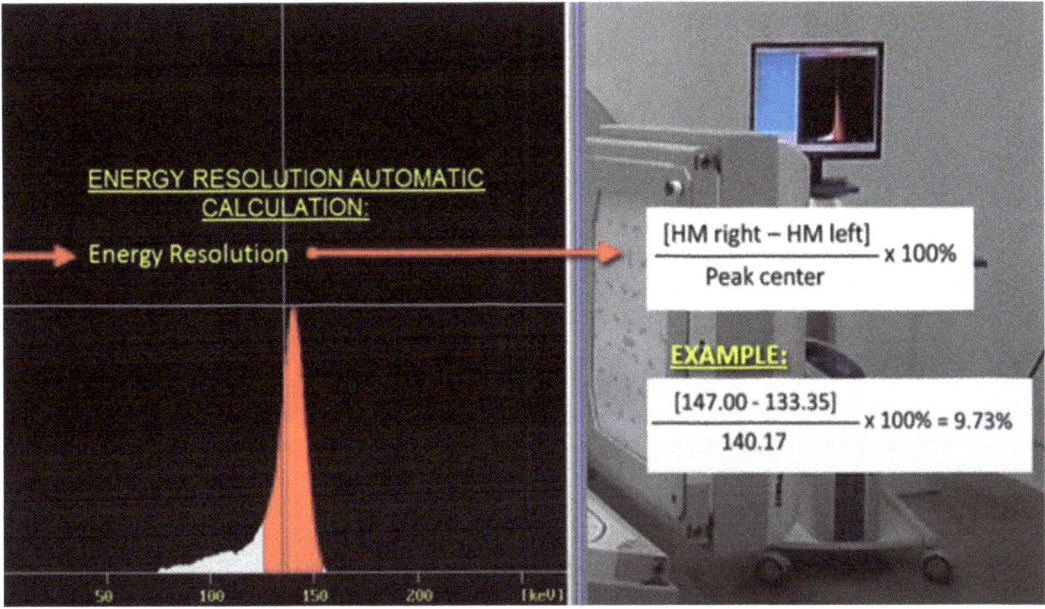

Fig. 16.28 Energy spectrum acquisition and estimation of energy resolution. (Image copyright, the IAEA, YouTube tutorial on quality control tests for SPECT systems [13]. Printed with permission)

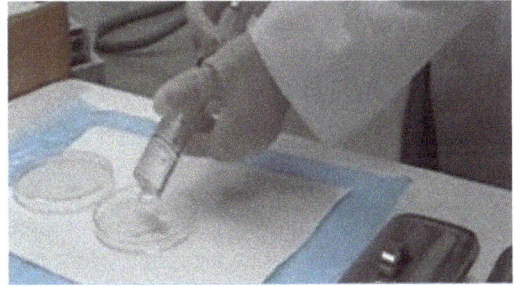

Fig. 16.29 A 2–3 ml layer of water in a 0.15 cm diameter flat plastic disc. (Image copyright, YouTube tutorial on quality control tests for SPECT systems [13]. Printed with permission)

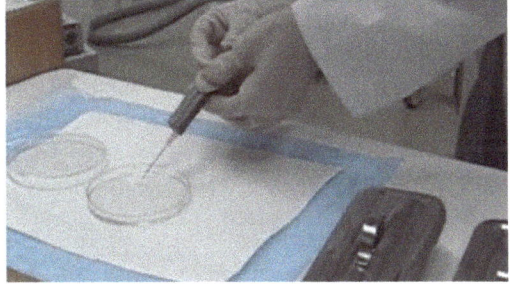

Fig. 16.30 In a 0.15 cm diameter flat plastic disc, activity is 20–80 MBq in 2–3 mL water. (Image copyright, the IAEA, YouTube tutorial on quality control tests for SPECT systems [13]. Printed with permission)

normally, it is 10 cm from the detector. The activity is 20–80 MBq in 2–3 ml water in a 0.15 cm diameter flat plastic disc known as a Petri dish. A low-attenuating source holder should be utilized to measure distance from the collimator indicated in Figs. 16.29, 16.30 and 16.31.

The size of the acquisition matrix is irrelevant. At least 1 min should be allowed for acquisition. All detectors, radionuclides and collimators utilized in the department should be subjected to a sensitivity test. Subtract the background from the sensitivity image counted for 60 s after removing the sensitivity source.

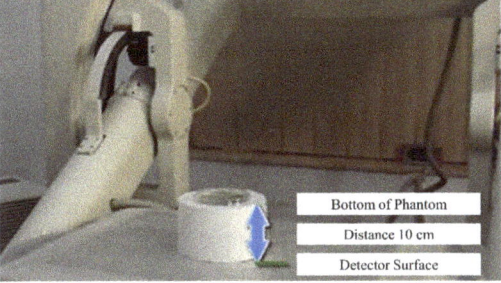

Fig. 16.31 0.15 cm diameter flat plastic disc keep 10 cm away from the detector. (Image copyright, the IAEA, YouTube tutorial on quality control tests for SPECT systems [13]. Printed with permission)

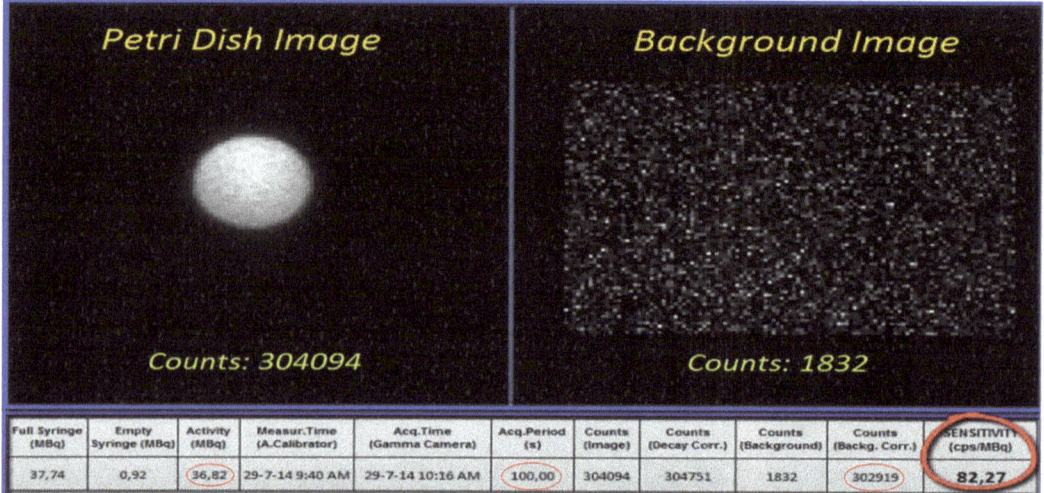

Full Syringe (MBq)	Empty Syringe (MBq)	Activity (MBq)	Measur.Time (A.Calibrator)	Acq.Time (Gamma Camera)	Acq.Period (s)	Counts (Image)	Counts (Decay Corr.)	Counts (Background)	Counts (Backg. Corr.)	SENSITIVITY (cps/MBq)
37,74	0,92	36,82	29-7-14 9:40 AM	29-7-14 10:16 AM	100,00	304094	304751	1832	302919	82,27

Fig. 16.32 The calculation of the sensitivity of the gamma camera. (Image copyright, the IAEA, YouTube tutorial on quality control tests for SPECT systems [13]. Printed with permission)

Image Analysis

Calculate the combined counts in the source and background pictures for each detector, radionuclide and collimator combination, using both the source and background images.

To get an accurate count, the whole picture matrix must be used. The sensitivity of one detector compared to the sensitivity of the other detector is shown in Fig. 16.32 for multi-detector systems.

$$\text{Sensitivity} = \frac{302919}{36.82 \times 100} = 82.27$$

Report

List all detector, radionuclide and collimator combinations evaluated to determine the system's sensitivity. Each pair of detectors in a gamma camera should be reported as a percentage difference in sensitivity.

Performance Specifications

As long as the camera system has more than one detector, two detectors should not have different system sensitivities by more than 5% for each radionuclide and collimator combination.

16.15.9 Performance of Intrinsic Count Rate

This test examines how many counts are lost due to dead time at high-count rates. Plot the count rate against the activity of the source. The count rate at which a 20% drop in counts is discovered should be noted. In gamma camera, each detector's count rate must be monitored.

The decay technique and two-point source approach are two ways of assessing count loss and dead time. One decay approach is to study the decay pattern of ^{99m}Tc during an overnight period since it has a half-life of 6 h. The alternative decay approach advised by NU 1-2012 is to insert successive attenuating layers of 0.1-cm-thick copper (Cu) sheets and watch for a decline in count rates. This is the preferred strategy for acceptability testing and may be accomplished in minutes. Figure 16.33 depicts the configuration for the NU-1-2012 method.

16.15.9.1 Two-Source Method

A two-source technique may also quantify system dead time if the camera system acts as if it is paralyzed, which is the case in most cases.

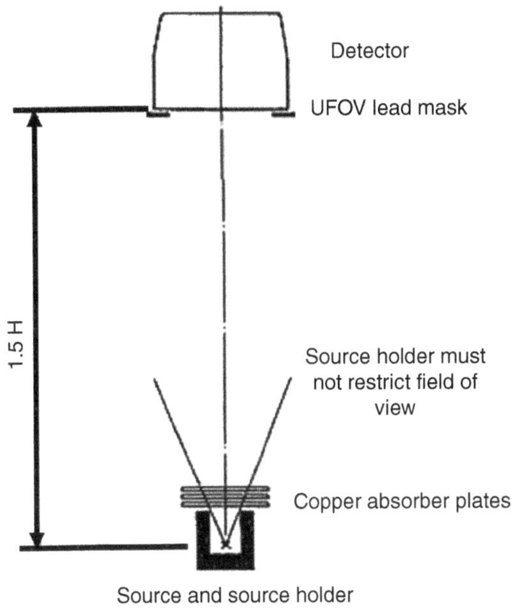

Detector

UFOV lead mask

1.5 H

Source holder must not restrict field of view

Copper absorber plates

Source and source holder

Fig. 16.33 Intrinsic count rate performance test positioning of the radiation source related to the detector. (Image copyright, the IAEA, TECDOC-602 (1991) [14]. Reprinted with permission)

When testing count rate in clinical settings, a collimator should be attached to the detector, and a scatter spectrum that matches the patient's scattering would be ideal. For testing reasons, creating these conditions might be a difficult task. This means that clinical count rate evaluations are unnecessary since the gamma camera system has a considerably greater capacity for count rates than is often found in the clinical context. On the other hand, a measurement with scattering would be necessary for clinical dynamic imaging circumstances where dead time count loss adjustments would be done, such as in cardiac study and renal dynamics.

Frequency

The highest peak count rate should be measured during acceptance testing. The paralyzed dead time interval may be calculated using the two-source technique, which considers all of the count rate properties for ^{99m}Tc.

Testing Procedure

16.15.9.2 Two-Source Method

Two nearly equal ^{99m}Tc (R1 and R2) sources with the activity of 50 µCi in each source holder should be made where each source results in about 10% count rate loss. Typically, such loss level occurs when a count rate of 80,000–100,000 cps is observed.

The intrinsic measurement is suggested. A measurement's activity is determined by factors such as detector size and distance from sources, which in this case may be as close as 300 mm. More than 75,000 counts per second may be achieved in the detector at 30 cm from the source activity at 35 µCi if the large field of view (LFOV) of the camera is 2000 cm^2 and is flushed with the photons. The activity would have to be scaled down by 2000 cm^2/detector area if the detector area is different from the 2000 cm^2.

The gamma camera should be configured to capture three static images simultaneously. Position source R_1 with activity 50 µCi at a 1-metre distance from the detector face (without using a collimator), and collect 10^6 counts in the first picture, noting down the time required to acquire 10^6 counts. Source R_1 should not be moved; instead, source R_2 with the same activity as in R_1 should be placed adjacent to it, and the image should be acquired for the same time as required for R_1; these counts are for source R_{12}. Set up is shown in Fig. 16.34 by removing source R_1 for the third image acquired the counts from source R_2 for the same period that was used for source R_{12}. The dead time for the two-source approach is calculated using the paralyze formula, which is as follows:

$$\tau = \frac{2 \times R_{12}}{(R_1 + R_2)} \times \ln \sqrt{\frac{(R_1 + R_2)}{R_1 \times R_2}}$$

The count rate (cps) for each source R_1, R_2 and R_{12} is computed separately. The dead time should be expressed in microseconds.

Calculate the input count rate, observe the output count rate for 20%, i.e. $R_{-20\%}$ and $C_{-20\%}$, respectively, and compare the results.

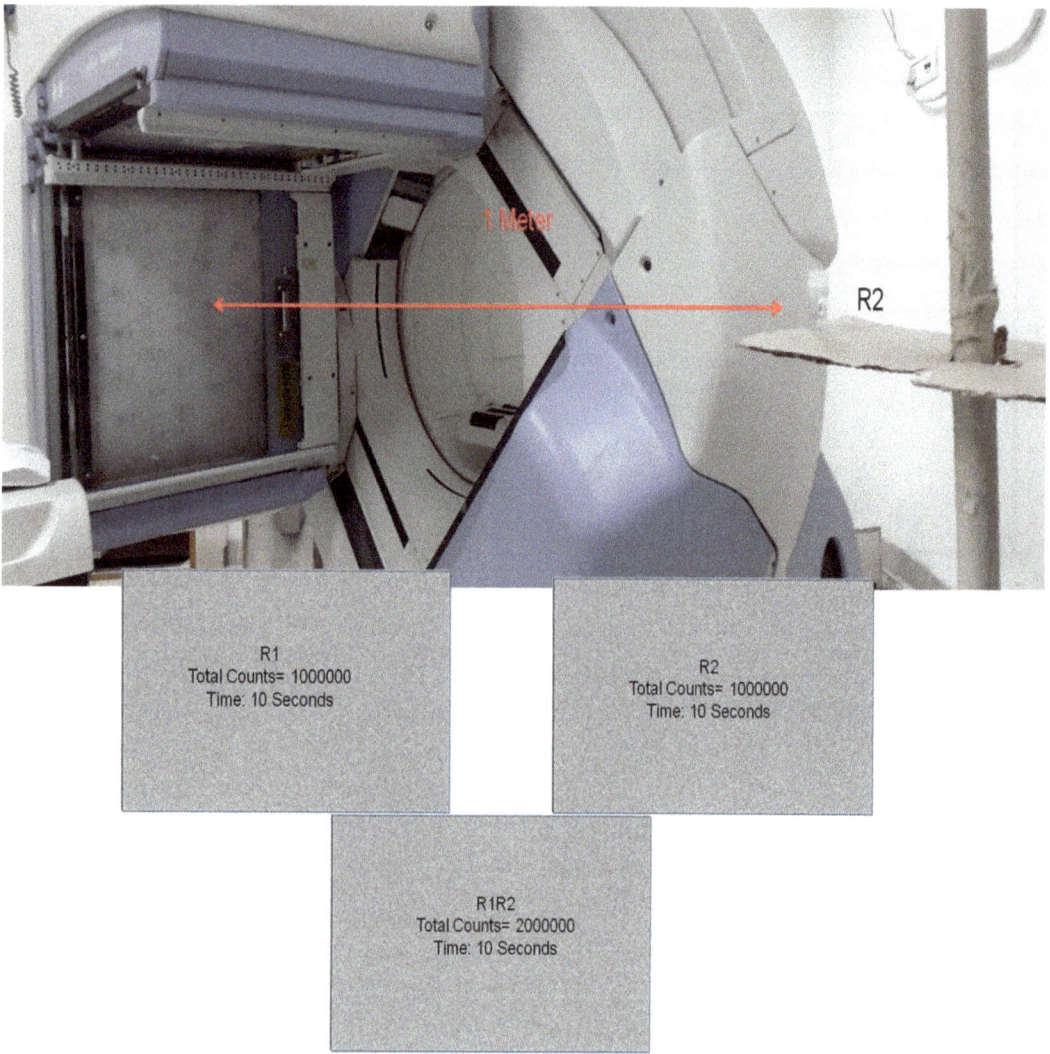

Fig. 16.34 Acquisition by using two source methods, R1 and R2

$$R_{-20\%} = \frac{1}{\tau} \times \ln\left(\frac{10}{8}\right) = \frac{0.2331}{\tau}$$

$$C_{-20\%} = 0.8 \times R_{-20\%}$$

Maximum Peak Count Rate Calculation

For this measurement, there should be a 37 MBq activity of ^{99m}Tc in the air. The first step is to determine the count rate of an airborne point source from a considerable distance away. Source are pushed closer to detectors until they achieve their maximum count rate, shown in Fig. 16.35. Then the count rate starts to fall.

16.15.9.3 Report

If any of these factors may be analysed, please indicate the maximum count rate, count rate at 20% loss and dead time. If the maximum count rate was achieved, regular or high-count rate mode must be indicated.

16.15.9.4 Performance Specifications

Refer to the manufacturer's requirements while doing acceptance testing. The acceptability value for maximum counts is within 20% of the reference value (+ or − 20%).

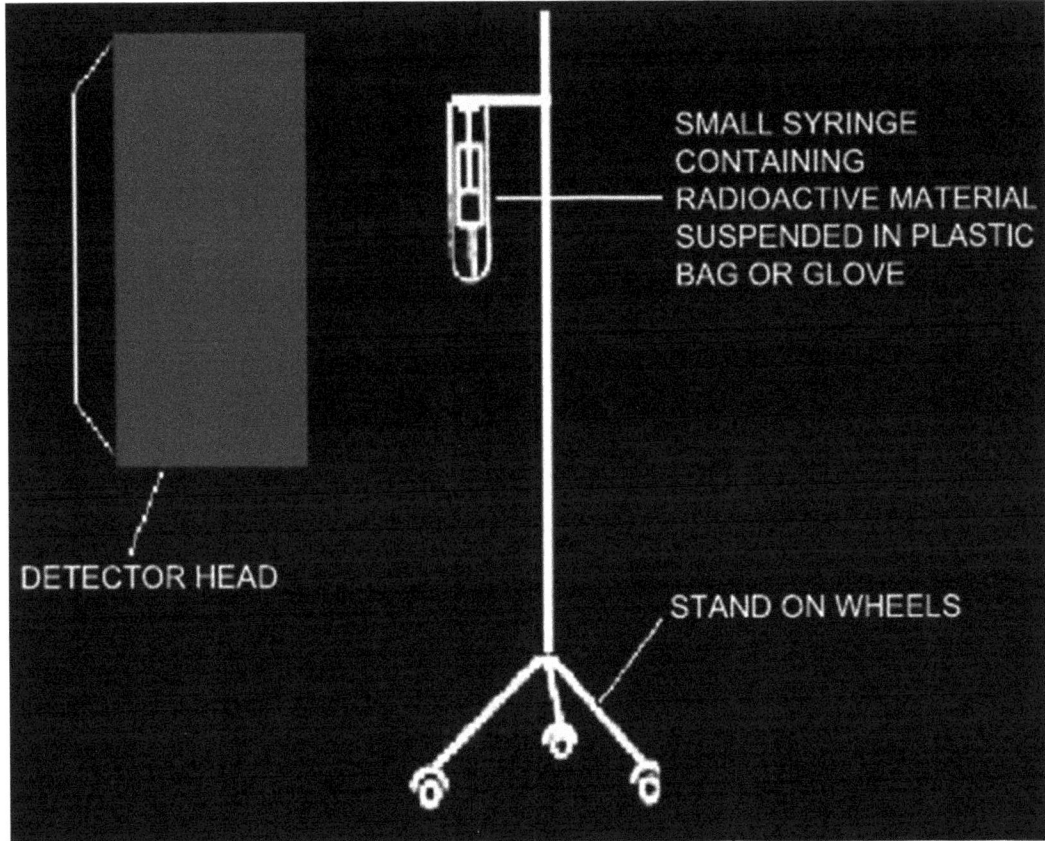

Fig. 16.35 Maximum peak count rate measurement using a point source of ^{99m}Tc. (Image copyright, the IAEA, TECDOC-602 (1991) [14]. Reprinted with permission)

16.15.10   Pixel Calibration

16.15.10.1   Purpose

The pixel size is utilized in various application programs for distance, surface area and volume calculations. The pixel size is determined by combining the matrix and zoom factors. To acquire precise organ or lesion size measurements, pixel calibrations must be performed. This is especially significant for calculating the absolute pixel size in the matrix employed in the reconstruction of tomographic images. This test must be carried out for all energies and collimators utilized in practice, and it must be computed for all matrix sizes and zoom settings.

16.15.10.2   Frequency

At the time of acceptance.

16.15.10.3   Testing Procedure

The test is carried out by putting two-point sources of ^{99m}Tc activity of 2 mCi each on the camera surface and X direction, approximately 10 cm apart, as illustrated in Fig. 16.36.

Configure the machine to execute a standard static acquisition of around 50 K counts using the smallest feasible matrix size, e.g. 256 × 256. Make certain that no zoom is utilized. Obtain one planar picture. Duplicate the whole method, but place the point sources along the Y-axis this time.

The same approach should be followed for all tomographic zoom situations and all other radionuclides and collimators utilized in practice.

16.15.10.4   Image Analysis

As seen in Fig. 16.37, draw a profile over the two-point source image. To get the pixel size,

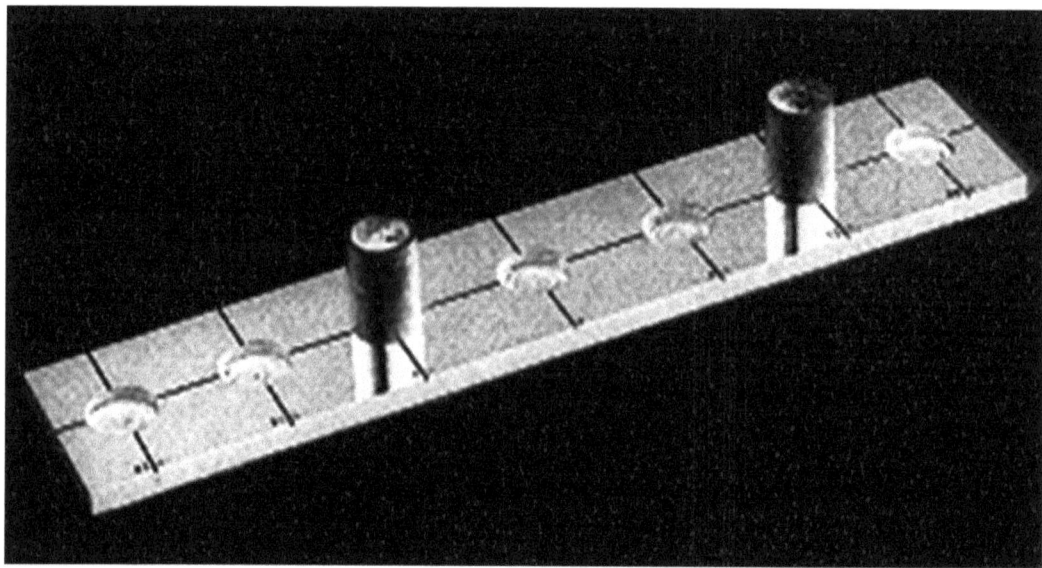

Fig. 16.36 Two-point sources of ^{99m}Tc with 2 mCi activity were maintained 10 cm apart

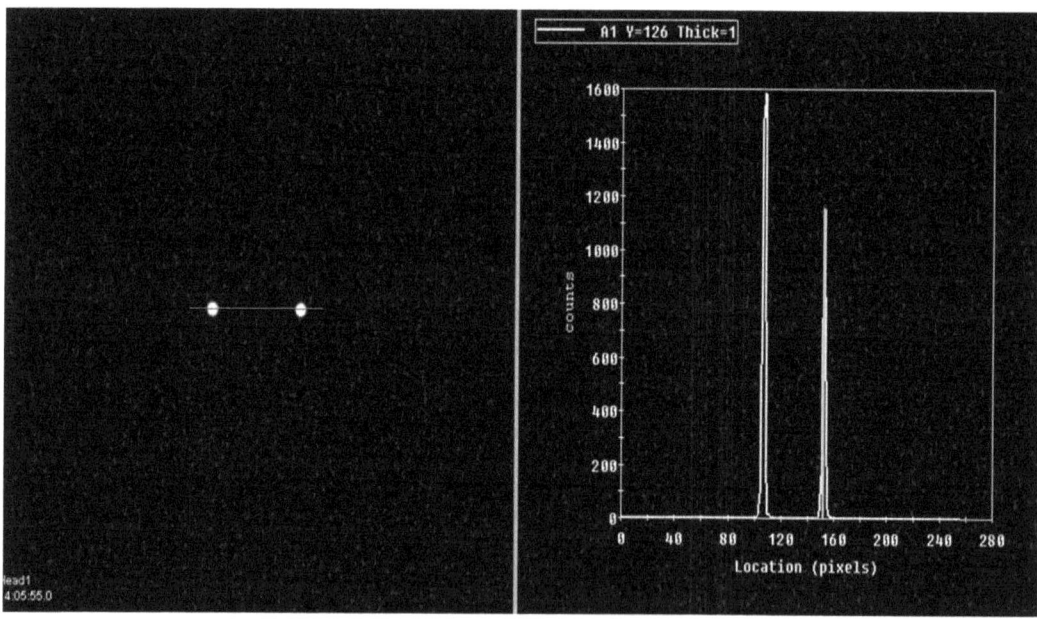

Fig. 16.37 Pixel size determination by drawing a profile across the point source image

keep the cursor on the peak of the line profiles of the two-point sources, and divide the mm distance between them by the pixel equivalent distance, and you'll have the answer. Pixel dimensions are expressed in millimetres.

$$\text{Pixel size} = \frac{\text{Distance between sources}}{\text{No.of pixels between two peaks}}$$

When calculating the pixel size of other matrix size, we must multiply the results by the neces-

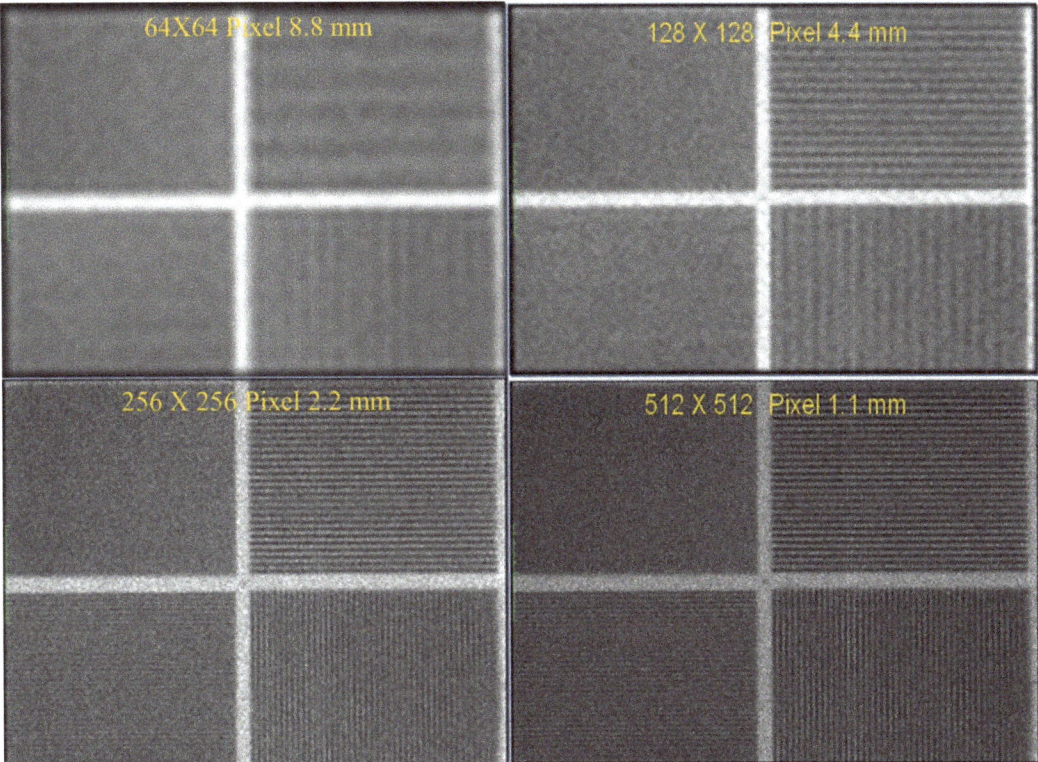

Fig. 16.38 According to the matrix size, the pixel size affects the resolution and noise in the image

sary factors. To get the 64 × 64 matrix pixel size from the 256 × 256 matrix size, we must multiply the 256 × 256 pixel matrix size; suppose it is 1 by a factor of 4, the pixel size is 4 mm. Additionally, if we computed pixel size of 64 × 64 matrix size, e.g. 4 mm, and wish to derive for 256 × 256 matrix size, we must divide by a factor of four to get pixels to size for 256 × 256, and it will be 1 mm. The pixel size affects the resolution and noise in a picture and the SPECT slice thickness, as seen in Fig. 16.38.

Performance Specifications

The difference between X and Y pixel size shouldn't be more than 5% of the whole.

16.15.11   Multiple Window Spatial Registration (MWSR)

The spatial resolutions of the individual energies are combined in the final image when multiple windows are utilized to form an image. If the spa-

tial registrations for the various energy windows are not correctly positioned, the resultant picture will lose spatial resolution, as demonstrated in Figs. 16.39 and 16.40. The spatial registration may be visually checked by imaging a source using multiple energy windows and comparing it to produced images independently by each window. Four different locations must be used, and image variances must be measured in millimetres for images taken from these four different locations.

16.15.11.1   Frequency

Optional acceptance testing.

16.15.11.2   Testing Procedure

Test Source

NU 1-2012 requires using ^{67}Ga point sources with 93, 184 and 300 keV photon energies. Other radionuclides, such as ^{111}In or ^{201}Tl, might be employed. As illustrated in Fig. 16.41, the point source is put in a container shielded with lead

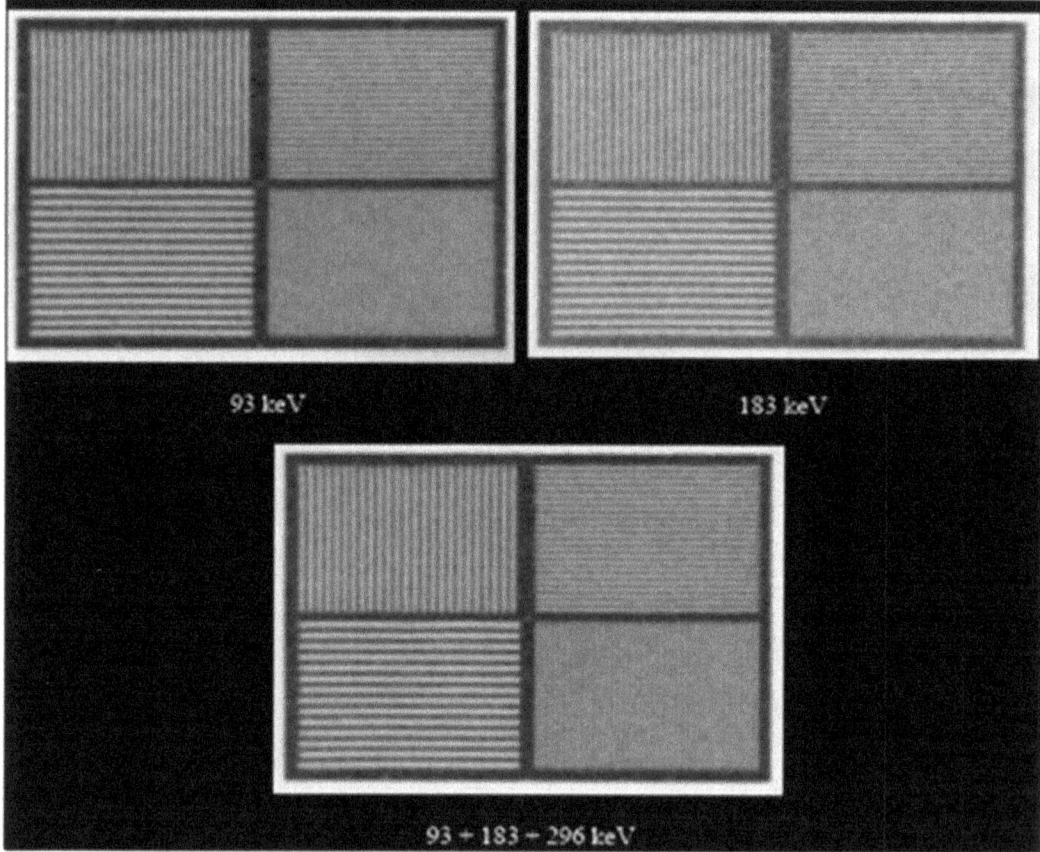

Fig. 16.39 Because MWSR is excellent, bar phantom images obtained with ^{67}Ga exhibit strong spatial registration. (Image copyright, the IAEA, Quality Control Atlas for Scintillation Camera Systems (2003) [15]. Reprinted with permission)

with a 0.3 cm diameter opening at the base. The lead container with the source may be mounted directly on the detector crystal covered by decoy or a LEAP collimator. On a parallel-hole collimator, five radionuclide sources of ^{67}Ga in five plastic vials may be used—one in the centre of CFOV and four at the camera's corners of the UFOV.

Image Acquisition

The test source is scanned nine times: one in the centre, four times along the X direction and four times across the Y direction. The off-central locations must be 0.4 and 0.8 times the distance from the central location to the border of the UFOV along the appropriate axes, as shown in Figs. 16.42 and 16.43. Create a 20 percent energy window for each image. Collect photos of point sources until the highest pixel count hits a thousand counts.

Image Analysis

You should draw lines 5–10 pixels wide in both the X and Y directions over each point source images you took. Centre the line profile on the point source and make it short. It should have an uneven number of pixels that contains the maximum counts' half-maximum on both sides. Use the X and Y direction line profiles to figure out the centroid of each point source's counts in millimetres. Using the method provided in Fig. 16.44, calculate and compare the shift in millimetres for every energy window pairing.

$$\sqrt{\left(x_1 - x_2\right)^2 + \left(y_1 - y_2\right)^2}$$

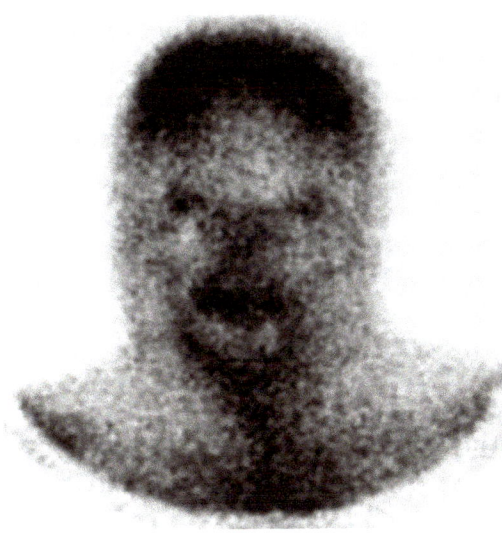

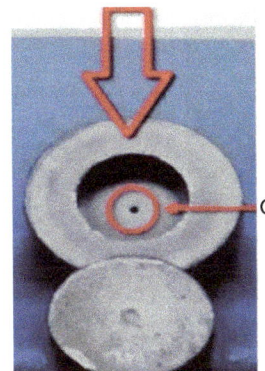

Fig. 16.41 A lead-lined cylinder with a 3 mm hole at the bottom. (Image copyright, the IAEA, YouTube tutorial on quality control tests for SPECT systems [13]. Printed with permission)

Fig. 16.40 Because the spatial registrations for the multiple energy windows are not adequately positioned, brain pictures produced with ⁶⁷Ga have poor spatial resolution. MWSR is not acceptable. (Image copyright, the IAEA, Quality Control Atlas for Scintillation Camera Systems (2003) [15]. Reprinted with permission)

Fig. 16.42 The test source is located in the middle of the detector. (Image copyright, the IAEA, YouTube tutorial on quality control tests for SPECT systems [13]. Printed with permission)

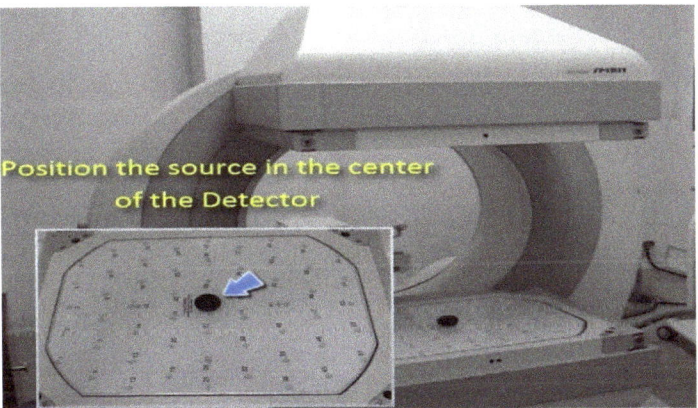

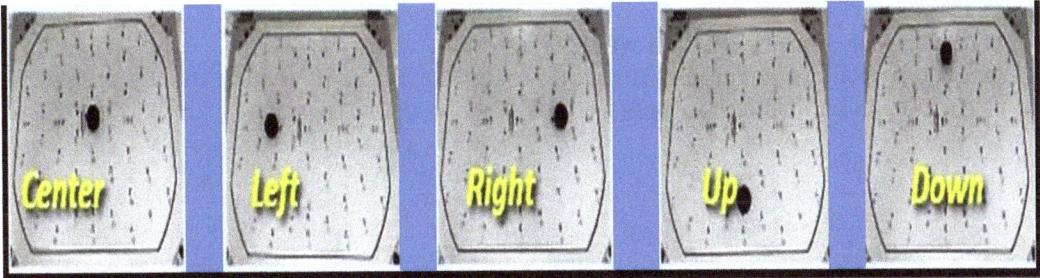

Fig. 16.43 The test source is positioned at the detector's off-off-centerition. (Image copyright, the IAEA, YouTube tutorial on quality control tests for SPECT systems [13]. Printed with permission)

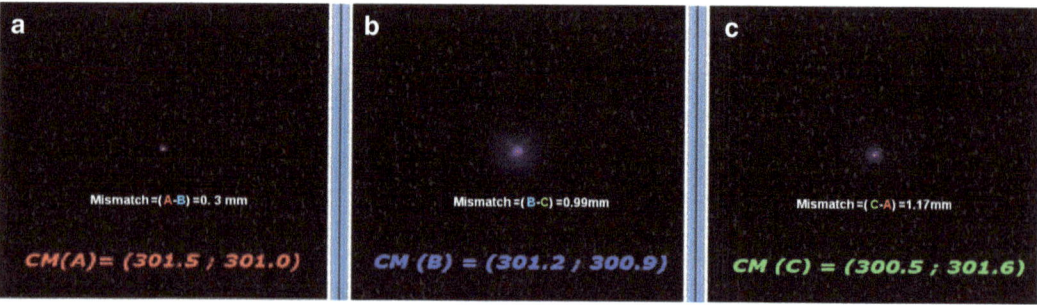

Fig. 16.44 Use the line profiles drawn in the X and Y directions to figure out the centroid of each point source's counts in millimetres. (Image copyright, the IAEA, YouTube tutorial on quality control tests for SPECT systems [13]. Printed with permission)

Report
It's important to say which radionuclide was used and how far it moved.

Requirements for Performance
Consult the manufacturer's instructions for the acceptance testing process.

16.15.11.3   Test of Collimator Hole Angulation

Purpose
To ensure that a slant hole collimator is set up appropriately for a tomographic acquisition for SPECT imaging, the angle of the holes in the collimator must be determined. This implies that the holes must parallel the rotational axis (AOR) direction.

Image Acquisition
- When you're ready to test the collimator, put it on the detector's head, and then turn it on. Turn the head to face the wall farthest away from the detector.
- It's best to put the point source of 2–3 mCi of ^{99m}Tc on a wall or other surface, at least 1.5–2.5 m away from the collimator. The source should be aligned with the centre of the collimator.

- Take a picture with a preset level of 5×10^6 counts and the largest possible matrix size with the system.
- Do the same thing again to take pictures at four more places on the collimator.
- Do the test for each of the parallel-hole collimators that you have. There is usually much damage to the foil collimators, but not to the cast collimators.

Image Analysis
After taking images of various locations, check them visually to determine the collimator septa's alignment and the holes' alignment. Whenever the image of the collimator is clean and free of artefacts (as seen in Fig. 16.45), the hole angulation is optimal.

Striping in the X and Y sides suggests major collimator hole angulation issues during fabrication. The axes of the holes are not straight and perpendicular to the surface of the collimator (Fig. 16.46).

16.15.12   Test of Collimator Quantitation Hole Angulation

Turn the detector so that it is towards the ground. Prepare a point source of 2–3 mCi of ^{99m}Tc to be used as a point source. Get a 50 K count on a

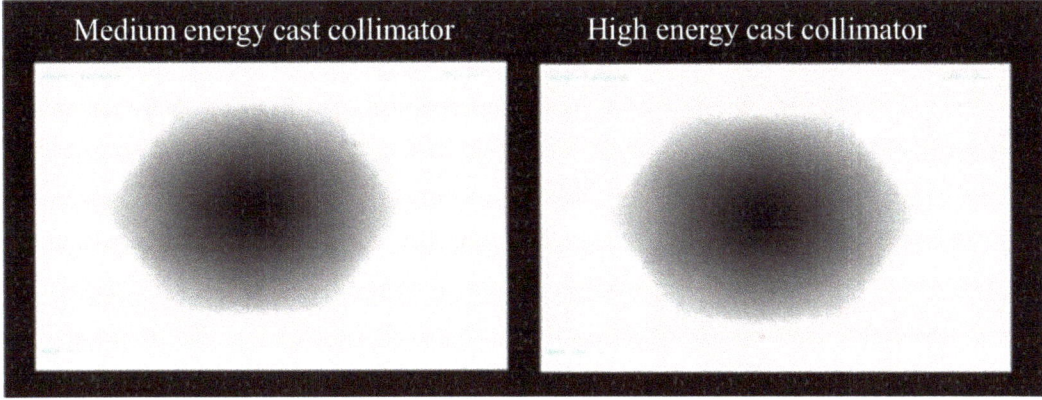

Fig. 16.45 Collimator septa and hole alignment is evaluated using a distant point source: cast and foil collimators indicate that the holes are perpendicular and parallel, which is acceptable. (Image copyright, the IAEA, Quality Control Atlas for Scintillation Camera Systems (2003) [15]. Reprinted with permission)

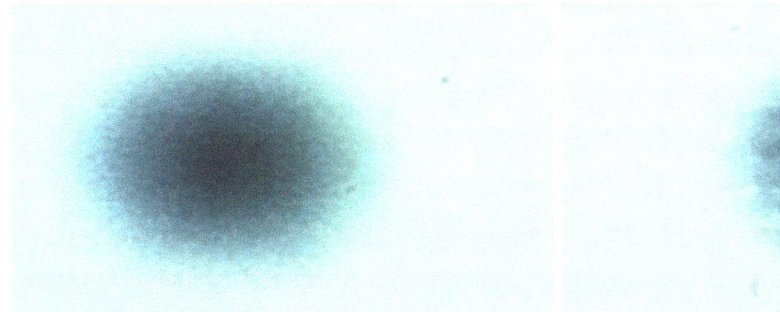

Fig. 16.46 The striped response in both the X and Y directions indicates serious collimator hole angulation difficulties during manufacture, i.e. the axes of the holes are not parallel to one another and perpendicular to the collimator's surface. (Image copyright, the IAEA, Quality Control Atlas for Scintillation Camera Systems (2003) [15]. Reprinted with permission)

standard planar image using the greatest possible matrix size. Image first at 5 cm and then lower the holder precisely 10 cm and image. Calculate the two-point image's x and y centroids (Fig. 16.47) by performing the following:

- for the lower position = (x_{lower}, y_{lower}),
- for the upper position = (x_{upper}, y_{upper}).

Using the formula below, get the separation between the two images (d).

$$d = \left(\left(x_1 - x_u \right)^2 + \left(y_1 - y_u \right)^2 \right)^{1/2} \times \text{Pixel size}$$

The hole angulations (HA) are calculated following the formula HA = arc tan (d/100).

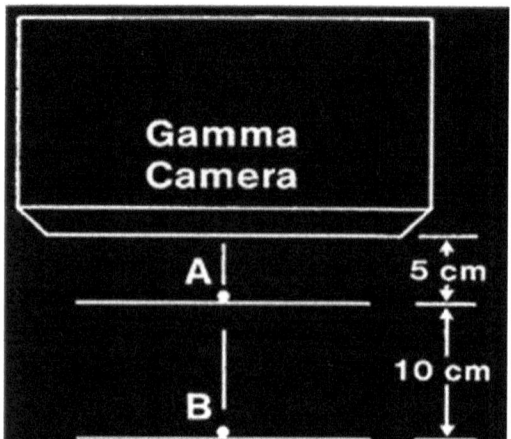

Fig. 16.47 The collimator hole angulation test configuration is shown here. Source location A is the place where the measurement is taken. After that, the source is moved vertically by 10 cm to point B, where a second measurement is taken from the source

16.16   Quality Control of SPECT System

Planar gamma camera imaging gives only a two-dimensional depiction of a three-dimensional distribution of activity within the body, but SPECT (single-photon emission computed tomography) delivers three-dimensional information about the organ by removing the overlapping underlying structures. The reconstruction approach amplifies image defects in planar views. To limit the detrimental influence of many artefacts on the quality of SPECT examinations, a strong quality assurance system is essential.

16.16.1   Flood Uniformity

Every SPECT camera has non-uniformity, even when properly adjusted; nevertheless, although these variances in-field uniformity are tolerated for planar pictures, they may cause severe artefacts in reconstructed SPECT images. Changes in local magnetic flux effects on PMT during detector rotating, PM tube deterioration, collimator shift and differential non-linearity in the computer interface might produce variations in

artefacts. This isn't a big deal with an improved version of PMT shielding designs. Defects in the collimator, as well as differences in sensitivity. Any non-uniformity flaw in an image is projected around the 360° (ROR) radius of rotation during reconstruction, forming a ring. This artefact is termed 'ring artefact' or 'bull's eye'. A region that is 'hot' might result in non-uniformity. If this is the case, the pattern that develops will be a ring of increased activity. The ring artefact is based on the following factors:

(a) The intensity and position of non-uniformity
(b) The object's size (proportional to the ratio of the object's size)
(c) Image counting statistics

16.16.2   Centre of Rotation (COR), Multiple-Head Registration and Head Tilt

The tomographic spatial resolution may be reduced because of the SPECT detector's inaccuracy, incorrect centre of rotation (COR) measurement, incorrect multiple-head registration (MHR) or axial head tilt. According to the manufacturer's standards, this approach is used as a standard quality control method monthly or as-needed basis.

When employing the COR calibration approach, ensure that the SPECT system's axis of rotation (AOR) is aligned with the projection image matrix's centre pixel column (Fig. 16.48). When used with multiple detector SPECT systems, the MHR acts as a calibration to ensure that each detector head samples the same volume. To maintain the detector heads parallel to the AOR at all times.

COR and MHR calibration is usually combined into a single measurement. The manufacturer's procedure handbook specifies the technique of measurement, which must be followed. Separate measurements are needed for various SPECT systems' 180- and 90-degree detector arrangements. Typically, the 90-degree arrangement is reserved for cardiac SPECT imaging.

Fig. 16.48 The camera obtains and shows the centroid, the source's position. Note how the two heads (180 degrees apart) position the two centroids differently

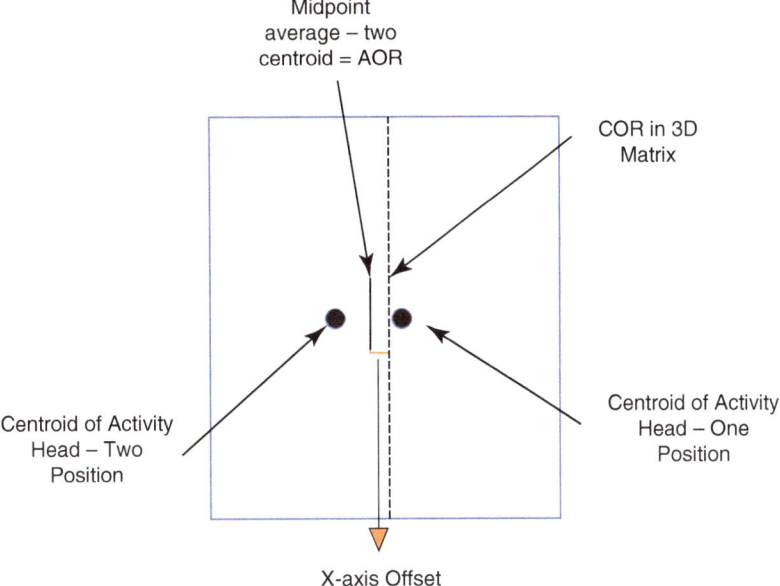

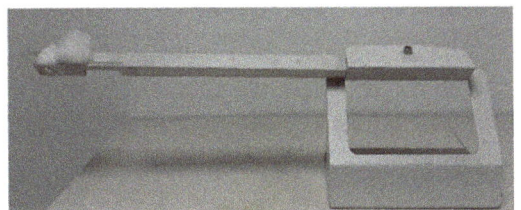

Fig. 16.49 Point source holder for COR

As shown in Fig. 16.49, point sources are placed on the imaging table or in a custom-built fixture to do the measurement. Each detector head must be rotated 360 degrees to acquire point sources using SPECT. As a function of detector angle, the calibration software determines and monitors the projected location of each point source in the sinogram of the camera pictures. Tomographic reconstruction is not necessary in the case of COR. The COR and MHR are calculated based on projected source locations.

Axial head tilt cannot be seen on the sinogram. It may be viewed in a cine presentation of the SPECT projection pictures. A head tilt is indicated by any sinusoidal oscillation in the axial (vertical) direction. Calibration of axial head tilt is not possible. Its repair necessitates using a service specialist to make a mechanical adjustment.

16.16.2.1   Preparing a Point Source

A single or several point source is used to check the calibration of the COR and multiple-head registration (MHR). The source should be put at the bottom of a V-shaped vial, although it might also be at the bottom of a syringe cap. The activity of the source is determined by manufacturer specifications, which vary from 74 to 111 MBq (2.0–3.0 mCi). Ensure that the activities of the different sources are within 10% of one another while assembling them.

Please refer to the gamma camera instructions or manufacturer manual for information on how many point sources to prepare and the best suitable activity. Testing and calibration might fail if the manufacturer's recommendations aren't followed.

16.16.2.2   Procedure for Checking

Point Source Placement

A compact ^{99m}Tc point source is used, as well as a method of keeping it in the air within the field of view, such as connecting the source to a long ruler or a specifically built supporting device (Fig. 16.49).

SPECT Acquisition

1. Confirm that the camera is properly aligned and that the head is parallel to the axis of rotation and not inclined using a spirit level. Keep the point source around 2 cm from the axis of rotation and around 2 cm from the field of view's centre in the air.
2. Carry out a standard tomographic acquisition using the finest digital matrix size possible, such as 256 × 256, gathering around 10,000 counts at each angular location. For this test, using 32 angles across 360° is sufficient.
3. Repeat step 1, but set the point source around 10 cm away from the rotation's centre.
4. Positioning the point source along the rotation axis but as far away from the centre slice as possible, for example, within 5 cm of the field of view's positive Y edge. During tomographic acquisition, it serves to retain the point source in the camera's field of view.

5. After measuring point 4, maintain the point source towards the field of view's edge in the opposite direction of the positive Y. Three-point sources may be used if the required software is available by doing a single set of measurements (Figs. 16.50 and 16.51).

Projection Image Processing and Analysis

Most manufacturers' systems come with software that calculates and includes the required correction into the regular tomographic acquisition and reconstruction method. The tests and processes used by each system vary significantly.

This test aims to pinpoint the centre of gravity of the point source picture angle by angle and to estimate the position of the centre of rotation.

COR Calculation (Infinia Hawkeye 4 SPECT/CT) (Figs. 16.52 and 16.53)

COR X (alpha) = COR X Offset + S × sin (alpha) + C × cos (alpha)

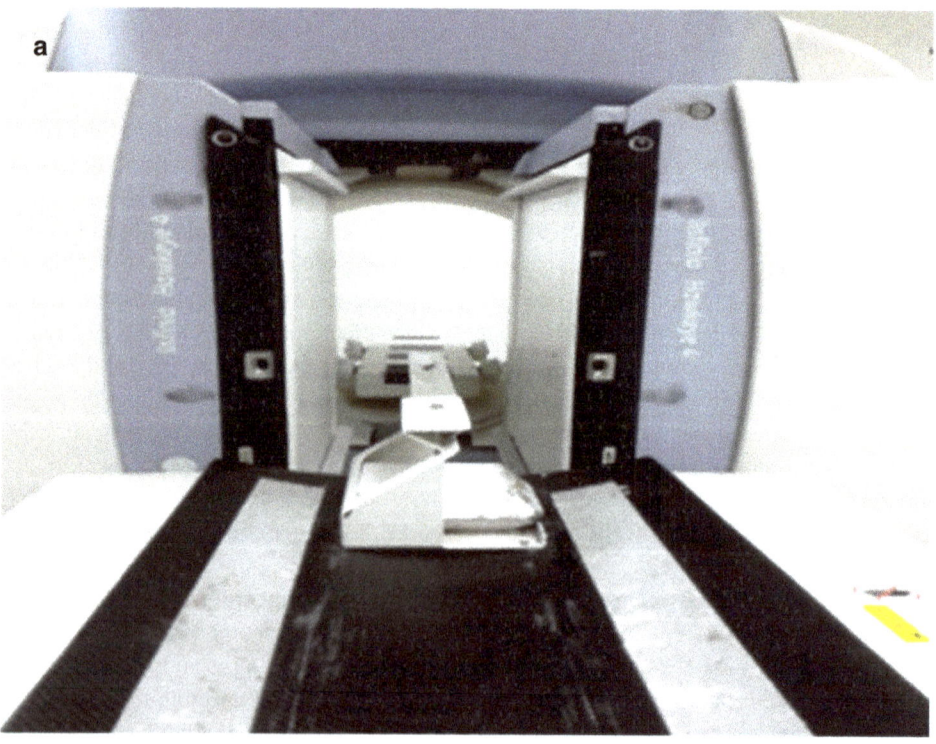

Fig. 16.50 (a) Position of point source ^{99m}Tc in the air for 180° COR. (b) Position of point source ^{99m}Tc in a red circle for both detectors in acquisition monitor for 180° COR

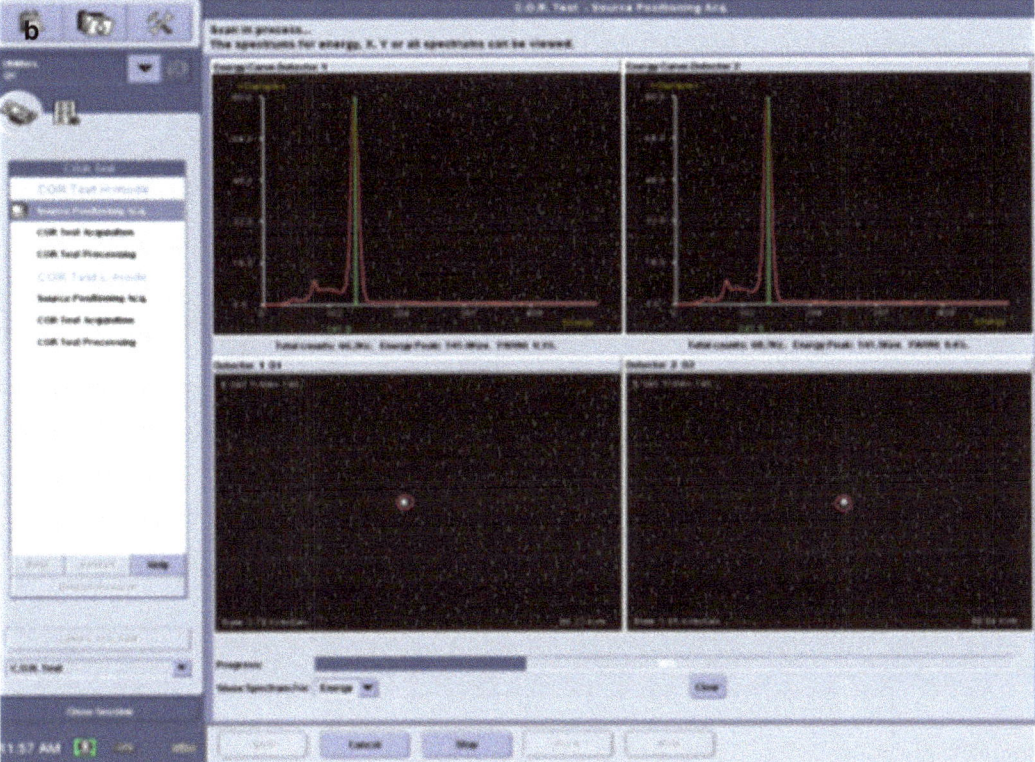

Fig. 16.50 (continued)

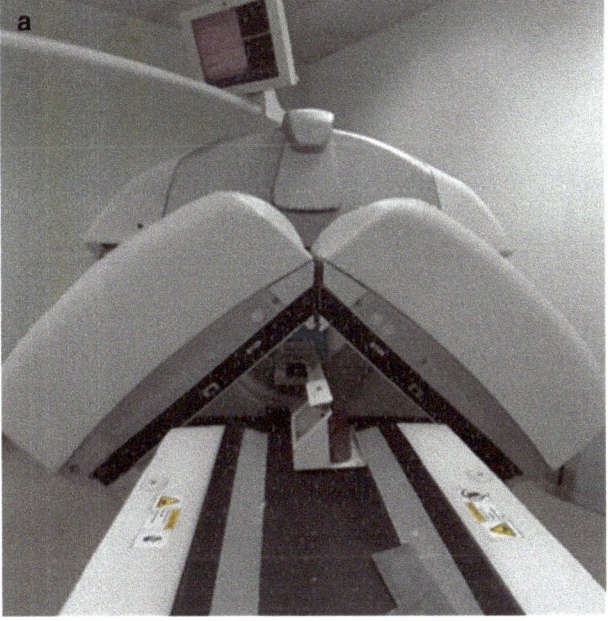

Fig. 16.51 (a) Position of point source ⁹⁹ᵐTc in the air for 90° COR. (b) Position of point source ⁹⁹ᵐTc in a red circle for both detectors in acquisition monitor for 90° COR

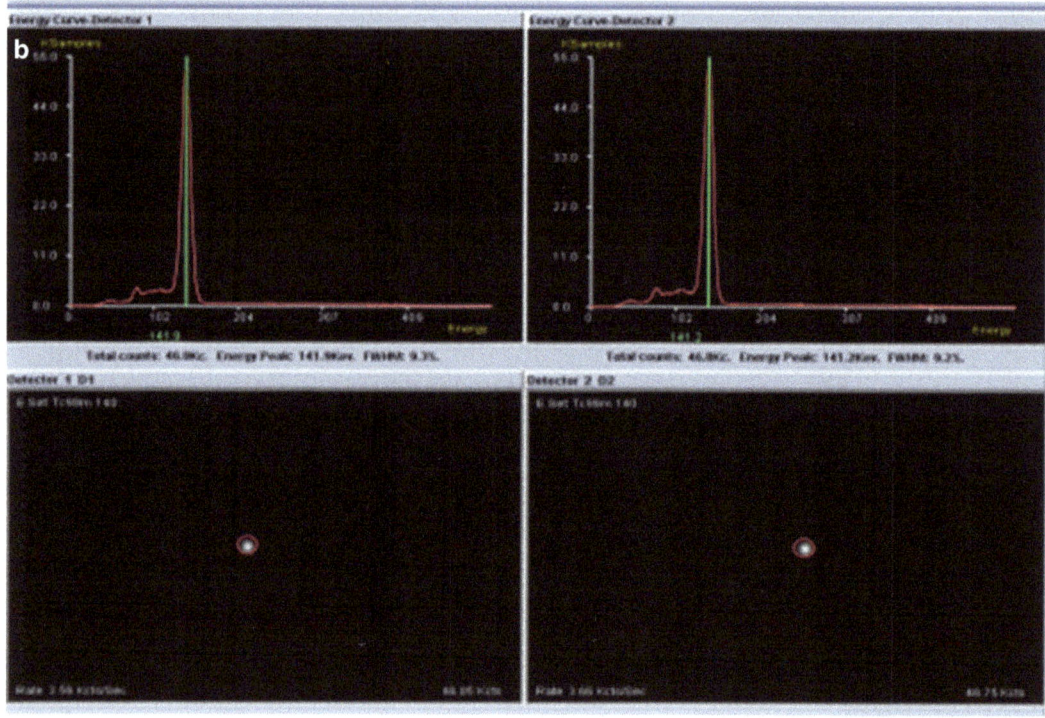

Fig. 16.51 (continued)

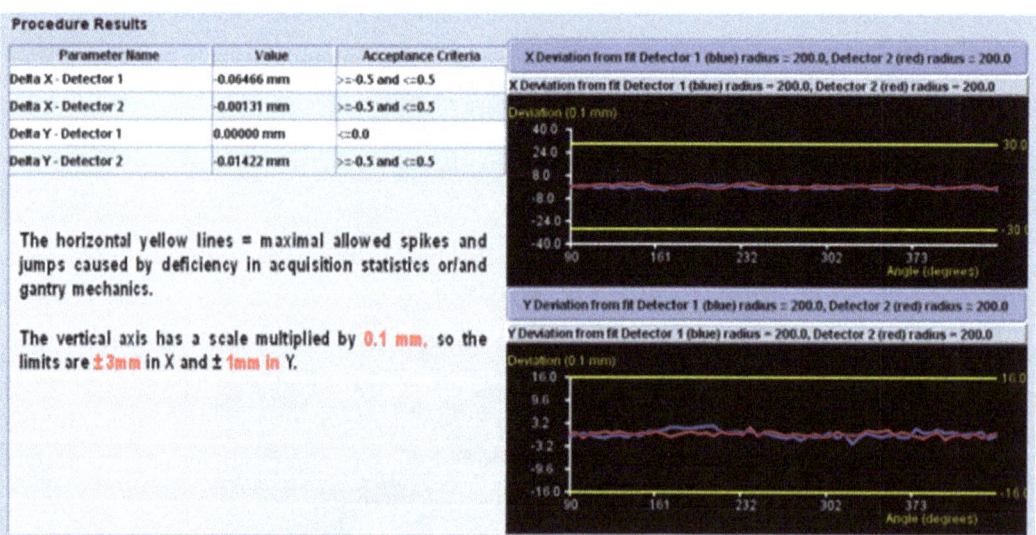

Fig. 16.52 COR processing for 180° in GE Infinia Hawkeye 4 (SPECT/CT)

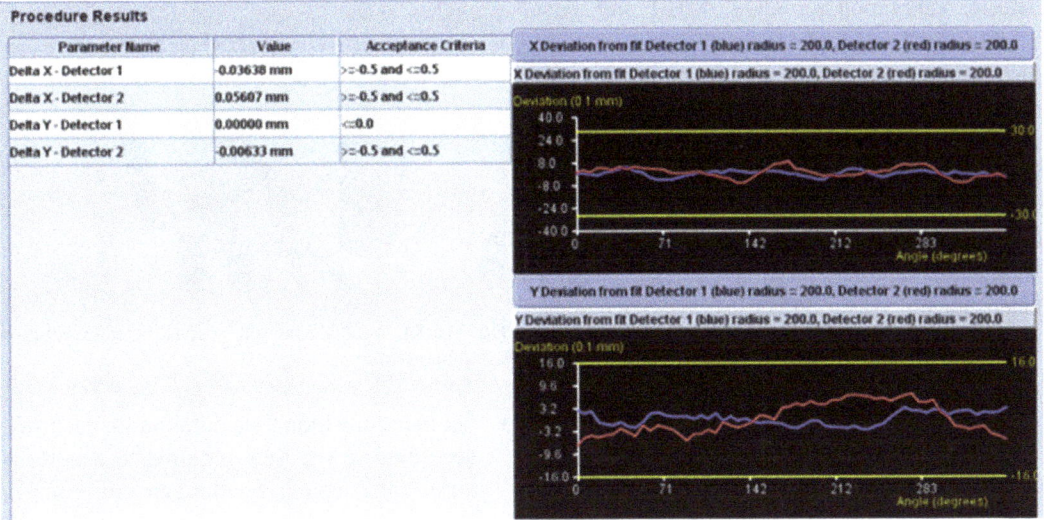

Fig. 16.53 COR processing for 90° in GE Infinia Hawkeye 4 (SPECT/CT)

- 'alpha' is the rotation angle, and the 'S' and 'C' are constants that reflect how far from the COR the point source was placed.
- The 'COR X Offset' constant reflects the offset between a virtual projection of the gantry axis of rotation on the transaxial axis (X) and the actual image. The same calculation is implemented for the Y-axis.
- COR X = A + B ×Radius; A and B are constant values stored in the system file per collimator type.

Report and Frequency

1. A 0.5-pixel offset reduces image quality, resulting in (1) a 30% drop in image resolution and (2) a 40% drop in contrast.
2. COR analysis is also essential for all collimators, especially when SPECT imaging is done with medium and high-energy collimators. Think about the added weight on your detector's gantry and how it can affect the rotating axis (AOR).
3. Put your detectors perpendicular to the image source to avoid a 3D matrix misalignment caused by a small tilt in the AOR. The offset will occur on the Y-axis if this occurs.
4. PMT (or ADC) fluctuation may exacerbate a fault as the camera spins around its axis, resulting in a COR error.

5. COR checks should be performed weekly as part of your usual QC. The allowable COR limit is ±1 mm. Some literature suggested that it should be monthly. You can decide the frequency to perform the COR according to your convenience and the manufacturer's advice.

16.16.3   Resolution of Tomographic Image

The following factors influence the reconstructed tomographic image's resolution:

1. System resolution for planar imaging with a choice of the collimator.
2. The orientation and radius of rotation of the detector.
3. Electronic COR, axial head tilt and MHR for several detectors.
4. Reconstruction algorithm—should be performed using FBP with a ramp filter.

A collimator intended for ^{99m}Tc, commonly a LEHR or parallel-hole collimator, is used to image a line source in the air at a set ROR.

For sources recorded at the same distance from the surface of the collimator as the radius of rotation, in this case, the reconstructed tomographic resolution should be substantially equivalent to the planar resolution. Inaccurate gamma camera detector calibration for COR, MHR and axial head tilt may reduce the tomographic resolution. Inspection of the SPECT line locations or point sources in the projection pictures generated during the line source's SPECT acquisition might reveal the mistakes. In a multi-detector SPECT system, a loss in tomographic resolution might be due to only one of the detectors. To get a full 360-degree acquired rotation for each detector independently, more SPECT acquisitions may be necessary.

16.16.3.1 Frequency

- Acceptance testing—An A-line or point source should be used to assess the tomographic resolution.
- If just one detector in a multi-detector system seems to be failing, separate 360-degree acquisitions for each detector may not be necessary.
- Weekly quality assurance.

16.16.3.2 Procedure for Checking

Line Source Placement

- When using ^{99m}Tc at an activity concentration of 1 mCi per cm^3 or the manufacturer's recommendation, the line source should be stretched to the end of the imaging pallet and over the gamma camera detectors.
- Position the source parallel to and close to the axis of rotation (AOR).
- Set the detectors' ROR to 20 centimetres. Reduce the ROR's radius to the smallest feasible value, and record it if 20 cm ROR is not possible. All acquisitions should have the same ROR.

SPECT Acquisition

- Figure 16.54: Acquire projection pictures using a circular orbit and step-and-shoot mode.
- Set the acquisition matrix to 128 × 128 and the number of views to 120 across 360 degrees.
- If required, employ an image magnification factor (zoom) to get the pixel size between 3.0 and 3.5 mm.

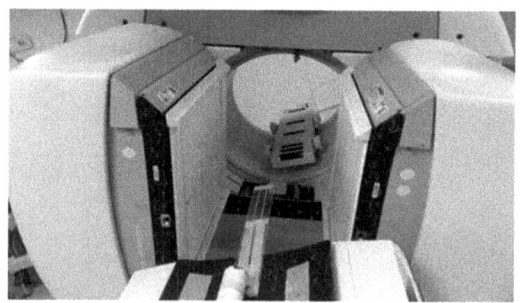

Fig. 16.54 Acquisition for reconstructed resolution using line source

- Set the acquisition time duration for each stop according to the time required to acquire at least 100 K counts for the first projection or image.

SPECT Image Reconstruction

Reconstruct the SPECT data for all data sets, including additional acquisitions for 360-degree SPECT for each detector, within the volume covered by the line source. FBP and the ramp filter were used to recreate the reconstructed SPECT image. An iterative reconstruction method may also be used. All resolution-enhancement settings for iterative reconstruction must be turned off which means no smoothening filter should be used. Only ramp filter is used while using iterative reconstruction methods.

Planar Image Acquisition

The detector should be positioned at the same distance as the ROR used during the SPECT imaging to get a planar picture. This can be accomplished without modifying the line source. Rotate the detectors clockwise to return the detectors and gantry to the start angle of the previous SPECT imaging or the detector at 180°.

Use the same acquisition matrix and magnification factor as the SPECT imaging, and ensure that each image counts at least 10,000.

Image Processing and Analysis for Spatial Resolution

The axially reconstructed projections of the line source are shown as point source distributions.

The FWHM of associated point-spread functions (PSF) is calculated during image processing by drawing a line profile (count density) across the point source image. All SPECT data sets are analysed simultaneously, including those from each detector's full 360-degree collecting arc.

Reconstructed transaxial images may be shown in Figs. 16.55 and 16.56.

Examine the point source pictures in three transaxial slices, one in the centre of the line source and the other two at 1 cm from each end of the line source. To get the PSFs, draw a 1-pixel-

Fig. 16.55 FWHM measurement in X and Y directions using LSF (line spread function)

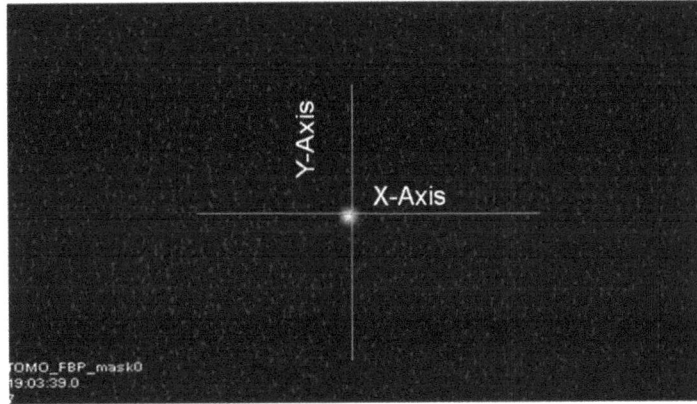

Fig. 16.56 FWHM measurement in X and Y directions using LSF (line spread function)

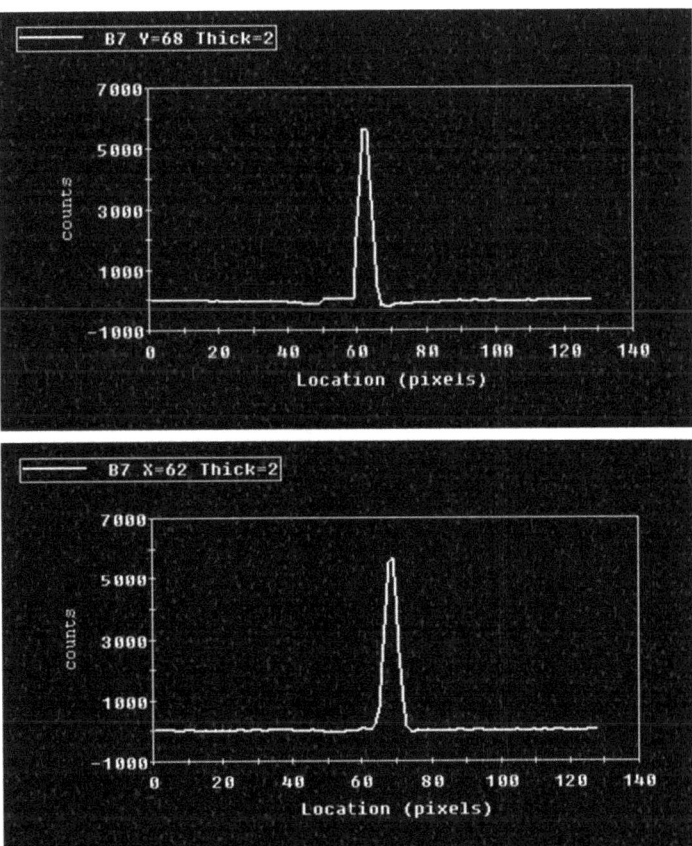

wide line profile in the X and Y directions across the brightest pixel in the reconstructed picture, and use curve fitting to get the FWHM. Neither the X nor the Y orientations are necessary when the line source is in the AOR.

Calculate the FWHM in mm for all PSFs obtained in axial slices and keep track of it.

Finding the maximum and half-maximum points in the PSF may need interpolation.

Each detector's picture of a planar line source is as follows:

- In the planar picture, calculate the FWHM (mm) of the LSF obtained at the three slice points of the line source.
- Determine and record the FWHM. The average FWHM of all detectors is then calculated. Compare each reconstructed data set's spatial resolution.
- Measure all detectors' relative planar spatial resolution to the average tomographic spatial resolution.
- Determine why the spatial resolution of SPECT-reconstructed images is 10% greater than planar resolution. It should be within 10%

MHR, COR and Head Tilt Error Analysis

Evaluate each SPECT data set's sinogram and lipogram. The line source's trajectory is projected onto the rotating gamma camera detector in the sinogram. The curve is sinusoidal. It should seem straight if the line source is at the AOR. Inspect the sinogram image for distorted curves and gaps. A detector that sags during rotation might produce a wavy curve. A gap or discontinuity may be seen in the sinogram in multi-detector systems. Estimate the size of the break or discontinuity, and it should be within a half-pixel distance.

Report

- Report the technique used to measure reconstructed tomographic resolution and if a set of data were gathered for each detector's entire 360-degree rotation. Report the image reconstruction technique utilized if it was not FBP.

- Provide the average FWHM values for the SPECT and planar.
- Describe any major COR, MHR or detector head tilt errors.

Performance Specifications

- If the reconstructed FWHM is less than 10% of the planar resolution, the SPECT/planar ratio should not exceed 1.1.
- The sinogram and lipogram should be within the manufacturer's standards.

16.16.4 Tomographic Uniformity and Contrast

- The Jaszczak SPECT phantom images are evaluated for spatial resolution, contrast and uniformity using ^{99m}Tc radionuclide. Other radionuclides may be tested. The uses of the SPECT phantom are described below and shown in Fig. 16.57:
- The collimator, artefacts, calibration and reconstruction parameters are all evaluated for their performance on the system.
- Evaluation of the inaccuracy in the centre of rotation.
- Evaluation of an artefact with non-uniformity.
- The effect of modifications in the radius of rotation on spatial resolution is investigated.
- The effect of reconstruction filters on spatial resolution.
- Analysing the effects of attenuation and scatter correction.
- Sensitivity of single slice volume.
- Sensitivity of total system volume.
- Delectability of the lesion.

16.16.4.1 Frequency

Quality assurance and control of Jaszczak phantom image quality study will serve as a benchmark for further annual and monthly quality assurance testing of overall SPECT performance. Quality assurance is done regularly, and it is done every 3 months.

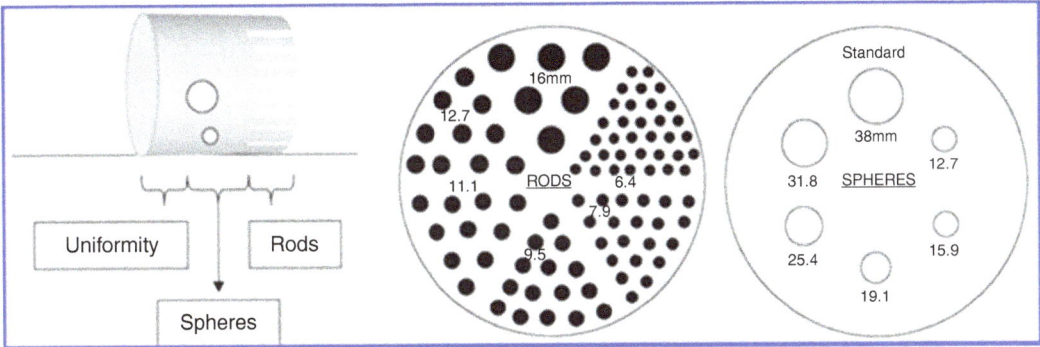

Fig. 16.57 Jaszczak SPECT phantom (image quality phantom) and its parts

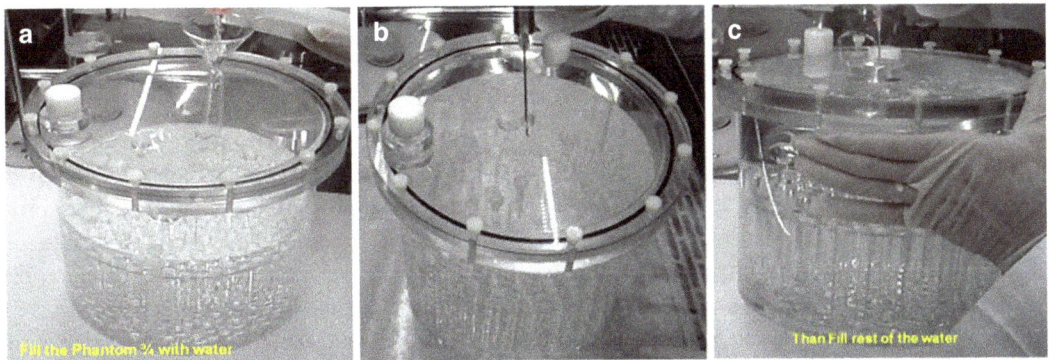

Fill the Phantom with ¾ of water   Insert the 10 mCi (400 MBq) of   Fill rest of the water
                                                 99mTc by syringe

Fig. 16.58 Steps for preparing a SPECT phantom. (Image courtesy of the International Atomic Energy Agency's YouTube lesson on quality control checks for SPECT systems [13]. Printed with permission)

16.16.4.2   The Procedure of the Test

How to Prepare the Phantom

The SPECT phantom (Jaszczak phantom) has one region of uniform activity, one area of the cold lesion and one rod sector to be recognized in a full performance phantom. It is also preferable to get a rough estimate of the resolution. As illustrated in Fig. 16.58, the phantom's activity level should be about 555–740 MBq (15–20 mCi) of ^{99m}Tc.

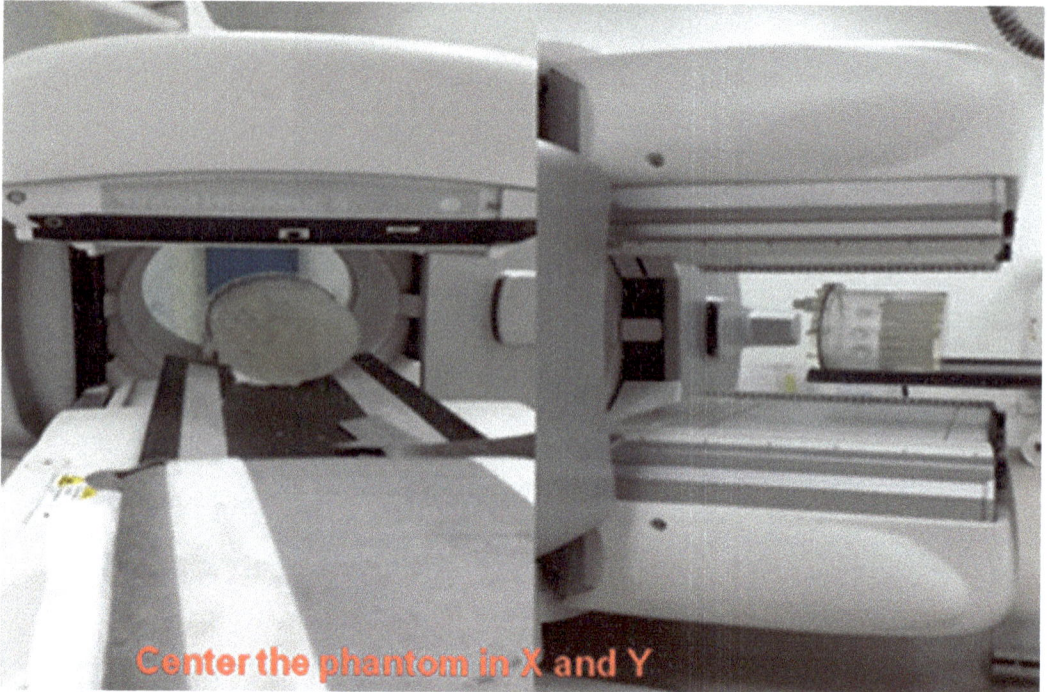

Fig. 16.59 Position of SPECT phantom

Phantom Positioning

- Install the highest resolution collimator possible for the radionuclide to be analysed for imaging.
- Place the activity-loaded phantom in the axial range of view by placing it lengthwise on the imaging table in the head-foot direction. As seen in Fig. 16.59, the phantom's centre is aligned with the rotating axis.
- Rotate the phantom such that the largest spheres and rods are on top and facing away from the table pallet. For acquisition, a circular orbit might be employed. Set the rotational radius to be as near as 20 cm as possible. For a circular orbit, the table on certain cameras restricts the closest radius to roughly 25 cm, so use 25 cm for them. This means the radius of rotation should be closed to the phantom with the precaution that the detector can rotate freely without touching the phantom.

Image Acquisition

- For ^{99m}Tc, set the symmetrical analyser window to 20%.
- To get the size of the pixel, about 2–3 mm, use a 128 × 128 matrix with a zoom factor. Over 360-degree projections, get 120 or more projections. Because two views are captured simultaneously by a dual-headed camera, 60 stops (or angles) will provide 120 views.
- The acquisition should be done by acquiring a total of 32 million counts. To calculate the time per view and count rate, divide 32 million by the number of views. A short static acquisition for 1 min with the phantom in place for the SPECT scan is a useful way to examine the count rate if the camera does not display the count rate.

Reconstruction of Image

Reconstruct the slices using FBP and a low-pass filter Butterworth or Metz filter with the order

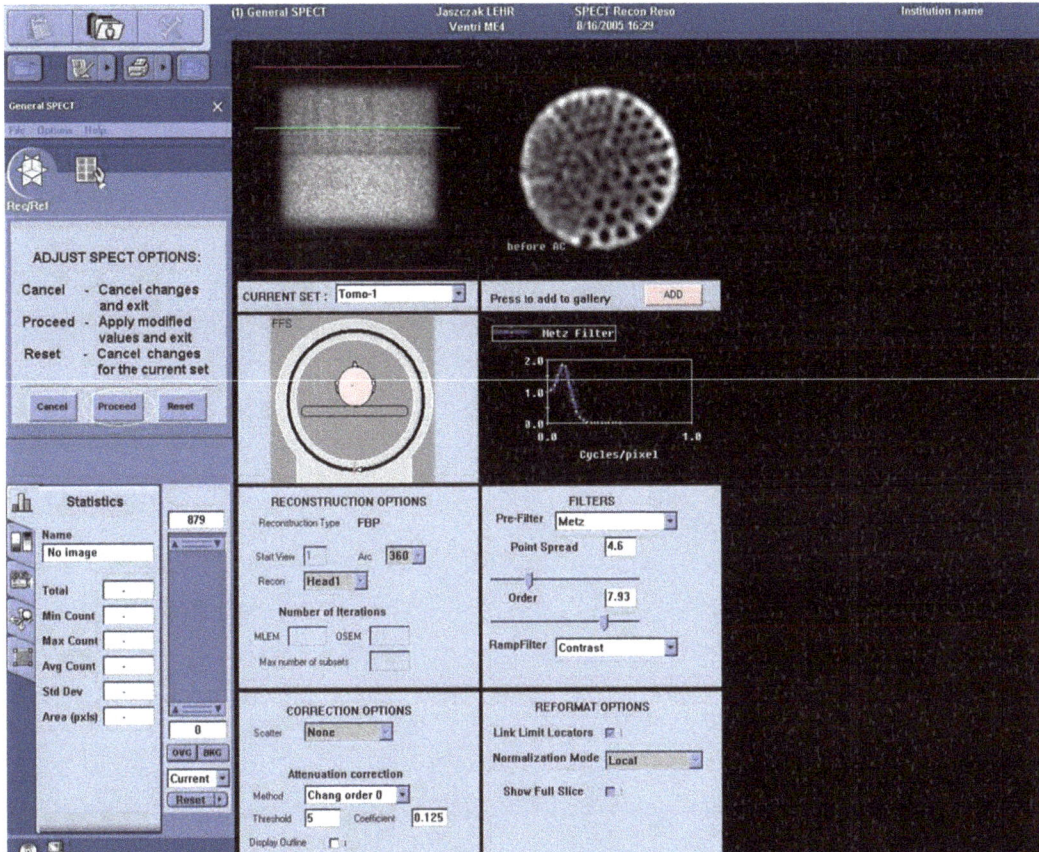

Fig. 16.60 Reconstruction using FBP and Metz filter

and cut-off frequency utilized in clinical investigations. The cut-off frequency may be adjusted to produce the desired outcome of images with the maximum resolution and contrast with less noise level. As demonstrated in Fig. 16.60, do not over-smooth the images.

Using the vendor's software, perform attenuation correction on the reconstructed transaxial slices using Chang's method for SPECT. The phantom is treated as a uniform body from all sides so that the attenuation will be the same from all sides. The coefficient for ^{99m}Tc is between 0.11/cm and 0.12/cm. This quantity may be modified to make the count density in the corrected slices' centre equal to the count density at the periphery. As a result, the attenuation will be consistent from all angles (Fig. 16.61).

Image Analysis

Figure 16.62 depicts how the reconstructed phantom slices' resolution, contrast and uniformity are visually assessed. Reduce the thickness of the reconstructed slices to around 1-cm-thick slices before analysing them.

Spatial Resolution

Identify the smallest rod sector visible in the slices and note the appropriate rod diameter. When more than 50% of a sector's rods can be identified, the sector is visualized, and it is the resolution of the SPECT. Note how many of the tiniest rods are visualized or how many are viewed with low contrast. As shown in Fig. 16.63, a total sum of 10–12 slices may be the best way to determine how good the spatial resolution is.

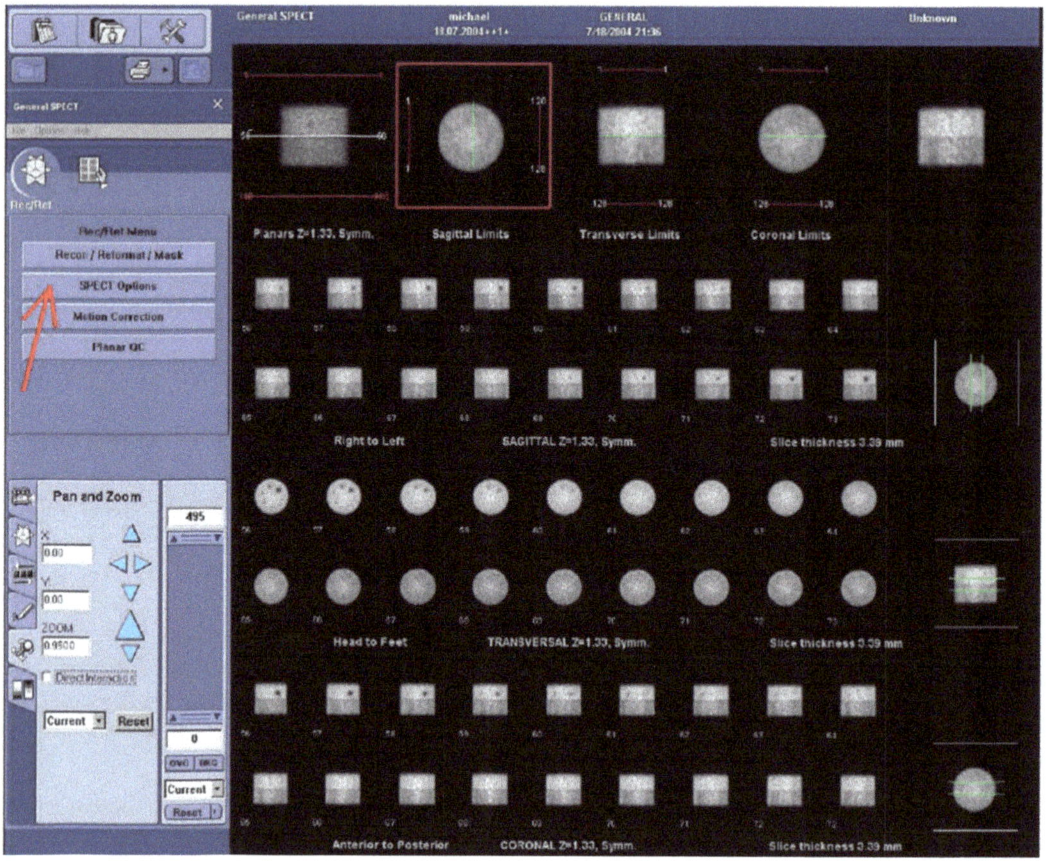

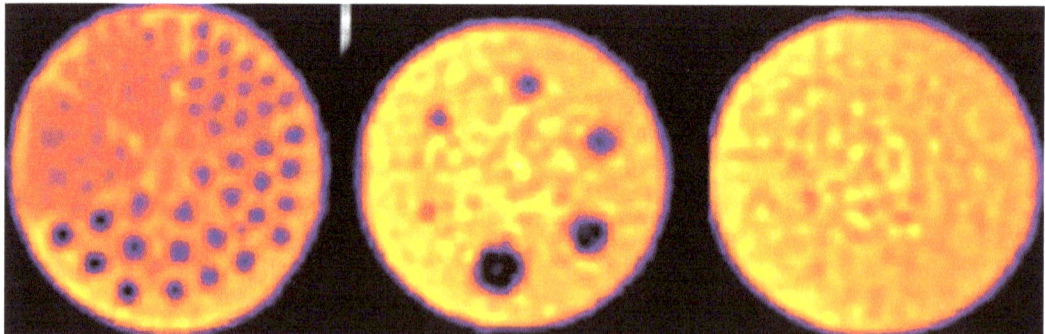

Fig. 16.61 Reconstructed images using FBP and Metz filter

Fig. 16.62 SPECT phantom slices reconstructed for spatial resolution, contrast and uniformity

Contrast Delectability

Figure 16.64 and Table 16.6 illustrate the matching sphere diameter for the smallest sphere observed in the image slices. Take note of whether the contrast of the smallest visible sphere is larger than or less than the noise. Also, as indicated below, compute the contrast for each sphere discovered in the reconstructed image slice:

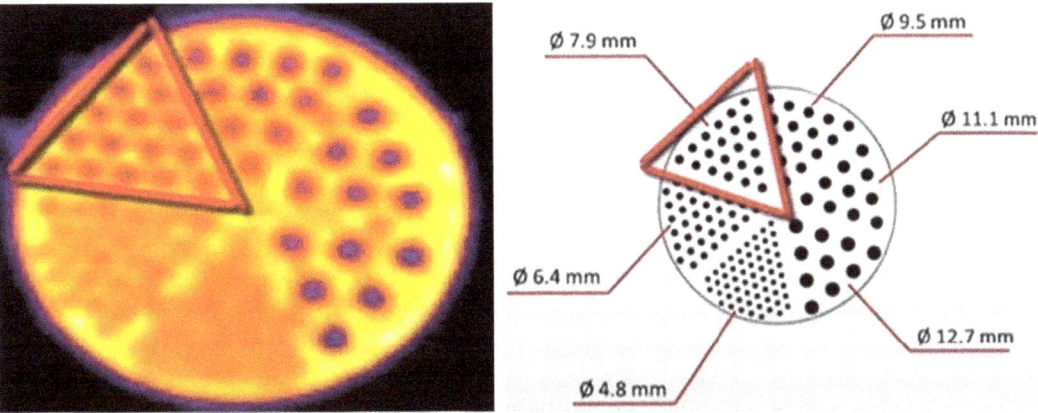

Fig. 16.63 Find the smallest rod section visible in the image slices, and write down the rod diameter

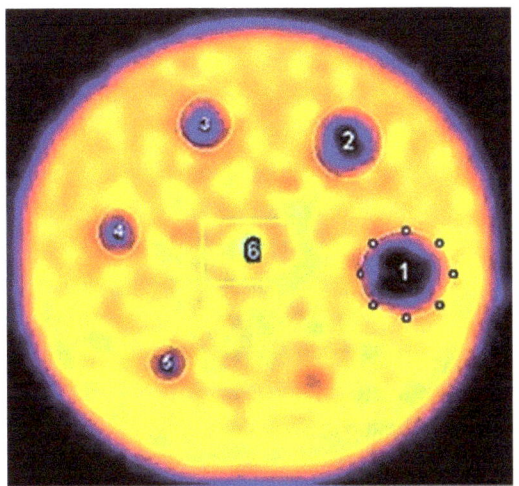

Fig. 16.64 The contrast was measured in a SPECT phantom with flood correction

Table 16.6 SPECT phantom contrast was measured with flood correction

Spherical diameter	Value ($V_{S_{Ph}}$)	Contrast
31.8	1	0.99
26.4	2	0.97
19.1	41	0.58
15.9	46	0.54
12.7	83	0.34
Background = 156 (V_{Bg})		

$$\text{Sphere contrast} = \frac{\text{mean pixel count from uniform section minimum sphere pixel count}}{\text{mean pixel counts from uniform section back ground}}$$

$$\text{Contrast} = \frac{V_{sph} - V_{Bkg}}{V_{sph} + V_{Bkg}}$$

Uniformity

- For SPECT, no uniformity index is computed. Instead, the image slices are examined for particular artefacts.
- Attenuation—either no attenuation artefacts are present, or they are under-corrected, or they are over-corrected.
- Around the reconstruction centre, there are rings with alternating high- and low-count densities. Only a central ring (solid or doughnut) at the axis of rotation may be distinguished in certain circumstances. Compare the size of the ring circle to the size of the noise. Ring artefacts with a magnitude smaller than the noise may be clinically inconsequential. Count the number of slices where the magnitude of the ring artefacts is greater than the noise.
- Focal area with increasing or decreasing count density regions might be mistaken for rods or spheres.

Analysis of Tomography Uniformity

- Examine each picture for artefacts, and display the whole set of reconstituted pictures that have been flood corrected.
- Minor artefacts may be present at the higher densities of counts utilised in the research, but significant ring artefacts should not be present.

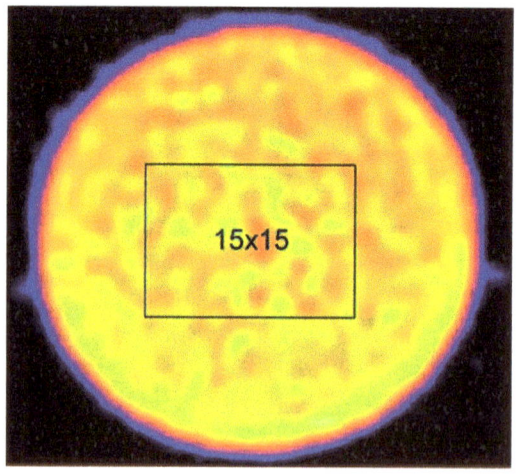

Fig. 16.65 Measurement of tomography measurement by drawing 15 × 15 ROI

- Draw a 15 × 15-pixel rectangular shape region (ROI) centred on the image(s) shown in Fig. 16.65 using one or more uniform slices.

Get the average counts per pixel, the highest and lowest pixel counts inside the ROI and the standard deviation of the mean counts. If the ROI's maximal and smallest pixel counts are not shown, solitary ROIs should have been employed to get the findings.

Calculate the integral uniformity of the reconstructed picture by using the following equation:

$$\text{Integral uniformity}\,(\%) = \frac{\text{maximum pixel counts} - \text{minimum pixel counts}}{\text{maximum pixel counts} + \text{minimum pixel counts}} \times 100$$

Root mean square noise:

$$\text{rms noise}\,(\%) = \frac{\text{standard deviation}}{\text{mean pixel value}} \times 100$$

Report

Report on all of the observations noted throughout the image analysis segment.

Performance Specifications

The following are the requirements that would be considered adequate in the case of [99m]Tc:

- Spatial resolution: 7.9 mm rods are completely resolved.
- Contrast: A sphere of 15.9 mm is shown.

There should be no ring artefacts with magnitudes larger than the noisy background, or if ring artefacts are seen in some slices, they should not be regarded as clinically significant. Acceptable values should be in one of the following ranges:

- 10.5–18.5% integral uniformity
- Noise (RMS): 3.2–7.0%.

16.16.5   SPECT/CT Spatial Registration

- Spatial registration tests assess the registration accuracy between the reconstituted SPECT and CT images. The spatial registration of SPECT/CT images is required for effective SPECT reconstruction using CT-based attenuation correction and the display of fused images for clinical interpretation.
- Problems with SPECT and CT co-registration are common, and they could be because of a system error or the patient's motion.
- To ensure the system is working properly, it must be checked and calibrated often, and then a transform matrix must be made to co-registered SPECT and CT data.
- Co-registration errors may occur suddenly or gradually, indicating a fault with the bed-moving mechanism.

16.16.5.1   Frequency

- Acceptance—Consider comparing the registration standards to those of the manufacturer. Acceptance testing findings should be used as a benchmark for future assessments.
- Annually, the values are verified to the manufacturer's registration criteria or baseline SPECT/CT registration measurements.
- Perform the registration test when the mechanical positioning of the SPECT and CT parts is modified or recommended by the manufacturer (e.g. when the SPECT and CT gantries have been serviced).

16.16.5.2   Procedure for Checking

- It is recommended that the spatial alignment of SPECT and CT images be assessed utilizing the manufacturer's registration phantom and analysis tools.
- The spatial registration of the reconstructed SPECT and CT images may be evaluated. To test for table displacements, an additional weight of roughly 70 kg (equivalent to the weight of a normal patient) should be placed on the table alongside the SPECT phantom during SPECT/CT acquisition or registration test.

Assessment of Spatial Registration

To evaluate the positional registration accuracy, display both SPECT and CT imaging sets from either the manufacturer's test method or the SPECT phantom acquisition utilizing available clinical SPECT and CT merging tools. Inside the CT images, the location of items in the test phantoms should be monitored, and the hot or cold spheres in the SPECT phantom images should be spatially aligned with the hot or cold spheres in the CT images. The hot spots of point sources should be centred inside the corresponding CT objects in the manufacturer's test phantom and vice versa. The cold spheres within the SPECT phantom should match the CT scan spheres.

If the locations of each source on the CT and SPECT pictures are not identical, the fusion software may be used to shift (translate) the CT images to match them with the SPECT images. For each of the evaluated locations (x, y, z), determine the absolute size of the offset in each direction in terms of pixel shifts or millimetres, and write it down.

The mean deviation was estimated in each direction using SPECT and computed tomography (CT) images collected in the same direction (x, y or z).

Report

For the images to be co-registered, give the size of each voxel and how much it moved along the three axes.

Performance Criteria

According to the manufacturer's specification, the mean variation along any axis should be <5 mm.

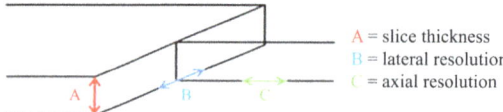

A = slice thickness
B = lateral resolution
C = axial resolution

Fig. 16.66 The thickness of a slice in the field of view's centre

Slice Thickness

To find out how thick a tomo slice is in the middle of the field of vision. To ensure that the tomographic resolution in Z-axis is acceptable (Fig. 16.66).

16.16.5.3 Procedure for Checking

- The results of this test are identical to those obtained during the air reconstructed resolution test (Sect. 16.16.3) and may be used to compute slice thickness. The data analysis is the main difference; otherwise, the same raw data might be utilised for this test.
- Place the point source in the air, 10 mm from the centre of rotation.
- Set the radius of 200 mm or as small as feasible if this is not practicable. Use a circular rotational orbit.
- Perform a tomo acquisition with the 128×128 matrix size with 120 angles and around 250 K counts per projection.
- Use FBP to reconstitute the data with a ramp filter or the sharpest filter the hardware permits. Instead of combining transaxial slices, use the slice thickness of each pixel.

Data Analysis

- Locate the transverse section with the clearest view of the point source. Create a one-pixel area of interest on the highest counting pixel.

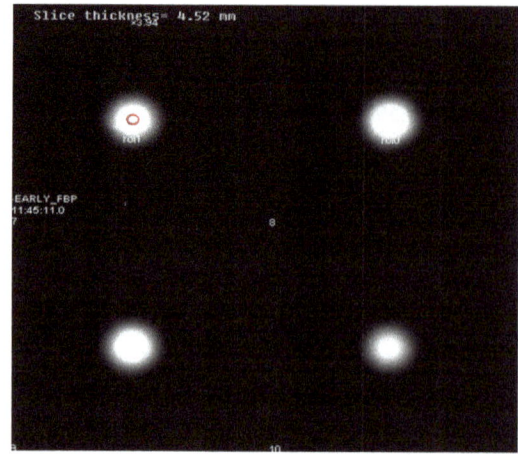

Fig. 16.67 One-pixel ROI on the pixel with the highest number of counts

This ROI creates a curve between the count and the transverse slice.
- Calculate the FWHM and FWTM as illustrated in Figs. 16.67 and 16.68.

Report

Assume the thickness of the slice is evaluated in the air. In such an instance, the full width at half maximum (FWHM) should be the same as the normal transversal tomographic resolution in the air for the same rotational radius in the air.

Performance Criteria

The centre slice thickness should be within 10% of the tomographic resolution.

These are the above quality control and assurance we should perform to know the status of our SPECT camera. Quality controls help us to give the artefacts free clinical image.

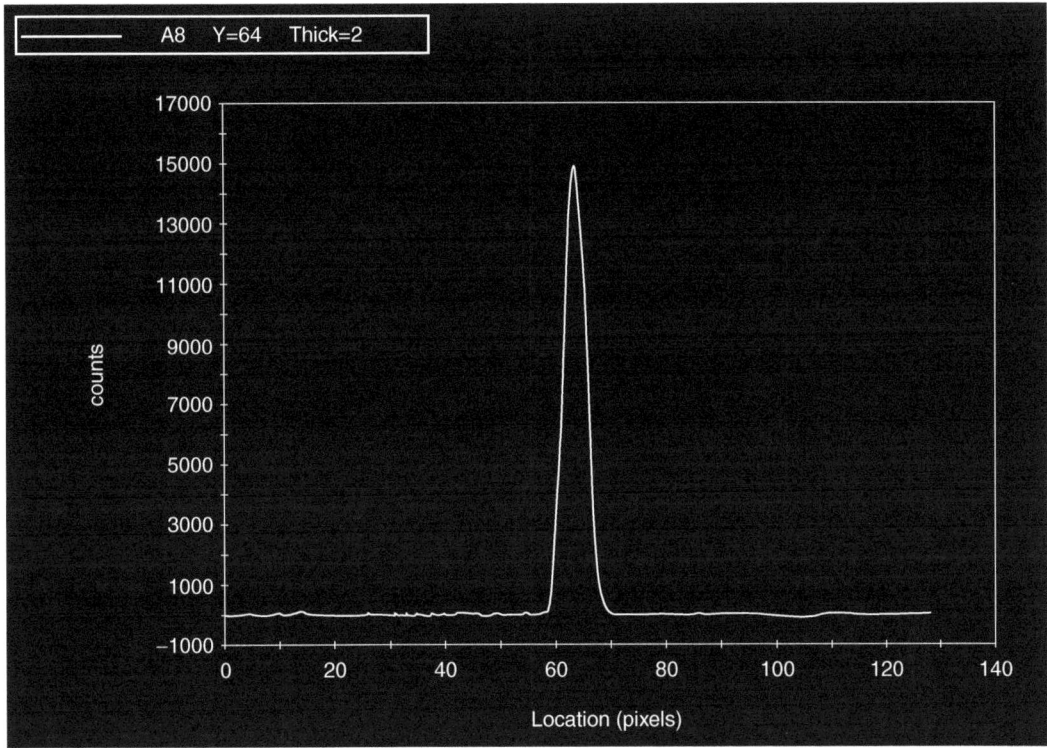

Fig. 16.68 To compute the FWHM, get a count versus transverse section curve

References

1. National Electrical Manufacturers Association. NEMA NU 1. Performance measurement of scintillation cameras. Rosslyn, VA: National Electrical Manufacturers Association; 2001, 2007.
2. International Electrotechnical Commission. IEC 60789. Medical electrical equipment – characteristics and test conditions of radionuclide imaging devices – anger type gamma cameras. Geneva: International Electrotechnical Commission; 2005.
3. International Electrotechnical Commission. IEC 61675-2. Radionuclide imaging devices – characteristics and test conditions – Part 2: single-photon emission computed tomographs. Consolidated Edition 1.1. Geneva: International Electrotechnical Commission; 2005.
4. International Electrotechnical Commission. IEC 61675-3. Radionuclide imaging devices – characteristics and test conditions – Part 3: gamma camera-based whole-body imaging systems. Ed 1. Geneva: International Electrotechnical Commission; 1998.
5. National Electrical Manufacturers Association. NEMA NU 2. Performance measurements of positron emission tomographs. Rosslyn, VA: National Electrical Manufacturers Association; 1994, 2001, 2007.
6. Bergmann H, Dobrozemsky G, Minear G, Nicoletti R, Samal M. An inter-laboratory comparison study of image quality of PET scanners using the NEMA NU-2 2001 procedure for assessment of image quality. Phys Med Biol. 2005;50:2193–207.
7. National Electrical Manufacturers Association. NEMA NU 3. Performance measurements and quality control guidelines for non-imaging intraoperative gamma probes. Rosslyn, VA: National Electrical Manufacturers Association; 2004.
8. National Electrical Manufacturers Association. NEMA NU 4. Performance measurement of small animal positron emission tomographs. Rosslyn, VA: National Electrical Manufacturers Association; 2008.
9. Ehsan S, et al. Assessment of display performance for medical imaging systems: executive summary of AAPM TG18 report. J Med Phys. 2005;32(4):1205–25.
10. Technical standard for electronic practice of medical imaging. ACR-AAPM-SIIM Technical Standard. 2012. www.acr.org/~/media/ACR/Documents/PGTS/standards/ElectronicPracticeMed-Img.pdf.

11. Chawla A, Samei E. Ambient illumination revisited: a new adaptation-based approach for optimizing medical imaging reading environments. Med Phys. 2007;34(1):81–90.
12. Dorbala S, Ananthasubramaniam K, Armstrong IS, Chareonthaitawee P, DePuey EG, Einstein AJ, et al. Single photon emission computed tomography (SPECT) myocardial perfusion imaging guidelines: instrumentation, acquisition, processing, and interpretation. J Nucl Cardiol. 2018;25(5):1784–846. https://doi.org/10.1007/s12350-018-1283-y.
13. https://humanhealth.iaea.org/HHW/MedicalPhysics/e-learning/Tutorial_videos_on_Quality_Control_tests_for_SPECT_systems/index.html.
14. International Atomic Energy Agency. Quality control of nuclear medicine instruments 1991, IAEA-TECDOC-602. Vienna: IAEA; 1991.
15. International Atomic Energy Agency. IAEA quality control atlas for scintillation camera systems. Vienna: IAEA; 2003, ISBN 92–0–101303–5.

Quality Assurance in Positron Emission Tomography-Computed Tomography (PET-CT)

17

Abstract

Nuclear medicine quality control starts with acceptance testing to verify that the equipment and software packages meet the manufacturers' specifications specified at the procurement tender, and these become reference values. Then a periodic assessment of the system performance of the scanners on an annual, semi-annual, quarterly, weekly, and daily basis evaluates the scanner performance deviations from its initial assessment.

A sufficient devotion of time for the testing process is important to identify any problems or malfunctions of the instrument and to complete the tests after calibration, repair, replacement, or upgrades to ensure the proper performance of the equipment before patient use. Quality control is important due to optimizing patient exposure and image quality during nuclear medicine imaging examinations.

17.1 Quality Assurance (QA) Program

Any QA program begins with the work done during installation to set up the system properly. The purpose here is to validate the performance characteristics of the tender process rather than to establish a complete collection of performance measurements. It serves as a baseline for subsequent measurements. Periodic service visits and frequent calibration will be part of the following maintenance schedule. The calibration will adjust the gain of the photomultiplier tubes (PMTs) to provide suitable signals for 511-keV photons within a certain energy window. These signals are needed to appropriately identify specific detector components. A timing calibration or alignment is necessary for coincident events. Other calibrations are required with time-of-flight capable scanners due to the possibility of deviation from ideal settings. For PET systems, normalization refers to the accuracy with which uniformity is monitored and employed in data reconstruction. The absolute calibration of a PET system is critical so that findings may be stated quantitatively in terms of SUV (kBq/ml). A known source of ^{18}F is carefully measured in a dosage calibrator for this purpose.

17.2 Routine Quality Assurance

Routine QA is performed to confirm the scanner's proper functioning and the integrity of the obtained images [1]. Routine testing monitors system stability and identifies any changes in scanner operation. The QA and calibration method of a scanner includes the adjustment of the PMT gain, energy maps, and coincidence timing calibration. The calibration correction converts reconstructed image pixel values into activity concentration (SUV) values and may be used to adjust for the scanner's axial sensitivity fluctuation. The manufacturer recommends that the PET QA be separated into daily, weekly, and quarterly operations. The daily QA software objectively evaluates the scanner's image quality over time. Single events, coincidence, dead-time, and the peak energy spectrum of the detectors are all monitored during daily QA. During weekly QA, all detectors are irradiated, and detector outputs are corrected. Calibrations are performed quarterly to give the system a standard for counting deviations and maximize system performance. These include single event position, update gain, and energy. It is possible to conduct detector coincidence timing characterization, 2D normalization, 3D geometric, and 2D/3D well-counter calibrations.

17.3   Performance Assessment

The objective of performance evaluation is to establish an experimental setting that allows for the measurement of image properties, comparison of various scanners, and prediction of scanner behaviour for patient investigations. The National Electrical Measurement Association (NEMA) published a standard, NU 2-2007 & 2012 (1), to evaluate PET systems' performance. The majority of manufacturers' standards are based on NEMA guidelines. Acceptance testing should adhere to the NEMA standard.

17.4   Responsibilities for Quality Control Tests

The trained nuclear medicine physicist is in charge of overseeing the QC program. In addition to acceptance testing and quality control, the physicist in nuclear medicine often has other tasks such as assisting diagnostic and therapeutic operations, training, radiation safety, computer system administration, and development [2]. However, they should be immediately accessible onsite for rapid consultation on QC issues.

The physicist performs acceptance tests after the equipment is installed. The physicist also needs to do post-service testing to ensure that changes made by the service engineer don't hurt image quality or patient doses.

The physicist also has to keep an eye on the equipment when it needs to be repaired, calibrated, or maintained (preventative maintenance, for example).

17.5   Important Points

While all QC tests are required to assure optimal image quality and patient radiation dosage, two processes, artefact assessment and QC checks for equipment utilized, demand special attention; image artefacts might be caused by a mismatch between the CT transmission attenuation map and the PET data, mobility, external radioactive contamination, the patient's prostheses and implants, or detector failure. Artefacts degrade image resolution and might lead to inaccurate diagnosis or staging. During acceptance and quality control testing, inspect each image for visible artefacts. When an artefact in a PET/CT reconstructed picture is detected, it is always best to seek a non-attenuated PET image to identify whether the issue is of PET or CT origin.

The QC test of the scan localization lasers, i.e. the lasers on the CT gantry used to position the patient relative to the image data volume, is the most important. The correct position of the image data volume is required for transmission to the PET-CT planning computer system. This daily exam consists of a fast and easy visual inspection.

17.6   Quality Control Records

Keeping a full set of records is a vital component of any QC. The key components are a record of the tests done, the time and date, a description of the results or notes concerning odd discoveries, and the performer's name. Images and important data should be kept in a binder for easy access and review.

17.7 Preventive Maintenance

The manufacturer's representative or skilled in-house personnel such as a biomedical engineer should perform preventive maintenance regularly. This procedure should ensure that the instrument is in the best possible functioning condition and that any faults are identified before major failures occur. Staff must also report any possible safety issues to the physicist, such as frayed wires or strange sounds.

17.8 Acceptance Test Procedures

Before beginning acceptance testing, all calibrations necessary for the scanner's installation and commissioning must be completed to confirm that the scanner is performing as intended. It should also be confirmed that the system passes the daily QC and the sinograms are in good working order.

It is advisable to conduct the tests in collaboration with the installation or service engineers. They know how the system software works and can give you access to service options if you need them.

17.9 Spatial Resolution

A system's spatial resolution is its capacity to differentiate between two points after image reconstruction. The measurement is carried out by imaging point sources in the air and then reconstructing images without any smoothing or apodization. Although this does not reflect the circumstance of imaging a patient in which tissue dispersion and a limited number of captured events necessitate a smooth reconstruction filter.

17.9.1 Purpose

This measurement aims to describe the width of the reconstructed picture point spread functions (PSF) of compact radioactive sources. The spread function's width is defined as its full width at half its maximum amplitude (FWHM) and its full width at a tenth of its maximum amplitude (FWTM).

17.9.2 Material Requirements (Table 17.1)

17.9.3 Activity Requirements

3 mCi (111 MBq) of F-18 in 0.1 mL in a 1 mL Luer-lock syringe, calibrated for T_0 (time 0), i.e. time of acquisition T_0 (time 0). The activity in the 1 mL Luer–lock (3 mCi) should remain 0.2–0.4 mCi (7.5–15 MBq) at the time of test T_0 in the resin of the capillary tube.

17.9.4 Activity in Point Source at the Start of Data Acquisition

0.2–0.4 mCi (7.5–15 MBq)

17.9.5 Source Distribution (Fig. 17.1)

1. Measure 3 mCi activity in a 1 mL syringe in a dose calibrator to confirm the activity in the syringe.
2. Fill the syringe with about 1 mL of air.
3. Connect the resin (F-18 absorber) end of the capillary tube to the 2 cm flexible tube and secure it with a Luer lock.
4. Remove the needle from the syringe and secure it with the Luer lock.

Table 17.1 Material required for spatial resolution

Item	Quantity
1 mL syringe	1
3 mL syringe	1
Luer lock	1
Tube holder	1
Tubing	2 cm
Capillary tube with F18-absorbing resin	1
Resolution L-fixture	1

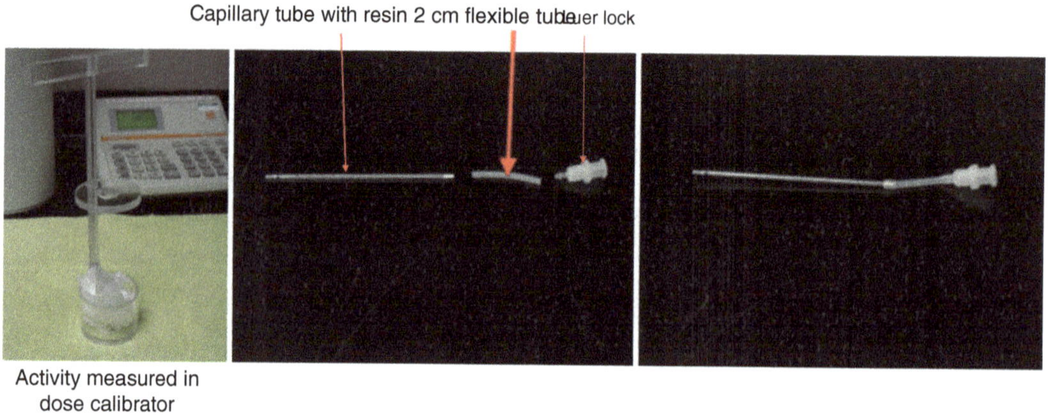

Capillary tube with resin   2 cm flexible tube   Luer lock

Activity measured in
dose calibrator

Fig. 17.1 Point source preparation

5. Push activity via a flexible tube into the capillary tube until it reaches the middle of the length of the capillary tube to absorb the activity in resin and draw all activity back into the syringe. The activity absorbed by the resin serves as the point source in this test.
6. Seal the capillary tube with wax to prevent activity from leaving at the other ends.
7. If F-18 droplets remain in the capillary tube, fill a 3 mL syringe with water, attach the syringe to the Luer-lock, push water into the capillary tube, and draw water back into the syringe.
8. Take the capillary tube out of the flexible tube and insert it into a dosage calibrator.
9. Repeat steps 5–8 until the point source reaches 0.2–0.4 mCi (7.5–15 MBq). If the point source activity level is greater than 0.4 mCi, fill a 3 mL syringe with water, attach the syringe to a Luer-lock, and force water into the capillary tube. Pull the water back into the syringe and insert a capillary tube into the dosage calibrator to measure the activity. To repeat Step 5, if the point source's activity is less than 0.2 mCi and more activity is needed to complete the test, repeat Step 5.
10. The point source activity must be 0.2–0.4 mCi (7.5–15 MBq) at the commencement of data collection.
11. Three-point sources are produced in the same manner.

17.9.6   The Positioning of the Source (Figs. 17.2, 17.3 and 17.4)

The sources shall be fixed parallel to the tomograph's long axis and located at six points as follows:

1. In the axial direction.
2. (a) one-fourth of the axial FOV from the centre of the FOV, (b) one-fourth of the axial FOV from the centre of the FOV
3. In the transverse direction, the source should be positioned (a) 1 cm vertically from the centre (to represent the centre of the FOV), (b) at $x = 0$ and $y = 10$ cm, and (c) at $x = 10$ cm and $y = 0$

    Data should be collected at the following places in the PET FOV:

    $(X, Y, Z) = (0, 1, 1/2FOV\ Z)$, $(10, 0, 1/2FOV\ Z)$, $(0, 10, 1/2FOVZ)$, $(0, 1, 3/4FOV\ Z)$, $(10, 0, 3/4FOV\ Z)$, and $(0, 10, 3/4FOV\ Z)$, as shown in Fig. 17.5.
4. Measurements must be taken at each of the six locations listed above. In each response function, at least 100,000 counts must be collected. Measurements may be collected from many sources.
5. The source arrangement is shown in Fig. 17.4.

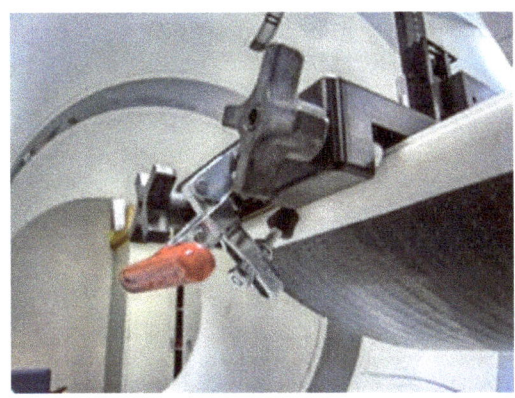

Fig. 17.2 Clamp the fixture to the PHS using the red clamps

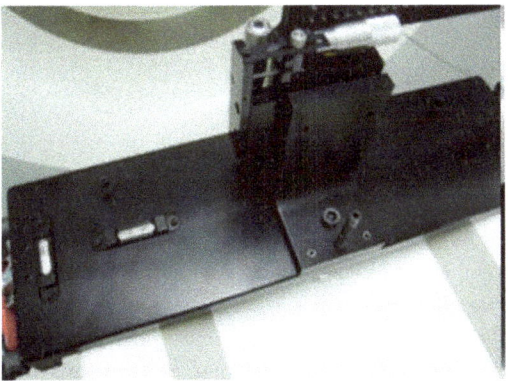

Fig. 17.3 Level the fixture using the three screws on the top of the fixture

17.9.7 Data Processing

All spatial resolution data must be reconstructed using filtered back-projection without smoothing or apodization.

17.9.8 Analysis

The spatial resolution (FWHM and FWTM) of the point source response function in all three dimensions for all six places must be established by creating one-dimensional response functions and profiles across the image volume in three orthogonal directions via the distribution's peak. The width of the response functions in the two directions perpendicular to the measurement direction must be about twice the FWHM.

Each FWHM (and FWTM) must be calculated using linear interpolation between adjacent pixels at half (or one-tenth) of the highest value of the response function (see Fig. 17.6). A parabolic fit of the peak point and its two closest surrounding points will be used to find the greatest value. Values must be translated into millimetres by multiplying by the pixel size. The pixel with the most counts in each one-dimensional response function is the source location that can be seen.

Fig. 17.4 Source positions for measuring resolution. (Image copyright, National Electrical Manufacturers Association, NEMA NU 2-2007 [1])

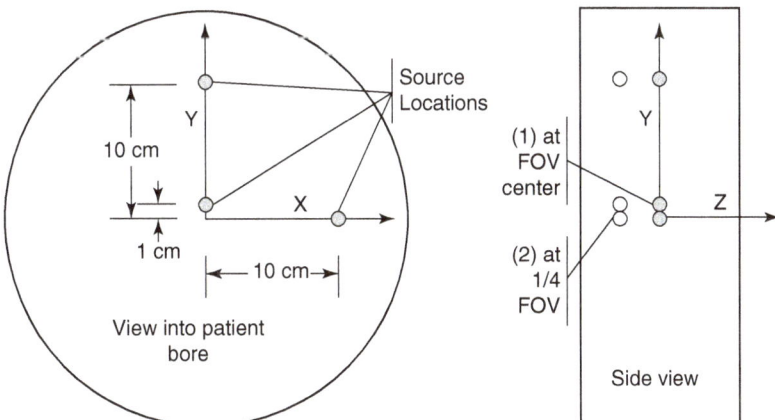

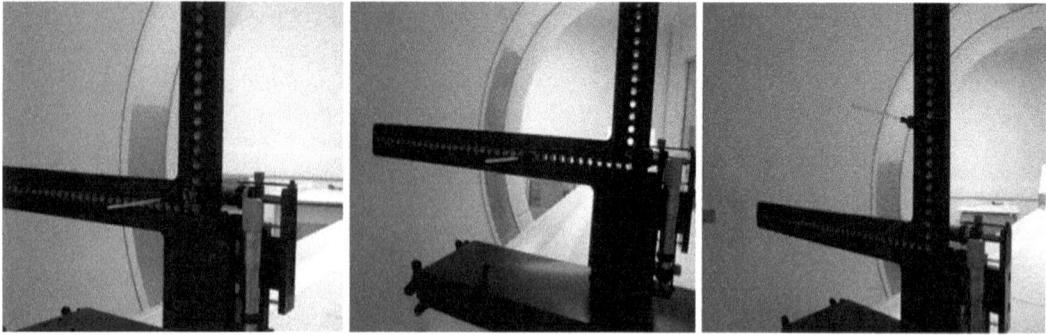

Fig. 17.5 Capillary tube holder to position (0,1,1/2FOV Z), (10,0,1/2FOV Z), (0,10, 1/2FOV Z) on the resolution L-fixture and insert the capillary tube (opposite end of resin) into the tube holder on the fixture. Z = Axial Direction

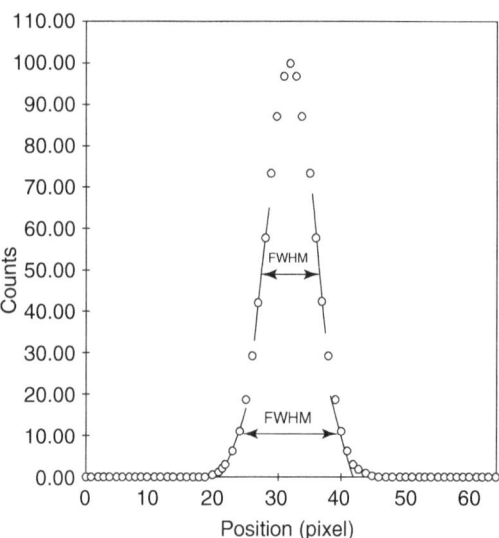

Fig. 17.6 Interpolation is used to generate a response function with FWHM and FWTM

17.9.9   Report

Axial, radial and tangential resolutions (FWHM and FWTM) for each radius (centre and 10 cm) averaged across both axial locations must be computed and reported as system resolution values according to Table 17.2.

17.9.10   Suggested Tolerances

The calculated FWHM values must not exceed the vendor's standard. The FWTM is often not mentioned. However, a predicted value may be inferred by remembering that the ratio between the FWTM and the FWHM for a theoretical Gaussian curve is 1.82. As a result, the estimated ratio between FWTM and FWHM for a real PET scanner should be 1.8–2.0. The user should establish reference values, tolerances, and action levels (i.e. to trigger the decision to place a call for maintenance). A suitable tolerance threshold for FWHM is

$$FWHM_{observed} < 1.05\,FWHM_{expected}.$$

17.9.11   Corrective Action

If the FWHM tolerance requirements are exceeded, action should be taken. The findings should be reviewed, and the testing method should be repeated to ensure the source was properly prepared. Then, the manufacturer should be told, and remedial action should be requested.

Table 17.2 Spatial resolution report values computation formulas. (RESx, RESy, and RESz refer to the spatial resolution measured in the x, y, and z-directions). (Table originally printed in NEMA NU 2–2007 [1]. Reprinted with permission)

	Description	Formula
At a 1 cm radius		
Transverse	Average x & y for both z positions (4 numbers)	$$RES = \left(\begin{array}{c} RESx_{x=0,y=1,\,z=centre} + RESy_{x=0,y=1,\,z=centre} + \\ RESx_{x=0,y=1,\,z=\frac{1}{4}FOV} + RESy_{x=0,y=1,\,z=\frac{1}{4}FOV} \end{array} \right) / 4$$
Axial	Average of $2z$ positions (2 numbers)	$$RES = \left(RESz_{x=0,y=1,\,z=centre} + RESz_{x=0,y=1,\,z=\frac{1}{4}FOV} \right) / 2$$
At a 10 cm radius		
Transverse radial	Average 2 transverse for both z positions (4 numbers)	$$RES = \left(\begin{array}{c} RESx_{x=10,y=1,\,z=centre} + RESy_{x=0,y=10,\,z=centre} + \\ RESx_{x=10,y=1,\,z=\frac{1}{4}FOV} + RESy_{x=0,y=10,\,z=\frac{1}{4}FOV} \end{array} \right) / 4$$
Transverse tangential	Average 2 transverse for both z positions (4 numbers)	$$RES = \left(\begin{array}{c} RESy_{x=10,y=0,\,z=centre} + RESx_{x=0,y=10,\,z=centre} + \\ RESy_{x=10,y=0,\,z=\frac{1}{4}FOV} + RESx_{x=0,y=10,\,z=\frac{1}{4}FOV} \end{array} \right) / 4$$
Axial resolution	Average 2 transverse for both z positions (4 numbers)	$$RES = \left(\begin{array}{c} RESz_{x=10,y=0,\,z=centre} + RESz_{x=0,y=10,\,z=centre} + \\ RESz_{x=10,y=0,\,z=\frac{1}{4}FOV} + RESz_{x=0,y=10,\,z=\frac{1}{4}FOV} \end{array} \right) / 4$$

17.10 Sensitivity

Like that of other imaging technologies, PET sensitivity is principally influenced by the detector system's absorption efficiency and solid angle of coverage of the imaged object.

The genuine coincidence rate, R_{true}, of a positron-emitting source placed between two coincidence detectors in an absorbing medium is given by

$$R_{True} = E_\varepsilon^2 g_{ACD} e^{-\mu t}$$

Where E is the rate of emission from the source (positrons/s); is the intrinsic efficiency of each detector, that is, the fraction of incoming photons detected; μ and T are the linear attenuation coefficient and the total thickness of the object, respectively. The geometric efficiency of the detector pair, g_{ACD}, is the fraction of annihilation events in which both photons are emitted in a direction that the detectors intercept.

17.10.1 Purpose

The sensitivity connects the count rate detected by the instrument to the quantity of radioactivity inside the FOV. As a result, sensitivity measurement calculates the rate of recorded real coincidence events per unit of radioactivity concentration for a standard source, such as a cylindrical phantom with specific dimensions. These results should be devoid of attenuation, scattering, and count rate aberrations to compare sensitivity measures across scanners. PET, or attenuation-free radioactivity measurement, is hampered by positrons needing a specific route in absorbing materials to be transformed into annihilation radiation. The test was carried out by taking measurements as the thickness of the absorbent material increased and extrapolating the results to zero absorption (9). The test necessitates specialized equipment, which consists of a series of aluminium tubes of various sizes.

17.10.2   Frequency

The competent medical physicist must execute the sensitivity test during acceptance testing or anytime it is believed that the detector system performance has changed considerably. Table 17.3 lists the materials needed for this exam. The sensitivity phantom depicted in Fig. 17.7 and Table 17.3 is the test equipment needed for this measurement.

17.10.3   Material (Table 17.4)

17.10.4   Activity Requirements

At the commencement of the acquisition, 0.125 mCi (4.6 MBq) ^{18}FDG is needed in the

Vsens-tube. Take activity in a 3 mL Luer-lock syringe in such an amount that it remains 0.125 mCi (4.6 MBq) at time T_0 while preparing the source.

17.10.5   Phantom Positioning

In the PET gantry tunnel, place foam fittings. Foam#1 should be placed at the rear of the PET gantry on the border of the PET tunnel. As indicated in Fig. 17.8, Foam#2 should be 30 cm apart from Foam#1 and towards the front of the PET gantry.

17.10.6   Activity Preparation

Place three Kimwipes at the bottom of the calibrator's plunger. Measuring the activity of the

Table 17.3 Material required to perform the sensitivity test

Item	Quantity
3 ml syringe	1
50 ml Pyrex beaker	1
Luer lock	1
Tube plug 1/16″	2
NEMA 2018 Sensitivity Sleeves and tubing	1
Foam fixture	2

Table 17.4 Sensitivity measurement phantom of different thickness used for sensitivity

Sleeves number	Diameter (inside) (cm)	Diameter (outside) (cm)	Length L (cm)
1	0.39	0.64	70
2	0.70	0.95	70
3	1.02	1.27	70
4	1.34	1.59	70
5	1.66	1.9	70

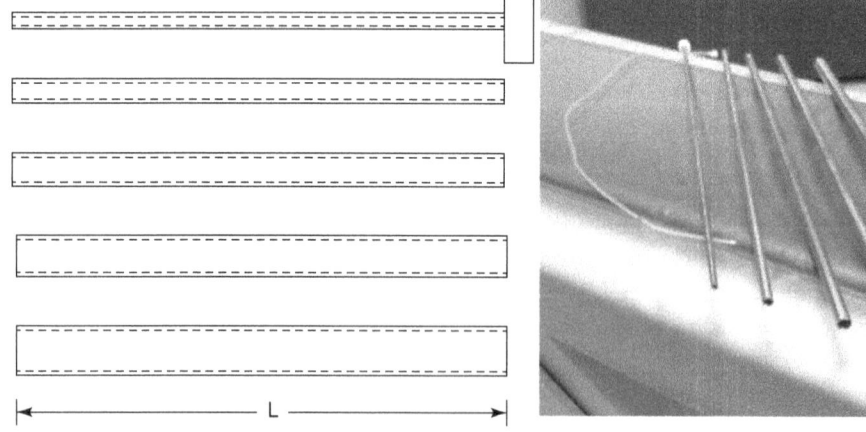

Fig. 17.7 Sensitivity measurement phantom. (Image source: Image copyright National Electrical Manufacturers Association, NEMA NU 2-2007 [1]. Reprinted with permission)

syringe in a dose calibrator if the activity and volume are larger than 0.18 mCi (6.66 MBq) + 10%, 3 mL ± 0.5 mL in Vsens-tube, transfer the surplus activity and volume from the syringe to a 50 mL beaker to acquire the requisite activity and volume. If the activity level is less than 0.18 mCi − 10%, additional activity is necessary to complete the test effectively. Take note of test

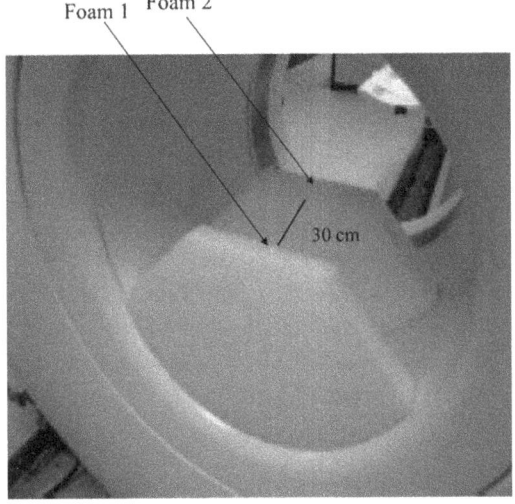

Fig. 17.8 Phantom positioning for Sensitivity Test

activity and time, and then inject about 1 mL of air into the syringe to allow for residual dose pull.

17.10.7 Line Source Preparation

- Connect the tube to the Luer-lock. Clean out tubes with compressed air if available (Fig. 17.9).
- Remove the needle from the syringe and attach it to the Luer lock.
- Fill an 80 cm tube with activity. The 70 cm extent of activity must be positioned in the 80 cm tube to provide a 0.5 cm air gap at the tip of the tube while assuring no air bubbles. If air bubbles are present, take the required steps to eliminate them. Plug the tube at the end opposite the Luer-lock.
- Remove Luer-lock and plug tube.
- Determine the length of activity in the tube. The extent of activity in the tube must be between 68 and 72 cm.
- Correct assay for line source length.

$$\text{Activity}_{corrected} = 70 \times \left(\frac{\text{Activity}}{\text{Length}} \right)$$

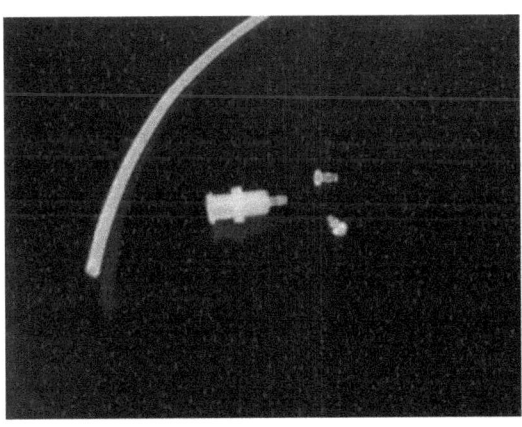
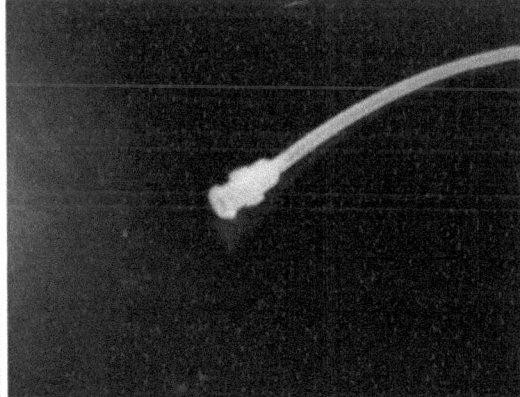

Fig. 17.9 Connect the tube to the Luer-lock

Place the tube in a shielded container and bring it to the scanning room.

17.10.8   Data Acquisition (for all Five Sleeves), Processing, and Analysis

For the 0 cm offset acquisition

- Insert the tube into a phantom. Place the tube so that the range of activity corresponds to the length of the sleeves (70 cm extent of FDG lined up with 70 cm long phantom). Place the sleeves in the small indentation in the foam (0 cm from the centre of the transverse FOV) and axially place the 5 aluminium sleeves between the foam fixtures. As shown in Fig. 17.10.
- Start acquisition at 0.125 mCi (4.6 MBq).
- Start the next acquisition by removing the sleeve with the largest diameter.
- Repeat until all five acquisitions for all sleeves are finished.
- Once data acquisition is complete for the 0 cm offset test, then analyse the data to calculate the sensitivity at the centre of FOV.

For the 10 cm offset acquisition

- As shown in Fig. 17.11, axially centre the five aluminium sleeves between the foam fixtures. The sleeves should be positioned 10 cm away

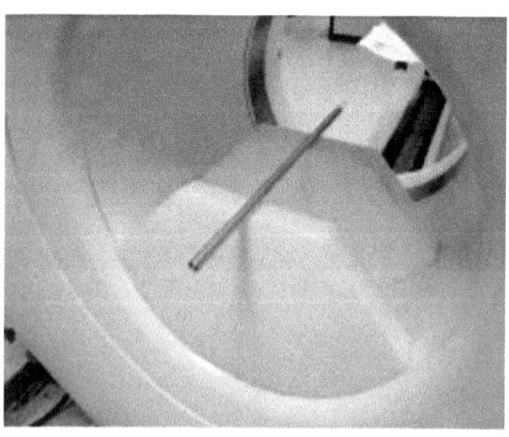

Fig. 17.10   Phantom positioning at 0 cm for Sensitivity Test

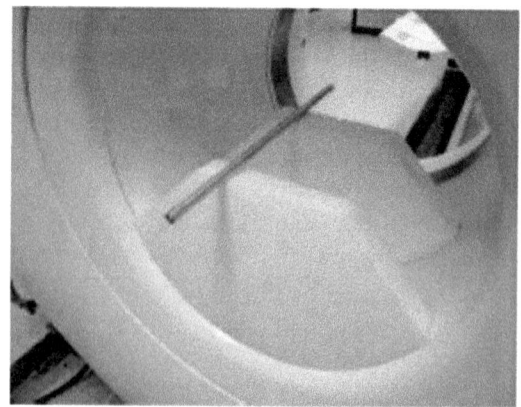

Fig. 17.11   Phantom positioning at 10 cm for Sensitivity Test

from the centre of the FOV (10 cm offset from the centre of the transverse FOV).

- Start acquisition. Remove the greatest diameter sleeve, and start the next acquisition. Repeat until only all five acquisitions have been completed.

17.10.9   Calculations and Analysis

17.10.9.1   System Sensitivity

Use the following calculation to adjust the count rate for isotope decay for each measurement linked with all five sleeves and each slice:

$$R_{\text{CORR}, j,i} = R_{j,i} \cdot 2^{(T_j - T_{\text{cal}})/T_{1/2}}$$

Once the isotope decay has been rectified, compute $R_{\text{CORR}j}$ by adding $R_{\text{CORR}j}$, I from each slice. Regression technique is then used to fit the data to the following equation:

$$R_{\text{CORR}, j} = R_{\text{CORR},0} \cdot \exp\left(-\mu_M \cdot 2 \cdot X_j\right)$$

Where $R_{\text{CORR},0}$, and μ_M are unknowns, X_j is the cumulative sleeve wall thickness. The count rate with no attenuation is represented by the word $R_{\text{CORR},0}$. To compensate for the limited quantity of scattered radiation, the value for attenuation in metal, μ_M, is permitted to vary.

The same approach must be followed for sensitivity measurements performed at 10 cm from the tomograph's centre. The system sensitivity will be calculated using the formula:

$$S_{tot} = \frac{R_{CORR,0}}{A_{cal}}$$

A_{cal} = Activity calculated

Calculate average sensitivity:

$$\text{Sensitivity}_{average} = \left(\text{Sensitivity}_{0cm\ acquisition} + \text{Sensitivity}_{10cm acquisition}\right)/2$$

17.10.9.2 Axial Sensitivity Profile

Compute the sensitivity for each slice using the data collected for the thinnest tube, $C_{1,j}$, at the 0-centimetre radial offset:

$$S_i = \frac{R_{CORR,1,i}}{R_{CORR,1}}.S_{tot}$$

17.10.10 Tolerances Suggestions

The sensitivity of the system in 2-D and 3-D modes should be equivalent to or higher than the

vendor's standards. Figure 17.12 depicts a typical 2D and 3D axial sensitivity profile.

The user must establish reference values, tolerances, and activity levels. A suitable tolerance threshold for system sensitivity may be:

$$S_{tot,measured} > 0.95\,S_{tot,expected}$$

17.10.11 Corrective Measures

If the sensitivity test tolerance requirements are not satisfied, the manufacturer should be contacted, and remedial action should be taken.

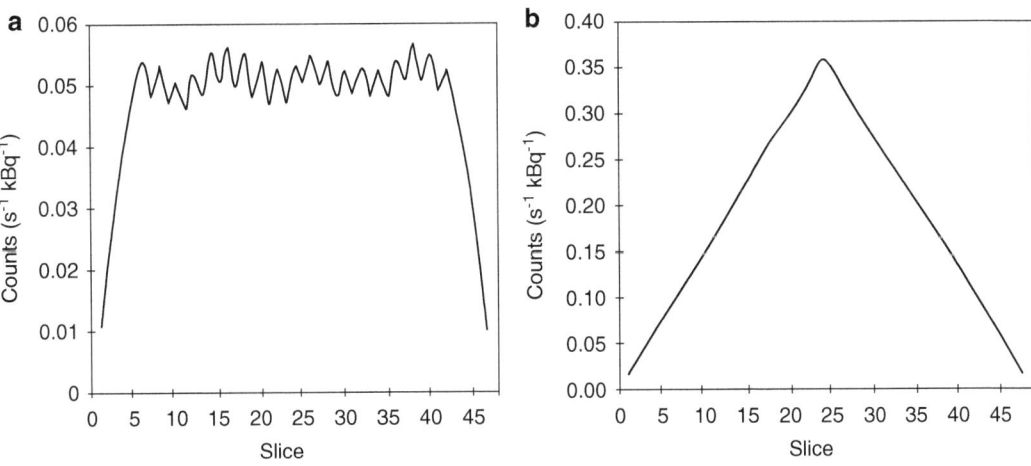

Fig. 17.12 Profiles of typical axial sensitivity in (**a**) 2-D and (**b**) 3-D. (Image copyright International Atomic Energy Agency (IAEA), Quality Assurance for PET and PET/CT Systems (2009) [2]. Reprinted with permission)

17.11   Scatter Fraction, Count Losses, and Randoms Measurement

The scatter of gamma rays released by positron annihilation causes coincidence occurrences to be mislocated. The sensitivity of positron emission tomography to scatter radiation varies due to design and implementation differences.

The approach requires the measurement of random coincidences, whether through a delayed event channel or a computation based on single-detector event rates. This approach is recommended because it allows for estimating scattering fractions as a function of count rate. It is essential for instruments with an inherent background that cannot achieve a random-to-true ratio of less than 1.0 percent.

17.11.1   Purpose

Scattering, count losses, and randoms affect image quality and quantitation accuracy. Scattering and randomness both introduce incorrect events. When random event coincidences are insignificant, i.e. at low count rates, the scatter fraction is defined as the ratio of scattering coincidences to the total of scattered and true coincidences. This may vary depending on the tomography's energy resolution, 2-D vs 3-D mode, and coincidence time windows length. The capacity of a tomograph to properly assess high and low radioactivity sources is reflected in its count rate performance. This is especially important since clinical investigations are routinely conducted with radioactivity levels at which count losses owing to system dead time are not trivial, yet the probability of random coincidences rises with activity concentration. The noise equivalent count (NEC) rate is utilized to represent tomography count performance as a function of radioactivity concentration. The corresponding radioactivity concentration may be used as a reference to select the best radioactivity to provide to patients. The NEC determines a scanner's useable count rates by considering, assuming Poisson statistics, the contribution of true events, scattered events, and randoms to the overall coincidence rate.

17.11.2   Frequency

During acceptance testing and any suspected changes in detecting system performance.

17.11.3   Material Requirements

The materials required to perform the Scatter Fraction, Count Losses, and Randoms Measurement test are listed in Table 17.5.

17.11.4   Activity

17.11.4.1   Requirements

35 mCi (1.2 GBq) ^{18}FDG required in $V_{scat-tube}$ in a 10 mL Luer-lock syringe at the start of acquisition. While preparing the source, take activity in a 10 mL Luer-lock syringe in such quantity that it should remain 35 mCi (1.2 GBq) at T_0.

17.11.5   Phantom Positioning

Place the phantom holder on the patient's bed and then the phantom on the holder. As indicated in Fig. 17.13, arrange the holders towards the phan-

Table 17.5 Material required to perform the scatter fraction, count losses, and random measurement

Item	Quantity
10 mL syringe	1
50 mL pyrex beaker	1
Tubing Inner diameter (ID): 1/8″ Outer diameter (OD): 3/16″	80 cm
Tube plug 1/8″	1
Luer-lock	1
NEMA scatter phantom: 203 mm diameter polyethene cylinder of length 700 mm	1
Tape	
Compressed air (dust remover spray)	2 bottles

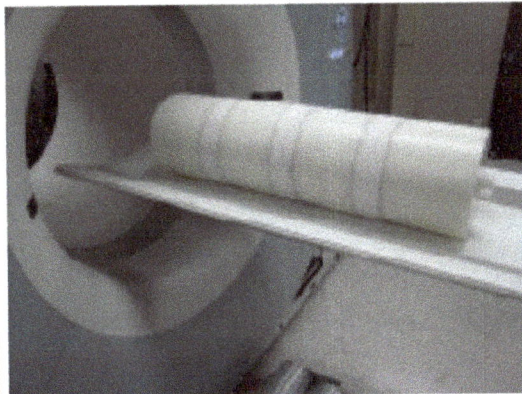

Fig. 17.13 Phantom positioning using a laser

tom's edge such that they are not in the PET field of vision.

Adjust the vertical PHS (Patient Handling System or Table) settings to line the phantom's centre with the lateral gantry lasers, as indicated in Fig. 17.13. Adjust the lateral location of the phantom on the PHS to align the phantom's centre. With gantry, lateral lasers and lasers to be aligned with phantom tube openings for the line source. The tube should be placed at the centre of the phantom, and the phantom should be placed on the holder to the PHS, as illustrated in Fig. 17.14. To ensure that the assay time is constant, synchronize your clock (such as a watch or cell phone) with the clock on the scanning computer.

17.11.6 Activity Preparation

- Place 3 Kimwipes in the bottom of the plunger of the dose calibrator.
- Place the syringe in the dose calibrator to record the amount of activity. Take the activity at around 51.5 mCi (1.0955 GBq) by assuming that it will take 1 h to prepare the source, so at the time of the scatter acquisition, the activity will be 35 mCi (1.2 GBq), which is required for the scatter acquisition.
- Record test activity and time, considering residual dose.
- Pull the syringe with roughly 1 mL of air.

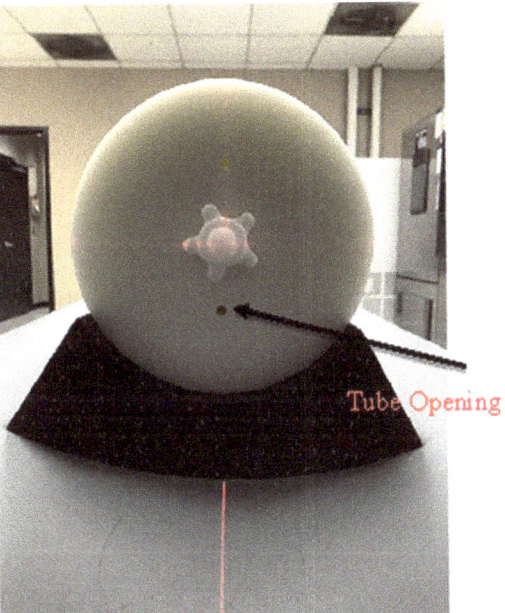

Fig. 17.14 Immobilize Phantom on PHS using tape

- If the activity level and volume in Vscat-tube ± 0.5 mL are larger than 51.5 mCi (1.9055 GBq) + 10%, transfer the surplus activity/volume from the syringe into a 50 mL beaker to achieve the requisite activity/volume, then suck water into the syringe from the beaker to obtain the Vscat-tube volume. If the activity level is less than 51.5 mCi (1.9055 GBq) - 10%, additional activity is necessary to complete the test properly. The activity should be 35 mCi (1.2 GBq) at the acquisition time. Calculation of the activity (example) shown in Fig. 17.15 using an Excel sheet:

- Record test activity and time, taking into consideration residual dose. Pull the syringe with about 1 mL of air.

17.11.7 Line Source Preparation as Shown in Fig. 17.16

- Connect the tube with the Luer-lock.
- If available, use compressed air to clean out the tube.
- Remove the needle from the syringe and attach it to the Luer-lock-on tube.

- Place activity in an 80 cm tube. The 70 cm extent of activity must be centred lengthwise in the 80 cm tube, with no air bubbles present. All air bubbles should be removed.
- Insert a plug into the end of the tube opposite the Luer lock.
- Remove the Luer-lock from the tube and plug it.
- Determine the length of activity in a tube. The range of activity in the tube must be between 68 and 72 cm.
- Use the correct assay for line source length using the following formula:

$$\text{Activity}_{\text{Corrected}} = 70 \times \left(\frac{\text{Activity}}{\text{Length}} \right)$$

Place the tube in a shielded container and bring it inside the scanning room.

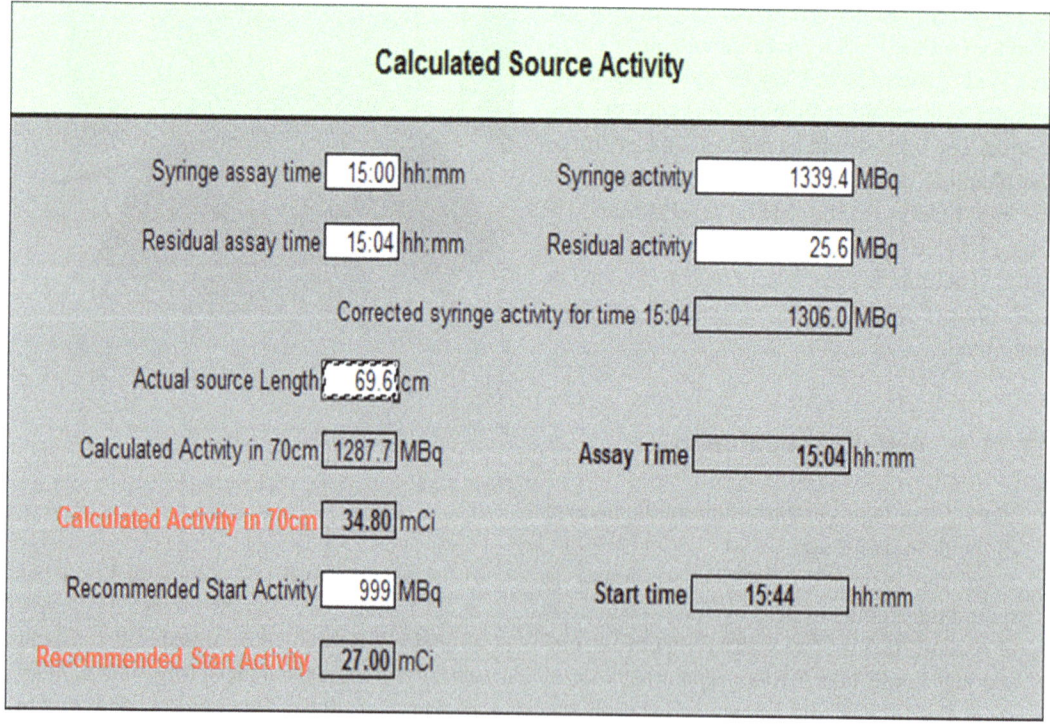

Fig. 17.15 Calculation of the activity for scattering phantom using Excel Sheet

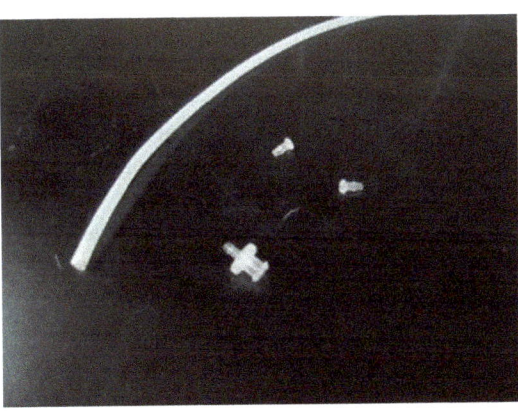
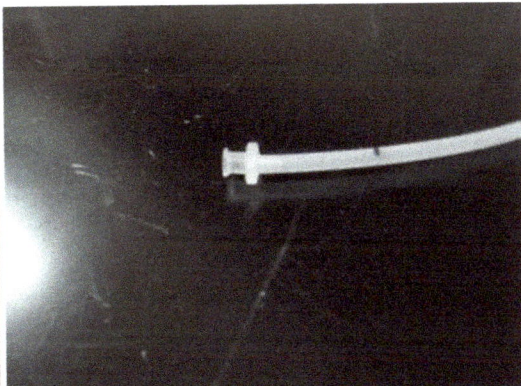

Fig. 17.16 Line source preparation

17.11.8 Data Acquisition

Place the tube into the phantom (70 cm extent of FDG lined up with 70 cm long phantom) as shown in Fig. 17.17.

A minimum of 500 K prompt counts should be included in each acquisition. It is also critical that the measurements surrounding the peak count rate be performed with sufficient frequency so that the peak rate can be reliably established. As a result, manufacturers are likely to prescribe a procedure for their scanners that includes beginning activity, acquisition timings, and acquisition durations.

Normally data acquisition was performed by acquiring 45 frames of 15 min each and 20 min pause between two acquisitions.

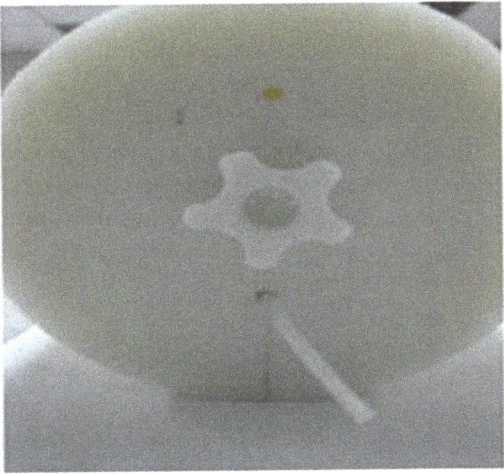

Fig. 17.17 Tube positioning in scatter phantom

17.11.9 Data Processing

Prompt and random sinograms must be produced for each acquisition *j* of the slice *i* on tomographs with an axial field of view of 65 cm or less (if no random estimate is available, only prompt sinograms are generated). For each acquisition, sinograms must be created for slices inside the centre of 65 cm of tomographs having an axial field of view greater than 65 cm. The measurements must not be corrected for fluctuations in detector sensi-

tivity or detector movements such as wobbling, randoms, scatter, dead time, or attenuation.

17.11.10 Symbols

Scatter fraction (SF)—a dimensionless ratio of scattered coincidence events to the total of scattered and true coincidence events in a given ROI of the scanner field-of-view. The radionuclide to be utilized for this measurement is 18F. The quantity of radioactivity must be sufficient to enable the following two rates to be measured:

(a) Rt, peak-peak true count rate.

(b) $R_{NEC, peak}$-peak noise equivalent count rate.

17.11.11   Analysis

Each acquisition j's prompt sinogram i is handled in the following manner:

1. As illustrated in Fig. 17.18, any pixels positioned more than 12 cm from the phantom's centre must be set to zero.
2. The centre of the line source response must be found for each projection angle inside the sinogram by locating the pixel with the highest value. Each projection must be moved such that the pixel with the highest value aligns with the centre pixel of the sinogram.
3. Following alignment, a sum projection must be generated in which a pixel in the sum projection is the sum of the pixels in each angular projection with the same radial offset as the pixel in the sum projection:

$$C(r)_{i,j} = \sum_{\phi} C\left(r - r_{max}(\phi),\phi\right)_{i,j}$$

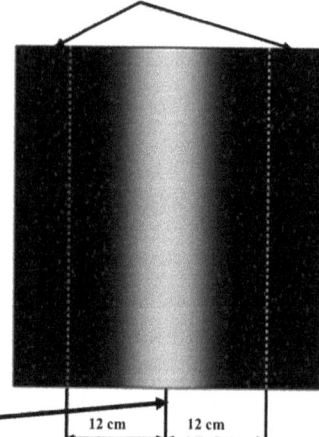

All pixels at a distance > 12 cm from the axis are set to 0

Field of view and phantom axis   12 cm   12 cm

Fig. 17.18 The ROI is shown schematically as a sinogram sum profile. (Image copyright International Atomic Energy Agency (IAEA), Quality Assurance for PET and PET/CT Systems (2009) [2]. Reprinted with permission)

where

(a) r is the projection's pixel number,
(b) The projection number in the sinogram (i.e. the row of the sinogram) is represented by the φ.
(c) $r_{max}(a)$ denotes the position of the greatest value in the projection φ.

The sum projection should provide the count's $C_{L,I,j}$, and $C_{R,i,j}$, the left and right pixel intensities at the margins of the 40 mm wide strip at the centre of the sinogram (see Fig. 17.19). Linear interpolation will determine the pixel intensities at 20 mm from the projection's centre pixel. The average of the two-pixel intensities $C_{Li,j}$ and $C_{R,i,j}$ is multiplied by the number of pixels, including fractional values, between the edges of the 40 mm wide strip, and the result is added to the counts in the pixels outside the strip to provide the number of random plus scatter counts C_{r+s}, i, j for slice i of acquisition j.

$C_{TOT, i, j}$ is calculated as the sum of all pixels in the sum projection for slice i of acquisition j.

The undermentioned formula should be used to compute the average activity $A_{ave,j}$ for each acquisition j. The following analysis is dependent on whether or not a randoms estimate is available:

$$A_{ave} = \frac{A_0}{\ln 2}\left(\frac{T_{1/2}}{T_{acq}}\right)\left\{1 - \exp\left(\frac{-T_{acq}}{T_{1/2}}\ln 2\right)\right\}$$

To calculate the average radioactivity during a certain acquisition, the activity, A_0, measured at the start of the acquisition, must be multiplied by half the radionuclide's half-life, $T1/2$, and the acquisition length T_{acq}, respectively.

17.11.12   Analysis with Randoms Estimate

It is necessary to set to zero any pixels in each randomly generated sinogram i of acquisition j that are more than 12 cm away from the centre of the phantom. The number of random counts, $C_{r, i, j}$, is calculated by adding all of the leftover counts in the sinogram i of acquisition j.

Fig. 17.19 Integration of background counts inside and outside 40 mm strip. (Image copyright International Atomic Energy Agency (IAEA), Quality Assurance for PET and PET/CT Systems (2009) [2]. Reprinted with permission)

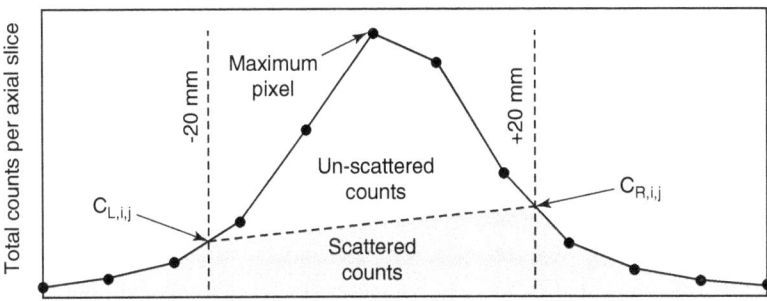

17.11.12.1 Scatter Fraction

Estimate the scatter fraction ($\mathrm{SF}_{i,j}$) for each slice i and acquisition j using the formula:

$$\mathrm{SF}_{i,j} = \frac{\sum_{j'} C_{r+s,i,j} - \sum_{j'} C_{r,i,j}}{\sum_{j'} C_{\mathrm{TOT},i,j} - \sum_{j'} C_{r,i,j}}$$

The scatter fraction SF of the system is calculated as follows:

$$\mathrm{SF}_j = \frac{\sum_i \sum_{j'} C_{r+s,i,j} - \sum_i \sum_{j'} C_{r,i,j}}{\sum_i \sum_{j'} C_{\mathrm{TOT},i,j} - \sum_i \sum_{j'} C_{r,i,j}}$$

17.11.12.2 Count Rates and NECR

Compute the following for each acquisition j:

For each slice i, the total event rate $R_{\mathrm{TOT},i,j}$ is calculated as follows:

$$R_{\mathrm{TOT},i,j} = \frac{C_{\mathrm{TOT},i,j}}{T_{\mathrm{acq},j}}$$

For each slice i, the real event rate $R_{t,I,j}$ is as follows:

$$R_{t,i,j} = \frac{\left(C_{\mathrm{TOT},i,j} - C_{r+s,i,j}\right)}{T_{\mathrm{acq},j}}$$

For each slice i, the random event rate $R_{r,ij}$ is as follows:

$$R_{t,i,j} = \frac{C_{r,t,j}}{T_{\mathrm{acq},j}}$$

as well as the scatter event rate $R_{s,I,j}$ for each slice i

$$R_{s,i,j} = \frac{C_{r+s,i,j} - C_{r,i,j}}{T_{\mathrm{acq},j}}$$

In where $T_{\mathrm{acq},j}$ denotes the acquisition time for frame number j.

Calculate the noise equivalent count rate $R_{\mathrm{NEC},i}$ and j for each slice i of each acquisition j on all systems except those that perform direct randoms subtraction:

$$R_{\mathrm{NEC},i,j} = \frac{R_{t,i,j}^2}{R_{\mathrm{TOT},i,j}}$$

$R_{\mathrm{NEC},i,j}$ for each slice i should be computed as follows in systems that employ direct randoms subtraction:

$$R_{\mathrm{NEC},i,j} = \frac{R_{t,i,j}^2}{R_{\mathrm{TOT},i,j} + R_{r,i,j}}$$

To calculate total system event rates, add up the slice event rates for all of the slice i:

$$R_{\mathrm{TOT},j} = \sum_i R_{\mathrm{TOT},i,j}$$

$$R_{t,j} = \sum_i R_{t,i,j}$$

$$R_{r,j} = \sum_i R_{t,i,j}$$

$$R_{s,j} = \sum_i R_{s,i,j}$$

$$R_{\mathrm{NEC},j} = \sum_i R_{\mathrm{NEC},i,j}$$

17.11.13 Alternative Analysis with no Random Estimate

The scatter fraction will be determined using the sequence's final acquisitions with count loss rates and random rates less than 1.0 percent of the true rate. For these acquisitions, it is assumed that $C_{j'r+s,i,j'}$ contains no random counts and only scattering counts and that $C_{TOT,i,j'}$ contains only true and scattering counts.

For each slice, the scatter fraction SF_i is derived by adding the low activity acquisitions together as follows:

$$SF_i = \frac{\sum_{j'} C_{r+s,i,j'}}{\sum_{j'} C_{TOT,i,j'}}$$

The count-weighted average of the SF_i values yields the system scatter fraction SF:

$$SF = \frac{\sum_i \sum_{j'} C_{r+s,i,j'}}{\sum_i \sum_{j'} C_{TOT,i,j'}}$$

17.11.14 Count Rates and NECR

- Calculate the following for each acquisition j:

For each slice, the total event rate $R_{TOT,i,j}$:

$$R_{TOT,i,j} = \frac{C_{TOT,i,j}}{R_{acq,j}}$$

For each slice i, the true event rate $R_{t,i,j}$:

$$R_{t,i,j} = \frac{\left(C_{TOT,i,j} - C_{r+s,i,j}\right)}{R_{acq,j}}$$

For each slice i, the random event rate $R_{r,i,j}$:

$$R_{r,i,j} = R_{TOT,i,j} - \left(\frac{R_{t,i,j}}{1-SF_i}\right)$$

for each slice i, the scatter event rate $R_{s,i,j}$:

$$R_{s,i,j} = \left(\frac{SF_i}{1-SF_i}\right) R_{t,i,j}$$

$T_{acq,j}$ denotes the acquisition time for frame j. Calculate the noise equivalent count rate $R_{NEC,i,j}$ for each slice i of each acquisition j for all systems except those that execute direct randoms subtraction:

$R_{NEC,i,j}$ should be computed as follows for each slice i in systems that employ direct randoms subtraction:

The sum of the relevant slice event rates across all slices i is determined as the total system event rate:

$$R_{NEC,i,j} = \frac{R_{t,i,j}^2}{R_{TOT,i,j}}$$

$$R_{NEC,i,j} = \frac{R_{t,i,j}^2}{R_{TOT,i,j} + R_{r,i,j}}$$

$$R_{TOT,j} = \sum_i R_{TOT,i,j}$$

$$R_{t,j} = \sum_i R_{t,i,j}$$

$$R_{r,j} = \sum_i R_{t,i,j}$$

$$R_{s,j} = \sum_i R_{s,i,j}$$

$$R_{NEC,j} = \sum_i R_{NEC,i,j}$$

17.11.15 Results

17.11.15.1 Plot of Count Rate

As illustrated in Figs. 17.20, 17.21, 17.22, and 17.23, plot the following five values as a function

Fig. 17.20 Plot scatter fraction vs activity concentration

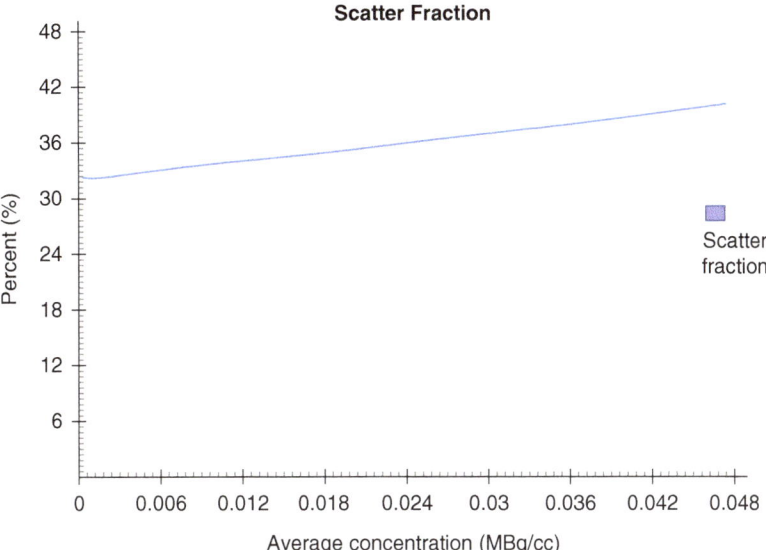

Fig. 17.21 Plot between true scatter ratio and activity concentration

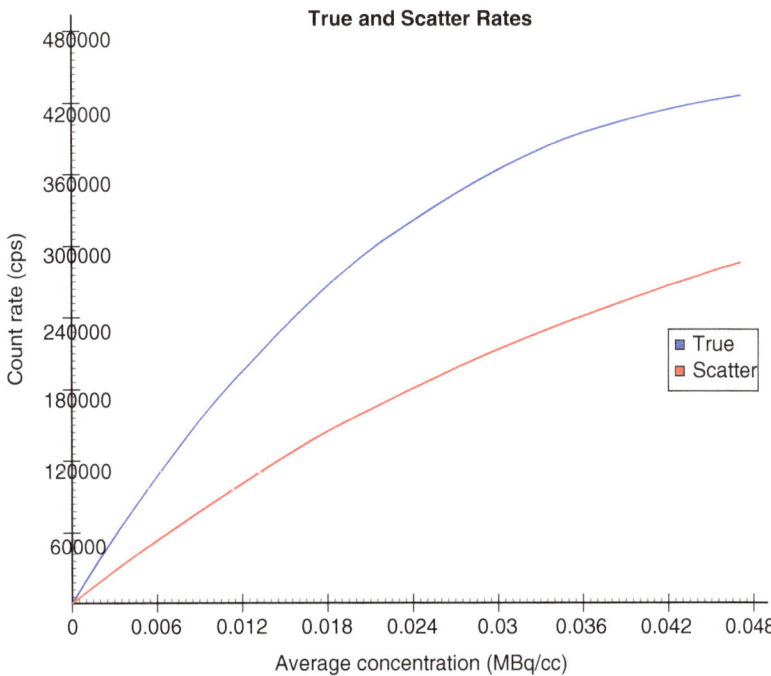

of the average effective radioactivity concentration $a_{ave, j}$, where volume V is the total volume of the cylindrical phantom (22,000 cm3).

(a) $R_{t,j}$ —True event rate for system.
(b) $R_{r,j}$—Random event for rate system.
(c) $R_{s,j}$—Scatter event rate for system.

(d) $R_{NEC,j}$ —noise equivalent count rate of system.
(e) $R_{TOT,j}$ —Total event rate of the system.

If a technique for estimating randoms was employed in the measurement, provide that as well.

Fig. 17.22 Plot between total, random, and activity concentration

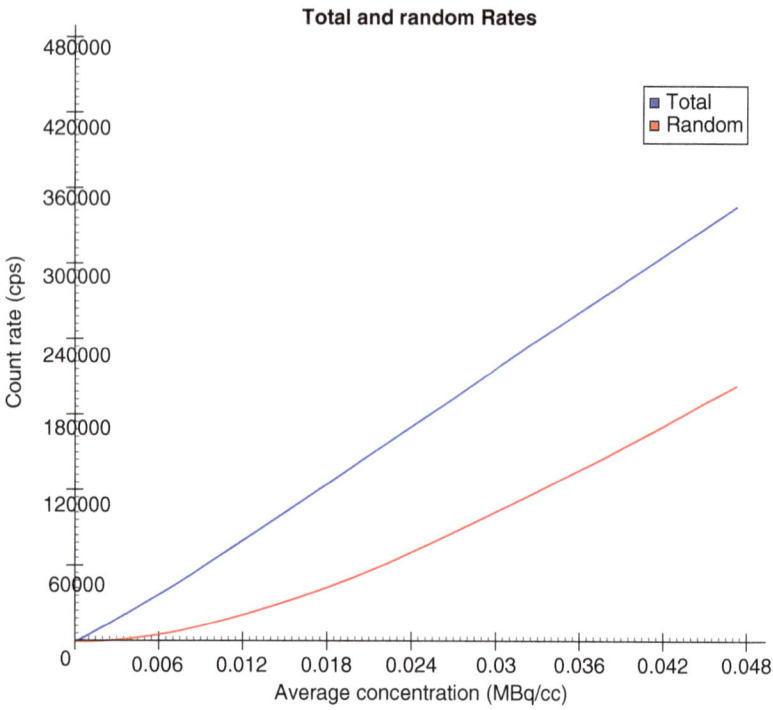

Fig. 17.23 Plot between NEC rate and activity concentration

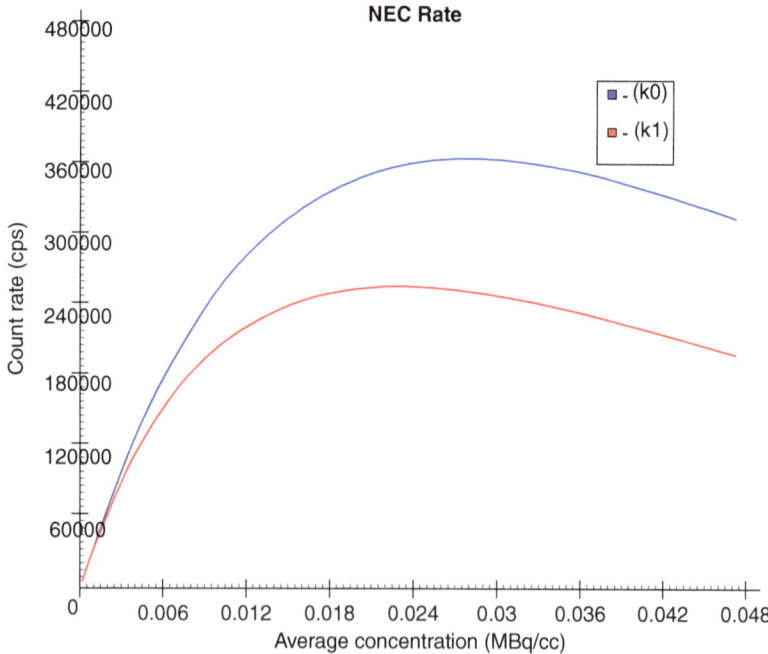

17.11.15.2 Values for Peak Count

The values retrieved from the above plot are as follows:

(a) $R_{t, peak}$ —true count rate of peak.
(b) $R_{NEC, peak}$ —peak NECR.
(c) $a_{t, peak}$ —the concentration of activity at which R_t, the peak, is attained,
(d) $a_{NEC, peak}$ —the concentration of activity at which R_{NEC} reaches its peak.

17.11.15.3 System Scatter Fraction

Report the value of SF at peak noise equivalent count rate if a random estimate was employed in the measurement, and plot system scatter fraction SF_j vs activity $a_{ave,j}$ as described in section Count rate plot. Report the value SF if no random estimate was utilized.

17.11.16 Tolerances Suggested

Scatter fraction, peak NEC, and radioactivity concentration for peak NEC should all meet or exceed the vendor's standards. The user should define reference values, tolerances, and action levels (i.e. to trigger the decision to place a call for maintenance). The following are suitable tolerance criteria for SF:

$$SF_{observed} < 1.05\, SF_{expected}$$

A record of the NEC curve, its peak value, and its peak radioactive concentration will be kept for future comparison.

17.11.17 Taking Corrective Action

Failure to meet tolerance standards necessitates further investigation to ensure that the source is accurately constructed and that all processes were successfully followed. It is important to notify the manufacturer and take remedial measures if the reason is not found.

17.12 Energy Resolution

17.12.1 Purpose

Only tomography equipment that uses singles-based attenuation correction and calibration is appropriate for this test. The energy resolution measurement enables accurate photomultiplier calibration to be determined and guarantees that the light collecting efficiency meets the requirements.

17.12.2 Frequency

If relevant, during acceptance testing and if there is a reasonable suspicion, the detection system's performance has altered considerably.

17.12.3 Materials

A point source of ^{18}F with a diameter of less than 1 mm is used in both the transaxial and axial directions. A capillary tube with less than 1 mm of internal diameter and an outer diameter of less than 2 mm may be used to provide an appropriate source. A source similar to that used for spatial resolution measurements may be employed for this test. To reduce the influence of dispersed radiation, the source should be hung in the air near the centre of the FOV. As for the spatial resolution test, the radioactivity of the source must be such that the percent dead time or random loss is less than 5%.

This is typically accomplished using radioactivity of about 37 MBq. The radioactivity concentration is calculated using the manufacturer's suggested activities for this test.

be translated into energy units using this factor (keV).

17.12.4   Data Gathering

Follow the manufacturer's instructions for energy testing or spectra collection and presentation. An example of such a spectrum is illustrated in Fig. 17.24. It should be taken a long enough period to get less than 10,000 counts at the top of the energy distribution.

17.12.5   Analysis

Determine the percent energy resolution of the system using the manufacturer's technique for energy testing. If no preset process is provided, the energy spectra should be evaluated to determine the FWHM of the energy peak distribution.

Calculating the peak location and fitting a parabolic curve to the top of the peak yields an estimated energy calibration factor. The FWHM may

17.12.6   Tolerances Suggested

The energy resolution values should not exceed those given in the vendor's specification.

The user should set reference values, tolerances, and action levels (i.e. to trigger the decision to place a call for maintenance). An appropriate tolerance criterion for FWHM is:

$$R_{measured} < 1.05 R_{Eexpected}$$

17.12.7   Taking Corrective Action

If the tolerance requirements for this test are not fulfilled, make sure that all processes are followed appropriately. If the source of the problem cannot be determined, the manufacturer should be alerted, and remedial action should be done.

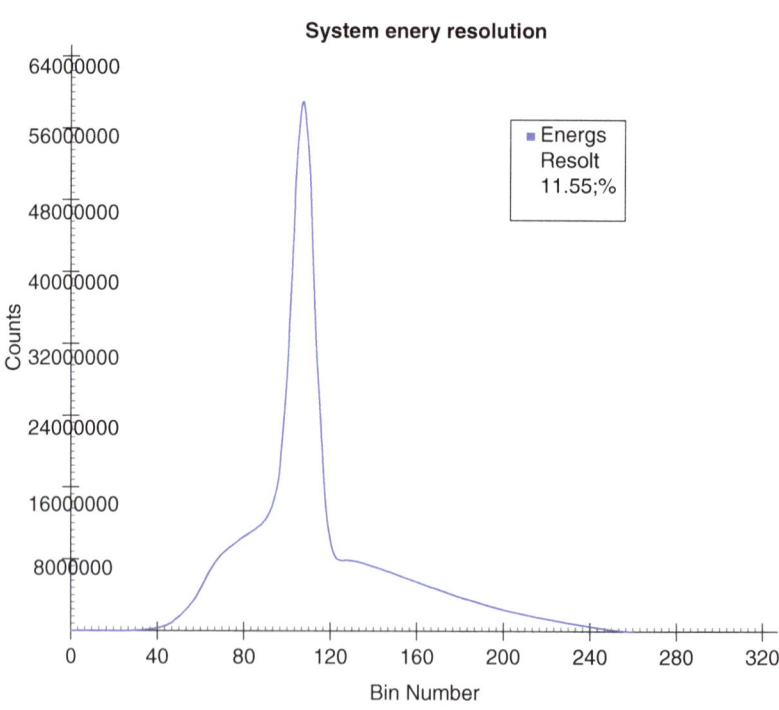

Fig. 17.24   An example of an acquired energy spectrum; energy resolution can be calculated from the FWHM of the energy peak distribution

17.13   Image Quality, the Accuracy of Attenuation, and Scatter Corrections

The gamma scattering rays released by positron annihilation cause coincidence occurrences to be incorrectly localized. The capacity of a PET to measure highly radioactive sources may be expressed by counting losses and random rates. The analysis and reporting of this measurement are discussed using two ways. The first technique requires the detection of random coincidences through a delayed event channel or a computation based on single-detector event rates.

This approach is favoured because it enables scattering fractions to be estimated as a function of the count rate. It's essential for instruments that can't reach a randoms-to-true ratio of less than 1.0 percent because of their inherent background. The second technique is for systems that don't have a random measurement capability.

17.13.1   Purpose

The initial goal of this approach is to determine the relative susceptibility of the system to scattered radiation. The scatter fraction, SF, for the complete tomograph represents scatter. The second goal of this approach is to assess the impacts of system dead time and random event creation at various degrees of source activity. The true event rate equals the overall coincident event rate minus the scattered and random event rates.

17.13.2   Frequency

Annually and whenever it is believed that the performance of the detection system has changed considerably, at the time of acceptance testing, as part of end-of-warranty tests, and whenever there is a suspicion that the performance of the detector system has changed significantly.

17.13.3   Material Requirements

Table 17.6 lists the materials needed for image quality, attenuation accuracy, and scatter corrections

A NEMA image quality (IQ) phantom has an internal capacity of 9.7 L and contains six spherical inserts with 10, 13, 17, 22, 28, and 37 mm internal diameters.

17.13.3.1   Activity Requirements

- Syringe 1: 0.1432 mCi or ($5.3 \times V_{background}$) MBq of ^{18}FDG in 5 mL, in a 10 mL syringe, calibrated for T_0, delivered 2 h before T_0.
- Syringe 2: ($0.1432 \times V_{background}$) mCi or ($5.3 \times V_{background}$) MBq of ^{18}FDG in 5 mL, in a 10 mL syringe, calibrated for T_0, delivered 2 h prior to T_0.
- Syringe3: 3 mCi (110 MBq) of ^{18}FDG in Vscat-tube, delivered 2 h before T_0 in a 10 mL Luer-lock syringe calibrated for T_0.
- Syringe4: 6 mCi (220 MBq) of ^{18}FDG in Vscat-tube, in a 10 mL Luer-lock syringe, calibrated for T_0, given 2 h before T_0.
- At the start of both sets of data collections, an inline activity source was used 3 mCi (110 MBq).
- Activity concentration in the phantom background 0.1432 mCi/L (5.3 MBq/mL) at the start of both sets of data collection.

Table 17.6 Lists the materials needed to improve image quality, attenuation accuracy, and scatter correction

Item	Quantity
10 mL syringe	6
2000 mL pyrex beaker	1
100 mL pyrex cylinder	1
Tubing (80 cm) "ID = 1/8" "OD = 3/16"	1
Tube plug "1/8"	1
Funnel	1
NEMA image quality phantom	1
NEMA scatter phantom	1
Industrial tape	1
Compressed air bottle (dust remover spray)	4 bottles

17.13.4   Phantom Positioning

- Position the image quality phantom and IQ foam fixture 30 cm away from the PHS palette tip on the PHS. Place the scatter phantom next to the foaming fixture on the PHS, as indicated in Fig. 17.25.
- Align the scatter phantom's lateral centre with the gantry lateral lasers by aligning the scatter phantom's lateral centre with the lateral gantry lasers. The tube is immediately under the phantom's centre (as shown in Fig.17.25).
- Centre the lung equivalent material on the vertical gantry lasers, as illustrated in Fig. 17.26, by adjusting the vertical PHS settings for the lung equivalent material on the image quality phantom.

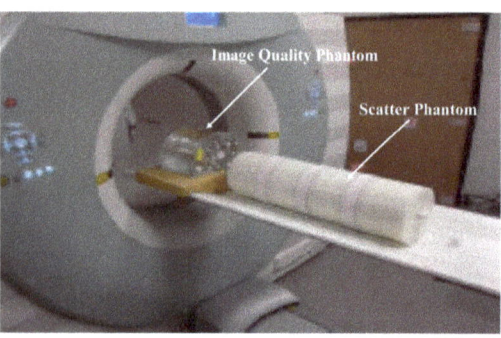

Fig. 17.25 Image quality phantom position with scatter phantom

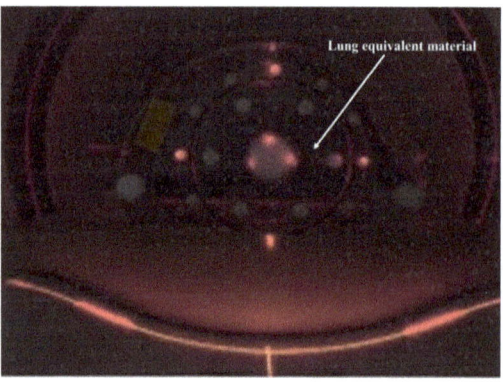

Fig. 17.26 Centring on image quality phantom

17.13.5   Activity Preparation

- Place a 3 Kimwipes in the bottom of the plunger of the dose calibrator.
- Place syringe 1 in dose calibrator and take note of activity. The activity level must be $(0.1432 \times V_{background})$ mCi or $(5.3 \times V_{background})$ MBq ± 5% in 5 mL ± 20%, when decay corrected to T_0.
- Place syringe 2 in dose calibrator and take note of activity. The activity level must be $(0.1432 \times V_{background})$ mCi or $(5.3 \times V_{background})$ MBq ± 5% in 5 mL ± 20%, when decay corrected to T_0.
- Place syringe 3 in dose calibrator and take note of activity. The activity level must be 3 mCi (110 MBq) ± 15% in $V_{scat\text{-}tube}$ ± 0.5 mL, when decay corrected to T_0.
- Place syringe 4 in dose calibrator and take note of activity. The activity level must be 6 mCi (220 MBq) ± 15% in $V_{scat\text{-}tube}$ ± 0.5 mL, when decay corrected to T_0.
- Pull approximately 1 mL of air into the syringe.

17.13.6   Preparation of Phantoms (8:1 Activity Concentration Ratio)

- The phantom background volume $V_{background}$ is necessary for the procedure outlined below. Because the volume will directly influence the recovery outcomes, it must be measured with great precision.
- The tube volume $V_{scat\text{-}tube}$ is necessary for the approach outlined below. Because the extent of the Image quality, accuracy of attenuation, and scatter corrections activity in the tube must equal 70 cm ± 2 cm, the volume must be determined with substantial precision. It's estimated that the volume will be between 4.8 and 5.5 mL.
- To fill the backdrop volume, remove the centre cap of the image quality phantom, as illustrated in Fig. 17.27.

- Measure ¾ $B_{ackground}$ in a beaker and pour water into the phantom background. To fill the spheres, remove the sphere cap from the image quality phantom, as illustrated in Fig. 17.28.
- Before injecting activity, make sure the insides of the four smallest spheres are dry since any remaining liquid will cause a contrast recovery mistake.
- Fill the two largest spheres with water.
- Using the graduated cylinder, measure 1/8 Background and pour into the 2000 mL beaker.
- Stir the contents of syringe 1 into the 2000 mL beaker.
- Fill the syringe with beaker solution that will be Syringe 5.

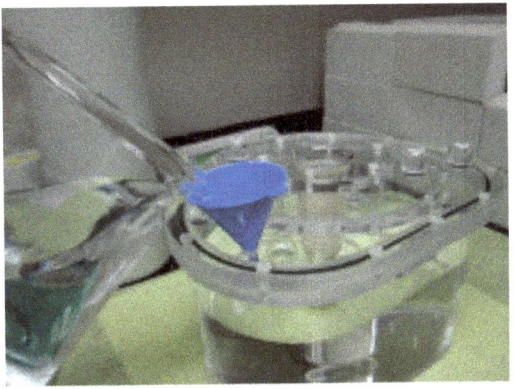

Fig. 17.29 Pour the contents using the funnel image quality phantom

- Using the image quality needle lengths, fill spheres 1 to 4 (4 smallest spheres) with the contents of the syringe 5.
- On the image quality phantom, reattach the top spherical cover.
- Fill the 2000 mL beaker with the remainder of syringe 5.
- Using the funnel, pour the beaker's contents onto the image quality phantom's background area, as illustrated in Fig. 17.29.
- Fill the phantom background with water to fill the remaining volume.

Seal the background by reattaching the plug to the image quality phantom.

17.13.6.1   Prepare Line Source Such as Stated Below

- Place the syringe in the dose calibrator to take note of the activity level.
- If activity level and volume are greater than 35 mCi (1.3 GBq) + 10% in $V_{scat-tube}$ ± 0.5 mL, transfer excess activity/volume from the syringe into a 50 mL beaker to obtain the required activity/volume and pull water into the syringe to obtain a volume of $V_{scat-tube}$. If the activity level is less than 35 mCi − 10%, more activity is required to complete the test successfully.
- Take note of assay activity and time, accounting for residual dose.
- Pull approximately 1 mL of air into the syringe.

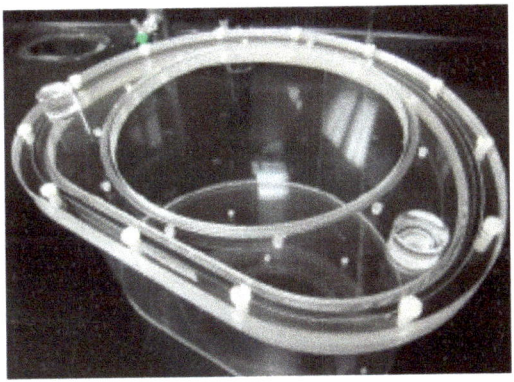

Fig. 17.27 Image quality phantom without centre cap

Fig. 17.28 Fill the spheres with water after removing spheres cap in image quality phantom

- Connect the tube to the Luer-lock.
- Clean out tube with compressed air, if available.
- Disconnect the needle from the syringe and connect the syringe to Luer-lock on the tube.
- Fill 80 cm tube with activity. The 70 cm extent of activity must be centred lengthwise in the 80 cm tube while ensuring no air bubbles. If air bubbles exist, perform necessary measures to remove bubbles.
- Plug the tube at the end opposite the Luer-lock.
- Remove Luer-lock from the tube and plug the tube.
- Measure the length of activity in the tube. The extent of activity in the tube must be between 68 cm and 72 cm.
- Correct essay for line source length.

$$\text{Activity corrected} = 70 X \left(\frac{\text{Activity}}{\text{Length}} \right)$$

- Place the tube into a shielded container and transport the container and image quality phantom to the scanner room.

17.13.6.2   Data Acquisition (8:1 Activity Concentration Ratio)

- Place the tube into scattering phantom and image quality phantom on foam.
- Select the AC (Attenuation Correction) Wholebody protocol under PET protocols.
- Modify the topogram length to 1024 mm.
- Acquire Topogram.
- Set acquisition to 1 bed.
- Align the centre of acquisition FOV with the centre of the spheres.
- Acquire a CT scan.
- In the PET recon card, set the following parameters:

  - 200 × 200 image matrix size
  - Iterative reconstruction, 3 iterations, 24 subsets.
  - 5 mm Gaussian filter.

The total axial imaging distance to be simulated and the axial distance the bed is translated between positions in a complete body study (usually less than the axial field-of-view of the scanner) will determine the data collection time. The imaging duration must be chosen to approximate an entire body scan in 60 min, with a total axial imaging distance of 100 cm. If attenuation correction is done at each bed position, this time should include both emission and transmission imaging timings. Total scan time for emission and transmission scans $T_{T,E}$ is computed as follows:

$$T_{T,E} = \frac{60}{\text{dist}} \times \text{axial step}$$

- In a complete body study, distance = 100 cm axial step is the distance the bed is moved.
- This time must include the lengths of both the emission and transmission scans and any transition periods, such as when shifting the transmission source or uploading data.
- Additional measurements may be conducted with extended imaging period to mimic a longer overall imaging time or a shorter total axial imaging distance. In particular, an imaging period of 60 min translates to a total axial imaging distance of 50 cm. The whole axial imaging distance is simulated, and the actual emission and transmission imaging periods must be recorded. Because the scans contain a limited number of counts, it is advised that three duplicate scans be obtained to increase the reliability of the data. To gather the same number of events, the time of successive replication scans should be adjusted for physical decay. Obtain a PET scan at T_0.

17.13.7   Preparation of Phantoms (4:1 Activity Concentration Ratio)

- Using syringe 5, extract 6 mL from the background using a plug of image quality phantom, as illustrated in Fig. 17.30.

Fig. 17.30 Preparation of the phantom

- Add the contents of syringe2 into the image quality phantom.
- Fill remaining volume of phantom background with contents of syringe 5.
- Reattach plug to image quality phantom.
- Remove plug from the tube and attach Luer-lock.
- Connect syringe 6 to Luer-lock on the tube.
- Remove the second plug.
- Remove the content of the tube.
- Prepare line source such as stated in Sect. 17.13.6.1.
- Place the tube in a shielded container and take it to the scanner room with the image quality phantom.

17.13.7.1   Data Acquisition #2 (4:1 Activity Concentration Ratio)

- Insert the tube into the scatter phantom and place the image quality phantom on the foam.
- In the PET procedures section, choose the AC Wholebody protocol.
- Increase the length of the topogram to 1024 mm.
- Obtain Topogram.
- Limit your acquisition to one bed.
- Align the centre of the spheres with the acquisition FOV.
- Obtain a CT scan.
- Set the following parameters in the PET recon card:
  - Image matrix size of 200×200 pixels.
  - 3 iterations, 24 subsets, iterative reconstruction
  - A Gaussian filter with a size of 5 mm.
- The total axial imaging distance to be simulated and the axial distance the bed is translated between positions in a complete body study (usually less than the axial field-of-view of the scanner) will determine the data collection time. The imaging duration must be chosen to approximate an entire body scan in 60 min, with a total axial imaging distance of 100 cm. If attenuation correction is done at each bed position, this time should include both emission and transmission imaging timings. Total scan time for emission and transmission scans $T_{T,E}$ is computed as follows:

$$T_{T,E} = \frac{60}{\text{dist}} \times \text{axial step}$$

- In a complete body study, distance = 100 cm axial step is the distance the bed is moved.
- This time must include the lengths of both the emission and transmission scans and any transition periods, such as when shifting the transmission source or uploading data.
- Additional measurements may be conducted with an extended imaging period to mimic a longer overall imaging time or a shorter total axial imaging distance. In particular, an imaging period of 60 min translates to a total axial imaging distance of 50 cm. The whole axial imaging distance is simulated, and the actual emission and transmission imaging periods must be recorded. Because the scans contain a limited number of counts, it is advised that three duplicate scans be obtained to increase the reliability of the data. To gather the same number of events, the time of successive replication scans should be adjusted for physical decay. Obtain PET scan at T_{0+110}.

17.13.8   Processing of Data

All slices must be recreated using all available corrections. Images for whole-body investigations must be rebuilt using the manufacturer's specified parameters (e.g. image matrix size, pixel size, slice thickness, reconstruction method, filters, or other smoothing techniques). These parameters of reconstruction must be reported.

17.13.9   Analysis

17.13.9.1   Image Quality

The analysis will be based on a transverse image centred on the cold and hot spheres. For all spheres, the same slice should be utilized. Draw regions of interest (ROIs) on each hot and cold sphere. The ROI must be circular, with a diameter equal to the sphere's measured inner diameter.

ROIs of the same size as the ROIs created on the hot and cold spheres should be drawn in the phantom's background on a slice centred on the spheres. At a distance of 15 mm from the phantom's border but no closer than 15 mm to any sphere, twelve 37 mm diameter ROIs must be created across the background (see Fig. 17.31). Smaller ROIs such as 10, 13, 17, 22, and 28 mm should be drawn concentric to the background ROIs of 37 mm. ROIs must also be drawn on slices near +/−1 cm and +/−2 cm on each side of the centre slice as possible. There will be 60 background ROIs of each size drawn, with 12 ROIs on each of the five slices. Between each

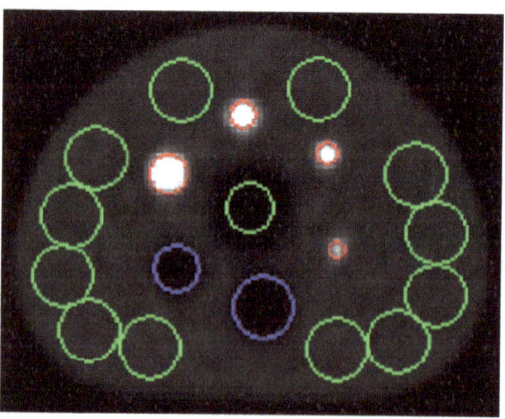

Fig. 17.31 For image quality analysis, place a background area of interest

measurement, the positions of all ROIs must be maintained (e.g. replicate scans). Each background ROI's average count should be recorded. For each hot sphere j, the percent contrast $Q_{H,j}$ is determined by:

$$Q_{H,j} = \frac{C_{H,j} / C_{B,j-1}}{a_H / a_{B-1}}$$

where

(a) $C_{H,j}$ is the average number of counts in the ROI for sphere j.
(b) $C_{B,j}$ Background ROI average counts for sphere j.
(c) a_H is the hot spheres' activity concentration and,
(d) a_B is the background activity concentration.

For each cold sphere j, the percent contrast $Q_{C,j}$ is determined as follows:

$$Q_{C,j} = \left[1 - \frac{C_{C,j}}{C_{B,j}}\right].100\%$$

where

(a) $C_{C,j}$ for sphere j, is the average count in the ROI.
(b) $C_{B,j}$ is the average of sphere j's 60 background ROI counts.

For sphere j, the percent background variability N_j is determined as follows:

$$N_i = \frac{SD_j}{C_{B,j}} \times 100\%$$

The standard deviation (SD) of the background ROI counts for sphere j is SD_j, which is computed as

$$SD_j = \sqrt{\frac{\sum_{k=1}^{K}\left(C_{B,j,k} - C_{B,j}\right)^2}{K-1, K}}, = 60$$

17.13.9.2 Accuracy of Attenuation and Scatter Corrections

A circular ROI with a diameter of 30 ± 2 mm on the lung insert should be centred. I record Clung,i, the average pixel value inside the ROI for each slice. Twelve circular background ROIs with a diameter of 30 ± 2 mm on each slice should be inserted at the positions given in Sect. 17.13.9.1. for the background ROIs.

The relative error $\Delta C_{\text{lung},i}$ in percentage units for each slice i will be computed as follows to quantify the residual error in scattering and attenuation corrections:

$$\Delta C_{\text{lung},j} = \frac{C_{\text{lung},j}}{C_{\text{B},j}} \times 100\%$$

where

(a) $C_{\text{lung},i}$ is the lung insert ROI's average count.
(b) $C_{\text{B},i}$ This is the mean of the 60 37-mm background ROIs used in the picture quality investigation. For each slice, i keep track of the average pixel values inside the ROIs ($C_{\text{B},i}$).

Figure 17.32 shows the result.

17.13.9.3 Accuracy of Radioactivity Quantitation

After all modifications have been made, this section of the study enables you to examine the scanner's quantification of radioactive concentration accuracy. The image quality phantom's backdrop compartment contained 5.3 ± 0.27 kBq/mL of radioactivity. As a result, the real radioactivity concentration is within 5% and is signified by A_{B}. The average radioactivity $C_{\text{B},i}$ of the twelve 3.7 cm background ROIs created for the image quality analysis in the slice I will be recorded in MBq/mL as $A_{\text{B},i}$ use the manufacturer's choice to show radioactivity concentration in MBq/mL, and the quantitation error A_i in slice i shall be computed as:

$$\Delta A_i = 100\left(A_{\text{B},i} - A_{\text{B}}\right)/A_{\text{B}}$$

17.13.9.4 Tolerances Suggested

The following information should be recorded in both circumstances (lesion-to-background ratios of 4:1 and 8:1):

(a) The phantom's initial background concentration, the time it was produced, and the injected radioactivity for a whole-body scan suggested by the manufacturer.
(b) The emission and transmission imaging periods, axial step size, and total axial distance simulated in this imaging study should all be standard values specified by the manufacturer for a whole-body scan.
(c) All reconstruction and correction settings used in this study are anticipated to be the

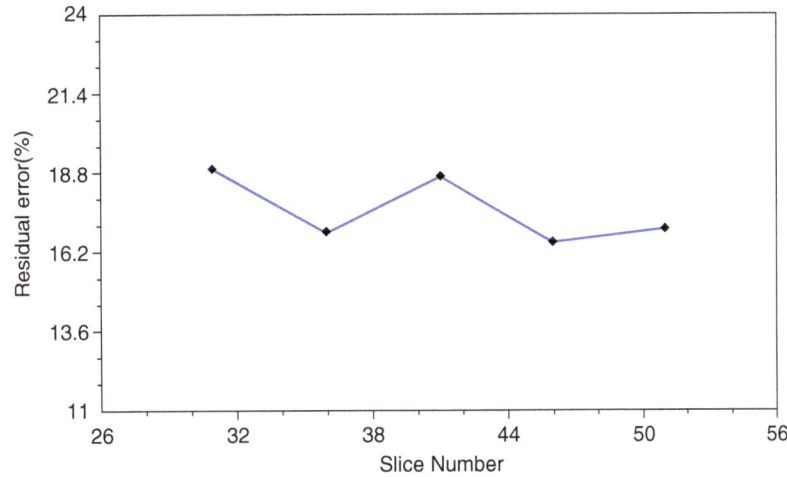

Fig. 17.32 Lung residual error

manufacturer's standard parameters for a whole-body scan, such as the reconstruction parameters (the number of subsets and iterations if using an iterative reconstruction technique and the reconstruction filters) and other smoothing used in the axial and transaxial directions, as well as the corrections made for scattering, randoms, attenuation, decay, dead time.

(d) For each sphere size, the percent contrast, and background variability. If repeat scans were taken, the average and standard deviation of % contrast and % background fluctuation over the replicates should be given.

(e) Each slice's values of ΔC_{lung}, I, and the average of these errors across all slices should be recorded.

Based on the three duplicate measurements, a tolerance of 5% is indicated for the baseline determined values for all image quality criteria.

17.13.9.5 Taking Corrective Action

Assume that the reconstructed images include image artefacts. The daily QC should be rechecked if lesion detectability is inadequate or the tolerance threshold is exceeded, and the system should be recalibrated. If the situation continues, the manufacturer should be contacted, and appropriate action should be taken.

17.13.10 Resolution of Coincidence Timing in TOF Positron Emission Tomography

17.13.10.1 Purpose

Only PET scanners in the TOF mode are subjected to this test. The capacity of the system to estimate the difference in time of arrival of the two coincidence photons is determined by the characterization of timing resolution. As a result, find out where the annihilation is most likely to happen along the LOR.

17.13.10.2 Frequency

A TOF scanner's temporal resolution, a critical feature for TOF scanners, must be measured consistently throughout acceptance testing.

17.13.10.3 Materials

Line source activity of 750μ Ci/1.0 mCi must span all rings and be located in the precise centre of the scanner.

17.13.10.4 Data Acquisition

F-18 is used to fill a line source. Use the same thinest alumunium sleeves for the sensitivity test you used for the line source.

As indicated in Fig. 17.33, do a 5-minute PET/CT scan on each bed. This would be accomplished by collecting time-of-arrival coincidences, histogramming time-of-arrival disparities, and estimating an FWHM as a measure of timing resolution.

17.13.10.5 Analysis

To calculate the timing FWHM, use the manufacturer's technique for measuring timing resolution.

17.13.10.6 Tolerances Suggested

Measured values of timing resolution, RT, should not exceed the specification given by the vendor.

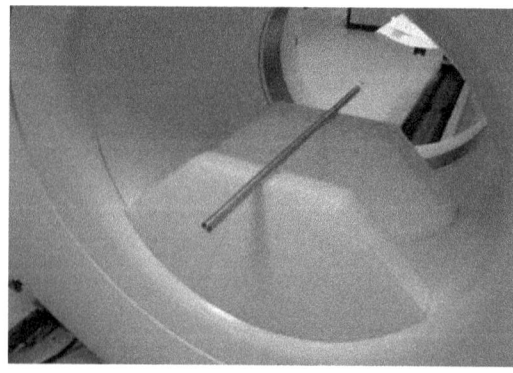

Fig. 17.33 The 1.0 mCi line source activity of ^{18}F must be centred in the scanner and span all rings

The user should set reference values, tolerances, and activity levels. An appropriate tolerance criterion for timing FWHM is

$$RT_{measured} < 1.05RT_{expected}$$

17.13.10.7   Taking Corrective Action

It is expected that the timing resolution would be a substantially consistent characteristic. If the tolerance limits are exceeded, the findings should be double-checked and the testing method repeated to validate the outcome. If the result remains outside the tolerance requirements, the system should be recalibrated by qualified service professionals.

17.14   Well Counter Correction

17.14.1   Purpose

To convert the count rate observed into radioactivity concentration. To calculate the calibration factor based on the intensity of image voxels to the real radioactivity concentration.

17.14.2   Frequency

Quarterly Acceptance

17.14.3   Material

Activity (1.5 mCi) and fillable cylindrical ^{18}F phantom.

17.14.4   Data Acquisition: In Phantom, Fill it with Activity

- Use the excel sheet to calculate the activity as indicated in Fig. 17.34.
- Place the phantom at the centre of the PET FOV depicted in Fig. 17.35.
- Obtain images with a count of 100 million.
- Starting with the centre slice, draw a 12-slice ROI. Draw 6 ROI below the center ROI and 6 ROI above the center ROI.
- Make a note of the average activity in each slice.
- Use the formula in Fig. 17.34 to calculate the CF.

17.14.5   Result

Calibration factor = _____

17.14.6   Tolerance

Manufacturer's specifications

Biograph Cross Calibration Worksheet	TruePoint and mCT systems

F-18 Phantom Information:

Measured Dose:	1.81	mCi	or	66.96	MBq	Time:	12:37:00 PM
Residual Dose:	0.01	mCi	or	0.33	MBq	Time:	12:38:00 PM
Net Dose:	1.80	mCi	or	66.62	MBq	Time:	12:37:00 PM
Phantom Volume:	6283	ml					

Scan Information:

Scan Start Time: 12:53:00 PM   Delay To Scan Start: 16.0 min

Calculated Specific Activity:

Calculated Specific Activity @ Scan Start: 9595.07   Bq/ml

ROI	Activity		ROI		
ROI #1	9255.40	Bq/ml	ROI #2	9170.30	Bq/ml
ROI #3	9278.30	Bq/ml	ROI #4	9244.10	Bq/ml
ROI #5	9219.10	Bq/ml	ROI #6	9245.00	Bq/ml
ROI #7	9182.90	Bq/ml	ROI #8	9213.20	Bq/ml
ROI #9	9192.20	Bq/ml	ROI #10	9167.60	Bq/ml
ROI #11	9188.10	Bq/ml	ROI #12	9133.40	Bq/ml

Measured Specific Activity @ Scan Start: 9207.47   Bq/ml

Dose Cross-Calibration Factor:

CCCF: 1.04

IMPORTANT! For Biograph TruePoint Systems, remember to update and save the PETCT_QualityCheck protocol with the Dose Cross-Calibration Factor. For Biograph mCT systems, remember to enter the Cross-Calibration factor on the Configure QC tab of the PET Qua

ECF Information:

Original ECF Value: 2.917E+07 Bq/ml   Corrected ECF Value: 3.040E+07   Bq/ml

Reviewed ECF Value: 3.038E+07 Bq/ml

Percent Difference Between the Corrected and Reviewed ECF Value: -0.06%

Fig. 17.34 Example of calculating the activity and cross-calibration factor in PET

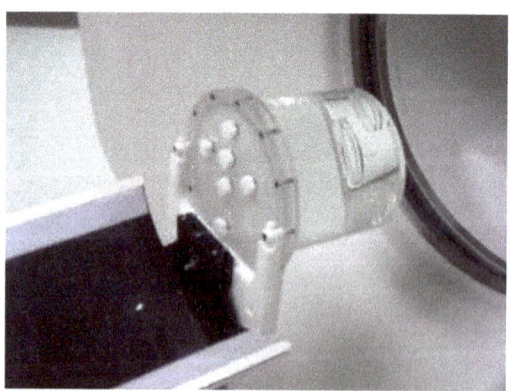

Fig. 17.35 Positioning of the Cross Calibration Phantom in the centre of the PET FOV

17.15 Calibration of Activity Concentration in 2D OR 3D17.15.1. Purpose

17.15.1   Test Purpose & Frequency

In order to rectify obtained sinograms for detector non-uniformities, data on scanner efficiency must be collected. Annually.

17.15.2   Material

The ^{18}F phantom is a fillable cylindrical phantom that is used to calibrate well counters with activity ($\leq$0.5 mCi).

17.15.3 Data Acquisition

- Ensure a backup copy of the prior calibration file before commencing the acquisition, and make sure there are enough total counts of providing accurate data statistics.
- At least 20 million counts should be acquired if no indicators are supplied or easily accessible.
- Scanning a phantom with a known quantity of radioactivity and volume/weight utilising a multi-bed clinical routine is a good way to assess SUV accuracy. Assume that the phantom's radioactivity, the radioactivity's calibration time, and the weight of the phantom volume are all recorded as part of the patient's information. The measured SUV should be 1 in this scenario.

$$SUV = \frac{\text{Decay corrected activity}\left(\dfrac{kBq}{mL}\right)}{\text{Injected dose}\dfrac{(kBq)}{\text{Patient Weight}(g)}}$$

17.15.4 Analysis of Data

Compare the results of the calibration with the manufacturer's recommendations.

17.15.4.1 Result

Standard Uptake Value=

17.15.5 Tolerance

Ten percent from expected results.

Further Reading

Bailey DL, Jones T, Spinks TJ. A method for measuring the absolute sensitivity of positron emission tomographic scanners. Eur J Nucl Med. 1991;18(6):374–9.

Bergmann H, Dobrozemsky G, Minear G, et al. An interlaboratory comparison study of image quality of PET scanners using the NEMA NU 2-2001 procedure for assessment of image quality. Phys Med Biol. 2005;50:2193–207.

Cherry SR, Sorenson JA, Phelps ME. Physics in nuclear medicine. Philadelphia, PA: Saunders; 2003.

Food and Agriculture Organization of The United Nations, International Atomic Energy Agency, International Labour Organisation, OECD Nuclear Energy Agency, Pan American Health Organization, World Health Organization. International basic safety standards for protection against ionizing radiation and for the safety of radiation sources, IAEA safety series no. 115. Vienna: IAEA; 1996.

International Atomic Energy Agency. Applying radiation safety standards in nuclear medicine, safety reports series no. 40. Vienna: IAEA; 2005.

International Atomic Energy Agency. Applying radiation safety standards in radiotherapy, safety reports series no. 38. Vienna: IAEA; 2006a.

International Atomic Energy Agency. Applying radiation safety standards in diagnostic radiology and interventional procedures using X rays, safety reports series no. 39. Vienna: IAEA; 2006b.

International Atomic Energy Agency. Nuclear medicine resources manual. Vienna: IAEA; 2006c.

National Electrical Manufacturers Association. Standards publication NU 2-2001—performance measurements of positron emission tomographs. Rosslyn, VA: National Electrical Manufacturers Association; 2001.

Williams NR, et al. Guidelines for the provision of physics support to nuclear medicine: report of a joint working group of the British Institute of Radiology and the British Nuclear Medicine Society. Nucl Med Commun. 1999;209:781–7.

References

1. National Electrical Manufacturers Association (NEMA). Standards publication NU 2-2007. Performance measurements of positron emission tomography.
2. Quality Assurance for PET and PET/CT Systems. International Atomic Energy Agency. Human health series no. 1. Vienna: IAEA; 2009.

Radiation Emergencies in Nuclear Medicine and Preparedness

Abstract

Nuclear Medicine is a field where sealed and unsealed radioactive sources are used for diagnosis, treatment and quality assurance. The unsealed sources make it more susceptible to radiation emergencies. Various situations include radioactive spills, loss or theft of radioactive source, medical events (misadministration), the incidental release of radioactive gases, medical emergency or death of a patient administered with therapeutic radiopharmaceuticals may encounter in routine practice. Also, a condition where the nuclear medicine department is the consignor of radioactive material, e.g. in case of a supply of cyclotron-produced radiopharmaceuticals to the other centres or in case of disposal of disused sources. If the vehicle carrying radioactive material meets with an accident, it becomes a radiological emergency for the department. This chapter guides licensees and radiological safety officers (RSOs) about how to prevent such situations and also, if occurred, then how to deal with them.

18.1 Introduction

The unsealed radionuclides in the form of radiopharmaceuticals are used to diagnose and treat diseases in nuclear medicine. The unsealed nature increases the risk of radiation emergencies compared to other radiological modalities such as radiology and radiotherapy. Any error, carelessness or absence/failure may lead to an unusual incident/accident. Every nuclear medicine facility should have a comprehensive emergency management system that envisages all possible emergency exposure situations and defines a clear allocation of responsibilities and actions to mitigate the consequences of such events.

The emergency management plan should be reliable and efficient, offer effective cooperation among emergency workers and their safety, provide training and retraining, arrange individual and environmental monitoring, etc. It should address the need to provide various instruments and consumables such as decontamination kits and radiation monitoring instruments, viz., survey meter and contamination monitor, personal protective equipment, etc. Its standard operating procedures should have arrangements for transition from an emergency exposure situation to an existing exposure situation, including recovery and remediation (Annexure XI of Euratom 14) [1].

The nuclear medicine practice is a planned exposure situation, and unintended or accidental exposures of patients or staff are still planned exposure situations as these incidents may occur at times. A plan to mitigate such situations should be prepared in advance [2]. All unusual incidents/accidents must be documented and reported to the regulatory authority as per existing law.

18.2   Prevention of Radiological Emergencies

Preventing accident is always the best option, and therefore a discussion on this prompted me to bring it first. A good work practice prevents accidents or unintended events and keeps radiation exposures minimal for occupational workers, patients, the public and the environment. The following minimum points be practised during work practice:

- There is no food or drink, cosmetics or smoking inside a radioactive area except that required for the patient's study.
- Eatables should not be kept in the refrigerator for storing cold kits in radiopharmacy.
- Mobile phones and handkerchiefs should not be used while handling radioactivity. Paper tissue may be used instead of handkerchiefs.
- Before entering the radioactive area, any cuts or wounds should be covered with a waterproof dressing.
- All radiation workers should wear full sleeve long aprons while working in *controlled areas* (discussed in para 18.3).
- To avoid contamination of hands, removing gloves should be based on surgical techniques.
- When leaving the controlled area, employees should wash their hands and check their hands, clothing and bodily contamination for any residual.
- Elbow operated taps, and liquid handwash should be used preferably.
- Pipetting by mouth is not allowed in nuclear medicine.
- Syringes used for radioactive liquids should be appropriately shielded.
- When working with radioactive liquids, needles should be recapped to ensure containment. The needle stick injuries can be avoided with the use of recapping tools. (IAEA Safety Standard SSG-46, para 4.114, 2018, [3]).
- The work place should be kept clean and clear of non-essential items.
- All radioactive material should be properly shielded and clearly labelled with the name and activity of the radiopharmaceutical at a particular date and time.
- Periodic surveys and contamination monitoring with wipe tests should be done for controlled and supervised areas.
- Inventory, administrative and predisposal waste management records should all be kept.

18.3   Design and Layout of Nuclear Medicine Department

The planning of the emergency management plan starts from the stages of sitting, design and layout of the department till decommissioning. Nuclear medicine facilities should be located away from general patient wards and public spaces [4]. The standalone nuclear medicine diagnostic centres should *NOT* be planned in residential complexes [4]. The site of the nuclear medicine department within the hospital needs consideration for easy transport of radioactive material from vehicle to the department, which allows minimal contact with general patient/public area. The patient entry and exit routes should also be planned to minimize movements.

The design and layout plan should consider the proposed workload and patient flow. Generally, a Nuclear Medicine facility is designed so that the workflow is from low active area to high active area. The radiopharmaceutical preparation area, source storage area and waste storage area should be located in such a place where movements can be restricted. There should be sufficient area for radiopharmacy, washable and leak-proof floors and walls and a decontamination area with adequate safety and shielding tools. The source storage area should have cupboards with a locking facility to ensure multilayer safety of sources.

The entrance of the high active area should have separate access control, which allows only authorized persons to enter. After passing through charcoal and high-efficiency particulate activity filters, fume hood exhaust must be released straight into the open (HEPA).

Signs and warning lights should be used in appropriate places, especially outside the radio-pharmacy room, source preparations and storage rooms, hybrid imaging rooms and high-dose therapy rooms. Pregnancy and breastfeeding signage should be placed in prominent areas. Preferably, it should be prepared in a bilingual language that allows everyone to understand, including regional ones. A sample of such posters is shown in Figs. 18.1 and 18.2.

The areas of the nuclear medicine department are to be properly delineated into *controlled, supervised and unsupervised areas*. *Controlled areas* are places where specific protection and safety measures are required for controlling exposure or preventing contamination spread under normal working conditions and minimiz-

Are you
pregnant or likely to be
pregnant? If yes,
then please inform us
before you are
given any radioactive medicine.

क्या आप **गर्भवती** या गर्भवती होने
की संभावना? यदि हां,
तो रेडियोधर्मी दवा लेने से पहले
निश्चित रूप से हमें सूचित करने की
कृपा करें।

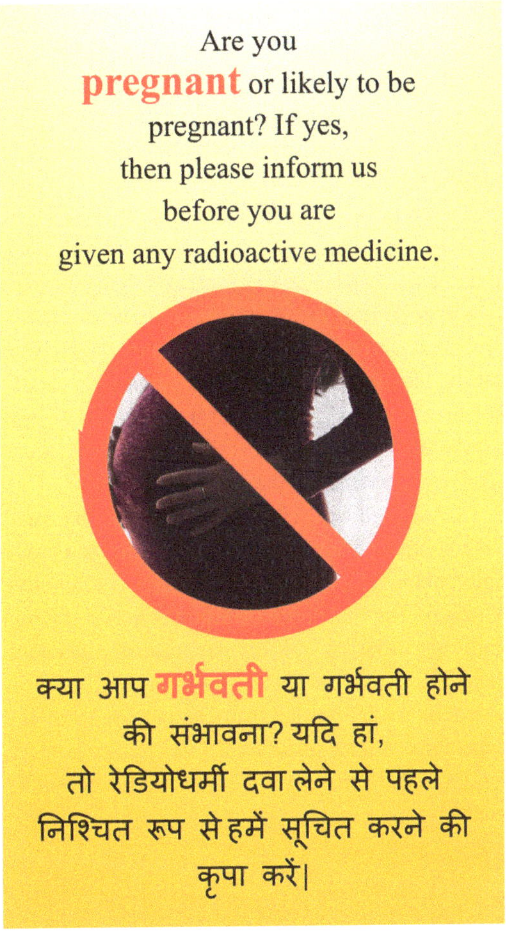

Fig. 18.2 Pregnancy caution poster

Are you
Breast feeding your child?
If yes,
then please inform us
Before you are
given any radioactive medicine

क्या आप अपने बच्चे को **स्तनपान**
करा रही हैं? यदि हां,
तो रेडियोधर्मी दवा लेने से पहले
निश्चित रूप से हमें सूचित
करने की कृपा करें।

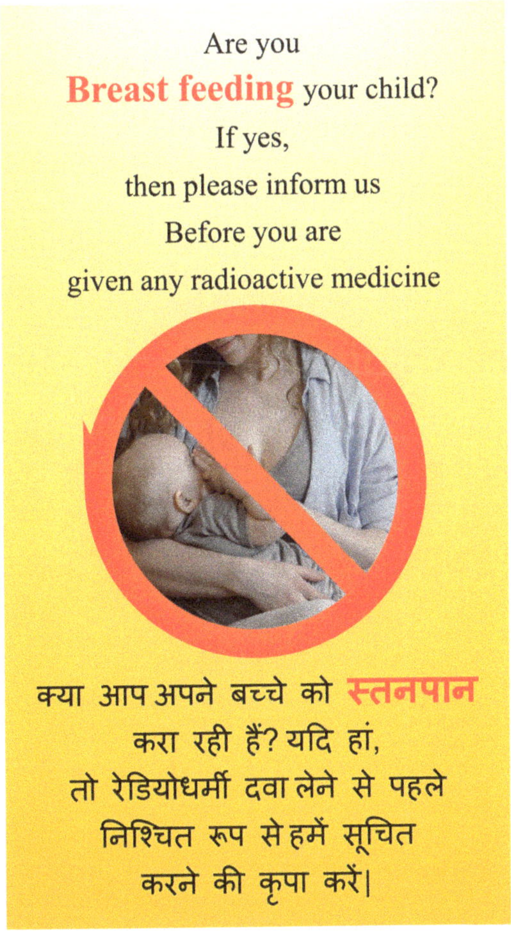

Fig. 18.1 Breastfeeding caution poster

ing the likelihood and severity of exposures in accidental situations [2]. Examples of the controlled area is radiopharmacy (hot lab), injection room, radioactive waste storage room, radioactive source storage room, imaging rooms having hybrid machines such as X-ray components (PET–CT or SPECT–CT), imaging room housing radiopharmaceutical dispensing equipment, e.g. PET radiopharmaceuticals or radioactive gas and aerosol dispenser devices, PET-waiting cubicles and high dose therapy rooms.

A *supervised area* is one which is not yet been recognized as a regulated area but where occupational exposure conditions must be monitored, even though special protection and safety measures are not generally required [2]. The examination room with probes, the Gamma camera room with no CT component, corridors, other areas where patients have been administered radiopharmaceuticals, and the areas around hybrid imaging equipment such as PET–CT and SPECT–CT are examples of supervised areas in nuclear medicine.

The entry into controlled and supervised areas are to be restricted, and no visitor shall be allowed to enter without an authorized attendant.

Unsupervised areas are the ones that are not part of the above but part of areas in the hospital or the diagnostic facility. Examples can be a reception area, visitors' area, general waiting area, staff rooms, etc. Unsupervised locations include areas nearby but not part of the radiation facility.

Drains of the Nuclear Medicine facility shall have separate connectivity to the main sewerage line unless other drains also carry radioactive material. The discharge of liquid effluents from a high-dose therapy facility is permitted through delay tanks only, and the plumbing shall be provided in a direct line from active areas to the delay tank. A leak-proof delay tank and corrosion-resistant effluent are essential while planning high dose therapy facility.

18.4   Emergency Management Plan

The 'Emergency Management Plan' should contain all the operational information on handling radiation and possible accidents/incidents with their management steps. The action of personnel who observe these events first (first responders) and contact details of RSO, licensee and other responsible persons should be displayed in the hot lab and other such areas. A sample display is shown in Fig. 18.3. Points to be included in Emergency Prepared Plan or a Radiation Safety Manual:

- The objectives of the radiation protection program.
- Regulatory compliances, including dose limits and dose constraints.
- Defined responsibilities of individuals and their contact details.
- Description of areas and procedures.
- The technique and equipment to optimize radiation exposures.
- Structural layout to facilitate decontamination when needed.

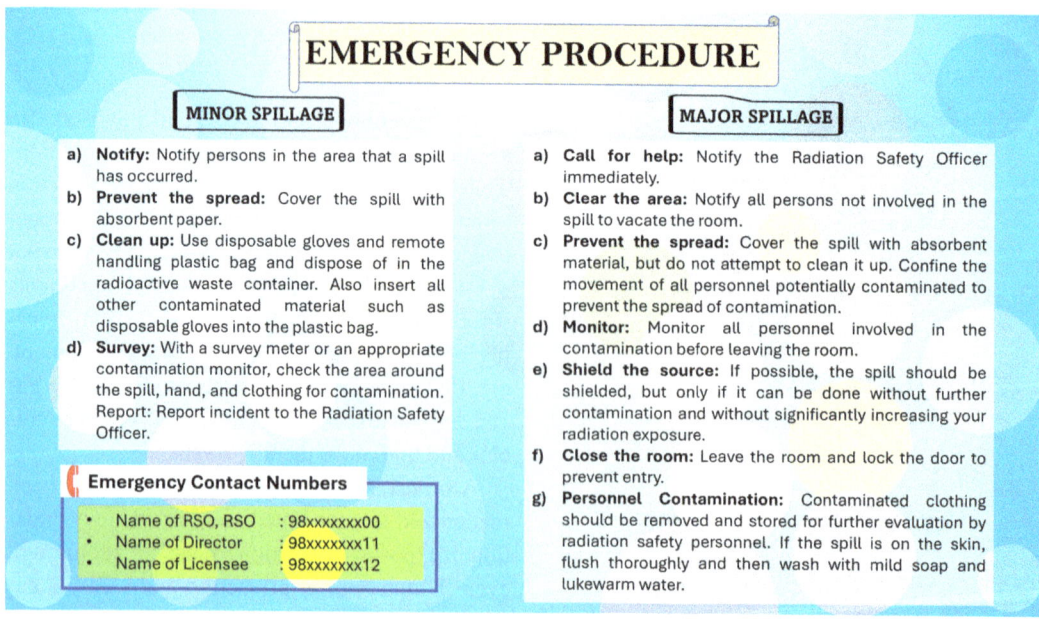

Fig. 18.3 Sample display of emergency procedure handling

- Types of radiation sources are handled in the department with their inventory and safety protocol.
- The procedure, frequency and record-keeping of survey and wipe test.
- List of personal protective equipment (PPEs).
- Contents of decontamination kits.
- Decontamination procedures.
- Personal monitoring and doses details.
- Use, maintenance, calibration and Quality Assurance (QA) program for radiation measuring and generating equipment.
- Procedures for the preparation, quality control and dispensing of radiopharmaceuticals.
- Procedures to prevent erroneous administration of radiopharmaceuticals.
- Methods to prevent inadvertent irradiation of an embryo/foetus or a child from breastfeeding.
- Standard operating procedures for therapy administration.
- Potential types of incidents/hazards and how to handle them, including immediate measures to avoid unnecessary radiation exposures to patients, employees, the general public and the environment.
- Actions in case the staff or the public exceed the dose constraints.
- Recording and reporting systems.
- Sources and categorization of radioactive waste.
- Radioactive waste disposal procedures (when, how, who authorizes the disposal, etc.)
- Protection of environmental issues.
- General requirements of record-keeping.
- Details of training and retraining with their frequency.
- Mechanisms for the periodic review of the emergency management plan or radiation safety manual.

18.5 Emergency Management Kit

The emergency management kit should possess all the items needed to manage the radioactive spills and include personal protective equipment, radiation survey meters and articles required to undertake surface and personal decontamination. A list of such items is mentioned below:

- *Personal Protective clothing:*
  - Lab coats.
  - Disposable waterproof gloves.
  - Safety glasses.
  - Shoe covers and.
  - Surgical caps.
- Portable Radiation survey and contamination monitors such as low-range Geiger-Muller counters. They are to be maintained operational and kept ready for use.
- Small and medium-sized plastic bags for waste collection, adhesive tape, labels, marker pen and rope.
- Warning signages such as 'Caution Radioactive Material' and 'Caution Radiation Area' and radioactive symbol stickers for pasting on wastes collected.
- *Surfaces Decontamination materials.*
  - Absorbent paper, long-handle tongs and forceps.
  - Soap/detergent.
  - Plastic sheets and bags.
  - Adhesive tape.
  - Buckets and mops.
  - Shielded containers for storage of radioactive waste.
  - *Radioiodine decontamination:* A solution containing 25 g of sodium thiosulphate, 2 g of sodium iodide in 1 L of 0.1 N sodium hydroxide, with a small amount of detergent [5].
  - *Floor decontamination:* Sodium hypochlorite solution or bleach (Calcium hypochlorite solution) freshly prepared or within 24 h. It is most effective in removing chemical agents from the surfaces [6]. However, it should *NEVER* be used on *Radioiodine* spills as it oxidizes it and makes it free to be volatile, and the contamination becomes airborne.
  - *Decontamination of machine surfaces:* The equipment manufacturers lay down processes for cleaning their machines on their website. Some recommend careful use of bleach solution wipes followed by alcohol-

based wipes. They also recommend the cautious use of warm water with soap wipes depending upon the part of the machine where the contamination has taken place. Each equipment has little different recommendations. GE Healthcare publishes its recommendations on https://cleaning.gehealthcare.com/
- *Some contaminations may not be cleaned to a satisfactorily level on different surfaces.* The following chemicals may be used depending upon the surface material:
  - *For metal objects*: Dilute nitric acid, 10% solution of sodium citrate or ammonium bifluoride. If this fails, Dilute Hydrochloric acid may be used on stainless steel.
  - *For glass and porcelain articles*: Chromic acid, Nitric acid, ammonium citrate, trisodium phosphate, glass cleaning solution or ammonium bifluoride.
  - *Plastics*: Ammonium citrate, dilute acids or other appropriate solvents. This technique may damage some plastic materials, so careful selection is advisable.
- *Personal Decontamination materials:*
  - Decontamination room with taps, shower, lukewarm water, soap dispenser operable without direct hand contact, shampoo for hair wash and disposable towels.
  - An emergency eyewash installed near the handwashing sink.
  - Swabs to collect contamination samples.
  - Plastic bags to keep individual contaminated clothing.
  - Marker pens, stickers to write name, date and time of collection.

18.6 Various Emergency Situations and Their Management

The emergency management system specific to nuclear medicine facilities may be subjected to the following incidents/accidents of an emergency nature:

- Spillage of radiopharmaceuticals.
- Incidental release of radioactive dust, fume and gases.
- Medical events (misadministration).
- Medical emergency, including surgery to the patient administered with therapeutic doses of radiopharmaceuticals.
- Death of a patient with a therapeutic dose of the radiopharmaceutical in the body.
- Unauthorized access to a nuclear medicine facility.
- Loss or theft of radioactive source.
- Fire, bomb threats, natural disasters such as tornados, floods and explosions.
- Accident of the vehicle carrying radioactive material of which consignor is nuclear medicine facility. Generally, various vendors supply radioactive materials, and the nuclear medicine department acts as the consignee. However, in the conditions where on-site cyclotron despatches positron-emitting radionuclides to other hospitals or centres or, in the disposal of spent radioactive generators or disused sources, nuclear medicine department also acts as a consignor.

18.6.1 Spill Management and Decontamination Procedure

Despite strict adherence to all safety practices, radioactive contamination or spillage can occur during work practice. Any inadvertent release of radioactive material, irrespective of its level (low or high) or place of release at a workplace, is identified as a radioactive spill. Depending upon incident-specific variations such as type and amount of radioisotope involved, the number of individuals affected, spread of contamination, types of surfaces involved, major and minor spills can be determined. The United States Nuclear Regulatory Commission (USNRC) provides a dividing line for some short-lived radionuclides spills [7], shown in Table 18.1, to differentiate between minor and major spills. However, it is

Table 18.1   Limit of minor spills for various radionuclides

Radionuclide	Activity level	Radionuclide	Activity level
Fluorine-18	370 MBq (10 mCi)	Technetium-99 m	3700 MBq (100 mCi)
Iodine 123	370 MBq (10)	Phosphorus-32	370 MBq (10 mCi)
Iodine-125	37 MBq (1)	Indium-111	370 MBq (10 mCi)
Iodine-131	37 MBq (1)	Ytterbium-169	370 MBq (10 mCi)
Chromium-51	3700 MBq (100)	Carbon-14	370 MBq (10 mCi)
Cobalt-57	370 MBq (10)	Thallium-201	3700 MBq (100 mCi)
Strontium-89	3700 MBq (100)	Samarium-153	3700 MBq (100 mCi)
Cobalt-60	37 MBq (1)	Selenium-75	370 MBq (10 mCi)
Gallium-67	3700 MBq (100)	Yttrium-90	3700 MBq (100 mCi)

always recommended to restrict the entry and allow the decay of short-lived radioisotopes.

18.6.1.1   Minor Spill Management

(a) *Notify*: Notify others in the vicinity that there has been a spill.

(b) *Prevent the spread*: Using absorbent paper, cover the spill.

(c) *Clean up*: Wearing disposable gloves, fold the absorbent paper with the clean side out and place it in a marked plastic bag for transfer to a radioactive waste receptacle using a remote handling tong. Keep any contaminated gloves or other disposable materials in the bag as well.

(d) *Survey*: Check the area around the spill, your hands, clothing and shoes for contamination with a low-range and thin-window radiation detection survey metre or a contamination monitor. Repeat the procedure if necessary.

(e) *Report*: Report the incident to the Radiological Safety Officer.

(f) *Document*: Document the complete incident report and notify the licensee and regulatory authority.

18.6.1.2   Major Spill Management

(a) *Do not panic.*

(b) *Clear the area*: Notify everyone not contaminated with the spill to leave the area.

(c) *Prevent the spread*: Do not attempt to mop up the spill; instead, cover it with absorbent paper.

(d) *Shield the source*: If possible, shield the spill to prevent significant radiation exposure without further contaminating it.

(e) *Close the room*: Lock or prevent entry to the area.

(f) *Call for help*: Notify immediately the Radiological Safety Officer.

(g) *Personnel decontamination*: Clothing that has been contaminated should be removed and stored for further assessment by the RSO. If the spill is on your skin, use soap and lukewarm water to clean it up (details for personal decontamination are in subsequent paragraphs).

(h) *Clean up*: The RSO to take calls for clean-up and supervise further action.

(i) *Record and report* the incident to the licensee and regulatory authority.

18.6.1.3   Decontamination

Any method that reduces the level of contamination or completely cleans this is called decontamination. The first step of decontamination is to decide whether to decontaminate or not [8]. The decision may depend on the half-life of the radionuclide, area or location of contamination, level of radiation exposure, cost, availability of disposable materials, etc.

Several radioisotopes used in nuclear medicine have short half-lives. The quantity of radioactivity can be reduced to an acceptable level simply by storing or isolating the contaminated item. If a situation necessitates decontamination, it's usually best to start with soap/detergents and water.

18.6.1.4 Decontamination Monitoring

Decontamination is effective only if it is accompanied by careful monitoring. Direct and indirect are two methods for monitoring, and either or both may be used as necessary. The direct method measures contamination with a contamination monitor directly with a probe. In contrast, the indirect method is the collection of wipe swabs from the contaminated area and then detecting the radiation. The direct method can survey larger areas in less time, whereas it cannot be used in inaccessible areas. An indirect method can be used in inaccessible areas or likely having higher radioactivity nearby, e.g. in the fume hood.

18.6.1.5 Surface Decontamination

- If a spill occurs, be calm, use common sense, protect people and not spread the contamination.
- If there are high radiation levels or the possibility of airborne contamination, evacuate the area immediately and secure it to prevent entry.
- Gloves are always to be worn while decontaminating and, if necessary, wear protective clothing and breathing apparatus.
- First, to start with the decontamination, restrict access and survey the area to assess radiation level. Gather all equipment and decontamination kit, and wear suitable protective clothing to keep radiation as low as reasonably achievable within minimal time.
- The mildest cleaning agent, i.e. soap and water, is used as the first treatment.
- When using swabs for decontamination, discard them after each wiping operation as it can spread contamination further.
- If the first treatment is insufficient, mild acids or alkali detergents should be used before any harsh decontamination treatments are used as a last resort.
- The amount of contamination eliminated between each cleaning stage must be assessed regularly.
- When using swabs for the final control, surfaces should be completely dry.

- Further cleaning of glassware can be done with chromic acid solution or solution of detergents.
- Equipment can be decontaminated by swabbing and scrubbing.
- Using acid on metal surfaces might cause unwanted corrosion and make subsequent decontamination more difficult.
- Electronic components must be properly cleaned with suitable solvents.
- Machine surfaces and metal tools can be cleaned with a very weak solution of inhibited phosphoric acid—this does not ruin polished surfaces.
- For stainless steel surfaces, use an inhibited phosphoric acid formulation followed by weak nitric acid, if necessary.

18.6.1.6 Personal Decontamination

- Removing clothing and quickly washing exposed skin and hair removes 95% of contamination [9].
- As the first step, wash the contaminated areas with soap and water. If contamination continues, wash and scrub gently with warm water and mild soap.
- Repeat the technique if the majority of the contamination has been eliminated.
- If contamination remains after two wash-and-scrub procedures, the infected person must be sent to medical authorities for further supervision.
- A 0.5% hypochlorite solution can remove pollutants from the skin and soft-tissue injuries, including open lacerations. However, it should *NOT* be used on Radioiodine contaminations, open abdominal injuries (cause peritoneal adhesions), eyes (cause corneal opacities), open chest, open brain or spinal cord injuries. After using hypochlorite solution, irrigate with sterile saline solution [9].
- Contamination of hair can be removed by rinsing with shampoo and lukewarm water.
- If suspected of airborne contamination, nasal swabs must be taken by blowing the nose into a paper tissue. If found positive, the victim should be sent to the medical personnel for further management.

- All injuries and illnesses should be reported to medical authorities as soon as possible. Life-saving actions should always prevail over decontamination.
- In case of internal radioiodine contamination, thyroid blocking agents such as Potassium Iodide (stable iodine) should be given immediately or within four hours of exposure as per medical advice (see Tables 18.2 and 18.3).

18.6.2 Incidental Release of Radioactive Fume, Dust and Gases

- Inform everyone else in the room that they must leave immediately.
- Turn off all air circulating equipment and close all valves.
- In case of brief exposure, hold your breath, but utilize a face mask for a longer time.
- Close all doors leading to the area and put warning signs to prevent entry.

Table 18.2 Doses of Stable Iodine to various individuals and age groups post-event

All individuals above 12 years	Pregnant women and children	Children under 3 years
180 mg KIO3—Immediately 85 mg KIO3—Day 2 and 3 180 mg KIO3—2 weeks post-event if required	50% of adult quantities	25% of an adult quantity

Table 18.3 Effects of stable Iodine administration and percentage dose reduction on thyroid

Stable iodine intake	Percentage dose reduction
Before event	100%
1-hour post-event	80%
6 h post-event	50%
1-day post-event	Negligible

- If the fume hood stops working during the handling, close the hood door as soon as possible, hold your breath and exit the area.
- All re-entries are to be approved by the radiological safety officer.

18.6.3 Loss or Theft of Radioactive Sources

An up-to-date inventory determines the number of sources, their type, activity and the name of the last custodian. A regular check on all the inventories is to be carried out regularly. In case of loss or theft of radioactive source:

- Perform a local search.
- Verify and confirm that other sources are secure.
- Make a thorough examination of all other possibilities.
- If the lost radioactive material is still not located, notify the licensee, police and regulatory authority in writing.
- The licensee may file an FIR (first information report).

18.6.4 Damage to ^{99m}Tc Generators

A larger amount of radioactivity is contained within radionuclide generators. In case they are suspected to be damaged:

- Evacuate the area immediately and inform RSO to confirm the spillage.
- RSO should survey all the areas from where the generator is brought to the radiopharmacy.
- Once spillage is confirmed, actions may be followed as per spill management.
- Monitor the personnel involved in handling the generator. If contamination is found, start decontamination immediately.

- Inform supplier, licensee and the regulatory authority.
- The damaged generator may be kept in the source storage room to decay to the background level and disposed of as per the routine procedure.

18.6.5   Medical Emergencies Involving Radioactivity Administered to Patients

Any medical emergency may arise with high dose therapy administered to patients, and immediate medical attention may be needed. Following actions may be initiated in such cases:

- Medical personnel are to be educated and instructed on handling radioactive patients.
- Medical workers should handle emergencies with caution, considering the spread of contamination and limiting external exposure.
- All emergency team members should wear impermeable protective gloves to avoid direct contact with the patient's mouth.
- The medical management should be under the guidance of the RSO.
- If the surgical procedure is protracted, staff rotation may be required.
- The RSO is to record the doses of all of the participants.

18.6.6   Receipt of Broken FDG Vial

Bulk 18F-FDG doses are supplied in vials inside tungsten/lead containers. Sometimes, the vial may be found broken. A very high radiation field is observed in such incidents. Following measures may be taken:

- Evacuate the area immediately.
- Inform the RSO to supervise the proceedings.
- Isolate the area once spillage is confirmed.
- Alternatively, keep the FDG vial within the shielded container and isolate it in areas with no occupancy, such as waste storage room,

source storage room or other places felt appropriate. The short half of the F-18 isotope allows radiation exposure to fall to the background level within 1 day.

- Inform supplier, licensee and regulatory authority.
- Monitor personal doses, check for possible personal contamination and record the event.

18.6.7   Fire

The fire drill should be performed at regular intervals, which involves the safe evacuation of patients, visitors and staff. In case of a minor fire, it must be extinguished, keeping in mind the vicinity of radioactivity present in the area. If necessary, the decontamination is to be performed once the fire is extinguished. In case of a major fire, the fire department must be informed. When firefighting personnel arrive, they should be notified about the existence of radioactive material and advised to proceed with caution. No one should be allowed to go inside the building until it has been decontaminated or found that no decontamination is needed.

18.6.8   Unauthorized Access to Radiation Area

Access to the areas where radioactive materials are used and stored is strictly for authorized entry only. Only persons with proper training and experience should be provided with access. In case of unauthorized entry of a person, the following actions may be initiated:

- Inform the person that the area is restricted for radiation protection purposes and escort the person from the facility.
- Get the person's contact information and enquire about the reason for the attempt to access the facility.
- Report the matter to security/hospital administration in case of any suspicion.

- Security/hospital administration may take administrative actions as necessary.
- Record the event.

18.6.9 Medical Events (Formerly Mis-Administration)

Medical events, previously called mis-administration, are already discussed in Chap. 14. But for completeness, the same is discussed here too in little more elaborations. The terminology mis-administration was changed to Medical Events in 2002 [10]. As per the new definition (the *United States Nuclear Regulatory Commission, Code of Federal Regulations-10 CFR-35.3045)* [11], Medical events are:

A dose that differs from the prescribed dose

(a) *For diagnostic:* 50% or more.
(b) *For therapeutic:* more than 20%.

or dose that would have resulted from the prescribed dosage by more than 50 mSv (5 rem) effective dose equivalent, 500 mSv (50 rem) to an organ or tissue, or 500 mSv (50 rem) shallow dose equivalent to the skin from any of the following:

1. Administration of *wrong radiopharmaceuticals.*
2. Administration through the *wrong route.*
3. Administration of radiopharmaceuticals to the *wrong patient.*
4. Administration of the *wrong dosage.*

The code further says the radiopharmaceutical dose administration to a patient not intended to receive a dosage, e.g. patient's attendant or a foetus in a pregnant woman, who is administered without confirming the pregnancy or such other cases, would also be a medical event.

The following actions are to be taken once the medical event is established:

- Report to the RSO and licensee immediately.
- The licensee must inform the referring physician and the affected individual(s) within 24 h

of its discovery, unless the referring physician informs the licensee that he or she will notify the individual or based on medical judgement, notifies that the individual would be harmful if informed.

- The licensee is not required to intimate the individual without first consulting with the referring physician.
- If the licensee is unable to contact the referring physician or the affected individual within the 24 h, the licensee must intimate the individual as soon as feasible.
- Because of any delay in information dissemination, the licensee should not delay any appropriate medical support to the individual, including any necessary remedial care resulting from the medical event.
- Complete records on each event must be preserved and available for examination for a period of 10 years.

The reporting procedure to the regulatory authority (Extract of Nuclear Regulatory Commission Regulations 10 CFR Part 35.3045) [11]:

- The licensee must notify the regulatory authority before the next calendar day post medical incident discovery. Within 15 days, a detailed written report must be submitted to the regulatory authority.
- The written report must include.
  - the name of the licensee.
  - the prescribing physician's name.
  - a summary of the occurrence.
  - the reason for the occurrence.
  - the effect on the person(s) who received the administration.
  - certification that the licensee notified the individual (or the individual's responsible relative or guardian), and if not, why not; and,
  - what actions have been taken or are planned to prevent a recurrence.
- The report may not include the person's name or any other information that could lead to his or her identification.

18.6.10 Death of Patient Administered with Radiopharmaceuticals

In case of death of a patient with the therapeutic quantity of radiopharmaceuticals in the body, wrap the body in plastic, followed by cloth, and then again in plastic and put a label on it. The incident is to be reported immediately to the regulatory authority, and as per their direction, the incident should be handled. If required, the body can be kept in the morgue.

18.6.11 Emergencies During Transport of Radioactive Material

There are three stakeholders during the transport of any radioactive material: the consignor, the consignee and the carrier. The consignor is the one who sends the goods, and the consignee is the recipient. The responsibility to deliver the consignment in good condition following all rules and regulations lies with the consignor. They must comply with the current *Code for Safe Transport of Radioactive Material*. The International Atomic Energy Agency (IAEA) has published the document *'Regulations for the Safe Transport of Radioactive Material Specific Safety Requirement No. SSR – 6 (Rev.1)'* [12] to harmonize the transport regulations among different countries. The member nations of the IAEA follow these guidelines and prepare their codes complying with the same, which is applicable in a particular country. The recommendations mentioned in subsequent paragraphs are extracted from Atomic Energy Regulatory Board (AERB) Safety Code (AERB/NRF-TS/SC-1 (Rev.1), 2016 [13] and the Australian Safety Guide on Radiation Protection in Nuclear Medicine (Radiation Protection Series Publication no.14.2), 2008 [14].

The responsibilities of the consignor

- To deliver radioactive consignment to the consignee only if the competent authority duly authorizes them.
- Obtain necessary approvals.
- The package carrying the radioactive material must be built and manufactured to contain the radioactive material to prevent contamination and shield enough not to expose the cargo handlers and the general public, which can be avoided and not justified.
- The carrier and cargo personnel are adequately trained and instructed about their responsibilities for safe handling, accumulation and segregation of packages from workers and the public.
- They should be well trained to respond to emergencies during transport.
- Intimate consignor before the dispatch of the shipment.
- Intimate any noncompliance to the regulatory authority.

Transport of packages

- Do not transfer radioactive materials in taxis, motorcycles or public transportation.
- The radioactive material can be transported in the boot and away from the driver in vans and station wagons.
- A radioactive symbol placard should be placed on each side of the vehicle and one on the back (Fig. 18.4).
- The package may be transported in the institution's transport vehicle if the driver has been trained to handle and secure the package in the vehicle. They should also know what to do in an accident or emergency. It is also necessary to provide written instructions.
- The package must be addressed and delivered to a permitted recipient; it should not be addressed to a 'Department' or delivered to a specific 'area' or the 'front desk'. It should either be transferred to the posses-

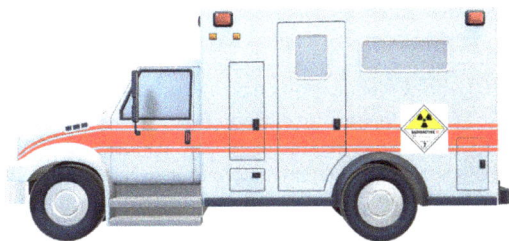

Fig. 18.4 Placard showing radioactive material inside

sion of an authorized person or left in a secure location without the authorized person's knowledge.

- All drivers transporting radioactive materials are to carry an emergency kit containing information about emergency management (called TREMCARD) and should contain the name and contact of the personnel to be called.
- The driver is to carry a mobile phone.
- When transporting radioactive material shipments, make sure no passengers are in the vehicle. If an authorized person responsible for the radioactive material is travelling in the vehicle, or if two or more people are required for off-site radionuclide procedures, they can all travel in the same vehicle.
- Remove the three yellow transport placards from the outside of the vehicle at the last destination. If there are no dangerous products in or on the vehicle, it is prohibited to show dangerous goods signs.

Actions in case of transport accidents.

In the event of an accident, remain calm, understand the following and act accordingly:

- The packaging is compliant with international standards and is built to withstand mishaps.
- The radioactive substance is unlikely to be damaged, and its container is unlikely to leak if the package is not severely damaged.
- It's critical to prioritize the needs of everyone who has been hurt.
- If a road vehicle transporting radioactive materials is engaged in a collision that results in a dangerous scenario (e.g. injury, road haz-

ard, leakage of radioactive material, fire, vehicle immobilization, etc.), the driver of the vehicle must.

- Call the emergency services such as police and ambulance.
- Intimate to the radiological safety officer of the institution or the licensee or the head of the department and.
- Support in the event of an emergency.

In addition to the above:

- If possible, exit the vehicle and examine the status of injury of others involved in the collision.
- If safe to assist, you may do so, else leave it to the emergency services if you're unsure.
- With little contact, assess the integrity of the radioactive package(s).
- If the situation permits, seek the help of passers-by to keep the traffic at a safer distance from the radiation.
- If the items are undamaged and the vehicle's damage does not need to be reported to the police, and the vehicle can still be driven safely, deliver the consignment to the consignees and inform them that the vehicle met with a minor accident on the way. On return, provide a detailed report to the radiological safety officer.

References

1. The Council of the European Union Directive 2013/59/EURATOM. Off J Eur Union. 2014.
2. IAEA Safety Standards for protecting people and the environment. Radiation protection and safety of radiation sources: International Basic Safety Standards. General Safety Requirements (GSR) Part 3. 2014.
3. IAEA Safety Standard Series (SSG) - 46. Radiation Protection and Safety in Medical uses of Ionizing Radiation 2018. ISSN 1020–525X.
4. AERB Safety code for Nuclear Medicine Facilities AERB/RF-MED/SC-2 (Rev. 2), Mar 2011.
5. Pant GS, Rajashekhar Rao B. Radiation safety in nuclear medicine. In: Basic physics of radiation and safety in nuclear medicine. 2018.
6. IAEA technical report series no. 395, state of art technology for decontamination and dismantling of nuclear facilities. 1999. ISBN 92–0–102499–1.

7. https://www.nrc.gov/docs/ML0827/ML082750235.pdf.
8. International Atomic Energy Agency. Manual of contamination of surfaces, Safety series 48. 1979, Para 6.1. ISBN 92-0-123079-6.
9. Kumar V, et al. Chemical, biological, radiological, and nuclear decontamination: recent trends and future perspective. J Pharm Bioallied Sci. 2010;2(3):220–38.
10. Ziessman HA et al. Nuclear medicine: the requisite. 4th edition. ISBN: 978-0-323-08299-0.
11. United States Nuclear Regulatory Commission Regulations 10 CFR Part 35.3045.
12. IAEA Safety Standards for protecting people and the environment. Regulations for safe transport of radioactive material. Safety standard series no.SSR-6 (Rev. 1). 2018 Edition. ISBN 978–92–0–107918–6.
13. Atomic Energy Regulatory Board (AERB) Safety Code (AERB/NRF-TS/SC-1 (Rev.1). Mar 2016.
14. Australian Radiation Protection and Nuclear Safety Agency. Safety guide: radiation protection in nuclear medicine. Radiation protection series publication no.14.2. 2008.

Nuclear Medicine Internal Dose Assessment

19

Abstract

Nuclear medicine dose assessment, informally called as 'Nuclear Medicine Dosimetry' is an important topic in managing nuclear medicine patients who are administered with radiopharmaceuticals, especially in therapeutic procedures. This chapter explains the need, As Low As Reasonably Achievable (ALARA) and As High As Safely Administrable (AHASA) concepts, the term 'absorbed dose' and its units, the formula for calculating absorbed dose, components involved in absorbed dose calculation, their resources, the concept of equivalent dose and effective dose, various systems of dose assessment including Medical Internal Radiation Dosimetry (MIRD), the International Commission on Radiation Protection (ICRP) and the Radiation Dose Assessment Resource (RADAR) methods, free resources for dose assessment in diagnostic nuclear medicine and a discussion on practical therapeutic nuclear medicine dose assessment. The chapter is prepared considering practical nuclear medicine dose assessment.

19.1 The Term 'Dosimetry'

Traditionally, dosimetry is the combination of 'Dose and metry'. The 'dose' means 'Absorbed Dose', whereas 'metry' implies 'measurement' (metrology). Metrology is the science of the measurement of physical quantities. Generally, in routine practice, external and internal dosimetry terms are used. External dosimetry has much to do with measurement, whereas in internal dosimetry, almost everything is based on theoretical calculations and no physical measurement at all. Saying 'internal dosimetry' cannot be an appropriate term, whereas using 'Nuclear Medicine Dose Assessment' suffices the need and describes assessing the dose.

19.2 The Need

Nuclear Medicine uses the properties of nuclear emissions to diagnose and treat diseases. These emissions deposit energy in the medium they interact and this property is utilized in radionuclide therapies (RNTs). However, they deposit their energies in non-target tissue too, which is a concern. A trade-off is always considered with a benefit versus risk, and here, 'Dose Assessment' comes to help in the decision. Diagnostic Nuclear Medicine uses a trace amount of radiopharmaceutical quantity; therefore, the doses are much lower. But in radionuclide therapies, the risk to certain critical organs may reach up to the threshold level for radiation-related complications.

Radiopharmaceutical behaviour differs in every individual because of different bio-kinetics, which becomes more significant in diseased subjects. Nuclear Medicine dose assessment has been performed with standardized phantoms or mathematical representations of human bodies. These are derived from the values obtained from the normal mean or median population. Ideally, they should not be applied to individuals with disease because of variabilities with normal populations.

In diagnostic nuclear medicine, mean absorbed dose estimates from standardized phantoms and models may be appropriate as the dose levels are far less. But, in therapeutic nuclear medicine, more individualized approaches to dosimetry should be preferred. Several scientific studies suggest personalized dose estimates can prevent unnecessary therapies, improve therapeutic efficacy and provide the foundation for significant advances in radionuclide therapies. Stabin and Flux summarize the need for individualized dosimetry in the chapter 'The Use of Dosimetry in the Planning of Patient Therapy' in the book 'Therapeutic Nuclear Medicine' by R.P. Baum (ed.), 2014 [1]. They have clearly shown the importance of personalized dosimetry in various radionuclide therapies.

Many articles suggest that the future of nuclear medicine lies in therapeutic nuclear medicine. However, the current clinical practice of nuclear medicine therapies applies a fixed dosage (empirical) approach, i.e. 'One-dose-fits-all approach', or with little modification on a case-to-case basis based on body surface area [2]. Dr. Michael G Stabin, in his book 'Fundamentals of Nuclear Medicine Dosimetry' (2008), says, 'many patients receive poor quality of care and therapy with no knowledge of radiation dose received and this way the knowledge of therapeutic nuclear medicine can never advance'. Strigari et al. (2014) [3] studied 92 papers on the term 'dosimetry', 79 studies were investigating dosimetry, and out of this, 48 showed a strong correlation between the absorbed doses delivered and the response indicating personalized dosimetry-based treatments would improve outcome and survival.

19.3   ALARA and AHASA Concepts in RNTs

For dose optimization in diagnostic and therapeutic nuclear medicine, ALARA and AHASA concept is used. ALARA i.e. 'As Low As Reasonably Achievable', concept is used since early days. It is the minimum amount of radionuclide required to diagnose or treat a patient, ensuring the least damage to non-target tissues. An example is a dose of I-131 carefully planned for the target absorbed dose to thyroid remnants as 300 Gy and metastatic disease as 80 Gy [4].

The AHASA is the acronym for 'As High As Safely Administrable', in which the 'Maximum Tolerable Absorbed Dose (MTAD)' concept is applied to non-target tissues or organs. This concept has been endorsed by the European Association of Nuclear Medicine in its procedural Guidelines [5] and says the treatment planning can depend on the maximum tolerable dose of critical non-target organs. An example is dose-limiting factors are 2 Gy for bone marrow and 23 Gy for kidneys in ^{177}Lu-DOTA-Octreotate treatment [6]. Konijnenberg et al. (2007) [7] argued for a limit of 29 Gy to the kidneys. The MTAD can be planned accordingly.

19.4   The Term 'Absorbed Dose'

The absorbed dose is defined as 'Energy absorbed per unit mass on any material'. This energy can be related to any kind of radiation being absorbed by any kind of matter. It is denoted as D. The equation for absorbed dose:

$$D = \frac{dE}{dm} \tag{19.1}$$

Unit: Système International (SI) unit is Joule/kilogram (J/kg) and special unit is Gray. The traditional unit is erg/g (erg/g) or rad (acronym of 'radiation absorbed dose').

1 Gy = 1 Joule/kg and 1 rad = 100 ergs/g
∵1 Joule = 10^7 ergs and 1 kg = 1000 g.
1 Gy = 100 rad or, 1 cGy = 1 rad.

The quantities rad, gray and sievert are collective nouns such as staff, team, army, etc., and should not be spoken as rads, sieverts, etc. However, this is seen in books and papers.

19.5 Dose Rate

Absorbed dose per unit time is called dose rate and is denoted as $\dot{D}$.

$$\dot{D} = \frac{D}{t} \qquad (19.2)$$

The unit is gray/time, i.e. gray/h or gray/s.

Short half-lives imply high dose rates and, conversely, long half-lives. The absorbed dose rate may affect the biological response of tissues to irradiation.

19.6 Absorbed Dose Calculation and its Components

To calculate the absorbed dose, we need to know about

(a) Radiation characteristics of administered radionuclide,
(b) Anatomy for which the dosimetry is being calculated and;

(c) The biokinetics of radiopharmaceutical.

Figure 19.1 shows various factors involved in radiation characteristics and bio-kinetics of tracer in the human body.

The radiation characteristics can be further divided into

- Half-life.
- Decay type and emissions.
- Yields.
- Radiation energies.
- Absorbed fraction.

The absorbed fraction is a *fraction of energy absorbed in a particular organ* and depends on the type and the energy of the radioisotope. The emitted particle/photon can be absorbed completely within the organ or escape depositing partial energy and irradiating other organs. Once the radioactivity is inside the body, each organ that acts as the source is also a target organ. The organs with no selective radiopharmaceutical uptake may be target organs. *In dosimetry*, particulate radiation (e.g. electrons, beta and alpha emissions) is considered to be completely absorbed in the source organ. The value of the absorbed fraction is assumed as *1.0* (Fig. 19.2a) when the source is the target. Whereas, for photons (gamma rays and X-rays), the value of the

Fig. 19.1 Components in absorbed dose calculation

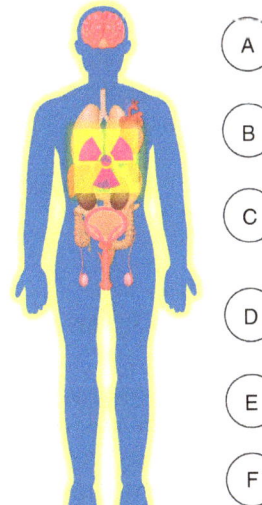

A Mass of the organ

B Radioactivity administered

C Radiopharmaceutical Uptake, residence and clearance pattern with respect to time (Biokinetics)

D Fraction of energy absorbed in the organ

E Mode and energy of decay

F Yield per nuclear transformation

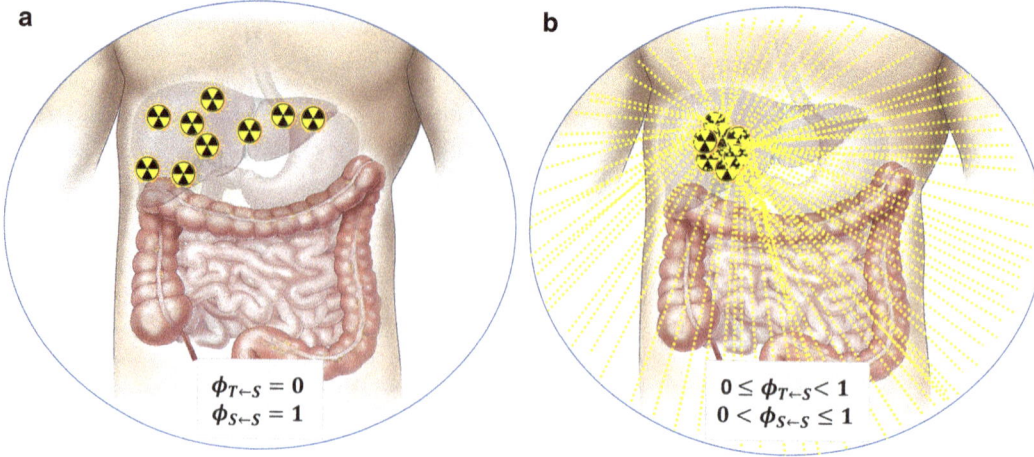

Fig. 19.2 (**a**) An absorbed fraction in particulate emission. (**b**) Absorbed fraction in photons

absorbed fraction is between 0 to 1 (Fig. 19.2b) as they can escape the objects. The absorbed fraction is denoted as φ ($T \leftarrow S$), e.g.: φ (*liver* ← *kidneys*).

The values of the absorbed fractions are calculated based on Monte-Carlo simulations and tabulated in many MIRD pamphlets such as pamphlet no. 3, 5, 8, 19, and the MIRD book on Head and brain dosimetry. Specific absorbed fraction (SAF) is explained in Sect. 19.9.

19.7 Assigning Numerical Values

Numerical values are assigned to all the quantities involved in the energy and mass terms to estimate the absorbed dose. The formula for absorbed dose is

$$D = \frac{k \, \tilde{A} \Sigma_i \, y_i E_i \phi_i}{m} \qquad (19.3)$$

where

D = Absorbed dose measured in gray or rad.
k = The proportionality constant and used for consistency of various units. Its value is 1.6×10^{-13} *Gy kg/Bq s or 2.13 rad. g/μCi hour* where isotope energy is in MeV.
$\tilde{A}$ = Time integrated activity (MIRD pamphlet 21, 2009), previously known as cumulative

activity (MIRD Primer, 1991) and measured in μCi-h or MBq-sec.
y_i = Yield per radiation or number of radiations with energy E_i emitted per nuclear transition.
E_i = Energy per radiation.
ϕ_i = Absorbed fraction.
m = Mass of the target region.

When we calculate the absorbed dose rate $\dot{D}$, the formula becomes

$$\dot{D} = \frac{kA\Sigma_i y_i E_i \phi_i}{m} \qquad (19.4)$$

where A = Time-integrated activity at time t.

The factor $y_i \times E_i$ can be written as Δ_i and is called the mean energy emitted by each ith nuclear transition. The values of Δ_i has been tabulated in various publications designed for MIRD calculations. The Eq. (19.4) can be rewritten as

$$\dot{D} = \frac{kA\Sigma_i \Delta_i \phi_i}{m} \qquad (19.5)$$

19.8 Organ Mass

Organ masses are different in every individual. Though it can be estimated using Computed Tomography (CT) and Magnetic Resonance Imaging (MRI) data, but it has its own challenges and requires high processing power [8]. For diagnostic Nuclear Medicine, a representative set of anatomical data is usually sufficient. Various reference models have been developed in the past to facilitate organ masses, e.g.

(a) Cristy and Eckerman (1987) developed phantoms for various age groups such as 0-year, 1-year, 5-years, 10-years, 15-years, Adult female and adult male (Table 19.1) [9].
(b) ICRP Publication 89 (2003) [10] provides anatomic and physiologic data for newborn, 01-year, 5-years, 10-years, 15-years male and female, adult male and female.
(c) ICRP Publication 110 (2009) [11] shows the relationship of CT-derived volume with the mass of various organs in adult male and female (Table 19.2), and broadly it can be considered $1 \text{ cm}^3 \approx 1$ g except in lungs and bones. This can be useful in patient-specific dose assessment if CT volume data is present.
(d) Stabin et al. (1995) [12] provided the mass of nonpregnant adult female and at the end of each trimester of pregnancy.

Table 19.1 Masses of six phantoms developed by Cristy and Eckerman

Age	0-year	1-year	5-years	10-years	15-years and adult female	Adult male
Mass in kg	3.6	9.7	19.8	33.2	56.8	73.7

Table 19.2 CT volume concerning masses of reference male and female as per ICRP 89, 2002. (The table was originally printed in ICRP publication 110 (2009) [11], reproduced with permission)

Organ	Male Volume (cm³)	Mass (g)	Reference mass (g)	Female Volume (cm³)	Mass (g)	Reference mass (g)
Adrenals	13.6	14.0	14	12.6	13	13
Blood (segmented vessels)	973.7	1032.1	5600	807.4	855.8	4100
Brain	1381	1450	1450	1238.1	1300	1300
Breast	25.6	25	25	511.9	500	500
Eyes	14.3	15	15	14.3	15	15
Eye lenses	0.4	0.4	0.4	0.4	0.4	0.4
Gall bladder	66	68	68	54.3	56	56
Gall bladder wall	13.5	13.9	10	9.9	10.2	8
Gall bladder contents	52.5	54.1	58	44.4	45.8	48
Gastro-intestinal tract						
Stomach wall	144.2	150	150	134.6	140	140
Stomach contents	240.2	250	250	221.2	230	230
Small intestine wall	625	650	650	576.9	600	600
Small intestine contents	336.6	350	350	269.2	280	280
Right colon wall	144.2	150	150	139.4	145	145
Right colon contents	144.3	150	150	153.8	160	160
Left colon wall	144.2	150	150	139.4	145	145
Left colon contents	72.1	75	75	76.9	80	80
Recto-sigmoid colon wall	67.3	70	70	67.3	70	70
Sigmoid colon contents	72.1	75	75	76.9	80	80
Heart	795.4	840	840	587.2	620	620

(continued)

Table 19.2 (continued)

Organ	Male Volume (cm³)	Mass (g)	Reference mass (g)	Female Volume (cm³)	Mass (g)	Reference mass (g)
Heart wall	314.3	330	330	238.1	250	250
Heart contents (blood)	481.1	510	510	349.1	370	370
Kidneys	295.3	310	310	261.9	275	275
Liver	1714.3	1800	1800	1333.3	1400	1400
Lungs	2891.3	1200	1200	2300.8	950	950
Lymphatic tissue	134	138	730	76.8	79.1	600
Muscle tissue	27,619	29,000	29,000	16666.7	17,500	17,500
Oesophagus	38.8	40	40	34	35	35
Ovaries				10.6	11	11
Pancreas	133.3	140	140	114.3	120	120
Pituitary gland	0.6	0.6	0.6	0.6	0.6	0.6
Prostate	16.5	17	17			
Residual (adipose) tissue	21535.2	20458.4	18,200	24838.3	23596.4	22,500
Salivary glands	82.5	85	85	68	70	70
Skin	3420.2	3728	3300	2496.8	2721.5	2300
Skeleton	7725.3	10,450	10,450	5767.4	7760.1	7760
Cortical bone	2291.7	4400	4400	1666.7	3200	3200
Trabecular bone	572.9	1100	1100	416.7	800	800
Cartilage	1000	1100	1100	818.2	900	900
Active marrow	1135.9	1170	1170	872.9	899.1	900
Inactive marrow	2530.6	2480	2480	1836.8	1800.1	1800
Miscellaneous	194.2	200	200	155.3	160	160
Spleen	144.2	150	150	125	130	130
Teeth	18.2	50	50	14.6	40	40
Testes	33.7	35	35			
Thymus	24.3	25	25	19.4	20	20
Thyroid	19.2	20	20	16.4	17	17
Tongue	69.5	73	73	57.1	60	60
Tonsils	2.9	3	3	2.9	3	3
Ureters	15.5	16	16	14.6	15	16
Urinary bladder wall	48.1	50	50	38.5	40	40
Urinary bladder contents	192.3	200		192.3	200	
Uterus				77.7	80	80
Total body	71,109.9	73,000	73,000	59,258	60,000	60,000

19.9 Specific Absorbed Fraction (SAF)

When the absorbed fraction is normalized to the mass of the target organ, it is called a specific absorbed fraction. It is denoted as Φ.

$$\Phi_i = \frac{\Phi_i}{m}$$

The unit of SAF is the inverse of mass.

Ready to use SAF values have been published in several MIRD, ICRP and scientific publications for various emissions and target organs.

The equation for absorbed dose rate, i.e. Eq. (19.5), can be written as

$$\dot{D} = kA\Sigma_i \Delta_i \Phi_i \quad (19.6)$$

And absorbed dose formula can be written as

$$D = k\,\tilde{A}\Sigma_i\,\Delta_i \Phi_i \quad (19.7)$$

In Eq. (19.7), all we need is to calculate is time-integrated activity and other variables are available in the literature.

19.10   Various Systems of Dose Assessment Calculations

Absorbed dose calculations can be very tedious, especially when multiple emissions and contributions from penetrating radiation. Except time-integrated-activity $\left(\tilde{A} \right)$ in Eq. 19.7, all other factors can be considered constant with some exceptions. The equation for absorbed dose can be simplified to time-integrated activity multiplied with a precalculated constant factor. Three systems, viz. the Medical Internal Radiation Dosimetry (MIRD), the International Commission on Radiation Protection (ICRP) and the Radiation Dose Assessment Resource (RADAR) group, have proposed this factor with different names with very comparable values. The MIRD group called it '*S-factor*', the ICRP called it Specific Effective Energy (SEE) and the RADAR group used '*Dose Factor*'.

19.10.1   Medical Internal Radiation Dosimetry (MIRD) Formalism

The Medical Internal Radiation Dose Committee was formed in 1965 under the Society of Nuclear Medicine to standardize internal dosimetry calculations [13]. The first publication, '**MIRD Pamphlet No. 1**', was published in 1968. The revision of this was published in 1975, 1988 and 1991. The formalism provides a broad framework for assessing the absorbed dose to whole organs, tissue subregions, voxelized tissue structures and individual cellular compartments for diagnostic and therapeutic nuclear medicine. In the publication MIRD pamphlet 21 (2009), the committee standardized nomenclatures used in dosimetry, formally adopted equivalent dose and effective dose and identified dosimetry quantities for deterministic effects relevant to targeted radionuclide therapy [14].

The committee gave an equation for absorbed dose calculation:

$$Dr_k = \sum_h \tilde{A}_h S\left(r_k \leftarrow r_h \right) \qquad (19.8)$$

where $S\left(r_k \leftarrow r_h \right)$ value contains emitted energy E with its probability Y, the absorbed fraction ϕ and the mass of the target region m and popularly known as S-value or S-factor. r_k is target region, and r_h is source region.

Naming subscripts with '*h*' and '*k*' for 'source' and 'target' and not having '*s*' and '*t*' was unusual. FORTRAN programmers did the early work for this system, and the letters '*i*' and '*j*' were already in use for other variables, so the letters '*h*' and '*k*' were used [2]. However, in the 2009 publication, the formula was rewritten as

$$D\left(r_T, T_D \right) = \sum_{r_s} \tilde{A}\left(r_T, T_D \right) S\left(r_T, r_s \right) \quad (19.9)$$

r_s and r_T are the source and target regions, T_D is dose-integration period. S-factor is characteristic of the radionuclide and anatomic models (age- and sex-specific) chosen to represent the patient or tissue of interest.

S-values are calculated using Monte Carlo radiation transport simulation codes for individual radionuclides in Source (r_s) and target regions (r_T). It depends on the shape, size, mass of the source and target regions, the distance and type of material between the source and the target regions, the type of radiation emitted from the source and the energy of the radiation. The unit of S-value is Gy/MBq s and seen in tables in format mGy MBq^{-1} s^{-1}.

In the MIRD system, there are 26 source regions, out of that, 25 are specific tissues, organs or contents of organs, and all other tissues are grouped into a 'remaining tissues' category (Table 19.3). There are 24 target regions, the whole body is noted as a target region. The S-factors for the whole body provide the absorbed dose to the full body, i.e. total energy absorbed by the whole body divided by the total body mass. It should not be confused with the effective dose, which is a stochastic risk-weighted sum of organ absorbed doses [15].

Table 19.3 Source and target regions in the MIRD system. (Copyright Springer Nature [14]. Reprinted with permission)

Source Regions		Target regions	
Adrenal glands		Adrenal glands	
Brain		Brain	
Breasts		Breasts	
Bone	Cortical	Gall Bladder wall	
	Trabecular		
Gallbladder contents		Gastrointestinal tract walls	Stomach
Gastrointestinal tract contents	Stomach		Small Intestine
	Small Intestine		Upper Large Intestine
	Upper Large Intestine		Lower large intestine
	Lower large intestine	Heart wall	
Heart Contents		Kidneys	
Heart wall		Liver	
Kidneys		Lungs	
Liver		Muscles	
Lungs		Osteogenic Cells	
Muscles		Ovaries	
Ovaries		Pancreas	
Pancreas		Red Marrow	
Red Marrow		Skin	
Spleen		Spleen	
Testes		Testes	
Thymus gland		Thymus Gland	
Thyroid gland		Thyroid gland	
Urinary bladder contents		Urinary bladder wall	
Uterus/Uterine wall		Uterus	
Remaining tissues		Whole Body	

19.10.1.1 Strengths and Inherent Limitations in the Formalism

The MIRD formalism is based on two assumptions:

(a) Uniform activity distribution in the source region.
(b) Calculates mean absorbed dose to the target region.

Its strength is *simplicity and ease* of use, whereas the weakness is that the *absorbed dose may vary throughout the region.*

19.10.2 The International Commission on Radiation Protection (ICRP)

The ICRP began in 1928 as an independent organization that advances radiological protection science by providing recommendations and guidance on all aspects of protection against ionizing radiation [16]. In ICRP 30 publication series, applicable for occupational workers, the commission used dose equivalent (*H*) term instead of absorbed dose and

$$H_{50,\,T} = 1.6 \times 10^{-10} \sum_S U_s \times \text{SEE} \left(\text{T} \leftarrow \text{S} \right) \quad (19.10)$$

Where 50 denotes the time period of 50 years over which dose is integrated for adult radiation workers, T, is the target tissue, 1.6×10^{-10} are the proportional constant k, which produces cumulative dose equivalents in sievert, from activity in becquerels, mass in grams and energy in MeV, U_s is time-integrated activity and S is source region. SEE is Specific Effective Energy and is equal to

$$\text{SEE} = \frac{\sum_i y_i E_i \phi_i (T \leftarrow S) Q_i}{m_T} \quad (19.11)$$

It is the same as the S-factor of the MIRD system, which is multiplied with Q_i, i.e. quality factor, the term used earlier by ICRP for radiation weighing factor w_R.

However, in publication 128 (2015) [17], which recommends radiation doses for patients from radiopharmaceuticals, the commission advocated using the MIRD formula, it being a straightforward method. They said to use the formula

$$D(T \leftarrow S) = \tilde{A}_s \times S(T \leftarrow S) \quad (19.12)$$

Where $\tilde{A}$ is the time-integrated activity and $S(T \leftarrow S)$ is the absorbed dose in target region T per unit of time-integrated activity in source region S.

The commission mainly differed in two terms, equivalent dose and effective dose. They account for the type of radiation and organ or tissue type, respectively. Both of these quantities were later adopted by the MIRD committee in pamphlet 21 (2009) [14].

19.10.2.1 Equivalent Dose

The mean equivalent dose H_T in a target organ or tissue T is given by

$$H_T = \sum_R w_R D_{T,R} \quad (19.13)$$

where $D_{T,R}$ is the mean absorbed dose from radiation R in tissue or organ T, and w_R is the radiation weighting factor. Unit of equivalent dose is Sievert, and it is equal to 1 Joule/kg.

The value of the radiation weighting factor (Table 19.4) roughly accounts for the ability of different radiations to cause biological effects

Table 19.4 Radiation weighing factors for various ionizing radiation

Ionizing radiation	ICRP 103 recommendations
Photons, electrons, positrons and β-particles	1
Alpha	20
Protons and charged pions	2 (In ICRP 60 it was 5)
Neutron	A continuous function of neutron energy varies between 2.5 and 20.7
Auger electrons (in nuclear medicine, emitters such as ^{99m}Tc, ^{123}I, ^{125}I and ^{201}Tl)	Varies from 1 to 20 on a case to case basis, e.g. 1 when ^{125}I is localized in the cytoplasm, 4 in the nucleus, 7–9 when bound to DNA. It can be up to 20 depending on the proportion of DNA-bound emitters (MIRD pamphlet 21)

based on their linear energy transfers (LET). It is used to relate absorbed dose to the probability of stochastic (cancer/heritable) health effects. It is not intended for use in predicting deterministic effects (harmful tissue reactions). If used as such, it may result in an overestimation of their occurrence and severity in irradiated tissues.

19.10.2.2 Effective Dose

Effective dose is a derived quantity that accounts for different relative radio-sensitivities of the various organs and tissues in the human body from all types of radiation at low energies. The concept applies only to stochastic effects for radiation protection planning and not to individual subjects for risk assessment.

The term radiation weighting factors, w_T, was introduced in Publication 26 (ICRP 1977) [18] as radiation detriment-adjusted nominal risk coefficients for six organs/tissues, and others were kept as remainder. In publication 60 (ICRP 1991) [19], numbers were increased to 12 + 1 (remaining), and in publication 103 (ICRP, 2007) [20], one more organ, salivary gland, was added to the list, and the total became 13 + 1 (Table 19.5). These coefficients are calculated by averaging estimates of the radiation-associated lifetime risk for cancer incidence for a combined population of equal numbers of males and females. The det-

Table 19.5 Tissue weighting factors as per ICRP 103 [19]. (Copyright ICRP. Reprinted with permission)

Tissue	w_T	No. of tissues	Summation
Red bone-marrow, colon, lung, stomach, breast, remainder tissue[a]	0.12 each	06	0.72
Gonads	0.08	01	0.08
Urinary bladder, oesophagus, liver, thyroid	0.04 each	04	0.16
Bone surface, brain, salivary glands, skin	0.01	04	0.04
Total			1.00

[a]Remainder tissues include adrenals, extra-thoracic region, gall bladder, heart, kidneys, lymphatic system, muscles, oral mucosa, pancreas, prostate, small intestine, spleen, thymus and uterus/cervix

riment is modelled as a function of life loss, lethality and loss of quality of life (ICRP 103).

19.10.2.3 Determination of Tissue Weighing Factors and Effective Dose

Tissue weighting factors are determined based on risks of radiation-induced stochastic effects in males and females. Sex-specific radiation detriments are calculated and averaged to find the values of w_T. These w_T values and the sex-averaged organ and tissue doses are used to calculate the effective dose.

$$E = \sum_T w_T \left[\frac{H\left(r_T, T_D\right)^{Male} + H\left(r_T, T_D\right)^{Female}}{2} \right]$$

(19.14)

19.10.2.4 Use of Effective Dose

- It allows the comparison of different *diagnostic* medical exposures (even from different modalities, such as X-rays and nuclear medicine).
- It allows taking dosimetry-based decisions to choose radiopharmaceuticals, e.g. use of ^{201}Tl chloride with ^{99m}Tc Sestamibi for use in myocardial imaging studies.
- Effective dose allows the summing of doses from radiopharmaceuticals to those received from other procedures outside of nuclear medicine.
- It permits relation to radiation risks.

19.10.2.5 Limitations of Effective Dose

- Limitations stem from the fact that w_T is both sex- and age-averaged.
- The w_T for the thyroid overestimates the risk of thyroid cancer in adult patients and underestimates the risk in children. For young children, the risk may be higher by a factor of two or three [18].
- It is appropriate for assessing stochastic risk in diagnostic exposures to populations of patients whose age and sex distribution do not significantly differ from w_T derivation conditions. At higher ages, the risks are less, perhaps by a factor of three [19].
- The effective dose may NOT be used in situations involving radiation therapy or to any single person. Its application is to populations of individuals.

19.10.3 Radiation Dose Assessment Resource (RADAR) Task Force Method

The Radiation Dose Assessment Resource (RADAR) task group of the Society of Nuclear Medicine and Molecular Imaging devised a simpler expression of the dose equation:

$$H_T = \sum_S N_S \times DF\left(T \leftarrow S\right) \quad (19.15)$$

Where N_S is the number of disintegrations per second, and it is the same as $\tilde{A}$ converted into disintegration per second (dps), and DF contains the rest of the terms of the absorbed dose formula with unit Gy/dps.

$$DF = \frac{k \sum_i y_i E_i \phi_i \left(T \leftarrow S\right)}{m} \quad (19.16)$$

Tables of DFs are embedded in the popular Internal Dose Assessment software OLINDA/EXM (Organ Level Internal Dose Assessment/

Exponential Modelling), a revised version of the MIRDOSE software, both coded by Dr. Michael G Stabin and Dr. Richard Sparks. Several important information and various kinds of data related to Dose Assessment are freely available on https://www.doseinfo-radar.com. Dose factor (DF) tables for some radionuclides and phantoms are available by email at stabinmg17@gmail.com [21].

19.11 Free Websites for Diagnostic Dose Estimation

The dose from diagnostic Nuclear Medicine is very small compared to therapeutic nuclear medicine, and so are the chances of stochastic effects. A properly performed diagnostic nuclear medicine procedure can never reach the threshold level of deterministic effects. Some approximation may be considered while evaluating doses in such procedures. Ready-to-use dose calculation tools based on standard phantoms are available online for free and may be considered for estimating approximate value. However, these estimations may vary significantly on individual patients in many cases because of patient-specific parameters.

The Society of Nuclear Medicine and Molecular Imaging (SNMMI) provides an online tool on their website to give effective doses and doses to the critical organ for various diagnostics radiopharmaceuticals. However, there is scope to increase more radiopharmaceuticals currently available (Fig. 19.3).

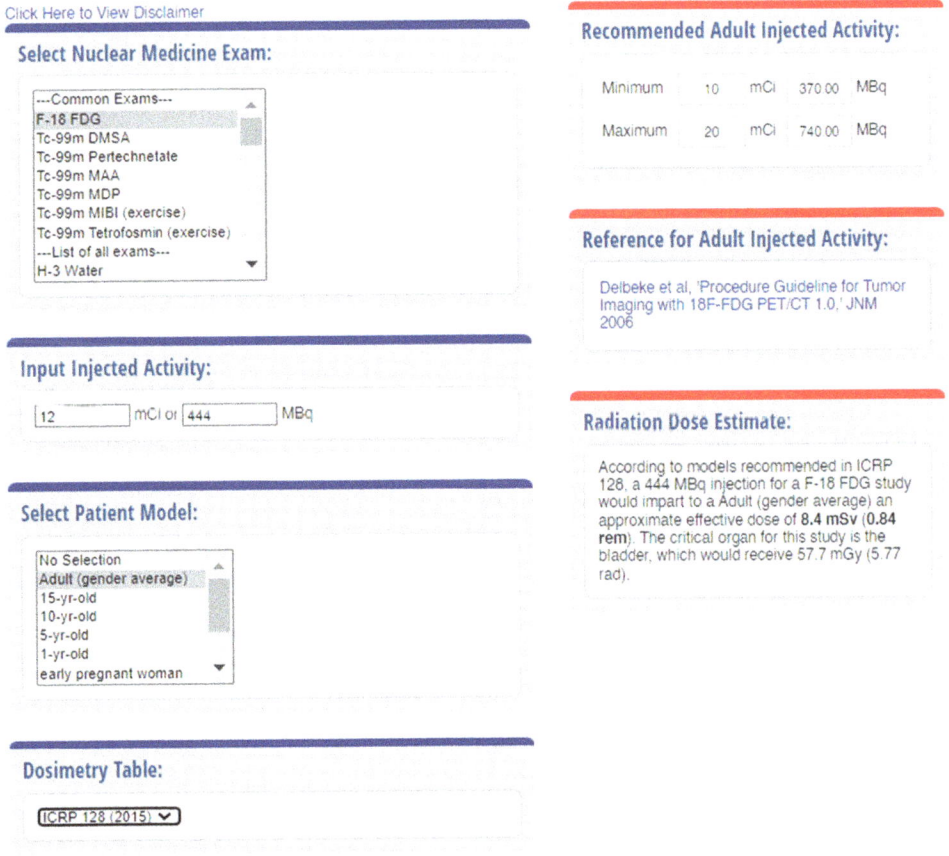

Fig. 19.3 Screenshot of SNMMI website radiation dose calculator. (Accessed on 23 Feb 2022 from http://www.snmmi.org/clinicalpractice/dosetool.aspx?itemnumber=1. Printed with permission)

The RADAR group on their website https://www.doseinfo-radar.com/RADARDoseRisk-Calc.html also provides effective dose calculations for various diagnostic radiopharmaceuticals. They also provide a short statement generated online that may be useful as part of a patient consent form document, explaining the radiation doses in numerical values and comparison with the equivalent number of days of exposure to natural background radiation (Fig. 19.4).

19.12 Dose Assessment in Therapeutic Nuclear Medicine

From Eq. (19.3), the formula for absorbed dose calculation is:

$$D = \frac{k\,\tilde{A}\Sigma_i\,y_i E_i \phi_i}{m}$$

In this equation, $\tilde{A}$ (time-integrated activity) is time-dependent with some exceptions where mass also changes with time and needs to be calculated for each therapeutic patient. Other factors such as specific absorbed fraction, SAF ($\phi_i = \frac{\phi}{m}$), and nuclear decay data $\Sigma_i\,y_i \times E_i$ values are combined in the form of the S-factor and are made available in the literature. In situations where organ mass changes, e.g. in hyperthyroidism or the case of tumour mass reduction, the S-factor needs additional calculations [22]. Estimating an individual patient's organ or tumour mass, 'm' may reduce significant errors in dose assessment (refer to Sect. 19.8).

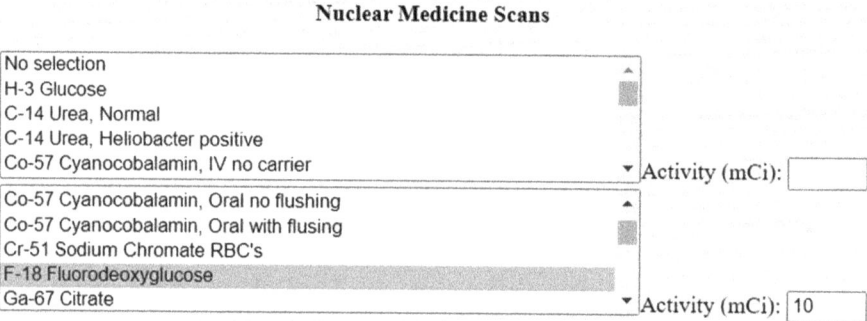

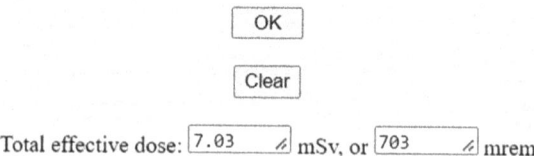

Fig. 19.4 Screenshot of RADAR website effective dose calculation. (Accessed on 23 Feb 2022 from https://www.doseinfo-radar.com/RADARDoseRiskCalc.html. Courtesy: Dr. Michael G Stabin. Personal communication, M Stabin, 2022)

19.12.1 Time-Integrated Activity Estimation

Absorbed dose estimation requires the kinetics of radiotracer to be characterized and quantified. Several imaging methods such as planar conjugate view scintillation imaging, single-photon emission computed tomography (SPECT) and positron emission tomography (PET) imaging have been tried. Non-imaging methods such as external radiation monitoring, tissue sampling (blood or biopsy) and excreta counting have also been used for dose assessment, and the most frequently used sample is blood [23]. Figure 19.5 shows a generalized time-activity curve for the whole body/organ. The area under the curve provides time-integrated activity estimation.

Dual-headed gamma cameras provide simultaneous whole-body acquisition of anterior-posterior scans, called conjugate view imaging. However, single-headed gamma cameras may also be used. SPECT and PET provide 3-dimensional details of under- or overlying regions. However, PET data has been used very limited in dose assessment and is an area of research. SPECT imaging can improve the accuracy of planar imaging measurements as it allows activity measurement of adjacent overlying, underlying or adjacent structures and enables the determination of tissue activity concentration in

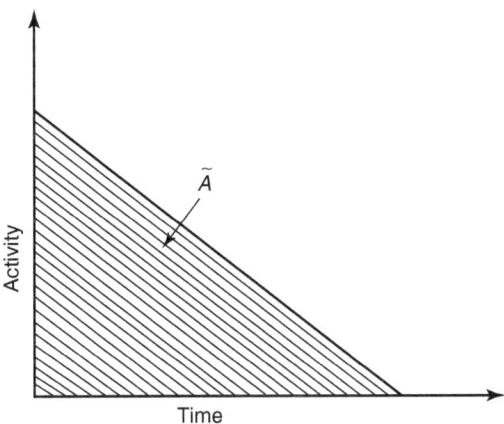

Fig. 19.5 Generalized whole body/organ time-activity curve. (Image copyright Springer Nature [2]. Reprinted with permission)

associated volume [24]. Although its accuracy is limited because of attenuation correction, non-optimal statistics, collimator resolution and scatter [24].

Absorbed dose calculations require mostly software assistance. The most widely accepted software is OLINDA/EXM (Organ Level Internal Dose Assessment/Exponential Modelling). The manual calculation can be very lengthy and requires many man-hours. There are several other software which include DOSE3-D [25], OEDIPE [26], the RMDP [27], PEREGRINE code [28], Planet Dose by Dosisoft [29], 3D-RD (3-Dimensional Radiobiologic *Dosimetry*) [30], RAPID (Radiopharmaceutical Assessment Platform for Internal Dosimetry) [31], BIGDOSE (Biomedical Imaging Laboratory BIG) [32] and many more. In these software, 3-D anatomical data from Computed Tomography (CT) or Magnetic Resonance (MR) images are fused with 3-D activity distributions from Single-Photon Emission Computed Tomography (SPECT) or Positron Emission Tomography (PET).

But software has its limitations, and it can give significantly erroneous results if quality input data is not provided. Good quality data requires an adequate number of data points obtained with carefully selected time points. As the number of measurements increases, the confidence in the data increases. One needs two data points per phase of exponential clearance. The reference books and papers have abstained from writing about it. MIRD publication 16 says, 'at least as many data points should be obtained as the number of initially unknown variables in the mathematical curve-fitting function(s). It further says that each exponential term in a multiexponential curve-fitting function requires two data points to be adequately characterized'. The same is adopted in Dr. Michael G Stabin in his book 'Fundamentals of Nuclear Medicine Dosimetry'. He also said, 'a typical human study for dosimetry will have perhaps three to seven-time points'.

The IAEA (International Atomic Energy Agency), in its document 'Nuclear Medicine Physics: A Handbook for Teachers and Students', says three data points per exponential phase should be considered the minimum data required to determine the pharmacokinetics, and it should follow for at least two to three effective half-lives.

This can be complicated in designing dosimetry studies, especially for new radiopharmaceuticals, because, at the beginning of dose assessment, we may not know the number of effective half-lives to expect for a particular patient or a compound. Here, the physical half-life can play an important role. We may understand that one or two data points will not give any information about the kinetic behaviour of radiopharmaceuticals as the graph cannot be formed with one or two data points (unless there is only one exponential phase of clearance, in which there are two unknowns, and two data points can define the curve, although three may be better). Collecting three data points can provide reasonable data. Four or more data points will, of course, always be better, but we are always balancing the desire to reduce uncertainty in our evaluation with the cost and inconvenience to the patients being imaged.

Now the question is, choosing the time points. These four to five data points may be divided in such a way that they should cover at least one physical half-life of the radionuclide. For a short-lived Tc-99m agent like DTPA, we may need several data points in the first 30–60 min. For an iodine-131 therapy agent, we may collect first at 30–60 min, second at 4 h, third at 24 h, fourth at

3–4 days and fifth at around the eighth day. For a lutetium-177 agent, we can modify the fourth and fifth data points at 2–3 days and 6–7 days. This collection method should provide information on both uptake and clearance phases, which is essential to reduce errors (Fig. 19.6).

For software-assisted dose assessment, all we need is to calculate the system calibration factor (C) or system sensitivity for planar dual-head anterior-posterior imaging or SPECT-CT or PET-CT data. However, the single whole-body image or SPECT only data requires many other corrections such as scatter correction, attenuation correction, background activity corrections, deadtime corrections and corrections for dose nonuniformities. Though background corrections are required in any case and correction for overlapping organs may also be needed if felt necessary for planar imaging. In planar conjugate imaging, the software uses the geometric mean image of anterior-posterior data to make it depth-independent. For SPECT-CT or PET-CT data, the software uses CT data for attenuation and scatter corrections and does it automatically.

There are several very good references available for the calculation of correction mentioned above which, including Dr. Michael G Stabin's book on 'Fundamentals of Nuclear Medicine Dosimetry (2008) [2]', Siegel et al. 'MIRD pamphlets 16 (1999) [24]' and Dewaraja et al. 'MIRD

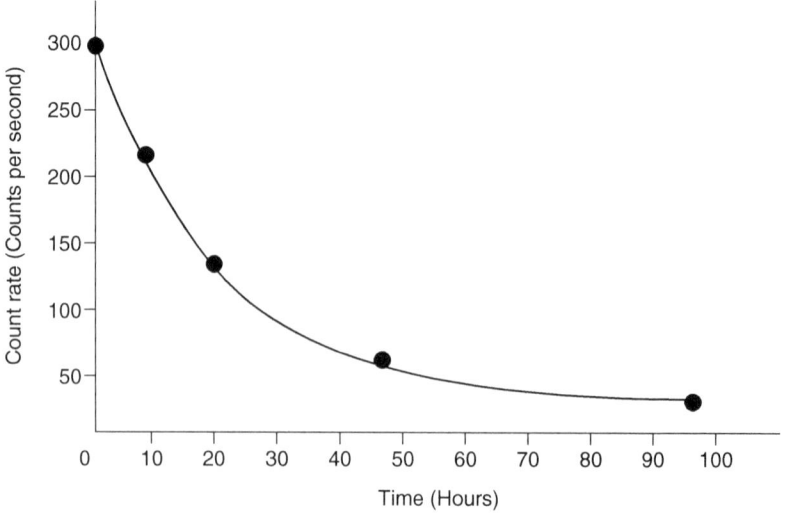

Fig. 19.6 Data points for time-activity curve. (Image courtesy: Dr. Michael G Stabin. Personal communication, M Stabin, 2022)

pamphlet 23(2012) [33]'. But just for completeness, an abstract is mentioned in forthcoming paragraphs.

19.12.2 Obtaining System Sensitivity

To obtain system sensitivity or calibration factor, represented as C, the same radionuclide to be administered to the patient is to be counted for a known activity (usually in a few tens of MBq) for a fixed time (e.g. 5 min) at a source-to-collimator distance. This approximates the patient midline distance used during the imaging time. The count rate per unit activity (in units of, e.g. cpm/MBq) represents the calibration factor.

19.12.3 Image Quantification

For image quantification from anterior-posterior data, the source activity A_j is given as

$$A_j = \sqrt{\frac{I_A I_P}{e^{-\mu_e t}}} \frac{f_j}{C} \qquad (19.17)$$

where

I_A and I_P = Anterior and posterior counts in given of interest (ROI) (count/time).
t = Patient thickness over the ROI.
μ_e = Attenuation coefficient correction for the source region.
C = System sensitivity (counts/time per unit activity).
f = Self-attenuation factor, and in most cases, it is considered as unity as normal variation in body thickness across individual ROI is small, and so a single attenuation factor may be used to calculate the activity in the entire ROI.

This expression assumes parallel hole collimators are used during anterior–posterior imaging, with no scatter (Fig. 19.7).

In conjugate imaging, the depth of interaction becomes independent of the depth of origin of the count.

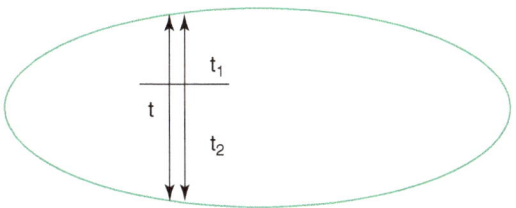

Fig. 19.7 Patient thickness. (Image courtesy: Dr. Michael G Stabin. Personal communication, M Stabin, 2022)

$$e^{(-\mu_e t_1)} \times e^{(-\mu_e t_2)} = e^{(-\mu_e (t_1 + t_2))} = e^{(-\mu_e t)}$$

19.12.4 Scatter Corrections

Activity in the region of interest has counts of Compton scattered events from the actual photopeak and possibly from higher energy peaks. As lower and upper scatter windows equal that of photopeak, two adjacent windows are drawn (Fig. 19.8), and counts in all three regions are noted. The corrected photopeak counts (C_T) are given by

$$C_T = C_{PP} - (C_{LS} + C_{US})$$

where C_{PP} is the total count recorded within the photopeak window, C_{LS} and C_{US} are the counts within the lower and upper scatter windows, respectively. The subtraction of the adjacent windows is assumed to compensate for the high-energy photon scatter tail upon which the true photopeak events ride. This may be determined by measuring a source of known volume submerged to a real depth in a water phantom whose dimensions are similar to that of a human subject.

19.12.5 Corrections for Background Activity

A background region of interest (ROI) must be drawn and subtracted from an ROI of any organ or tumour on a pixel-by-pixel basis. This corrects for activity present in overlying and underlying tissues. The area of drawing back-

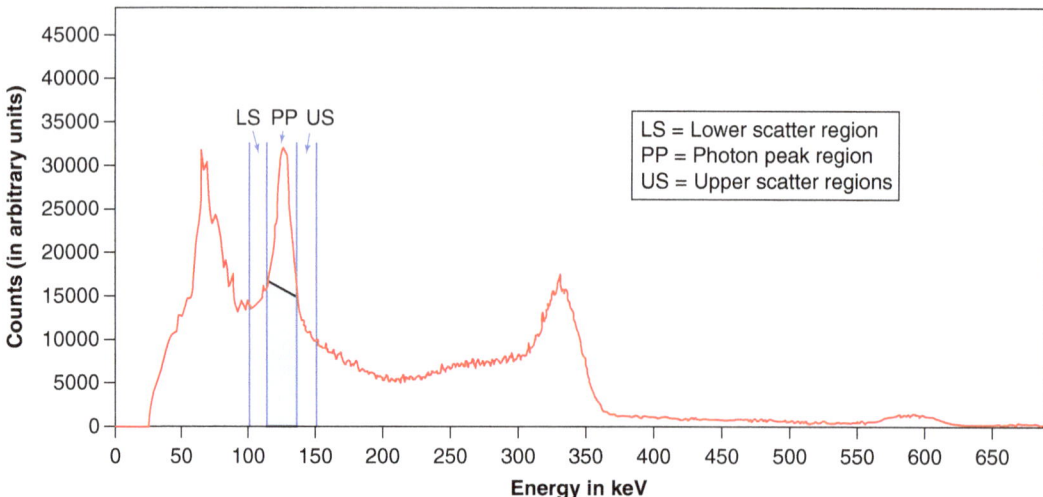

Fig. 19.8 Compton scattered events contributing to actual photon peak

ground ROI is carefully selected and should be shown somewhere on a screenshot for reproducibility. It should be drawn away from areas where major vessels or streak artefacts are present.

19.12.6   Tomographic Imaging

The quantification of voxels is much more difficult, as *volumes of interest* (VOIs) are established by drawing regions on multiple *slices* of the tomographic reconstruction. Dewaraja et al., in MIRD pamphlet 23 (2012), provided a detailed procedure needed to obtain quantitative SPECT data. This includes choice of the collimator, choice of energy windows, reconstruction methods, attenuation and scatter correction, deadtime corrections, compensation for other image-degrading effects, choice of target regions, corrections for dose nonuniformities and other aspects to be considered to obtain quantitative information in individual voxels used to define source and target regions. However, the SPECT-CT or PET-CT data need only system sensitivity data, and the rest are processed from CT data.

19.13   Conclusion

All the present methods to estimate nuclear medicine doses have merits and demerits. But a good dose assessment can be carried out with little effort, especially in the era of software. It is essential to bring nuclear medicine dose assessment into routine clinical practice to understand therapeutic nuclear medicine better. The national regulatory authorities can play an important role in it.

References

1. Stabin Michael G, Flux GD. The Use of dosimetry in the planning of patient therapy. In: Baum RP, editor. Therapeutic nuclear medicine, medical radiology. Radiation oncology. Berlin: Springer-Verlag Berlin Heidelberg; 2012. https://doi.org/10.1007/174_2012_739.
2. Stabin Michael G. Fundamentals of Nuclear Medicine Dosimetry (2008). ISBN: 978-0-387-74578-7.
3. Strigari L, et al. The evidence base for the use of internal dosimetry in the clinical practice of molecular radiotherapy. Eur J Nucl Med Mol Imaging. 2014;41(10):1976–88. https://doi.org/10.1007/s00259-014-2824-5.
4. Traino AC, Di Martino F. A dosimetric algorithm for patient-specific 131I therapy of thyroid cancer based

on a prescribed target-mass reduction. Phys Med Biol. 2006;51:6449–56.

5. Lassmann M, Haenscheid H, Chiesa C, Hindorf C, Flux G, Luster M. EANM dosimetry committee series on standard operational procedures for pre-therapeutic dosimetry I: blood and bone marrow dosimetry in differentiated thyroid cancer therapy. Eur J Nucl Med Mol Imaging. 2008;35:1405–12.

6. Sandstrom M, et al. Individualized dosimetry of kidney and bone marrow in patients undergoing 177Lu-DOTA-Octreotate treatment. J Nucl Med. 2013;54:33–41.

7. Konijnenberg M, Melis M, Valkema R, Krenning E, de Jong M. Radiation dose distribution in human kidneys by octreotide in peptide receptor radionuclide therapy. J Nucl Med. 2007;48:134–42.

8. McParland Brian J. Nuclear medicine radiation dosimetry, advance theoretical principles. Berlin: Springer Science & Business Media; 2010. ISBN 978-1-84882-126-2

9. Cristy M and Eckerman KF. Specific absorbed fractions of energy at various ages from internal photon sources. VII adult male. ORNL/TM-8381/V7. Oak ridge national Laboratory 1987.

10. Valentin J ICRP publication 89. Basic Anatomical and Physiological Data for Use in Radiological Protection: Reference Values. 2002. ISBN 008-044-2668.

11. Clement CH. ICRP 110. Adult reference computation phantoms. 2009. ISBN 0146–6453.

12. Stabin MG et al. Mathematical models and specific absorbed fractions of photon energy in the nonpregnant adult female and at the end of each trimester of pregnancy. 1995. ORNL/TM-12907.

13. Stelson AT, Watson EE, Cloutier RJ. A history of medical internal dosimetry. Health Phys. 1995;69:766–82.

14. Bolch WE, Eckerman KF, Sgouros G, Thomas SR. MIRD Pamphlet No. 21: a generalized schema for radiopharmaceutical dosimetry—standardization of nomenclature. J Nucl Med. 2009;50:477–84.

15. Brian J. McParland. 2010. Nuclear medicine radiation dosimetry advanced theoretical principles. Page 496–498/63.

16. Clarke RH, Valentin J. The History of ICRP and the Evolution of its Policies. ICRP publication 109. 2009.

17. Clement CH. ICRP 128. Radiation Dose to Patients from Radiopharmaceuticals: a Compendium of Current Information Related to Frequently Used Substances 2015. ISBN 978-147-393-9479.

18. ICRP Publication 26. 1977. Recommendations of the International Commission on Radiological Protection

19. Smith H ICRP 60. 1990. Recommendations of the International Commission on Radiological Protection.

20. Valentin J Publication 103.2007. Recommendations of the International Commission on Radiological Protection.

21. https://www.doseinfo-radar.com/RADARphan.html.

22. Traino AC, Di Martino F, Lazzeri M, Stabin MG. Influence of thyroid volume reduction on calculated dose in radioiodine therapy of graves' hyperthyroidism. Phys Med Biol. 2000;45:121–9.

23. Lathrop KA, Harper PV, Charleston DB, Atkins FB, Mock BH. Acquisition of quantitative biologic data in humans for radiation absorbed dose estimates. In: Cloutier RI, Coffey IL, Snyder WS, Watson EE, editors. Radiopharmaceutical dosimetry *symposium. Proceedings of a conference held at Oak Ridge, TN. April 26–29 1976*, vol. 1976. Washington: U.S.: Government Printing Office; 1976. p. 164–73.

24. Siegel JA, et al. MIRD Pamphlet No. 16: Techniques for Quantitative Radiopharmaceutical Biodistribution Data Acquisition and Analysis for Use in Human Radiation Dose Estimates. J Nucl Med. 1999;

25. Clairand I, et al. DOSE3D:EGS4 MonteCarloCodebased software for internal radionuclide dosimetry. J Nucl Med. 1999;40:1517–52.

26. https://www.irsn.fr/EN/Research/Scientific-tools/Computer-codes/Pages/OEDIPE-Personalised-dosimetric-evaluation-tool-3443.aspx.

27. Guy MJ, et al. RMDP: A Dedicated Package For 131I SPECT Quantification, Registration and Patient-Specific Dosimetry. Cancer Biother Radiopharm. 2004;18:61–9.

28. Lehmann J, et al. Monte Carlo treatment planning for molecular targeted radiotherapy within the MINERVA system. Phys Med Biol. 2005;50(5):947–58.

29. https://www.dosisoft.com/company-overview/company-history/.

30. Prideaux AR, et al. Three-dimensional radiobiologic dosimetry: application of radiobiologic modelling to patient-specific 3-dimensional imaging-based internal dosimetry. J Nucl Med. 2007;48(6):1008–16.

31. Besemer AE, et al. Development and validation of RAPID: a patient-specific monte carlo three-dimensional internal dosimetry platform. Cancer Biother Radiopharm. 2018;33:155–65.

32. Li T, et al. BIGDOSE: software for 3D personalized targeted radionuclide therapy dosimetry. Quant Imaging Med Surg. 2020;10(1):160–70.

33. Dewaraja YK, Frey EC, Sgouros G, Brill AB, Roberson P, Zanzonico PB, Ljungberg M. MIRD Pamphlet No. 23: quantitative SPECT for patient-specific 3-dimensional dosimetry in internal radionuclide therapy. J Nucl Med. 2012;53:1310–25.

Computed Tomography Dose Assessment

20

Abstract

Since the introduction of multi-modality imaging, i.e. Positron Emission Tomography-CT (PET-CT) and Single Photon Emission Computed Tomography-CT (SPECT-CT), the number of studies in Nuclear Medicine has grown tremendously. The previous chapter has discussed dose assessment from the radiation emitted by radionuclides in Nuclear Medicine. Computed Tomography (CT) has its inherent property of imparting radiation to the patient. The amount of radiation released by CT machines is significantly higher than the conventional diagnostic Nuclear Medicine procedures. Considering the radiation safety of patients, it becomes very important to discuss dose assessment from the CT component in a radiation safety book. CT dose is entirely different from the radiation doses imparted by radionuclides in nuclear medicine, and so are the dose assessment method and terms. The chapter is designed to cater to needs of Nuclear Medicine Professionals. It explains how the CT machine works, various terms used for CT dose assessment, calculation of effective doses from CT dose reports generated from machines during the scan, diagnostic reference levels and achievable doses.

20.1 Introduction

In the era of fusion imaging, Computed tomography (CT) has gained much importance in Nuclear Medicine. Since it has its inherent property of imparting radiation dose to the patient, talking about radiation safety becomes an important topic. Same as Nuclear Medicine Dose Assessment, CT dosimetry is also a 'dose Assessment' as there is no exact radiation measurement to the patient. Rather it is estimated with the help of phantoms, exposure parameters and scan length. CT dose assessment is required in Nuclear Medicine for PET-CT, and Single Photon Emission Computed Tomography-CT (SPECT-CT) scans. In radiology, CT scans are generally performed for parts of the body such as head, head and neck, chest, abdomen, abdomen and pelvis, etc. However, in PET-CT, scans are performed for the whole body, i.e. from vertex to mid-thigh, the base of the skull to mid-thigh, vertex to the knee or, in some cases, vertex to toe depending upon need, whereas, in SPECT-CT, the area of interest in images as per the need which can be anywhere in the body. Newer SPECT-CT systems allow the whole body (up to 200 cm in length) SPECT-CT if needed.

CT dose assessment is entirely different from those performed for radionuclides in Nuclear Medicine with different terms and methods to perform. Various terms such as Volume CT Dose Index ($CTDI_{vol}$), Weighted CT dose index

(CTDI$_w$), CTDI$_{100}$, Dose Length Product (DLP), Size Specific Dose Estimate (SSDE), Conversion Factors, Polymethyl Methacrylate (PMMA), Pencil Ionization chamber (IC), etc. will be discussed in detail and finally how effective doses can be calculated from the dose report obtained from the scanner will be shown.

20.2 Design and Working Principle of CT Scan Machines

The technology and capabilities of CT scanners have changed tremendously since their inception by Sir Godfrey Hounsfield in 1967 [1], especially with the introduction of multirow-helical CT scanners in 1998 [2].

The 'multiple-detector-row' CT scanner, also called MDCT, contains an arc of detectors and the X-ray tube(s). The multiple arrays (rows) of detectors are placed along the z-direction, which is perpendicular to the axial CT plane, as shown in Fig. 20.1. The scanner moves detectors and the X-ray tube together with the help of a slip-ring (set of parallel conductive rings to transfer high voltage to rotating device) technology power supply. The slip-ring technology allows helical acquisition as fast as 0.33 s for a full X-ray tube rotation about the isocentre in one direction continuously around the patients.

The major components of CT scanners are the Scanner and the Computer. The scanner consists of a power source, cooling devices, X-ray tube(s), filtration, collimators, detectors, rotating gantry and the data transfer system. Cooling devices include blowers, filters or devices that perform oil to air heat exchange. The X-ray tube produces X-ray photons, which help to generate the CT image. Tungsten (atomic number 74), a high Z material, is often used as an anode target material because it produces a higher intensity X-ray beam. Filters filter out the undesirable X-rays with low energy (soft X-rays) as they do not contribute to the image quality (Fig. 20.2).

Collimations restrict the X-ray beam to a specific area and help to reduce scatter radiation. The scatter radiation affects image quality and increases the radiation burden on the patients. Reducing the scatter radiation improves contrast resolution and decreases patient dose. Collimation also controls the slice thickness by narrowing or widening the X-ray beam. There may be pre-patient collimation and post-patient collimation.

Detectors are the components that collect information regarding anatomic structure attenuated by the beam. The detector arrays comprise multiple detector elements situated in an arc or a ring, each of which measures the intensity of transmitted X-ray radiation along a beam projected from the X-ray source to that detector element. Detectors can be made from different

Fig. 20.1 64-row multi-detector CT machines

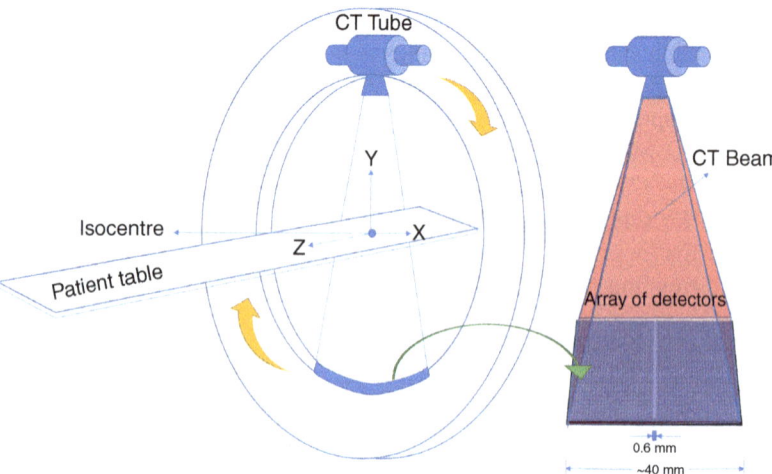

substances, each with advantages and disadvantages. The details of CT detectors are explained in Chap. 15. The computer systems consist of a data acquisition system, scanner console and reconstruction hardware.

20.2.1 PET-CT Scanners

PET-CT scanners are the combination of two machines housed as one machine. PET has an array of detectors that detects the radiation emitted from the patients, whereas CT emits the radiation to form images. The PET machine detectors are discussed in Chaps. 15 and 17. Typical dose reports generated from the CT component of PET-CT scanners are shown in Figs. 20.3, 20.4 and 20.5.

The dose reports from Figs. 20.3, 20.4 and 20.5 contain scan series, type of scan (scout, helical or axial), scan range, CTD_{vol}, DLP and the Phantom type, and they are explained in forthcoming paragraphs. Because of different types of CT machines, varying scan motions, X-ray beam collimation geometries and operating conditions, the dose index of CT contains various terms and discussed in subsequent paragraphs.

Fig. 20.2 Bow-tie filters

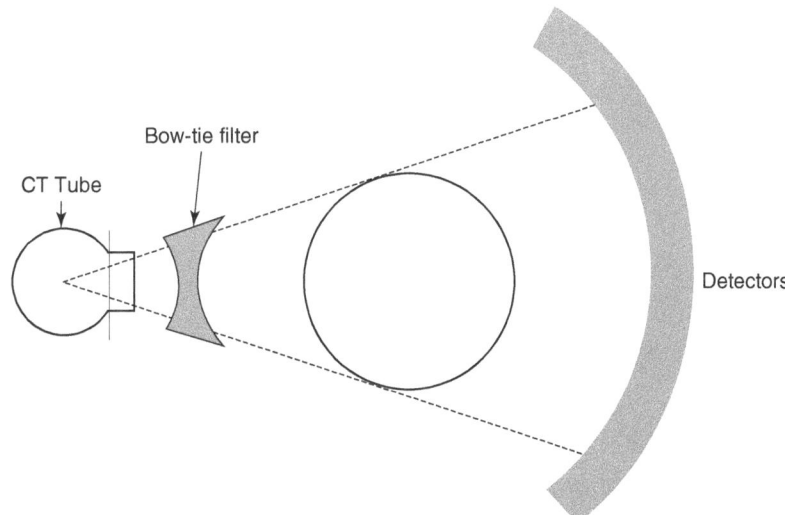

Dose Report

Series	Type	Scan Range (mm)	CTDIvol (mGy)	DLP (mGy–cm)	Phantom cm
1	Scout	–	–	–	–
2	Helical	1250.000–1580.000	7.60	276.20	Body 32
4	Helical	50.000–1971.480	11.68	1184.24	Body 32
			Total Exam DLP:	1460.44	

Exam Description: 18F-FDG WB PET CT

Fig. 20.3 Dose report generated by a 16-slice PET-CT on Vertex to mid-thigh protocol

		Dose Report			
Series	Type	Scan Range (mm)	CTDIvol (mGy)	DLP (mGy-cm)	Phantom cm
1	Scout	–	–	–	–
2	Helical	I173.750–I503.750	8.11	295.03	Body 32
4	Helical	I63.750–I800.510	13.35	1033.95	Body 32
16	Helical	S36.250–I113.710	42.62	713.51	Body 32
			Total Exam DLP:	2042.48	

Exam Description: 18F-FDG WB PET CT

Fig. 20.4 Dose report generated from the base of the skull to mid-thigh and Brain Protocol

		Dose Report			
Series	Type	Scan Range (mm)	CTDIvol (mGy)	DLP (mGy-cm)	Phantom cm
1	Scout	–	–	–	–
2	Helical	I195.500–I445.500	5.99	169.84	Body 32
4	Helical	I295.000–I645.000	11.59	444.65	Body 32
200	Axial	I285.562–I285.562	11.90	11.90	Body 32
6	Helical	I292.000–I543.020	13.52	382.15	Body 32
6	Helical	I292.000–I543.020	13.52	382.15	Body 32
6	Helical	S58.000–I926.520	8.82	905.19	Body 32
			Total Exam DLP:	2295.87	

Exam Description: 18F-FDG Triple Phase PET-CT

Fig. 20.5 Dose report generated from a Triple-Phase Vertex to Thigh Protocol

20.3 CTDI

CTDI term was proposed by Shope et al. in 1981 [3] and is an abbreviation of '"Computed Tomography Dose Index'. It is obtained from the dose distribution that occurs when the X-ray tube performs one single 360° rotation with no table motion. It measures radiation output from a CT tube when it rotates fully in one single rotation. The CTDI is also denoted as $CTDI_\infty$ and obtained by integrating the axial dose pro-file (Fig. 20.6) for a single CT slice divided by its beam width.

$$CTDI_\infty = \frac{1}{nT} \int_{-\infty}^{\infty} K(z)\,dz \qquad (20.1)$$

where

n = Number of data channels used in a particular scan*.

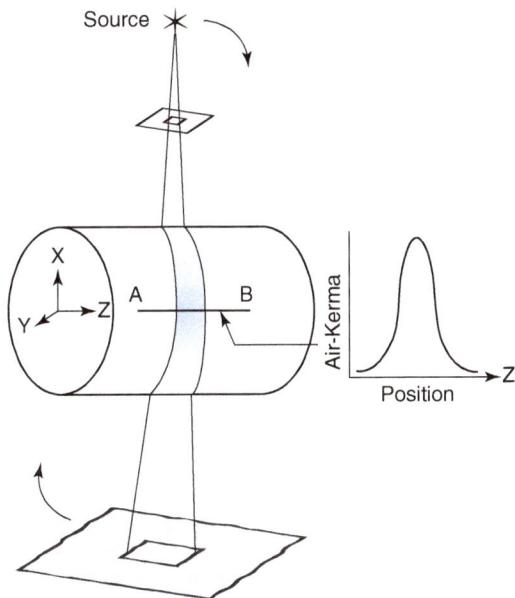

Fig. 20.6 Typical dose distribution from a single scan of CT system

Slices are axial images (two-dimensional cross-sectional) reconstructed following data acquisition and depend largely on the study protocol chosen. Rows are an arrangement of discrete elements across the detector, e.g. 4, 6, 8, 16, 64, 128, 320 row CT detectors. Each row contains hundreds (e.g. 800) [5] of detector elements. Data channels are electronics that convert the analogue signal produced by the detector to a digital value for image reconstruction. Channels are one of the limits of CT functionality.

T = Nominal width of the tomographic section along the z-axis imaged by one data channel (Fig. 20.1).

$K(z)$ = Air-kerma as a function of position on the z-axis (Fig. 20.6).

The parameter nT is described as '*Nominal beam width*', and in general, it is a very useful measure in CT acquisition protocols and dosimetry. Earlier, $K(z)$, i.e. Air-Kerma, was written as $D(z)$ [4], but since the term dose is not the absorbed dose to the phantom material but rather the absorbed dose to the air in the ion chamber, it is written as $K(z)$ [10].

CTDI values are measured in *mGy (milli-Gray)*. Values of CTDI predict the dose that results from a series of contiguous scans. When measurements are made at the *scanner isocentre* without any patient or phantom, it is called *CTDI*$_{air}$.

[* In Multi-detector CT machines (MDCT), the number of detector rows and data channels are better represented than the number of slices.

20.4 CTDI₁₀₀

Integration of the CT axial dose profile in a single full 2π rotation of CT tube using a *100 mm* long pencil-shaped ionization chamber in a 15 cm long Polymethyl Methacrylate (PMMA) cylindrical phantom is called *CTDI*$_{100}$. The 16 and 32 cm phantoms are the standard adult head and body CT phantoms. However, custom-made phantoms dedicated to paediatric studies averaged over body diameters such as at birth (10 cm), 1 year (13 cm), 5 years (16 cm), 10 years (20 cm) and 15 years (25 cm) have also been used [6] (see Fig. 20.7). The phantom is to be placed concentrically with the isocentre of the CT scanner and the centre of the phantom along the z-axis and located at $z = 0$. CTDI measurements include the energy deposited in the scatter tails.

$$\text{CTDI}_{100} = \frac{1}{nT} \int_{(-50\ \text{mm})}^{(50\ \text{mm})} K(z)\,\mathrm{d}z \qquad (20.2)$$

CTDI₁₀₀ represents the accumulated multiscan air-kerma at the centre of a 100 mm scan and underestimates the accumulated air-kerma for longer scan lengths (ICRU 87) [4].

The measurement in the ionization chamber is obtained by placing it in central and peripheral holes provided in phantoms. Some head and body phantoms include multiple holes in the phantoms at their periphery on various angles, and these values are typically averaged (see Sect. 5).

Fig. 20.7 Phantoms for measurement of CTDI$_{100}$ and 100 mm pencil ionization chamber (Image copyright John Wiley and Sons; Medical Physics 34 (7), 3018–3033 (2007) [6], reprinted with permission)

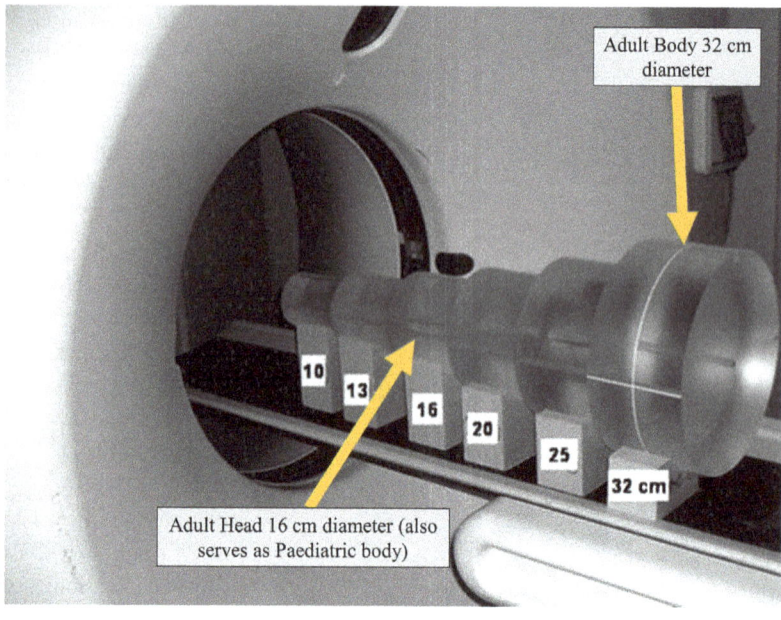

Adult Body 32 cm diameter

10
13
16
20
25
32 cm

Adult Head 16 cm diameter (also serves as Paediatric body)

20.5   CTDI$_w$

The absorbed doses vary between centre and periphery in scanned volume. In a 32-cm body phantom that mimics the adult body, surface doses in the periphery may be twice that of the central region. However, in head phantom (Fig. 20.7), doses do not vary much and are almost similar in central and peripheral regions. CTDI measured at the surface, i.e. periphery of the phantom, is called as *CTDI$_p$* whereas at the centre, it is called as *CTDI$_c$*. A term *CTDI$_w$* i.e. weighted CTDI is used to approximate the average dose in dosimetry phantom. The formula is:

$$CTDI_w = \frac{2}{3} CTDI_p + \frac{1}{3} CTDI_c \quad (20.3)$$

In a single full rotation of the CT tube, one-half of the energy is deposited directly into the irradiation volume, whereas the other half is deposited in scattered tails adjacent to the directly irradiated volume [7]. The intensity of scattered radiation falls rapidly with the inverse square law. However, tissues beyond the directly irradiated region are always exposed to scattered radiation.

20.6   CTDI$_{vol}$

In helical or spiral scanning, where the gantry rotates continuously with continuous patient table increment, the gaps or overlaps are accounted for with Volume CTDI (*CTDI$_{vol}$*). The distance travelled by the table in one 360° gantry rotation divided by beam width or beam collimation, *nT* is called pitch. If the table travels 5 mm in one rotation and the beam collimation is 5 mm, then pitch equals $\frac{5 \text{ mm}}{5 \text{ mm}} = 1.0$.

The formula for CTDI$_{vol}$ is:

$$CTDI_{vol} = \frac{CTDI_w}{pitch} \quad (20.4)$$

With pitch 1.0, CTDI$_{vol}$ is equal to CTDI$_w$, and doses are similar to those resulting from contiguous axial scanning. Pitch is inversely propor-

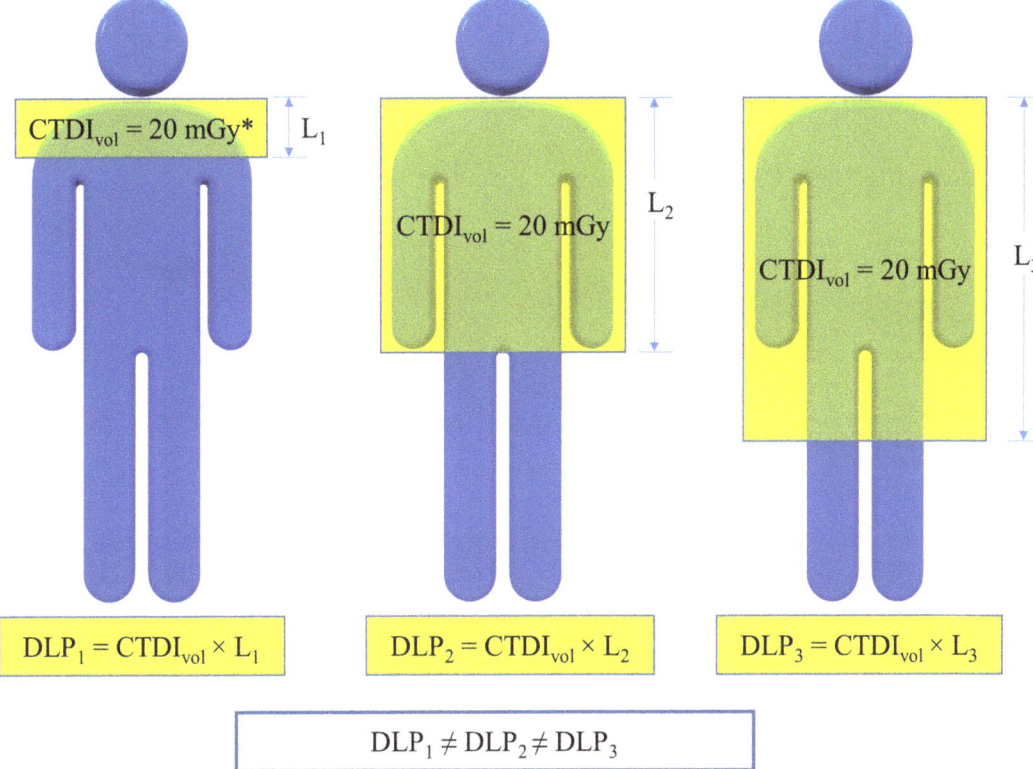

$$DLP_1 = CTDI_{vol} \times L_1 \qquad DLP_2 = CTDI_{vol} \times L_2 \qquad DLP_3 = CTDI_{vol} \times L_3$$

$$DLP_1 \neq DLP_2 \neq DLP_3$$

* An example value, to make reader understand. Actual value will be different depending upon mA, kV, pitch, tube rotation time and size of phantom.

Fig. 20.8 $CTDI_{vol}$ and DLP values

tional to the dose. If the pitch is <1.0, doses increase because of overlapping scans, and when the pitch increases to more than 1.0, doses decrease because the energy is spread out over a larger volume.

$CTDI_{vol}$ is independent of the total scan length and is a universal matrix that talks about the intensity of radiation incidents on patients and do not represent patient dose (Fig. 20.8). It depends on X-ray tube voltage (kV), tube current (mA), tube rotation time (sec), CT Pitch (P) and phantom size (Small or Large).

20.7 DLP

The dose length product (DLP) is the product of $CTDI_{vol}$ and scan length (L).

$$DLP = CTDI_{vol} \times \text{Scan length}(L) \qquad (20.5)$$

It is measured in mGy.cm.

The DLP is proportional to the total dose (energy) imparted to the patient and a good indicator of the total amount of radiation incident on a patient. It is directly related to the patient stochastic risk and may be used to set reference values for a given type of CT examination to achieve the ALARA (As Low As Reasonably Achievable) concept.

20.8 CT Dosimetry Phantoms

The report 31 (1990) [8] of The Association of American Physicists in Medicine (AAPM) describes two phantoms to obtain $CTDI_{100}$, one is

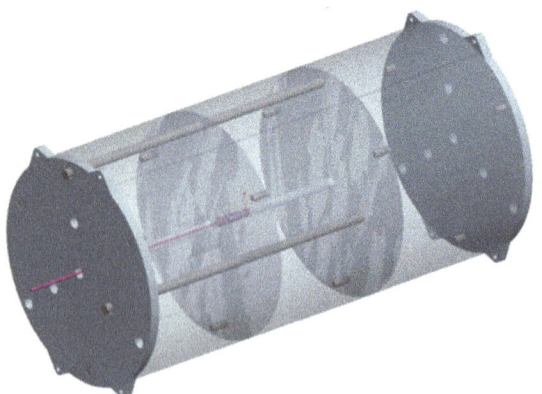

Fig. 20.9 ICRU/AAPM phantom for CT dosimetry (Image Courtesy: The AAPM report 200, reprinted with permission)

a 16-cm diameter adult head phantom which also serves as a paediatric body phantom with an equivalent cross-sectional area of the typical 2-year-old, and the other is a 32-cm diameter adult body phantom (Fig. 20.7), both are 15 cm in length. They are made up of Polymethyl methacrylate (PMMA) and traded with different names such as Lucite, Perspex and Plexiglass. It is a clear plastic with the chemical formula $(C_5H_8O_2)_n$ and has a density of 1.185 g/cm³. These phantoms have ~12 mm diameter holes at the centre and periphery to allow insertion of the ionization chamber.

The phantoms have been widely used for a long time and have proven very useful. However, they suffer from limitations such as excluding the dose that would accumulate for longer scans and is unsuitable for newer scanners with a very wide beam of more than 100 mm, etc. AAPM Report 111 [9] offered several recommendations for a new measurement methodology and suggested several phantom designs.

AAPM and The International Committee on Radiation Units and Measurements (ICRU) jointly developed a phantom called ICRU/AAPM phantom (Fig. 20.9), suitable for a wide range of CT scanners designs and scanning conditions. The same is discussed in AAPM report 200 (2020) [10].

This phantom comprises polyethylene (density—0.97 g/cm³), cylindrical in the shape of 30 cm in diameter and 60 cm in length. The density is chosen to mimic human adipose tissue's absorption properties closely. It is relatively light in weight, readily available and relatively expensive. To make it more manageable, it can be divided into three sections, with each section having a mass of around 13.7 kg and 20 cm long, with a total weight of 41.1 kg when fully assembled. The weighing of each section is about the same as a fully assembled 32-cm CTDI phantom.

These phantoms have shown good correlations and can be adopted as future standards (AAPM 200 and ICRU 87).

20.9 Size-Specific Dose Estimate (SSDE)

Since 2002, as a mandatory requirement, all CT scanners display $CTDI_{vol}$ and DLP before and after the scan as a 'dose report'. $CTDI_{vol}$ is determined for only 16-cm and 32-cm PMMA head and body reference phantoms. These system-generated CT dose report values estimate patient doses, but they would be for the phantom and not for the patients. In practice, there are wide variations in patient sizes. The size variability among patients may significantly change the absorbed doses (2–3 times) [7] then observed in phantoms, especially among paediatric patients. To account for size variabilities, the AAPM provided detailed guidelines on 'Size-Specific Dose Estimate (SSDE)' in its report 204 (2011).

The SSDE is a 'patient dose estimate' that considers corrections based on the patient's size, using linear dimensions measured on the patient or patient images. The AAPM creates two tables (Tables 20.1 and 20.2) (AAPM report 204) related to 32-cm and 16-cm diameter PMMA phantoms, which provide correction factors to multiply with $CTDI_{vol}$ reported by CT scanners.

$$SSDE = CTDI_{vol} \times \text{size specific conversion factor}$$
$$(20.6)$$

'Anterior-Posterior (AP)' and/or 'Lateral' dimensions can be obtained by electronic callipers (Fig. 20.8). Both tables provide four sets of data contained within sub-Tables A, B, C and D. In these tables, the AP dimension is the thickness of body part from anterior to posterior or vice versa, whereas the lateral dimension is the measurement from left to right or vice versa. An effective diameter can be obtained from these dimensions based on the formula shown in Fig. 20.10. The effective diameter is the diameter of a circle whose area is the same as that of patient cross-section.

The AAPM report 204 also provides a table (Table 20.3) to estimate the effective diameter based on the patient's age. This table can be used if the user does not have AP, lateral dimensions of the patient and only age are available. One can find the effective diameter based on age, and this effective diameter can be matched from

Table 20.1 Conversion factors based on 32-cm diameter PMMA phantom for $CTDI_{vol}$ (Table originally printed in AAPM 204, reprinted with permission from the AAPM)

Table 1A

Lat + AP Dim (cm)	Effective Dia (cm)	Conversion Factor
16	7.7	2.79
18	8.7	2.69
20	9.7	2.59
22	10.7	2.50
24	11.7	2.41
26	12.7	2.32
28	13.7	2.24
30	14.7	2.16
32	15.7	2.08
34	16.7	2.01
36	17.6	1.94
38	18.6	1.87
40	19.6	1.80
42	20.6	1.74
44	21.6	1.67
46	22.6	1.62
48	23.6	1.56
50	24.6	1.50
52	25.6	1.45
54	26.6	1.40
56	27.6	1.35
58	28.6	1.30
60	29.6	1.25
62	30.5	1.21
64	31.5	1.16
66	32.5	1.12
68	33.5	1.08
70	34.5	1.04
72	35.5	1.01
74	36.5	0.97
76	37.5	0.94
78	38.5	0.90
80	39.5	0.87
82	40.5	0.84
84	41.5	0.81
86	42.4	0.78
88	43.4	0.75
90	44.4	0.72

Table 1B

Lateral Dim (cm)	Effective Dia (cm)	Conversion Factor
8	9.2	2.65
9	9.7	2.60
10	10.2	2.55
11	10.7	2.50
12	11.3	2.45
13	11.8	2.40
14	12.4	2.35
15	13.1	2.29
16	13.7	2.24
17	14.3	2.19
18	15.0	2.13
19	15.7	2.08
20	16.4	2.03
21	17.2	1.97
22	17.9	1.92
23	18.7	1.86
24	19.5	1.81
25	20.3	1.76
26	21.1	1.70
27	22.0	1.65
28	22.9	1.60
29	23.8	1.55
30	24.7	1.50
31	25.6	1.45
32	26.6	1.40
33	27.6	1.35
34	28.6	1.30
35	29.6	1.25
36	30.6	1.20
37	31.7	1.16
38	32.7	1.11
39	33.8	1.07
40	34.9	1.03
41	36.1	0.98
42	37.2	0.94
43	38.4	0.90
44	39.6	0.87
45	40.8	0.83

Table 1C

AP Dim (cm)	Effective Dia (cm)	Conversion Factor
8	8.8	2.68
9	10.2	2.55
10	11.6	2.42
11	13.0	2.30
12	14.4	2.18
13	15.7	2.08
14	17.0	1.98
15	18.3	1.89
16	19.6	1.81
17	20.8	1.73
18	22.0	1.65
19	23.2	1.58
20	24.3	1.52
21	25.5	1.45
22	26.6	1.40
23	27.6	1.34
24	28.7	1.29
25	29.7	1.25
26	30.7	1.20
27	31.6	1.16
28	32.6	1.12
29	33.5	1.08
30	34.4	1.05
31	35.2	1.02
32	36.0	0.99
33	36.8	0.96
34	37.6	0.93
35	38.4	0.91
36	39.1	0.88
37	39.8	0.86
38	40.4	0.84
39	41.1	0.82
40	41.7	0.80
41	42.3	0.78
42	42.8	0.77
43	43.4	0.75
44	43.9	0.74
45	44.4	0.73

Table 1D

Effective Dia (cm)	Conversion Factor
8	2.76
9	2.66
10	2.55
11	2.47
12	2.38
13	2.30
14	2.22
15	2.14
16	2.06
17	1.98
18	1.91
19	1.84
20	1.78
21	1.71
22	1.65
23	1.59
24	1.53
25	1.48
26	1.43
27	1.37
28	1.32
29	1.28
30	1.23
31	1.19
32	1.14
33	1.10
34	1.06
35	1.02
36	0.99
37	0.95
38	0.92
39	0.88
40	0.85
41	0.82
42	0.79
43	0.76
44	0.74
45	0.71

Table 20.2 Conversion factors based on 16-cm diameter PMMA phantom for CTDI$_{vol}$. (Table originally printed in AAPM 204, reprinted with permission from the AAPM)

Table 2A

Lat + AP Dim (cm)	Effective Dia (cm)	Conversion Factor
12	5.7	1.50
13	6.2	1.47
14	6.7	1.44
15	7.2	1.42
16	7.7	1.39
17	8.2	1.36
18	8.7	1.34
19	9.2	1.31
20	9.7	1.29
21	10.2	1.26
22	10.7	1.24
23	11.2	1.22
24	11.7	1.19
25	12.2	1.17
26	12.7	1.15
27	13.2	1.13
28	13.7	1.10
29	14.2	1.08
30	14.7	1.06
31	15.2	1.04
32	15.7	1.02
33	16.2	1.00
34	16.7	0.98
35	17.2	0.97
36	17.6	0.95
37	18.1	0.93
38	18.6	0.91
39	19.1	0.89
40	19.6	0.88
42	20.6	0.84
44	21.6	0.81
46	22.6	0.78
48	23.6	0.75
50	24.6	0.72
52	25.6	0.70
54	26.6	0.67
56	27.6	0.64
58	28.6	0.62
60	29.6	0.60
62	30.5	0.57
64	31.5	0.55
66	32.5	0.53
68	33.5	0.51
70	34.5	0.49
72	35.5	0.47
74	36.5	0.46
76	37.5	0.44
78	38.5	0.42
80	39.5	0.41
82	40.5	0.39

Table 2B

Lateral Dim (cm)	Effective Dia (cm)	Conversion Factor
6	8.2	1.36
7	8.7	1.34
8	9.2	1.32
9	9.7	1.29
10	10.2	1.26
11	10.7	1.24
12	11.3	1.21
13	11.8	1.19
14	12.4	1.16
15	13.1	1.13
16	13.7	1.10
17	14.3	1.08
18	15.0	1.05
19	15.7	1.02
20	16.4	0.99
21	17.2	0.96
22	17.9	0.94
23	18.7	0.91
24	19.5	0.88
25	20.3	0.85
26	21.1	0.83
27	22.0	0.80
28	22.9	0.77
29	23.8	0.75
30	24.7	0.72
31	25.6	0.70
32	26.6	0.67
33	27.6	0.65
34	28.6	0.62
35	29.6	0.60
36	30.6	0.57
37	31.7	0.55
38	32.7	0.53
39	33.8	0.51
40	34.9	0.48
41	36.1	0.46
42	37.2	0.44
43	38.4	0.42
44	39.6	0.40
45	40.8	0.39
46	42.1	0.37
47	43.3	0.35
48	44.6	0.33
49	45.9	0.32
50	47.2	0.30
51	48.5	0.29
52	49.9	0.27
53	51.3	0.26
54	52.7	0.24
55	54.1	0.23

Table 2C

AP Dim (cm)	Effective Dia (cm)	Conversion Factor
6	5.8	1.50
7	7.3	1.41
8	8.85	1.33
9	10.2	1.26
10	11.6	1.19
11	13.0	1.13
12	14.4	1.07
13	15.7	1.02
14	17.0	0.97
15	18.3	0.92
16	19.6	0.88
17	20.8	0.84
18	22.0	0.80
19	23.2	0.76
20	24.3	0.73
21	25.5	0.70
22	26.6	0.67
23	27.6	0.64
24	28.7	0.62
25	29.7	0.59
26	30.7	0.57
27	31.6	0.55
28	32.6	0.53
29	33.5	0.51
30	34.4	0.50
31	35.2	0.48
32	36.0	0.46
33	36.8	0.45
34	37.6	0.44
35	38.4	0.42
36	39.1	0.41
37	39.8	0.40
38	40.4	0.39
39	41.1	0.38
40	41.7	0.37
41	42.3	0.36
42	42.8	0.36
43	43.4	0.35
44	43.9	0.34
45	44.4	0.34
46	44.8	0.33
47	45.2	0.33
48	45.6	0.32
49	46.0	0.32
50	46.4	0.31
51	46.7	0.31
52	47.0	0.30
53	47.2	0.30
54	47.5	0.30
55	47.7	0.30

Table 2D

Effective Dia (cm)	Conversion Factor
6	1.49
7	1.43
8	1.38
9	1.32
10	1.27
11	1.22
12	1.18
13	1.13
14	1.09
15	1.05
16	1.01
17	0.97
18	0.93
19	0.90
20	0.86
21	0.83
22	0.80
23	0.77
24	0.74
25	0.71
26	0.69
27	0.66
28	0.63
29	0.61
30	0.59
31	0.56
32	0.54
33	0.52
34	0.50
35	0.48
36	0.47
37	0.45
38	0.43
39	0.41
40	0.40
41	0.38
42	0.37
43	0.35
44	0.34
45	0.33
46	0.32
47	0.30
48	0.29
49	0.28
50	0.27
51	0.26
52	0.25
53	0.24
54	0.23
55	0.22

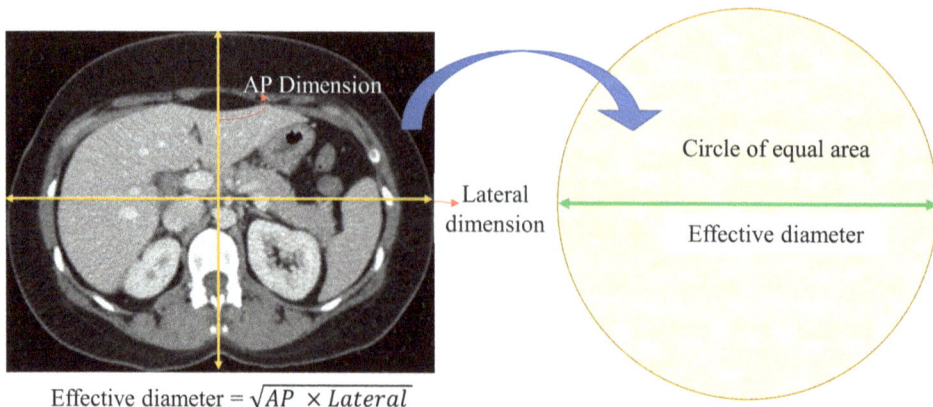

Effective diameter $= \sqrt{AP \times Lateral}$

Fig. 20.10 Patient size estimation from AP and lateral dimensions. (CT images used here are reproduced with the permission of RadiologyCafe.com)

Table 20.3 Estimation of effective diameter based on Age. (Table originally printed in AAPM 204, reprinted with permission from the AAPM)

Patient Age (years)	Effective diameter (cm)
0.0	11.2
0.2	12.1
0.4	13.1
0.6	13.9
0.8	14.6
1.0	15.1
1.2	15.6
1.4	16.0
1.6	16.3
1.8	16.6
2.0	16.8
2.5	17.3
3.0	17.6
3.5	17.9
4.0	18.1
4.5	4.5
5.0	18.5
6.0	19.0
7.0	19.6
8.0	20.2
9.0	20.9
10.0	21.6
11.0	22.4
12.0	23.2
13.0	24.1
14.0	25.0
15.0	26.0
16.0	27.0
17.0	28.1
18.0	29.2

Table 20.1D or 20.2D to get appropriate conversion factors to estimate SSDE. However, this table should be used only when patient sizes are not available, as there can be a large variation in size in same-age patients.

20.10 CT Doses

- To estimate the patient dose from $CTDI_{vol}$, we need to multiply it with the patient size-specific factor and scan length.

$$\text{Patient absorbed dose} = CTDI_{vol} \times \text{Patient size factor} \times \text{Scan length} \tag{20.7}$$

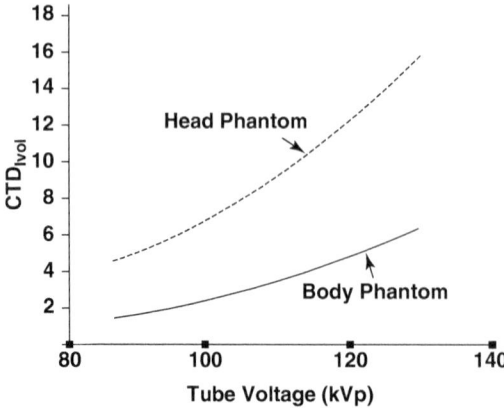

Fig. 20.11 Relationship between tube voltage (kVp) and CTD$_{vol}$. (Copyright Wolters Kluwer Health, Inc. Originally published in Walter Huda, Review of Radiologic Physics, 3rd edition (2010). Reprinted with permission)

- CT doses are directly proportional to the mA and the scan rotation time. However, increasing the tube voltage increases the dose significantly. Figure 20.11 shows the variation of CTDI doses with tube voltage. Body doses are lower because the larger phantom attenuates the X-ray beam much more than the head phantom. The relationship between CTDI$_{vol}$ and kV is,

$$CTDI_{vol} \propto \left(kV\right)^{2.6} \qquad (20.8)$$

   For example, if we increase the kVp from 80 to 140 kVp, the CTDIvol increases about five-fold [11].
- Tube current modulations around the patient can reduce patient doses without affecting image quality. Most CT manufacturers nowadays provide software-assisted tube current modulations.
- CT doses are inversely proportional to pitch. A Pitch of 0.5 doubles the dose, whereas a pitch of 2 halves the dose.
- The dose profile is not uniform along the patient axis. Doses at the surface may be higher

than the dose at the centre. In head scans, the dose ratio from surface to centre is approximately 1:1, whereas, in body scans, it is 2:1.
- The patient dose is directly proportional to the product of the acquired slice thickness and the total number of slices. Multiphase studies increase the patient dose substantially, may up to quadruple.

20.11   Estimating Effective Doses

The effective dose defined in ICRP 60 (1991) and modified in ICRP 103 (2007) is used to compare radiation detriments of various radiological and nuclear medicine procedures. The effective dose is neither a physical dose quantity nor intended to be used to assess an individual patient's radiation dose [12]. The computation of effective dose is generic and based on weighing factors derived from radiobiological considerations. It reflects the overall risk of any radiation exposure to a reference person, where the risk is averaged over all ages and both sexes.

   There are two common methods to calculate effective doses in CT scans [12]. The first, the gold standard method based on organ dose estimates, uses tissue-weighting factors specified by the ICRP. The second, a computationally simple method is based on the DLP and a 'dose conversion coefficient (k)' determined for different anatomic regions examined (Table 20.4) [12].

   Effective dose estimation based on organ dose estimates and tissue weighting factors uses NRPB (National Radiological Protection Board, UK) Monte Carlo data sets and calculated by The *ImPACT* Excel spreadsheet (Fig. 20.12) and other software. The spreadsheet can be downloaded from the impactscan.org website, and code data sets are provided by gov.uk [13] by email *medicalradiationdoses@phe.gov.uk* for free. The

Table 20.4 Conversion coefficients for effective dose calculation using the DLP. For head and neck, 16 cm diameter CT head phantoms are used, whereas other conversion factors are based on 32 cm diameter CT body phantom. (Table originally reprinted in AAPM 96, reprinted with permission)

Region of the body	k/[mSv/(mGy-cm)]				
	0-year-old	1-year old	5-year-old	10-year-old	Adult
Head and neck	0.013	0.0085	0.0057	0.0042	0.0031
Head	0.011	0.0067	0.0040	0.0032	0.0021
Neck	0.017	0.012	0.011	0.0079	0.0059
Chest	0.039	0.026	0.018	0.013	0.014
Abdomen	0.049	0.030	0.020	0.015	0.015
Pelvis	0.049	0.030	0.020	0.015	0.015
Chest, abdomen and pelvis (trunk)	0.044	0.028	0.019	0.014	0.015

spreadsheet can calculate effective doses according to ICRP 103 and ICRP 60. The NRPB data sets contain 23 series of Monte Carlo calculations that model the conditions of exposure for a range of common makes of CT scanners. Each data file contains 208 sets of normalized dose data for 27 organs of a mathematical phantom (Fig. 20.13) for every 5 mm thick transverse slab [13].

The European Commission guidelines published the dose conversion coefficients in year 2000 for adults, and later in 2004, Shrimpton (Shrimpton 2004), in its report NRPB-PE/1/2004 [14], provided dose conversion coefficients for paediatric patients of different ages. The Shrimpton coefficients were adopted by the European Commission and the AAPM (AAPM 96, 2008) [15]. The effective doses were compared with DLP values for the corresponding clinical exams. A set of coefficients, k, was determined for a different region of the body being scanned (head, neck, thorax, abdomen or pelvis). Using the following formula, the effective dose E can be estimated:

$$E \approx k \times \mathrm{DLP} \qquad (20.9)$$

where E is the effective dose in mSv if DLP has units of mGy.cm and 'k' has units of mSv/(mGy cm).

10 to 15% deviations in effective dose estimates have been reported using this method relative to the gold standard organ dose-based technique at 120 kV [15]. However, the use of DLP to estimate E is a reasonably robust method for estimating the effective dose (ICRU 87) [4].

The conversion factors mentioned in Table 20.4 were calculated based on ICRP 60 (1990) weighing factors. Huda et al. (2011) [16] calculated conversion factors based on ICRP 103 weighting factors for 16-cm and 32-cm phantoms. Saltybaeva et al. (2014) [17] provided conversion factors for lower extremities based on ICRP 103 (Table 20.5).

In the same paper, Huda et al. (2011) compared conversion factors obtained from weighing factors of ICRP 60 and ICRP 103. He concluded, 'ICRP 103 weighting factors increase effective doses in the head by ~11%, in the chest by ~20%, and decrease effective doses for pelvic scans by ~25%'.

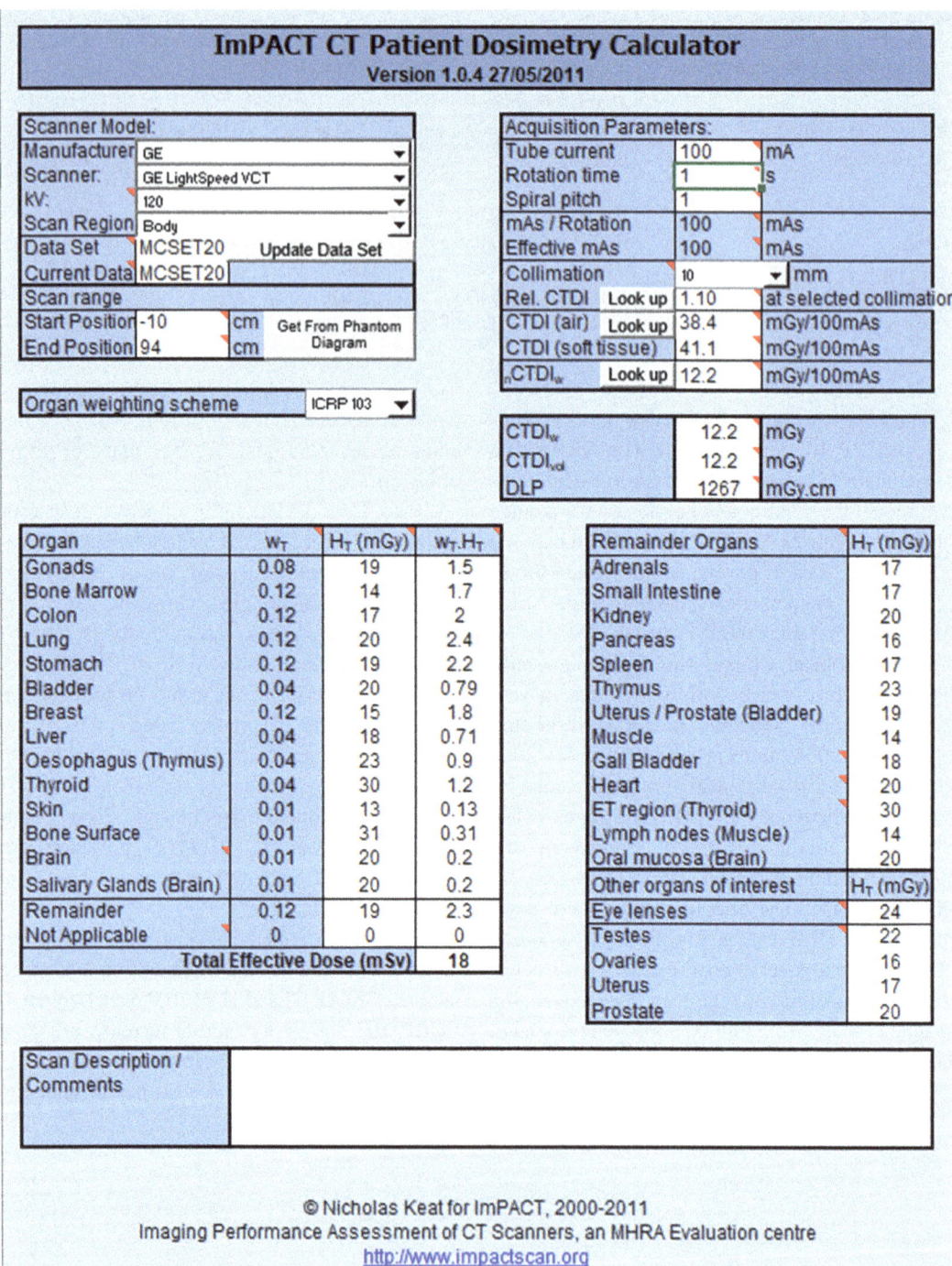

Fig. 20.12 ImPACT CT Dosimetry calculator

Fig. 20.13 Mathematical phantom used with *ImPACT* software to compute patient doses covering Vertex to mid-thigh

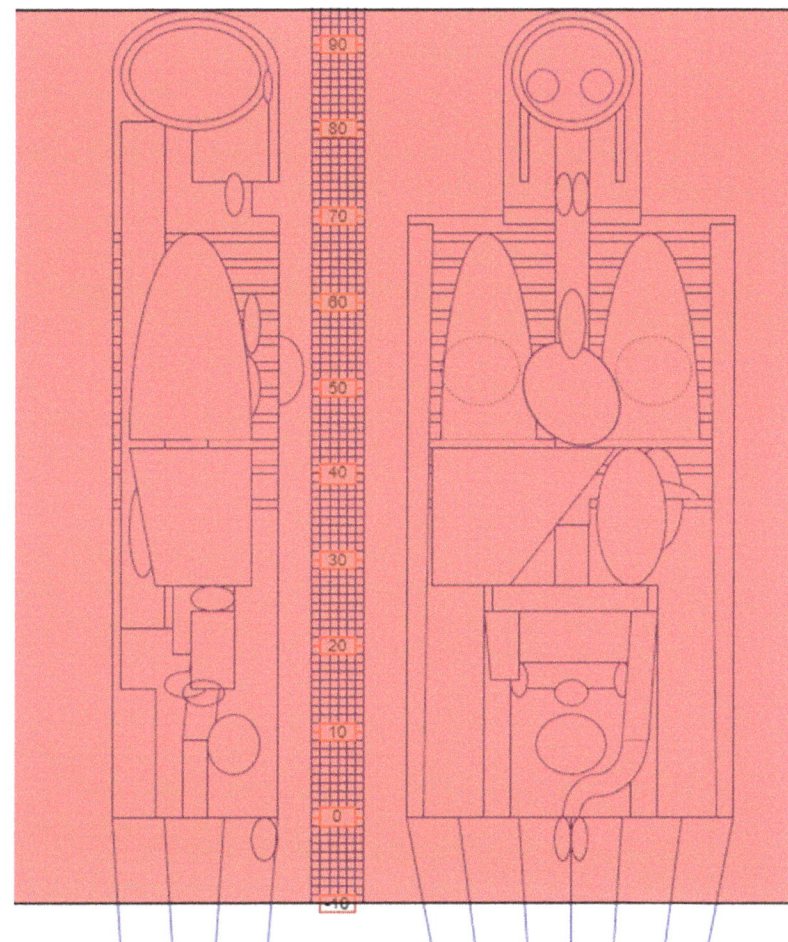

Table 20.5 Conversion factors based on ICRP 103 recommended tissue weighting factors

Region of the body	k/[mSv/(mGy-cm)]		Reference
	16-cm diameter phantom	32-cm diameter phantom	
Head	0.0024	0.0049	[16]
Neck	0.0053	0.0107	[16]
Head and neck	0.0045	0.0090	[16]
Chest	0.0102	0.0204	[16]
Abdomen	0.0082	0.0163	[16]
Pelvis	0.0071	0.0143	[16]
Abdomen + pelvis	0.0086	0.0171	[16]
Chest + abdomen + pelvis (trunk)	0.0093	0.0186	[16]
Whole body (head + torso)	0.0077	0.0154	[16]
Lower extremities		0.006 (male) 0.0073 (female)	[17]

20.11.1 Calculation of Effective Dose Using Dose Reports

Let us calculate the effective dose from the patient dose report (Figs. 20.3, 20.4 and 20.5) obtained from the PET-CT scanner using the DLP method. Conversion coefficients are used from Table 20.5 (based on ICRP 103) for the corresponding anatomic regions. In the absence of size dimensions, SSDE factor = 1 is assumed.

i. *Figure 20.3 Vertex to mid-thigh protocol*:
   DLP for HRCT thorax = 276.20 mGy-cm
   DLP for the whole body = 1184.24 mGy-cm
   Effective dose from HRCT thorax data = 276.20 × 0.0204 = 5.6 mSv
   Effective dose from whole body data = 1184.24 × 0.0154 = 18.2 mSv
   Total effective dose = 5.6 + 18.2 = 23.8 mSv

ii. *Figure 20.4: Eye to mid-thigh + Brain protocol:*
   Effective dose from HRCT thorax data = 295.03 × 0.0204 = 6.0 mSv
   Effective dose from the whole-body data = 1033.95 × 0.0154 = 15.9 mSv
   Effective dose from the brain data = 713.51 × 0.0049 = 3.5 mSv
   Total effective dose = 6.0 + 15.9 + 3.5 = 25.4 mSv

iii. *Figure 20.5: Triple phase Vertex to thigh protocol*:
   Effective dose from HRCT Thorax = 169.84 × 0.0204 = 3.5 mSv
   Total DLP from abdominal region scans = 444.65 + 11.90 + 382.15 + 382.15 = 1220.85 mGy-cm
   ∴ Effective dose from abdominal region = 1220.85 × 0.0163 = 19.9 mSv
   Effective dose from the whole-body CT = 905.19 × 0.0154 = 13.9 mSv
   ∴ Total effective dose = 3.5 + 19.9 + 13.9 = 37.3 mSv

When we compare effective doses from these three scan protocols, it is clearly illustrated that a multiphase scan significantly increases the absorbed dose to the patient. The same is concluded in many studies [11].

In general, effective doses in children are higher than in adults because of smaller organ sizes. Head CT examinations in infants and young children can be up to four times higher than those for adults using the same techniques [11].

20.12 Diagnostic Reference Levels (DRL) and Achievable Dose (AD)

Diagnostic reference levels (DRLs) and achievable doses (ADs) are used to optimize image quality and the dose. DRL is 75th percentile and an investigational level used to identify unusually high radiation doses, whereas Achievable Dose is 50th percentile and provides images with no diagnostic loss of data. The 75th percentile means 75% of institutions surveyed had exposure levels at or below the DRL, whereas the 50th percentile means half of the facilities are producing images at lower doses, and half are using higher doses. Doses higher than AD need to be justified by a corresponding improvement in diagnostic performance.

Various papers have been written on DRL and achievable doses for Nuclear Medicine procedures, including PET-CT, and summarized by Alkhybari et al. (2018) [18]. ICRP Publication 135 (2017) [19] says, 'despite wide variations between PET-CT systems (four-fold variations in $CTDI_{vol}$), CT DRL values of 8 mGy ($CTDI_{vol}$) and 750 mGy.cm (DLP) have been proposed for the whole-body PET-CT'. Alkhybari et al. (2019) [20] suggested a National DRL and achievable dose for 18F-FDG PET-CT (Vertex to thigh protocol) for the parts of Australia and the country New Zealand (Table 20.6).

Table 20.6 Suggested DRL and Achievable doses for 18F-FDG PET-CT (Vertex to thigh) procedure in New Zealand

Administered Activity (MBq)			CTDIvol (mGy)			DLP (mGy cm)		
DRL	Achievable dose	25th percentile	DRL	Achievable dose	25th percentile	DRL	Achievable dose	25th percentile
332.87	282.00	275.00	13.07	9.67	9.41	1319.05	970.49	686.87

References

1. https://www.isct.org/computed-tomography-blog/2017/2/10/half-a-century-in-ct-how-computed-tomography-has-evolved. Accessed Jul 2021.
2. Flohr, T. G., Schaller, S., Stierstorfer, K., Bruder, H., Ohnesorge, B. M., & Schoepf, U. J. (2005). Multi–Detector Row CT Systems and Image-Reconstruction Techniques. Radiology, 235(3), 756–773. doi:https://doi.org/10.1148/radiol.2353040037.
3. Shope TB, Gagne RM, Johnson GC. A method for describing the doses delivered by transmission x-ray computed tomography. Med Phys. 1981;8(4):488–95. https://doi.org/10.1118/1.594995.
4. Boone JM, Brink JA, Edyvean S, Huda W, Leitz W, McCollough CH, McNitt-Gray MF. Report 87. J ICRU. 2012;12(1):NP–NP. https://doi.org/10.1093/jicru/ndt006.
5. Goldman LW. Principles of CT: multislice CT. J Nucl Med Technol. 2008;36(2):57–68. https://doi.org/10.2967/jnmt.107.044826.
6. Brisse HJ, Madec L, Gaboriaud G, Lemoine T, Savignoni A, Neuenschwander S, Rosenwald J-C. Automatic exposure control in multichannel CT with tube current modulation to achieve a constant level of image noise: experimental assessment on pediatric phantoms. Med Phys. 2007;34(7):3018–33. https://doi.org/10.1118/1.2746492.
7. AAPM report no. 204. Size Specific Dose Estimate (SSDE) in paediatric and adult body CT examination. The Report of AAPM Task Group 204. 2011. ISBN: 978-1-936366-08-8.
8. AAPM report no. 31. Standardized methods for measuring Diagnostic x-ray exposures. Report of task group 8, Diagnostic x-ray imaging committee. July 1990. Isbn 0-88318-874-0.
9. AAPM report 111. The future of CT dosimetry. Report of AAPM task group III. 2010. ISBN 978-1-888340-94-5.
10. AAPM report no. 200. The Design and Use of the ICRU/AAPM CT Radiation Dosimetry Phantom: An Implementation of AAPM Report 111. The Report of AAPM Task Group 200. January 2020. ISBN: 978-1-936366-74-3.
11. Huda W. Review of radiologic physics. 3rd ed. Philadelphia, Pennsylvania: Lippincot Williams & Wilkins; 2012. ISBN 978-0-7817-8569-3
12. Christner JA, Kofler JM, McCollough CH. Estimating Effective Dose for CT Using Dose–Length Product Compared With Using Organ Doses: Consequences of Adopting International Commission on Radiological Protection Publication 103 or Dual-Energy Scanning. Am J Roentgenol. 2010;194(4):881–9. https://doi.org/10.2214/ajr.09.3462.
13. https://www.gov.uk/government/publications/computed-tomography-ct-data-analysis-software-to-assess-radiation-doses/normalised-organ-doses-for-x-ray-computed-tomography-calculated-using-monte-carlo-techniques-professionals. Accessed 29 Jul 2021.
14. Shrimpton PC, Hillier MC, Lewis MA, Dunn M. Doses from Computed Tomography (CT) examinations in the UK—2003 review. 2005. NRPB—W67. ISBN 0 85951 556 7.
15. AAPM Report no. 96. The Measurement, Reporting, and Management of Radiation Dose in CT. Report of AAPM Task Group 23: CT Dosimetry. 2008. ISBN: 978-1-888340-73-0.
16. Huda W, Magill D, He W. CT effective dose per dose length product using ICRP 103 weighting factors. Med Phys. 2011;38(3):1261–5. https://doi.org/10.1118/1.3544350.
17. Saltybaeva N, Jafari ME, Hupfer M, Kalender WA. Estimates of effective dose for CT scans of the lower extremities. Radiology. 2014;273(1):153–9. https://doi.org/10.1148/radiol.14132903.
18. Alkhybari EM, McEntee MF, Brennan PC, Willowson KP, Hogg P, Kench PL. Determining and updating PET/CT and SPECT/CT diagnostic reference levels: a systematic review. Radiat Prot Dosim. 2018;182(4):532–45. https://doi.org/10.1093/rpd/ncy113.
19. Vaño, E., Miller et al. (2017). ICRP publication 135: diagnostic reference levels in medical imaging. Ann ICRP, 46(1), 1–144. doi:https://doi.org/10.1177/0146645317717209.
20. Alkhybari EM, McEntee MF, Brennan PC, Willowson KP, Kench PL. Diagnostic reference levels for 18 F-FDG whole-body PET/CT procedures: results from a survey of 12 centers in Australia and New Zealand. J Med Imaging Radiat Oncol. 2019; https://doi.org/10.1111/1754-9485.12857.

Abstract

In Nuclear Medicine and Research application, transporting radioactive material from manufacturer to users' premises is essential. The sources generally used are short half-life, and the activity used ranges from a few kBq to GBq. As the transport of radioactive material is associated with hazard, adequate safety measures must be followed during its transportation. Therefore, the transport of radioactive material in India is regulated by Atomic Energy Regulatory Board (AERB) as per the regulations prescribed in AERB Safety Code AERB/NRF-TS/SC-1(Rev-1), 2016. The requirements prescribed in the code are based on the International Atomic Energy Agency (IAEA) regulations for the 'Safe Transport of Radioactive Material' SSR-6, 2012, which is superseded by SSR-6 (Rev. 1), 2018.

preparation, consigning, loading, and carriage, including in-transit storage, unloading, and receipt at the final destination.

The Regulations do not apply to:

(a) The means of transport, where radioactive material is an integral part of conveyance (e.g. aircraft wings).

(b) When the movement of radioactive material is not through public roads and railways, it's moved within an organization where appropriate transport safety is in force and established (e.g. from hot lab to imaging room for administration).

(c) A person or live animal implanted or incorporated with radioactive material for diagnosis or treatment. (e.g. treatment of the eye using eye plaque).

(d) Consumer products containing radioactive material approved by the regulatory authority after their sale to the end-user (e.g. gas mantle).

21.1 Regulatory Aspects

The regulatory requirements prescribed in AERB/NRF-TS/SC-1(Rev-1), 2016 [1] apply to all transport codes, i.e. land, water or air. The scope of the regulations applies to all stages of transportation, including design, manufacture, maintenance, and repair of packaging, and the

21.2 Definitions of the Terms Used

Term definitions are adapted from IAEA 'Regulations for the Safe Transport of Radioactive Materials SSR-6 (Rev.1), 2018' [2]. Reprinted with permission to English language only.

21.2.1 Radioactive Material

Radioactive material shall mean any material containing radionuclides. Both activity concentration and total activity in the consignment exceed the values given in the 'Regulations for Safe Transport of Radioactive Material' of AERB/IAEA regulations (Excerpt is enclosed as Table 21.1).

21.2.2 Special Form Radioactive Material

Special radioactive material shall mean either an indispensable solid radioactive material or a sealed capsule containing radioactive material.

21.2.3 A₁ and A₂ Values

'A$_1$' shall mean the maximum activity value of *special form radioactive material*, which is listed in the regulations of AERB/IAEA that can be transported in a Type A package.

'A$_2$' shall mean the maximum activity value of radioactive material, *other than special form*

radioactive material, which is listed in the regulations of AERB/IAEA that can be transported in a Type A package.

21.2.4 Contamination

Contamination shall mean the presence of a radioactive substance on a surface in quantities above 0.4 Bq/cm^2 for beta and gamma emitters and low toxicity alpha emitters (natural uranium, depleted uranium, natural thorium, uranium-235 or uranium-238, thorium-232, thorium-230 when contained in ores or physical and chemical concentrates, or alpha emitters with a half-life of fewer than 10 days), and 0.04 Bq/cm^2 for all other α emitters.

21.2.5 Exclusive Use

Exclusive use shall mean the sole use, by a single consignor, of a conveyance or a large freight container, in respect of which all initial, intermediate, and final loading and unloading is carried out following the directions of the consignor or consignee.

Table 21.1 A$_1$/A$_2$ values of some radionuclides. *Table originally published in IAEA 'Regulations for the Safe Transport of Radioactive Materials SSR-6 (Rev.1), 2018'* [2]. *Reprinted with permission to English language only*

Radionuclide (atomic number)	A₁	A₂	Activity concentration for exempt material	Activity limit for an exempt consignment
	TBq	TBq	Bq/g	Bq
Co-60	4×10^{-1}	4×10^{-1}	1×10^{1}	1×10^{5}
Co-57	1×10^{1}	1×10^{1}	1×10^{2}	1×10^{6}
Cs-137	2×10^{0}	6×10^{-1}	1×10^{1}	1×10^{4}
SR-90	3×10^{-1}	3×10^{-1}	1×10^{2}	1×10^{4}
Tl-204	1×10^{1}	7×10^{-1}	1×10^{4}	1×10^{4}
Tl-201	1×10^{1}	4×10^{0}	1×10^{2}	1×10^{6}
Ir-192	1×10^{0}	6×10^{-1}	1×10^{1}	1×10^{4}
C-14	4×10^{1}	3×10^{0}	1×10^{4}	1×10^{7}
Ga-67	7×10^{0}	3×10^{0}	1×10^{2}	1×10^{6}
I-125	2×10^{1}	3×10^{0}	1×10^{3}	1×10^{6}
I-131	3×10^{0}	7×10^{-1}	1×10^{2}	1×10^{6}
P-32	5×10^{-1}	5×10^{-1}	1×10^{3}	1×10^{5}
H-3	4×10^{1}	4×10^{1}	1×10^{6}	1×10^{9}
Mo-99	1×10^{0}	6×10^{-1}	1×10^{2}	1×10^{6}
Tc-99 m	1×10^{1}	4×10^{0}	1×10^{2}	1×10^{7}
Re-188	4×10^{-1}	4×10^{-1}	1×10^{2}	1×10^{5}

21.2.6 Surface Contaminated Object

Surface contaminated object (SCO) shall mean a solid object that is not itself radioactive but has radioactive material distributed on its surfaces. SCO shall be in one of two groups:

(a) SCO-I: A solid object on which:
  (i) the non-fixed contamination (removal) on the accessible surface averaged over 300 cm^2:
    • ≤ 4 Bq/cm^2 for beta, gamma, and low toxicity alpha emitters.
    • ≤ 0.4 Bq/cm^2 for all other alpha emitters.
  (ii) the fixed contamination (any contamination other than non-fixed contamination) on the accessible surface averaged over 300 cm^2.
    • $\leq 4 \times 10^4$ Bq/cm^2 for beta, gamma, and low toxicity alpha emitters.
    • $\leq 4 \times 10^3$ Bq/cm^2 for all alpha emitters.
  (iii) the no-fixed contamination plus the fixed contamination on the inaccessible surface averaged over 300 cm^2.
    • $\leq 4 \times 10^4$ Bq/cm^2 for beta, gamma, and low toxicity alpha emitter.
    • $\leq 4 \times 10^3$ Bq/cm^2 for all other alpha emitters.
(b) SCO-II: A solid object on which:
  (i) 4 Bq/cm^2 < non-fixed contamination ≤ 400 Bq/cm^2 for β, γ and low toxicity α emitters.
  (ii) 4 Bq/cm^2 < non-fixed contamination ≤ 40 Bq/cm^2 for all other alpha emitters.
  (iii) 4×10^4 Bq/cm^2 < fixed contamination $\leq 8 \times 10^5$ Bq/cm^2 for β, γ, low toxicity α emitters.
    4×10^3 Bq/cm^2 < fixed contamination $\leq 8 \times 10^4$ Bq/cm^2 for all other β, γ emitters.
  (iv) 4×10^4 Bq/cm^2 < non fixed plus fixed contamination $\leq 8 \times 10^5$ Bq/cm^2 for β, γ, low toxicity α emitters.
    4×10^3 Bq/cm^2 < non fixed plus fixed contamination $\leq 8 \times 10^4$ Bq/cm^2 for all other α emitters.

21.2.7 Package

The package shall mean the packaging with its radioactive contents as presented for transport. Following are the different types of packages used for the transport of the radioactive material:

1. Excepted Package.
2. Industrial (IP-1, IP-2, IP-3) Package.
3. Type A Package.
4. Type B(U)/(M) Package.
5. Type C Package.

21.2.7.1 Excepted Package

This type of package is the simplest one. This type of package is used to transport radioactive material of very trivial activity such as empty packages, consumer products manufactured of Naturally occurring radioactive materials such as natural uranium, depleted uranium or natural thorium; Limited content of radioactivity as defined in regulations, etc. Excepted packages are widely used in the medical field, e.g. while transporting RIA kits and sometimes during transport of radioisotopes used for research purposes in universities.

21.2.7.2 Excepted Limit of Activity of a Radioisotope

It is the activity of a radioisotope which is $10^{-3} A_1$ value if the source is in a special form, or $10^{-3} A_2$ value if the source is in other than a special form, or $10^{-4} A_2$ if the source is in liquid form. This is the maximum activity of a radioisotope that can be carried in an Excepted package provided the external radiation level does not exceed 0.5 mR/h.

21.2.7.3 Industrial Packages (Type IP-1, Type IP-2, Type IP-3)

This type of package is used to transport radioactive material in bulk quantities, like Low Specific Activity (LSA) material and Surface Contamination Objects (SCO).

21.2.7.4 Type A Package

The packages during transportation come across situations such as falling from a small height, meeting the rainwater, stacking of the packages during storage, and puncture due to the fall of a small, pointed steel rod onto the package. Such transport conditions are known as Normal Conditions of Transport. Type A packages are designed to withstand such normal conditions of transport.

Most of the transport packages used in medical applications, especially nuclear medicine and industrial applications, are Type A packages.

Requirements of Type-A Package

A Type A package is designed to withstand normal conditions of transport. The following are the requirements for a Type A package:

(i) Minimum dimension of the package should not be less than 10 cm.
(ii) Design of the Package shall take into account the temperature ranges of −40 °C to 70 °C for the packaging components.
(iii) The containment system should retain its radioactive contents under ambient pressure reduction to 60 kPa.
(iv) Subjected to water spray, free drop, stacking, and penetration tests.
(v) Designed to completely enclose the liquid contents and ensure their retention within the secondary outer containment, even if the primary inner component leaks.
(vi) Subjected to additional 9 m free drop and 1.7 m penetration tests.

Additional requirements for packages transported by air:

(i) For packages to be transported by air, the temperature of accessible surfaces should not exceed 50 °C at an ambient temp of 38 °C with no account taken for insolation.
(ii) The packages should be so designed that, if they were exposed to an ambient temperature ranging from −40 °C to +55 °C, the integrity of containment would not be impaired.
(iii) The packages should have a containment system able to withstand without leakage an internal pressure which produces a pressure differential of not less than maximum normal operating pressure plus 95 kPa.

21.2.7.5   Type B(U)/(M) Package

The packages during transportation can also be met with an accidental situation such as falling from a great height, catching fire, immersion in water, etc. Therefore, the packages designed to carry large activity of radioisotope beyond A1 and A2 values are required to withstand normal and accidental conditions of transport. The type B(U) and type B (M) packages are designed to withstand normal and accidental transport conditions.

The prior approval from the regulatory body of all the countries is not required, though, which type B (U) packages with radioactive material are transported. Thus, the type B(U) packages can be used worldwide. Whereas the prior approval of the regulatory body of countries, except the regulatory body of the country of origin of the design of the package, is required through, into, within which type B(M) packages with radioactive material are purposed to be transported.

Note: In the above definition of Type B(U) and Type B(M), 'U' means Universal, and 'M' means Multilateral.

21.2.7.6   Type C Package

This type of package is designed to withstand very severe accidental conditions of transport like the mid-air crash of planes. They are mainly used to transport very high activity of radioactive material by air.

21.3   Contamination Level for Packages

The non-fixed contamination on the external surfaces of any package shall not exceed (a) 4 Bq/cm^2 for β, γ and low toxicity α emitters; (b) 0.4 Bq/cm^2 for all other α emitters.

21.4 Categories of Packages

Radioactive packages other than the Excepted Package are categorized into three categories based on the maximum radiation levels on the external surface and at 1 m from the external surface of the package. The number equal to the maximum radiation level (expressed in mrem/h) at 1 m from the external surface of the package is called Transport Index (TI), as shown in Fig. 21.1. From the figure, the T.I. is 4.5 mR/h Table 21.2 gives the different categories of the package.

Though the Excepted packages are not categorized into any of the above three categories, the external radiation level of such packages is restricted to 0.005 mSv/h. Different categories of the label are given in Fig. 21.2.

Fig. 21.1 Transport index

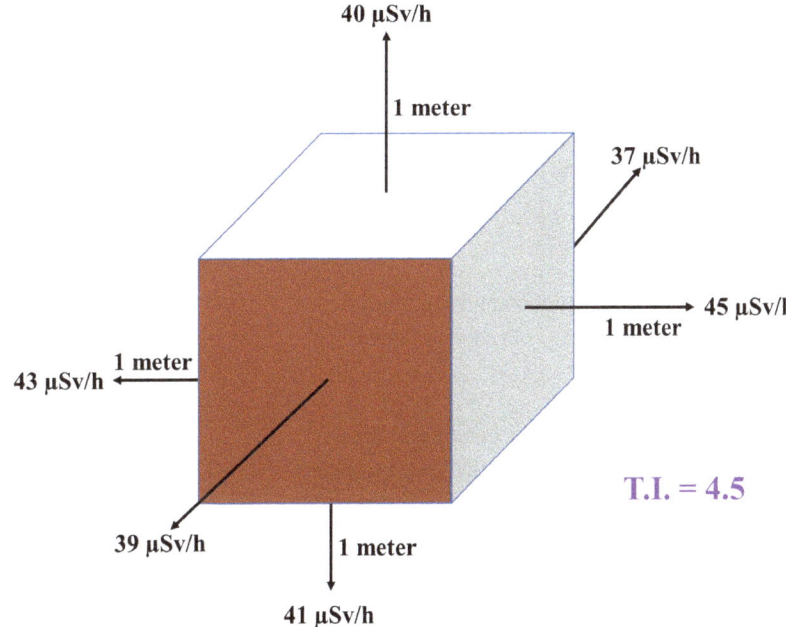

Table 21.2 Categories of package. *Table originally published in IAEA 'Regulations for the Safe Transport of Radioactive Materials SSR-6 (Rev.1), 2018'* [2]. *Reprinted with permission to English language only*

Transport index	The maximum radiation level at any point on the external surface	Category
0	Not more than 0.005 mSv/h	I-WHITE
More than 0, but not more than 1	More than 0.005 mSv/h, but not more than 0.5 mSv/h	II-YELLOW
More than 1, but not more than 10	More than 0.5 mSv/h, but not more than 2 mSv/h	III-YELLOW

Fig. 21.2 Different categories of label. *Figure originally published in IAEA 'Regulations for the Safe Transport of Radioactive Materials SSR-6 (Rev.1), 2018'* [2]. *Reprinted with permission to English language only*

21.5   Marking, Labelling, and Placarding

21.5.1 Marking

An Excepted package containing radioactive material should be durably marked the following:

1. Name and address of the consignor (sender).
2. Name and address of the consignee (receiver).
3. Appropriate UN number (list shown in Table 21.3).

Any package other than Excepted package containing radioactive material should bear the following markings:

1. Name and address of the consignor (sender).
2. Name and address of the consignee (receiver).
3. Type of Packages such as Type IP-1 or Type IP-2 or Type IP-3 or TYPE A or TYPE.
   - B(U) or B (M) or Type C. Type A package would contain a registration no. and for.

- Type B(U)/(M), Type C packages, a Competent Authority identification mark, and serial number.
4. Appropriate UN number and corresponding shipping name (list shown in Table 21.3). This depends upon the type of radioactive material being transported.
5. The gross weight of the package, if it is more than 50 kg.

21.5.2 Labelling

Appropriate transport category labels with respective columns duly filled, i.e. name of radioactive material, its activity, and its Transport Index, should be pasted at least on two laterally opposite sides of the package.

21.5.3 Placarding

Placards for large freight containers and rail, road vehicles carrying radioactive materials.

Table 21.3 Excerpts from list of United Nations Numbers, Proper Shipping names and descriptions. *Table originally published in IAEA 'Regulations for the* *Safe Transport of Radioactive Materials SSR-6 (Rev.1), 2018'* [2]. *Reprinted with permission to English language only*

UN No.	PROPER SHIPPING NAME[a] and description
2910	RADIOACTIVE MATERIAL, EXCEPTED PACKAGE—LIMITED QUANTITY OF MATERIAL
2911	RADIOACTIVE MATERIAL, EXCEPTED PACKAGE—INSTRUMENTS OR ARTICLES
2909	RADIOACTIVE MATERIAL, EXCEPTED PACKAGE—ARTICLES MANUFACTURED FROM NATURAL URANIUM OR DEPLETED URANIUM OR NATURAL THORIUM
2908	RADIOACTIVE MATERIAL, EXCEPTED PACKAGE—EMPTY PACKAGING
2912	RADIOACTIVE MATERIAL, LOW SPECIFIC ACTIVITY (LSA-I) non-fissile or fissile-excepted
3321	RADIOACTIVE MATERIAL, LOW SPECIFIC ACTIVITY (LSA-II) non-fissile or fissile-excepted[b]
3322	RADIOACTIVE MATERIAL, LOW SPECIFIC ACTIVITY (LSA-III) non-fissile or fissile-excepted[b]
2913	RADIOACTIVE MATERIAL, SURFACE CONTAMINATED OBJECTS (SCO-I or SCO-II) non-fissile or fissile-excepted[b]
2915	RADIOACTIVE MATERIAL, TYPE A PACKAGE, non-special form, non-fissile or fissile-excepted[b]
3332	RADIOACTIVE MATERIAL, TYPE A PACKAGE, SPECIAL FORM non-fissile or fissile-excepted[b]
2916	RADIOACTIVE MATERIAL, TYPE B(U) PACKAGE, non-fissile or fissile-excepted[b]
2917	RADIOACTIVE MATERIAL, TYPE B(M) PACKAGE, non-fissile or fissile-excepted[b]
3323	RADIOACTIVE MATERIAL, TYPE C PACKAGE, non-fissile or fissile-excepted[b]
2919	RADIOACTIVE MATERIAL, TRANSPORTED UNDER SPECIAL ARRANGEMENT, non-fissile or fissile-excepted[b]

[a]The 'PROPER SHIPPING NAME' is found in the column 'PROPER SHIPPING NAME and description' and is restricted to that part shown in CAPITAL LETTERS. In the case of UN 2909 and UN 2911, where alternative PROPER SHIPPING NAMES are separated by the word 'or', only the relevant PROPER SHIPPING NAME shall be used
[b]'FISSILE-excepted' applies only to those packages complying with applicable provisions in the regulations

21.6 Transport Documents

The consignor is responsible for proper labelling, marking and placarding, and removing any irrelevant displays. The following information must be included in the transport documentation by the consignor.

(i) Particulars of the consignment.
(ii) Consignors declaration 'I hereby declare that the contents of this consignment are fully and accurately described above by the proper shipping name and are classified, packed, marked and labelled, and are in all respects in proper condition for transport by (insert mode(s) of the transport involved, say if it is by road, you have to write, "by road", if it is by sea, you have to write "by sea") according to the applicable international and national government regulations'.

(Signature of the Consignor)

(iii) Instructions to the carriers: Information about how to store, handle, and restrictions on the mode of transport, emergency arrangements appropriate to the consignment.
(iv) TREMCARD: The transport documents should include a TREMCARD (i.e. Transport Emergency CARD), which should spell out the vehicle crew's action plan and the person at the scene of an accident.

Note: The radioactive consignment shouldn't be transported in a shared taxi, auto-rickshaw or passenger bus. An exclusive use vehicle shall be provided. The personnel involved with the transport shall be clearly instructed to sit in the vehicle as far away as possible from the container and not hold it very close to the body during the hand carriage of the package.

Disclaimer In multiple places, the chapter has used copyrighted material from the IAEA as transport regulations are guided internationally by the IAEA and adopted by the member nations. The contents have been reprinted with due permission from the IAEA.

References

1. IAEA Regulations for the Safe Transport of Radioactive Material, SSR-6 (Rev.1), 2018.
2. AERB Safety Code AERB/NRF-TS/SC-1 (Rev.1), 2016 for Safe Transport of Radioactive Material.

Legislation and Role of National Regulatory Authority in Nuclear Medicine

22

Abstract

In the last few decades, there has been tremendous growth in Nuclear Medicine (N.M) because of the technological advancement of SPECT-CT, PET-CT and Medical Cyclotron modalities. Moreover, the production of new radionuclides for diagnostic and therapeutic purposes has also made this field more demanding. About 90% of the diagnostic examinations are being carried out using ^{99m}Tc and ^{18}F-based radiopharmaceuticals, and only in a few cases, other radionuclides are used. Similarly, ^{131}I and ^{177}Lu are used more commonly than ^{32}P, ^{89}Sr, ^{90}Y and ^{186}Re for therapeutic procedures. To control the safe use of the radiation sources (radiation generating equipment and radioactive substances), the Government of India has promulgated Atomic Energy Act-1962 (The Atomic Energy Act, 1962 (33 of 1962)) and Radiation Protection Rules-2004 (The Atomic Energy Radiation Protection Rules, A.E. (R.P.R.)-2004) made thereunder. Atomic Energy Regulatory Board (AERB) implemented these rules in our country.

22.1 Introduction

All the radiation facilities using ionizing radiation for diagnostic and therapeutic purposes need to be regulated. Depending on the source, activity, use and handling procedures, they may have potential hazards. Thus, these sources is to be regulated, and there needs to be a statutory basis for its implementation and enforcement. In India, AERB was constituted in 1983, and the Chairman of the Board is the Competent Authority for the implementation of rules under the Act. As per the mandate of AERB, the Surveillance procedures, Codes, Standards and Guides are issued, and the same is available on the AERB website (www.aerb.gov.in) [3]. Thus, Radiation protection regulations evolve from a hierarchy of legal documents such as Code, Standard, Guide and Manual.

22.2 The Atomic Energy Act

As referred to earlier, Atomic Energy Act-1962 [1] governs the use of radiation sources. This Act consists of '32' Sections, where the Sections relevant to N.M practice are dealt with in this Chapter. Section '16' provides controls through written consent for handling radiation sources. Section '17' provides for safety-related issues involved and identifies the areas of concern and responsibility for those engaged in the applications of radioisotopes and the use of radiation generating plants. Sections '24', '25' and 26 deal with penalties prescribing fine, imprisonment, or both, depending on the offence. Section '27' deals with the power to appoint the Competent Authority to enforce the rules framed under the provision of the Act. Section '30' deals with the power of the government to make appropriate rules to enforce the provision of the Act.

22.2.1 Rules Issued Under the Act

The following rules have been issued under the Act:

(a) The Atomic Energy (Radiation Protection) Rule G.S.R-303, 2004 **superseding** the.
    Radiation Protection Rules, G.S.R-1601, 1971
(b) The Atomic Energy (Working of the Mines, Minerals, and Handling of Prescribed Substances) Rules, G.S.R-781,1984.
(c) The Atomic Energy (Safe Disposal of Radioactive Wastes) Rules, G.S.R-125, 1987.
(d) The Atomic Energy (Factories) Rules, G.S.R-253, 1996 and
(e) Atomic Energy (Radiation Processing of Food and Allied Products) Rules 2012.

The rules 'a' and 'c' are relevant for regulatory control in N.M. Practice.

Further, there are '35' rules in A.E. (R.P.R.)-2004 [2], but here we have explained the rules which are of direct relevance to Nuclear Medicine practice and are explained here. As per rule '3' of A.E. (R.P.R.), there is a need for a license for the facility's operation, and this license is required during all stages like siting, design, construction, commissioning, operation, recommissioning of Radiation installation. Further, a license in the form of authorization is required to handle radioactive substances in the N.M facilities.

The term handle means the manufacture, possession, storage, use, transfer, sale, export, import, transport or disposal. The issuance of Safety Standards and Safety Codes is governed by Rule '16' Under rule '19'. The Employer shall designate a person having appropriate qualifications as a Radiological Safety Officer (R.S.O) after the due approval of the Competent Authority. The rules '20' and '21' mention the responsibilities of the Employer and the licensee, respectively. Under rule '22', the responsibilities of R.S.O are specified. The inspection of radiation installations is covered under Rule '30'. In case of contravention of any of the rules, the regulatory authority can investigate, seal or seize radiation installation or radioactive material under rule '31'.

22.2.2 Surveillance Procedures Issued Under the Rules [2]

The Competent Authority issues the following surveillance procedures for nuclear medicine practice, i.e. (i) Safe Transport of Radioactive Materials and (ii) Medical Applications of Radiation.

22.3 Safety Code for Nuclear Medicine Facilities

The Board issued the AERB Safety Code for Nuclear Medicine Facilities, AERB/RF-MED/SC-2 (Rev.2) [4] in March 2011, which is still in force to implement regulatory measures to ensure the radiation safety in N.M practices. The Code covers the regulatory requirements for N.M facility, specification of equipment, preparation and use of radiopharmaceuticals, quality assurance related to radiation protection, patient protection, radiation protection and surveillance procedures, management.

As per the Code [4], the following responsibilities of Employer, Licensee, R.S.O, Nuclear Medicine Physician and Nuclear Medicine Technologist are given below.

22.3.1 The Employer

The employer shall

(a) Employ an adequate number of nuclear medicine physician(s), nuclear medicine technologist(s) and an R.S.O.
(b) Provide appropriate equipment and tools to concerned persons for safe handling of radioactive material.
(c) Assign responsibilities to individuals in accordance with this Code and establish lines of communication and authority.
(d) Provide personnel monitoring devices to radiation workers.
(e) Provide any special personnel monitoring device, including those for internal monitoring that may be required and report to the Competent Authority in the event of any qualified staff leaving the institution or remaining absent for a period exceeding sixty days.

22.3.2 Licensee

The licensee shall

(a) Constitute a Local Safety Committee to review the facility's safety in terms of operational safety, quality assurance and regulatory compliance. The R.S.O. shall be a member of the committee and Competent Authority shall be informed about the committee and its work.
(b) Constitute a radiation protection program depending on the type of nuclear medicine facility. The program shall be submitted to the Competent Authority before obtaining regulatory consent.
(c) Submit periodic safety status reports to the Competent Authority.
(d) Report to the Competent Authority on any change in the safety organization of the facility and.

(e) Report any unusual occurrence pertinent to radiological health and safety to the Competent Authority within 24 h of the event.

22.3.3 The Radiological Safety Officer

The R.S.O. shall

(a) Advise and assist the licensee in organizing a radiation protection program appropriate for the facility and ensuring that the staff observes safe work practices.
(b) Ensure safety, security and containment of radioactive sources, carry out radiation and contamination monitoring of work areas, patient waiting areas, radioactive waste disposal sites and public areas and maintain a record.
(c) Ensure that radiation monitoring instruments are kept in proper working condition and are calibrated at regular intervals.
(d) Establish procedures for the management of emergencies and conduct periodic drills to ensure their effectiveness.
(e) Report any unusual incident in writing to the licensee and take remedial measures to mitigate the consequences of the incident and to prevent a recurrence.
(f) Maintain records of the doses of workers, the inventory of sources received, used and disposed of, any unusual incident, cause of such incident and remedial measures taken.
(g) Ensure segregation and monitoring of the waste before interim storage or final disposal.
(h) Advise and assist the licensee in ensuring regulatory compliance for obtaining authorization from the Competent Authority for procurement, use, transport or disposal of radioactive material.
(i) Inform the Competent Authority of his/her leaving the institution and.

(j) Advise and assist the licensee in the transport of radioactive.

(k) Material/radioactive waste in the public domain.

(l) Ensure urgent processing of personnel dosimeters in cases of suspected overexposure.

22.3.3.1 In Addition to the Above, R.S.O. of High Dose Therapy Shall

(a) Ensure that patients administered with radioisotopes for in-patient therapy are hospitalized in the approved isolation wards.

(b) Carry out regular monitoring of therapy patients, patient areas and nurse station areas.

(c) Ensure that the effective dose to the patient's comforter should not normally exceed 5 mSv during a patient's treatment.

(d) Ensure that the radiation level at 1 m from the patient is below 30 μSv/h when the patients are discharged.

(e) Assess living conditions of the patient and ensure the dose to any family member other than comforter does not exceed 1 mSv before the discharge of the patient.

(f) Ensure management of cadavers in accordance with stipulations.

(g) Ensure sampling and monitoring of effluents from therapy wards before their release to public sewers.

(h) Ensure liquid effluents released to public sewers do not exceed authorized discharge limits.

(i) Maintain a separate logbook for data on monitoring of therapy patients from the time of hospitalization until discharge from the ward.

(j) Segregate and monitor patient linen before interim storage or reuse.

(k) Provide personnel monitoring to patient comforters, if required, and maintain appropriate records.

(l) Give appropriate instructions for radiation safety and precautions to patient comforters in managing therapy patients.

(m) Restrict the entry of visitors to isolation wards and.

(n) Decide, in consultation with the nuclear medicine physician-in-charge, the safety precautions to be followed regarding disposal of cadavers containing radionuclides not exceeding the limits prescribed by Competent Authority.

The responsibilities of a Nuclear Medicine Physician and Nuclear Medicine Technologist, as mentioned in AERB Safety Code [4] for N.M. facilities, are given below.

22.3.4 Nuclear Medicine Physician

The nuclear medicine physician shall

(a) Have the responsibility of dosage administration and maintenance of records providing the name of the patient, nature of procedure, radiopharmaceutical prescribed, quantity prescribed, name of the nuclear medicine physician with signature and date and name of the person administering the radiopharmaceutical with signature and date.

(b) Prevent any possibility of misadministration and promptly report to the licensee and the Competent Authority in the event of any misadministration, adverse reaction or death of a patient administered with radioactivity.

(c) Inform the patient on safety measures to be observed to avoid radiation exposure to the family members and others.

(d) Ensure that any patient hospitalized after administration of therapeutic quantity of radioactivity is kept isolated, the spread of contamination prevented and exposure of staff, other patients and public minimized.

(e) Instruct nursing and ancillary staff on radiation safety and precautions in nursing/management of therapy patients.

(f) Obtain informed consent from the relatives of the patient before the administration of therapeutic dose and.

(g) Ensure the efficacy of new nuclear medicine procedures from experience gained from previous clinical trials, animal experiments and published literature. Prior clearance from the institution's Ethics Committee shall be obtained for any new trial or deviation from established norms with implications from a radiological safety standpoint.

22.3.5 Nuclear Medicine Technologist

The nuclear medicine technologist shall

(a) Ensure the proper functioning of all nuclear medicine equipment, carry out periodic calibrations, quality assurance checks and maintenance.
(b) Ensure the radiopharmaceutical quality requirements, the route of administration and the dosage accuracy before giving it to a patient and take precautions to avoid misadministration.
(c) Avoid spillage of radioactivity or contamination of the patient, premises, persons and material by exercising care during dispensing/administration of radioactivity.
(d) Report to R.S.O and the nuclear medicine physician of any mishap in dispensing/administration of dosage to the patient or any unusual incident and.
(e) Assist the R.S.O in maintaining inventory records, use, waste disposal of sources and other safety matters.

Moreover, to have safe disposal of radioactive waste generated during the operation of the Nuclear Medicine facility, the Atomic Energy Safe Disposal of Radioactive Waste, Rules, GSR-125, 1987 [5] was made. As per these rules, the N.M. facility cannot dispose of the radioactive waste without valid authorization from the Competent Authority. These rules encompass the disposal limits for various radionuclides.

The fundamental principles of radiation protection as per ICRP [6] adopted in the country are:

- Justification of practices.
- Optimization of protection and safety.
- Compliance with dose limits.

It is the responsibility of the licensee to ensure the protection of the workers, the public and the safety of the environment, as he is obliged to set up and implement the technical and organizational measures needed. Any modifications to any practice or source for which the licensee is authorized should be notified to the 'AERB'.

The role of the national regulatory authority (AERB) in controlling radiation doses consists of three basic elements:

- Regulations.
- Licensing, Registration, Authorization, Consent.
- Inspection.

Where Licensing and Authorization are the two important terms used in Nuclear Medicine.

These regulations made by AERB are considered the milestone of the India regulatory program, which defines the basic requirements that must be followed by users of these Nuclear Medicine facilities, as well as the measures and the resources needed to achieve the protection and safety objectives.

They define such matters as:

- Principles of the optimization of protection.
- Compliance with dose limits.
- Classification of working areas.
- Categorization of workers.
- Monitoring requirements.
- Actions to be taken routinely and in the emergencies.
- Conditions under which licensing, registration, authorization and consent are issued.

To apply for License and Authorization, the Nuclear Medicine facility may apply through e-LORA (e-licensing of radiation application) to AERB by giving all the necessary details asked in the application form. The application should include the complete details of the facility, the infrastructure, the persons handling the radioactive materials, personnel monitoring services and radiation protection instruments available with the facility. The guidelines for filling the application form used in e-LORA are given on the AERB website (www.aerb.gov.in) [3].

These applications are reviewed in AERB, and after careful examination, the necessary permissions with conditions are accorded to the user facility within the stipulated time as mentioned in the Radiation Protection Rules, A.E. (R.P.R.)-2004 [2]. Any modifications of license conditions concerning use, processing, design, manufacture, construction, importing, exporting, siting, locating, commissioning, operation, distribution, loaning, selling, maintaining, repairing, hiring, transferring, decommissioning, disassembling, transporting, storing and disposal shall be notified to AERB. Any modification carried out without prior approval from AERB is viewed as a violation of regulatory norms. The regulatory authority may initiate necessary action. Further, the license renewal is done only after receiving the duly completed periodical safety status report.

In India, the regulatory inspection provides the most positive assurance that radiation protection and safety requirements are being met and provides the opportunity to enforce corrective actions. The licensee must permit authorized representatives of the 'AERB' to inspect their authorized activities to assess compliance and safety. The 'AERB' provides inspectors with a very good knowledge of radiation protection principles and practices and a good understanding of the radiation facility operation being inspected. Finally, it is the Employer's duty that an appropriate training programme for the occupational workers involved in the N.M. facility should be developed, and the same should be implemented and reviewed periodically.

To ensure radiation safety in N.M practices, AERB is playing an important role in the growth of N.M by giving radiation safety advice and implementing the rules and regulations laid down therein and keeping the effective dose under control of the occupational worker and general public at large. All the necessary details are available on the AERB website *www.aerb.gov.in* and updated periodically.

22.4 Conclusion

The Atomic Energy Regulatory Board has issued regulatory documents from time to time. The licensee should keep a complete set of current regulatory documents and apprise the concerned staff regarding the availability of such documents and their contents. The R.S.O of the N.M facility plays an important role in ensuring radiation safety in the institution. However, the ultimate responsibility lies with the Employer of the Nuclear Medicine Facility.

References

1. The Atomic Energy Act, 1962 (33 of 1962).
2. The Atomic Energy Radiation Protection Rules, A.E. (R.P.R.)-2004.
3. www.aerb.gov.in
4. AERB Safety Code for Nuclear Medicine Facilities, AERB/RF-MED/SC-2 (Rev.2).
5. Atomic Energy (Safe Disposal of Radioactive Waste) Rules, GSR-125, 1987.
6. International Commission on Radiological Protection (ICRP-103), 2007.

Radioactive Waste Disposal and Safe Management of Disused Sealed Radioactive Sources

Abstract

In nuclear medicine, mostly short-lived radio-nuclides are used. Therefore, radioactive waste arising from nuclear medicine procedures is easier to manage. The basic objective of waste management is to ensure that radiation exposure to the public and environment does not exceed the prescribed dose limits. In India, disposal of radioactive waste in the public domain shall be undertaken in accordance with the Atomic Energy (Safe Disposal of Radioactive Waste) rules, 1987, promulgated under Atomic Energy Act, 1962. The protection of the Public and Environment is best achieved by (a) facility design (Isolation Measures-Zones, discharge control) (b) licensing (discharge procedures, limits) and (c) monitoring and inspection.

23.1 Fundamental Radioactive Waste Management (RWM) Principles

To achieve the objective of safe radioactive waste management, the International Atomic Energy Agency (IAEA) has published Safety Series No. 111-F on Principles of Radioactive Waste Management Safety Fundamentals. This series emphasizes (a) Protection of human health; (b) Protection of the environment; (c) Protection beyond national borders; (d) Burden on future generations and (e) National legal framework.

23.2 Classification of Wastes

Radioactive waste can be classified in a number of ways like high-, intermediate- and low-level wastes according to activity level and according to the radiation dose at the surface for solid wastes. However, the revised classification of IAEA is in accordance with waste disposal strategies. The classification of waste outlined in SS 111-G-1.1 is (a) exempt waste/clearance levels; (b) low- and intermediate-level waste with sub-classification as short-lived and long-lived; and (c) high-level waste. IAEA also specifies that special consideration should be applied to waste containing long-lived natural radionuclides, heat generation and liquid/gaseous waste. Nuclear medicine waste comes under classification (b).

The waste may be in a large variety of forms like solid, liquid, gaseous, combustible or non-combustible, aqueous or non-aqueous, precipitates and contaminated equipment. In biological research work, the waste may consist of excreta, tissue specimens, or animal carcasses.

In nuclear medicine, the radiopharmaceutical vials, the disposable syringes used for injection (solid) and spent ^{99}Mo (liquid) form the major content of radioactive waste. The tissue papers used for lining the work tables and decontamination during spillage, etc. are also considered as solid waste. The management of radioactive waste is best done in two stages, viz. (a) Collection and (b) Disposal.

23.3   Radioactive Waste Collection

Radioactive waste should be segregated from non-active waste while collecting in each laboratory or individual area to reduce the volume of active waste. Foot-operated waste bins with disposable polythene lining should be used for collecting solid radioactive waste and polythene carboys for liquid waste. When both short-lived radionuclides such as ^{99m}Tc and longer-lived ones such as ^{131}I are being used, then in such cases, separate waste collection bins and waste storage bags should be provided.

23.4   Radioactive Waste Disposal

Basic guidelines for disposal are:

1. Dilute and disperse for low-level radioactive waste.
2. Delay and decay for waste containing short-lived isotopes.
3. Concentrate and contain intermediate- and high-level radioactive waste.

We do not come across high-level waste in Nuclear Medicine; therefore, the first two methods are employed for radioactive waste disposal.

23.4.1 Solid Waste

Wastes containing short-lived radioactive isotopes, such as from isotope generators, may be stored for ten half-lives and disposed of as ordinary waste and released into municipal dump provided the activity of any article, (e.g. vial or syringe should not exceed 1.35 µCi (50 KBq) subject to concentration limit of 135 µCi/m^3 (5 MBq m^{-3})) when linked with other normal waste from the hospital. Contaminated clothing and linen are allowed to decay for ten half-lives before reuse.

23.4.2 Liquid Waste

Low activity short-lived radioactive waste need not be stored (<µCi level). This may be disposed of into the sanitary sewage system with adequate flushing with water following the disposal. High activity short-lived radioactive waste collected in carboys should be stored for sufficient decay, depending on the initial activity, in the waste and then can be disposed into the main sewerage through a specially laid drainage system.

Only soluble and dispersible liquid waste should be disposed of through sewerage. The disposal limits for sanitary sewerage systems for various radionuclides as per the Atomic Energy (Safe Disposal of Radioactive Waste) Rules, G.S.R 125, 1987 are indicated in Table 23.1. The gross quantity of radioactive material released into the sewerage system by any hospital should not exceed 37 GBq/year.

For hospitals handling large activity of ^{131}I for treatment of Ca-thyroid, '*Delay and Decay Tank*' may be required, where the outlet from the Isolation ward is connected to the Delay and Decay tank and after a decay period, the effluents are released into the main sewerage after monitoring the concentration level prescribed by the National Regulatory Authority.

23.4.3 Incineration of Wastes

Insoluble liquid waste such as that from liquid scintillation counting may be disposed of by incineration. Likewise the combustible radioactive waste, like tissue papers, animal carcasses, etc. When diluted with the day's hospital waste going to an ordinary incinerator, it must not have an activity concentration above 135 µCi/m^3 (5 MBqm^{-3}).

As incineration releases part of the radioactivity to the atmosphere, it should be done in controlled conditions in a segregated place to ensure that the radioactivity released does not affect the

Table 23.1 Disposal limits for sanitary sewerage system

Radionuclide	Maximum limit on total discharge per day MBq	The average monthly concentration of radioactivity in the discharge MBqm^{-3}
^{3}H	92.5	3700
^{14}C	18.5	740
^{24}Na	3.7	222
^{32}P	3.7	18.5
^{35}S	18.5	74
^{45}Ca	3.7	10.1
^{99}Mo + ^{99m}Tc	3.7	185
^{125}I	3.7	22.2
^{131}I	3.7	22.2

immediate environment, and the ashes collected for separate disposal as solid waste. Since these are difficult to achieve in practice, incineration is not generally recommended.

23.5   Record Keeping

A logbook should be maintained recording the identity and quantity of each radioisotope disposed of, description of the waste, time of disposal, name of the person who has supervised the disposal operations and the date of radiation surveillance by the Radiological Safety Officer (RSO). Records, in respect of the disposal operations, shall be maintained and made available to the inspection team as and when carried out by the Regulatory Authority (RA). The facility needs to submit the Annual Periodic Safety Status Report to (RA).

23.6   Management of Cadavers Containing Radionuclides

Handling of cadavers containing radionuclides must be done under the supervision of the RSO. Cremation or burial of such cadavers must be undertaken in accordance with the procedures approved by the regulatory authority.

23.7   Disposal of Disused Sealed Radioactive Sources (DSRS)

Any radioactive material becomes a radioactive waste after its useful lifetime is over. Once these sources become unusable, they are known as disused radioactive sources. These disused sources need to be safely disposed of as mandated by the Atomic Energy (Safe Disposal of Radioactive Wastes) Rules, 1987, G.S.R 125, which are promulgated under the Atomic Energy Act, 1962. The radioactive wastes generated in nuclear medicine facilities are of two types, viz. sealed and unsealed (open) sources. Usually, during practice, it has been observed that generation of waste due to unsealed sources is minimum; therefore, all nuclear medicine facilities are equipped to dispose and manage the waste generated through the use of such unsealed sources as mentioned earlier. However, the major concern is for the radioactive waste generation due to the sealed sources.

The sealed sources like ^{22}Na and ^{68}Ge used in nuclear medicine facilities are generally long-lived sources and mainly used for calibration of imaging equipment used in nuclear medicine facilities. Such sealed sources are required to be returned to the original supplier/country of origin once their useful life is over. Such sources which are no longer in use and not intended to be used for the purpose for which it is acquired are called Disused Sealed Radioactive Sources (DSRS).

The recent experience has shown that these DSRS due to various reasons, such as the non-existence of original foreign suppliers, the exorbitant cost involved in the export of such sources back to the original supplier and non-availability of original documentation could not be returned back to the country of origin.

Regulatory Authority has incorporated various administrative measures and regulatory hold points in the consenting processes applicable for various radiation practices based on its hazard potential. The few of such administrative and regulatory processes incorporated are listed below.

- During the procurement of a radioactive source, the radiation facility is required to submit the undertaking from the supplier of the source about their confirmation to accept the source back once it becomes disused.
- Decommissioning plan incorporating the financial arrangements is obtained from the institute as part of licensing requirements. The radiation facility is required to submit a copy of the agreement (duly authenticated by the employer of the user institute and Indian Supplier) in which the Indian supplier has agreed to assist the end user taking into account the financial arrangements made by the end user.
- Agreement between the institute and source supplier about arrangements for the return of source, including the financial provision, is required to be submitted at the time of procurement.

- Incorporate the requirement of financial provisions for decommissioning in case of the unlikely event of bankruptcy or other constraints to meet the cost of safeguarding radioactive sources.

In addition, the Government has also established the mechanism to ensure the safe management of disused sources, wherein the radiation facilities are genuinely finding it difficult to return the source to its original supplier. However, the primary option for the safe management of such DSRS is to ensure its return to the original supplier.

In the safe management of such DSRS, the role of regulatory authority is to facilitate the radiation facility by issuing the shipment approval certificate from a radiation safety viewpoint for the transport of DSRS to its destination, i.e. either to the original supplier or to waste disposal agency in the country for its safe management.

Regulatory Authority (RA) always ensures the safety and security of such DSRS till they are in the possession of radiation facilities through routine regulatory inspections and review of the periodic safety status reports. RA also initiates enforcement actions against radiation facilities which are delaying the return of DSRS to its original supplier.

In nuclear medicine practice, an undertaking from the supplier of the source about their confirmation to accept the source back once it becomes disused is obtained at the time of issuance of procurement permission.

Model Questions for Radiological Safety Certification Examination in Nuclear Medicine

24

Abstract

This chapter can be helpful to students appearing for the Radiological Safety Officer (RSO) examination. It contains model 250 multiple choice questions, 100 true-and-false questions, 60 fill-in-the-blank questions, and 40 match-the-following questions. Similar questions may be asked in the RSO written examination whereas for viva-voce questions, many questions can arise from different chapters of this book.

24.1 Questions for RSO Exams

24.1.1 Section A

24.1.1.1 Multiple Choice Questions

1. Chronologically, which among the following is the slowest reaction?
   a. Physical
   b. Biochemical
   c. Biological

2. The following will help isotopes to stabilize
   a. As Z increases, N/Z will decrease to reduce the overall size of the nucleus
   b. As Z increases, N/Z will increase to increase the overall binding energy of the nucleus to compensate for p-p repulsion
   c. As Z increases, N/Z will strictly remain 1

3. Which of the following is **NOT TRUE** for a radioactive isotope?
   a. Theoretically, the activity of an isotope never reaches zero
   b. Among α, β and γ decay, change in Z will happen only with α and β emissions
   c. Every isotope when it undergoes decay, the daughter nuclide is invariably a stable isotope

4. Which of the following applications uses isotope origin radiation?
   a. X-ray fluorescence
   b. Proton induced X-ray fluorescence
   c. P.E.T. scanning

5. Which of the following is not correct?
   a. Two half-lives will lead to the complete decay of isotopes
   b. The energy of γ radiation does not change after 2 half-lives
   c. The energy distribution of β particles is not mono-chromatic, unlike γ radiation

6. S.I. unit of KERMA is
   a. Rad
   b. Sievert
   c. Gray
   d. Roentgen

7. Tissue weighting factor for lung as per ICRP-103 is
   a. 0.20
   b. 0.12
   c. 0.05
   d. 0.01

8. Radiation weighting factor for gamma rays is
   a. 20
   b. 10
   c. 1
   d. 5

9. If the physical half-life of a radionuclide is 2 days and the biological half-life is 4 days, the effective half-life of the radionuclide is
   a. 6 days
   b. 13 days
   c. 1.33 days

10. In the SI unit, 10 mCi of ^{131}I would correspond to
    a. 370 Bq
    b. 370 kBq
    c. 370 MBq

11. S.I. unit of effective dose is
    a. Becquerel (Bq)
    b. Gray (Gy)
    c. Sievert (Sv)

12. Radiation weighting factors (w_R) and Tissue weighting factors (w_T) are used to evaluate
    a. Absorbed dose (D)
    b. Equivalent dose (HT)
    c. Effective dose (E)

13. S.I. Unit of Derived Air Concentration (D.A.C.) is
    a. bq/cm^2
    b. bq/cm^3
    c. bq/m^3
    d. None of the above

14. The SI unit value of Annual Limit on Intake (A.L.I.), as defined in the ICRP publication, is
    a. Bq
    b. Bq/m^3
    c. µCi
    d. µCi/m^3

15. The absorbed dose for given energy in an organ is
    a. Indirectly proportional to the mass of the organ
    b. Indirectly proportional to activity present in the organ
    c. Indirectly proportional to the concentration of radioactivity in the organ

16. g—rad/µCi-h is the unit of
    a. Gamma-ray constant

b. Absorbed dose constant
c. Equilibrium dose constant
d. Attenuation constant

17. The effective half-life of ^{131}I in the thyroid tissue as compared to ^{125}I is
    a. Longer
    b. Shorter
    c. Same

18. Exposure rate constant (gamma-ray constant or K-factor) for ^{131}I is
    a. 22.0 R/h/mCi at 1 cm
    b. 2.2 R/h/mCi at 1 cm
    c. 0.22 R/h/mCi at 1 cm
    d. 8.0 R/h/mCi at 1 cm

19. S.I. unit of exposure is
    a. becquerel (Bq)
    b. Gray (Gy)
    c. Coulombs/kg (C/kg)

20. SI unit of Equivalent Dose (H_T) is
    a. Rad
    b. Gray (Gy)
    c. Sievert (Sv)

21. Exposure rate constant for ^{99m}Tc is
    a. 0.8 mR/h/mCi at 1 cm
    b. 8.0 mR/h/mCi at 1 cm
    c. 0.8 R/h/mCi at 1 cm
    d. 8.0 R/h/mCi at 1 cm

22. Organ absorbed dose is
    a. Directly proportional to the effective half-life of the radionuclide
    b. Inversely proportional to the effective half-life of the radionuclide
    c. None of the above

23. A.L.I., as defined in the ICRP publication, stands for
    a. Allowed limit of intake of radionuclide
    b. Annual limit of intake of radionuclide
    c. Average limit of intake of radionuclide

24. The Government notified the radiation protection rule of India in the year
    a. 1962
    b. 1971
    c. 1987
    d. 2004

25. If the activity of a radionuclide in the old unit is 15 mCi, in the S.I. unit, it would disintegrate at a rate of
    a. 5.55×10^6 dps
    b. 5.55×10^7 dps

c. 5.55×10^8 dps

d. 5.55×10^9 dps

26. In case of spillage of very high radionuclide activity, the steps taken are; (i) information to R.S.O., (ii) decontaminate the area, (iii) inform others about incidence and (iv) control the spread of contamination. The most appropriate order of steps taken is
   a. i, ii, iii, iv
   b. ii, iii, iv, i
   c. iii, iv, i, ii
   d. iv, i, ii, iii

27. Derived Air Concentration (DAC) is derived from ALI using the equation
   a. ALI/2400 Bq/m^2
   b. ALI/2400 kBq/cm^3
   c. ALI/2400 MBq/cm^3
   d. ALI/2400 Bq/m^3

28. Radiation Weighting Factor (W_R) is used to derive
   a. Effective dose from the absorbed dose and its S.I. unit is Sv
   b. Effective dose from equivalent dose and its S.I. unit is Gy
   c. Equivalent dose from the absorbed dose and its S.I. unit is Sv
   d. Equivalent dose from the absorbed dose and its S.I. unit is Gy

29. Effective dose of 20 mSv is equal to
   a. 2000 mRem
   b. 200 mRem
   c. 20 mRem
   d. None of the above

30. Sensitivity to radiation carcinogenesis is maximum for
   a. Senior citizens >60 years
   b. Children <20 years
   c. Adult females
   d. Adult males in the age group of 20–60 years

31. Probability of excess cancers/100 mSv in radiation workers estimated based on human data by ICRP 103 (2007) is
   a. 125/10^6
   b. 60/10^4
   c. 410/10^6
   d. 4.1/10^3

32. The estimated detriment for the whole population as per ICRP 103 is
   a. 4.2% Sv^{-1}
   b. 5.7% Sv^{-1}
   c. 7.3% Sv^{-1}
   d. 5.6% Sv^{-1}

33. The estimated detriment for the working (occupational) population as per ICRP 103 (2007) is
   a. 4.2% Sv^{-1}
   b. 5.7% Sv^{-1}
   c. 7.3% Sv^{-1}
   d. 5.6% Sv^{-1}

34. Basic principle of radiation protection is to
   a. Completely avoid deterministic effects and minimize stochastic effects well below detectable levels with principles of industry standards
   b. Only avoid deterministic effects
   c. Set dose limits based on actual epidemiological observations of occupational workers

35. Doubling dose for genetic effects is
   a. The dose needed to achieve double the mutation rate compared to the natural mutation rate
   b. The dose needed to double the total mutations
   c. The dose needed to observe double the frequency of recessive disorders in a human population

36. Radiation weighting factor definition is based on
   a. R.B.E. for cell killing
   b. R.B.E. for chromosomal aberrations at a high dose
   c. A conservative estimation of R.B.E. for stochastic effects or hazards assigned to a particular type of radiation

37. Tissue weighting factor is estimated based on
   a. The fractional contribution of a tissue based on relative detriment towards stochastic risk
   b. Total weight of tissue
   c. Threshold of the deterministic effect associated with the tissue

38. Combination of additive and multiplicative models has been used in ICRP 103 since
    a. Most of the life span study is now complete, and most of the cancer risks with short latent periods are already quantified
    b. Life span study is completed
    c. The additive model was found to be simpler and more efficient

39. Estimated DDREF is based on
    a. Completely human data from A.B.S.
    b. Experimental data from mice model with limited human evidence
    c. Cell culture experiments

40. Principle of dose limits for occupational workers is based on
    a. 1 in 10^6 probabilities of cancer in a lifetime
    b. 1 in 10^3 excess cancer per year weighed with detriment
    c. 1 in 10^4 excess cancer in a lifetime with detriment

41. Which of the following physical factors has linear relation in yielding personnel dose
    a. Time
    b. Distance
    c. Shielding

42. Intensity of radiation decreases:
    a. With the inverse of the square of the distance from a point source
    b. Inversely with distance
    c. Linearly with distance

43. Lower dose limits are defined to the public compared to occupational workers due to
    a. They are not the direct beneficiaries and are not monitored
    b. Sensitive population such as children and pregnant women are present in public
    c. Both a and b

44. Dose limits for individual organs such as limbs, skin and eye lenses are based on
    a. Absorbed dose
    b. Equivalent dose
    c. Effective dose

45. Tissue weighting factor for skin as per ICRP-103 is
    a. 0.20
    b. 0.12

    c. 0.05
    d. 0.01

46. Radiation weighting factor for alpha rays is
    a. 20
    b. 10
    c. 1
    d. 5

47. Among the three types of radiations given below, the Relative Biological Effectiveness (R.B.E.) is higher for
    a. Beta particles
    b. Gamma rays
    c. Alpha particles

48. The most hazardous radioisotope among the following in terms of internal hazard is
    a. ^{99}Tc
    b. ^{131}I
    c. ^{137}Cs
    d. ^{14}C

49. Which of the following radioisotope tends to give a more localized dose?
    a. ^{99}Tc
    b. ^{131}I
    c. ^{137}Cs
    d. ^{14}C

50. A sudden increase in background counts in gamma cameras indicates
    a. Possible damage to the crystal
    b. Possible contamination of crystal
    c. Deteriorating NaI(Tl) scintillator
    d. The shift in amplifier gain

51. The function of the fume hood in handling a large dose of ^{131}I is
    a. To contain surface contamination due to radioactivity
    b. To contain air contamination
    c. To suck out the vapors activity and release into the atmosphere

52. The maximum permissible exposure level in a controlled area, assuming 40 working hours/week, is
    a. 2.5 mR/h
    b. 1.0 mR/h
    c. 0.25 mR/h

53. Half-Value Thickness ^{131}I gamma rays in lead is
    a. 3.0 mm
    b. 3.0 cm
    c. 2.0 mm

54. The half-value thickness of lead for ^{99m}Tc gamma energy of 140 keV is
   a. 3.0 mm
   b. 0.3 mm
   c. 30 mm
   d. 0.03 mm

55. The tenth-value thickness of lead for ^{131}I gamma energy of 364 keV is
   a. 1.0 cm
   b. 0.3 cm
   c. 3.0 cm

56. Gamma ray constant for ^{131}I is 2.2 R-cm^2/mCi-hour, the exposure rate from 1 mCi of ^{131}I at 2 cm will be
   a. 4.84 R/h
   b. 1.1 R/h
   c. 0.55 R/h
   d. 0.3 R/h

57. Method(s) of control of radiation hazards from unsealed radionuclides is/are
   a. Containment of radionuclide
   b. Spend less time near the radionuclide
   c. Use of radiation shield
   d. All that are mentioned above

58. If H.V.T. of gamma rays emitted by a radionuclide is 0.3 mm in the lead, the thickness of the lead material required to reduce the exposure rate from 10 mR/h to 1 mR/h is approximately
   a. 0.3 mm
   b. 1.0 mm
   c. 10 mm
   d. None of the above

59. H.V.L. (half-value layer) is
   a. The thickness of an attenuator that reduces a monoenergetic beam's intensity (number of photons) by 90%
   b. The thickness of a monoenergetic beam attenuator reduces the intensity (number of photons) by 50%
   c. 0.693/H.V.L.
   d. None of the above

60. T.V.L. (tenth-value layer) is
   a. The thickness of an attenuator that reduces a monoenergetic beam's intensity (number of photons) by 90%
   b. The thickness of a monoenergetic beam attenuator reduces the intensity (number of photons) by 50%

c. 0.693/T.V.L
d. None of the above

61. Linear attenuation coefficient (μ)
   a. The thickness of an attenuator reduces a monoenergetic beam's intensity (number of photons) by 90%
   b. The thickness of a monoenergetic beam attenuator reduces the intensity (number of photons) by 50%
   c. 0.693/H.V.L
   d. None of the above

62. Exposure situations arising from natural background radiation can be considered as
   a. Planned exposure
   b. Emergency exposure
   c. Existing exposure

63. Spillage of radio-pharmaceutical and exposure arising thereof can be considered as
   a. Occupational exposure
   b. Emergency exposure
   c. Public exposure

64. An occupational worker or a radiation professional undergoing diagnostic treatment. The exposure during such a procedure is
   a. Part of their occupational exposure and should wear a personnel monitor during the procedure
   b. Personnel/medical exposure and cannot be considered as part of occupational exposure
   c. Emergency exposure

65. Conditions that apply to dose to patient comforters are
   a. No dose limits and no constraints
   d. Dose constraint of 20 mSv per episode
   e. No dose limit, but dose constraint of 5 mSv per episode, and 1 mSv per episode for children

66. Most of the radionuclides used in nuclear medicine tend to
   a. Get distributed uniformly in the body and continue to deliver dose even after several days of the procedure
   b. Get accumulated in target organs, but most of the activity gets eliminated through the biological route
   c. Since their biochemistry is different, they do not get absorbed in any tissue, and they remain in humeral media

67. Effective dose is measured using the following principle
    a. External dose multiplied by W_T.
    b. Internal dose for the tissue T multiplied with W_T.
    c. Summation of effective doses estimated for external and internal arising from all routes

68. Separate dose limits are defined to extremities as they are
    a. Highly sensitive to stochastic effects
    b. More likely to come in contact with radiation
    c. Having low W_T, allow limits to surpass the threshold of deterministic effects

69. Separate dose limits are defined for pregnant women. The criteria are
    a. 1 mSv per year for the pregnant women
    b. 1 mSv per year for the foetus
    c. 1 mSv to the foetus after the declaration of pregnancy for the rest of the pregnancy duration

70. The use of L-bench while handling radioactivity is
    a. To contain the surface contamination due to radioactivity
    b. To contain the air contamination
    c. To provide adequate shielding to the occupational worker
    d. To provide adequate shielding to the extremities

71. Personnel monitoring of nuclear medicine professionals is mandatory since
    a. They are high-risk and highly exposed professionals
    b. It is mandatory only due to protocol
    c. The chance of excess exposure does exist as open isotopes are handled

72. Critical target structures in living cells for the action of radiations
    a. Glucose
    b. Vitamins
    c. Lipoproteins
    d. D.N.A. molecules

73. The most important type of radiation (ionizing) damage in D.N.A. is
    a. Thymine dimers
    b. Double-strand break

    c. Single-strand break
    d. Interstrand cross-linking

74. Which is NOT a free radical?
    a. $H°$
    b. H_2O_2
    c. $OH°$
    d. $HO_2°$

75. Changes in the genetic information of a cell can be caused by
    a. A loss of chromosome
    b. Substitution of a base
    c. Deletion of a chromosome
    d. All of the above changes

76. Threshold for deterministic effects
    a. Depends upon the effect
    b. Depends upon the dose rate
    c. Depends upon the sensitivity of the target cells
    d. All of the above

77. Which of the following is not a stochastic effect?
    a. Leukaemia
    b. Cataract
    c. Thyroid cancers
    d. Genetic disorders

78. Among the children exposed to Chernobyl, the cancer type occurred prominently during the first decade after irradiation.
    a. Breast cancer
    b. Thyroid
    c. Leukaemias
    d. Multiple myeloma

79. Two of the most radioresistant type of cells in the human body
    a. Nerve cells and muscle cells
    b. Lymphocytes and granulocytes
    c. Spermatogonia and spermatocytes
    d. Liver cells and fibroblasts

80. The tissue weighting factors are based on
    a. Sensitivity of the tissue toward stochastic effects of radiation
    b. The total mass of the tissue
    c. The density of the tissue
    d. Total dose absorbed by the tissue

81. The risk estimation of radiation exposure is estimated using
    a. Epidemiological studies on the population exposed to high doses

b. Follow-up of radiological accidents and early radiation workers
c. Animal and model experiments
d. All of the above

82. The earliest radiation effect which appears after exposure to large acute doses is
a. Skin burns
b. Cataract
c. Temporary sterility
d. Radiation sickness

83. While planning a nuclear medicine laboratory, a public area is located
a. Near the control area
b. In the middle of the control and supervised area
c. At the entrance of the laboratory

84. ^{177}Lu is very effective in the treatment of
a. Liver cirrhosis
b. Thyroid Carcinoma
c. Polycythemia vera
d. Neuroendocrine tumour

85. The activity limit above which the patients are to be hospitalized is decided by
a. Licensee of the institution
b. The ethics committee of the institution
c. Local statutory body
d. Regulatory authority

86. The minimum permissible distance between the isolation ward and nearby residential premises is
a. 3 ft
b. 3 m
c. 3.5 m
d. 4 ft

87. Isolation ward is required for the use of
a. ^{177}Lu
b. ^{131}I
c. $^{177}Lu + {}^{131}I$
d. None of the above

88. The isolation ward with an adequate thickness of walls is required for reducing the radiation level for
a. The general public
b. The radiation worker
c. a and b
d. None of the above

89. Personnel monitoring is not required for workers where the plausibility of receiving an effective dose is
a. Less than 3/10th of the maximum yearly dose
b. Less than 1/10th of the maximum yearly limit
c. <1 mSv in a calendar year

90. The present maximum Annual Effective Dose Limit for occupational workers in India is
a. 20 mSv
b. 30 mSv
c. 50 mSv

91. The lowest whole-body dose that can be measured by chromosomal aberration analysis in human lymphocyte culture is
a. 100 mSv
b. 10 mSv
c. 50 mSv
d. 500 mSv

92. As per ICRP-103 recommendations, the equivalent dose to the foetus is
a. 20 mSv
b. 30 mSv
c. 2 mSv
d. 1 mSv

93. For the protection of the patient in nuclear medicine practice, the basic principle(s) enunciated by ICRP and adopted by B.S.S. is/are
a. Justification of practice with radiopharmaceutical
b. Optimization of the practice so that patient dose is as low as reasonably achievable
c. All the above
d. None of the above

94. The personnel monitoring badge worn at chest level records
a. The dose received by the skin
b. The dose received by the whole body
c. The dose received by the chest
d. None of the above

95. Test of radioactive contamination in a source storage room *(where background*

radiation level is high) is effectively done using

a. Direct method
b. Indirect method
c. Both (a) and (b)

96. The TLD badge worn at chest level monitors the
  a. X, β and γ
  b. X and β
  c. β and γ
  d. X, β, γ and neutron

97. The use of personnel monitoring devices while working with radioactive sources is to
  a. Estimate radiation absorbed dose rate
  b. Estimate radiation absorbed dose
  c. Estimate radiation effective dose rate
  d. Estimate radiation effective dose

98. It is called for an investigation of Overexposure if a person is receiving exposure >10 mSv during
  a. One month
  b. Monitoring period
  c. One year

99. Occupational annual dose limits for radiation workers are based on
  a. Zero risks to worker
  b. Acceptable risk to worker
  c. Just below the maximum risk to worker

100. The maximum permissible exposure level in a supervised area, assuming 40 working hours/week, is
  a. 2.5 mR/h
  b. 1.0 mR/h
  c. 0.25 mR/h
  d. 0.1 mR/h

101. The effective dose received by patient in a diagnostic procedure using ^{99m}Tc is in the range
  a. 2–8 mSv
  b. 20–80 mSv
  c. 2–8 Sv

102. Gamma ray constant for ^{99m}Tc is 0.8 rad-cm^2/mCi-h, the exposure rate from 1 mCi of ^{99m}Tc at 4 cm will be
  a. 1.84 rad/h
  b. 1.01 rad/h
  c. 0.05 rad/h
  d. 0.3 rad/h

103. Lead is used as shielding material for gamma rays because
  a. It has the high atomic number
  b. It has a high density
  c. It does not produce bremsstrahlung
  d. Both a and b

104. The wrist badge is worn below the hand gloves
  a. To avoid contamination of badge
  b. To reduce the dose received by the badge
  c. To reduce beta doses
  d. To reduce dose to the hands

105. For treatment of hyperthyroidism, the radioisotope of iodine ideally used is
  a. ^{125}I
  b. ^{123}I
  c. ^{131}I
  d. ^{127}I

106. In control of external radiation hazards, the factor(s) used is/are
  a. Time
  b. Distance
  c. Shielding
  d. All the above

107. A transparent lead glass having a lead equivalence of 3.0 mm will reduce the intensity of ^{99m}Tc gamma radiation by a factor of about
  a. 1
  b. 10
  c. 100
  d. 1000

108. The permissible level of contamination in general corridors of a nuclear medicine laboratory is
  a. 10^{-3} μCi/cm^2
  b. 10^{-1} μCi/cm^2
  c. 10^{-5} μCi/cm^2
  d. 10^{-2} μCi/cm^2

109. The most appropriate reason for the sudden decrease in detector response is due to
  a. Possible damage to the crystal
  b. Possible contamination of crystal
  c. Deteriorating NaI (Tl) scintillator
  d. The shift of the gain amplifier

110. The range of beta particles in tissue is of the order of
  a. Millimetres
  b. Centimetres
  c. Metres

111. Therapeutic radionuclides used in radiopharmaceuticals are generally with
    a. High energy gamma rays
    b. Positron emitters
    c. Beta emitter with no or low gamma energy
112. The $^{99m}TcO4$-the solution is placed in a thick lead pot and counted for quick estimation of the ^{99}Mo breakthrough. The role of the lead pot is
    a. To keep the vial steady without toppling
    b. To cut off the gamma rays from ^{99m}Tc
    c. To cut off the gamma rays from ^{99}Mo
113. In the supervised area, the radiation level
    a. Should not exceed 10 μSv/h
    b. Should not exceed 1 μSv/h
    c. May exceed 10 μSv/h
114. A nursing mother undergoing diagnostic studies with ^{99m}Tc should be advised not to breastfeed the child for a period of at least
    a. 4 hours
    b. 4 weeks
    c. 24 hours
115. If H.V.T. in lead for gamma rays emitted by a radionuclide is 3 mm. How much thickness is required to reduce? The exposure rate from 10 mR/h to 1 mR/h is approximately
    a. 1 cm
    b. 1.5 cm
    c. 5 cm
    d. 10 cm
116. The exposure rate constant for 18F is
    a. 0.56 mR/h/mCi at 1 cm
    b. 5.6 R/h/mCi at 1 cm
    c. 0.56 R/h/mCi at 1 cm
    d. 56 R/h/mCi at 1 cm
117. A transparent lead glass having a lead equivalent of 10 mm will reduce the intensity of annihilated photons from 18F by a factor of about
    a. 1
    b. 2
    c. 3
    d. 4

118. An absorber placed in front of a gamma source would reduce
    a. The intensity of radiation
    b. The energy of radiation
    c. Both (a) and (b)
    d. None of the above
119. The permissible surface contamination limit for beta-emitting radionuclides on work surfaces is
    a. 37 Bq/cm^2
    b. 3.7 Bq/cm^2
    c. 0.37 Bq/cm^2
120. During the operation of a medical cyclotron, the radiations present in the cyclotron vault are
    a. Only protons
    b. Only neutrons
    c. Protons, neutrons and prompt gammas
    d. Only (a) and (b)
121. A nursing mother undergoing therapy with ^{131}I should be advised not to breastfeed the child for a period of at least
    a. 4 hours
    b. 24 hours
    c. 6 weeks
    d. Complete cessation
122. In a case of sudden malfunctioning of a dose calibrator, equipment used to calculate the amount of activity of the radioactive source is
    a. Gamma camera
    b. Contamination monitor
    c. Direct reading dosimeter
    d. Radiation survey meter
123. In ^{99m}Tc labelled radiopharmaceuticals (R.P.) free TcO4$^-$ is
    a. Chemical impurity
    b. Radionuclide impurity
    c. Radiochemical impurity
    d. Pyrogens
124. Sterility of radiopharmaceuticals means being free from
    a. Pyrogen contamination
    b. Radiochemical contamination
    c. Radionuclide contamination
    d. Chemical contamination

125. The limit for ^{99}Mo breakthrough in ^{99m}Tc is
   a. 0.01%
   b. 0.1%
   c. 1%

126. Basic guideline(s) for disposal of radioactive waste in Nuclear Medicine are (i) dilute and discharge; (ii) delay, decay, discharge; and (iii) concentrate and contain
   a. Only (i) and (ii)
   b. Only (ii) and (iii)
   c. Only (i) and (iii)
   d. All of the above

127. The max limit on the total activity of ^{131}I that may be discharged per day by an institution into a sanitary sewerage system is
   a. 0.37 MBq
   b. 3.7 MBq
   c. 37 MBq
   d. 370 MBq

128. How frequently should a C.O.R. check be carried out?
   a. At least once a week
   b. Before each patient study
   c. Every 6 months
   d. At least once a month

129. Diagnostic radionuclides used in radiopharmaceuticals are
   a. Beta emitters with no or low gamma rays
   b. High energy gamma rays
   c. Positron emitters
   d. b and c

130. In a uniformity correction flood for a SPECT camera, what minimum number of counts should be obtained?
   a. 100 thousand counts
   b. 20 million counts
   c. 30 million counts
   d. 5 million counts

131. Which of the following collimators will magnify an image?
   a. Pinhole collimator
   b. Parallel hole collimator
   c. Converging
   d. Slant hole collimator

132. The most appropriate instrument to measure the exposure rate from a Ca-thyroid patient treated with radioiodine is
   a. An ionization chamber type survey meter (range 0 to 5 R)

b. A portable contamination monitor (0 to 20 mR)
   c. A G.M. type survey meter (range 0 to 20 mR)

133. In the P.E.T. scanner, the sensitivity response between the two detectors varies between
   a. 10–50%
   b. 50–100%
   c. 90–100%
   d. No variation

134. The detector used in a gamma camera is
   a. NaI (Tl) activated scintillator
   b. BGO scintillator
   c. Plastic scintillator
   d. HPGe

135. ^{137}Cs source is used for testing a gamma-ray spectrometer (G.R.S.) because
   a. It emits 2 beta energies
   b. It emits gamma rays of 662 keV energy
   c. It has a long half-life
   d. All of the above

136. The most appropriate instrument used for routine survey of the N.M. facility is
   a. Ionization type detector
   b. Proportional counter type detector
   c. G.M. counter type detector

137. Liquid scintillation detectors are used for measuring
   a. Gamma radiation
   b. Low energy beta particles
   c. None of the above

138. The most appropriate instrument to measure the surface dose rate from a patient with thyroid cancer treated with ^{131}I is
   a. An ionization chamber type of survey meter with a range of 0–5 R/h
   b. A portable contamination monitor with a range of 0–500 cps
   c. A G.M. type survey meter with a range of 0–20 mR/h
   d. A pocket dosimeter with a range of 0–9999 mR

139. In gamma-ray spectrometer (G.R.S.), the single-channel analyser is equipped with
   a. Coincidence circuit
   b. Anti-coincidence circuit
   c. None of the above

140. Which of the following must be done at least once every six months?
   a. COR
   b. Sealed source leak test
   c. Linearity test
   d. Hot lab survey

141. The fundamental imaging concept of the C.T. scanner is the following:
   a. The C.T. numbers
   b. Relative X-ray attenuation
   c. Determination of grayscale
   d. Voxel atomic number

142. Which detector is not used as detectors in C.T. scanners?
   a. Air ionization chambers
   b. B.G.O.
   c. $CdWO_4$
   d. NaI

143. Frequently used C.T. detectors:
   a. Ionization chambers
   b. CdWO4
   c. Caesium Iodide thallium activated CsI (Tl) crystals
   d. Gadolinium Oxysulfide (Gd2O2S)

144. Helical CT has the following advantages over axial CT:
   a. Spatial resolution
   b. Data acquisition rate
   c. Image quality
   d. Image reconstruction time

145. Radiation dose to the patient increases with:
   a. Increases with Patient size
   b. Increase in tube voltage
   c. Decrease in tube current
   d. Decrease in scan time

146. For decontamination of glassware, the commonly used agent is
   a. Organic solvent
   b. Chromic acid
   c. Sulphuric acid

147. To test the intrinsic uniformity of a camera used for planar imaging, a dosage of 99mTc is being prepared. What is the required level of activity for the image?
   a. 1 mCi
   b. 500 µCi

   c. 5 mCi
   d. 10 µCi

148. The radiation dose to an adult euthyroid per MBq of ^{131}I administered is about
   a. 500 mGy
   b. 50 mGy
   c. 5 mGy

149. When ^{131}I is administered for the treatment of thyroid diseases, the dose to the thyroid is
   a. Mostly due to beta radiation
   b. Mostly due to gamma radiation
   c. Equally due to both beta and gamma radiations

150. If a radionuclide used in N.M. examination emits non-penetrating radiation from the same organ, the absorbed fraction (ϕ) for the radiation in the target organ would be
   a. 10
   b. 1
   c. 0

151. The unit of Specific Absorbed fraction (Φ), used in the MIRD method of internal dose calculation, is
   a. MeV
   b. Rad
   c. $gram^{-1}$
   d. No unit

152. The benefits of employing a 57Co sheet source over a fluid-filled flood source with 99mTc is:
   a. It does not need to be prepared every day
   b. It is cost-effective
   c. It has a shorter half-life and
   d. None of the above

153. $LD_{50}(60)$ dose is
   a. 3–5 Gy
   b. 3–5 mGy
   c. 3–5 mSv
   d. 1–2 Gy

154. Important characteristics of stochastic effects of radiation are that it
   a. Occurs in high frequency in human beings mostly due to many other agents
   b. May occur after several years
   c. May occur in very few exposed
   d. All the above

155. The earliest radiation effect that appears after exposures to acute localized doses is
    a. Skin erythema
    b. Cataract
    c. Temporary sterility
    d. Radiation sickness

156. Lethal damage induced by acute whole-body irradiation in the range of 3–6 Gy results from
    a. Bone marrow damage
    b. Intestinal damage
    c. C.N.S. injury
    d. Skin damage

157. The effective decay constant(λ_e) is related to the biological decay Constant (λ_b) and the physical decay constant(λ_p) by the relation
    a. $\lambda_e = \lambda_b + \lambda_p$
    b. $\lambda_e = (\lambda_b \times \lambda_p)/(\lambda_b + \lambda_p)$
    c. $\lambda_e = (\lambda_b \times \lambda_p)/(\lambda_b - \lambda_p)$

158. For calculation of radiation dose due to internally deposited radionuclide, the half-life to be considered is
    a. Biological half-life
    b. Physical half-life
    c. Effective half-life

159. How does the image resolution change when the distance between a patient and a parallel hole collimator is reduced?
    a. It improves
    b. It decrease
    c. It remains the same
    d. None of the above

160. Stochastic effects of radiation are
    a. Caused by non-lethal type cellular effects
    b. Lethal type cellular effects
    c. Sure to occur in the exposed person
    d. None of the above

161. The example of stochastic effect is
    a. Skin erythema
    b. Cataract
    c. Leukaemia
    d. Sterility

162. In internal dosimetry, the radiation absorbed dose to an organ is
    a. Inversely proportional to the mass of the thyroid
    b. Directly proportional to the effective energy of beta particles
    c. Directly proportional to the effective half-life of the radionuclide
    d. All of the above

163. If the physical half-life ($T_{1/2}$) of a radionuclide is 8 days, the physical decay constant (λ_p) of a radioisotope is
    a. $8.66\,\text{days}^{-1}$
    b. $0.866\,\text{days}^{-1}$
    c. $0.086\,\text{days}^{-1}$
    d. $8\,\text{days}^{-1}$

164. The radionuclide which has the shortest physical half-life is
    a. ^{11}C
    b. ^{13}N
    c. ^{15}O
    d. ^{18}F

165. When is a photon absorbed, and which portion of the gamma camera emits light?
    a. PMT
    b. PHA
    c. Scintillation crystal
    d. Collimator

166. Which collimator will be most useful for obtaining high-quality images of a small organ, gland or joint?
    a. LEAP
    b. LEHR
    c. Diverging
    d. Pinhole

167. In MIRD, to calculate the dose, the Reciprocity Principle is defined/used
    a. To calculate cumulative activity from activity in the source organ
    b. To calculate the absorbed fraction in a target organ from a specific absorbed fraction of other target organs
    c. To calculate equilibrium absorbed dose constant
    d. None of the above

168. When a gamma photon interacts with a radiation detector, the time in which it is incapable to detect another gamma is called:
    a. Temporal resolution
    b. Spatial resolution
    c. Dead time
    d. Efficiency

169. The unit of 'Time-integrated Activity (cumulated activity)' used in the calculation of internal dose by the MIRD method is
    a. Kg-Gy/Bq-Sec
    b. Bq-Sec
    c. Bq

170. The Z component of the signal of P.M.T. is processed by
    a. Pulse height analyser
    b. CRT
    c. PMT
    d. All of the above

171. The energy spectrum of a patient injected with 99mTc shows a peak around 90–140 keV. This part of the spectrum is called:
    a. Photo peak
    b. Energy peak of 99mTc
    c. Iodine peak
    d. Compton scatter

172. The time-integrated activity in an organ depends on
    a. Administered activity
    b. Fractional uptake in organ
    c. Effective half-life
    d. a and c

173. The S-factor depends on
    a. Number of emissions and its energy
    b. Organ activity
    c. Distance to the target organ
    d. All the above

174. The effective dose for a C.T. examination does not consider:
    a. Mean organ doses
    b. Radiation L.E.T.
    c. Patient age
    d. Organ sensitivity
    e. All exposed organs

175. A transport index is a number that expresses the distance between two points
    a. The maximum radiation level in mrem/h at 1 metre from the package's centre
    b. The average radiation level in mrem/h at 1 metre from the source's exterior surface
    c. The maximum radiation level in mrem/h at 1 metre from the package's external surface

176. Type 'A' package is designed to withstand
    a. The normal condition of transport
    b. Accident condition of transport
    c. Both (a) and (b)
    d. None of the above

177. An excepted package can have radioactive content having activity
    a. Less than that in a Type A package
    b. Greater than that of a Type A package
    c. Equal to that in Type A package
    d. None of the above

178. As per U.N. classification of dangerous goods, the radioactive material is represented by
    a. 1
    b. 3
    c. 6
    d. 7

179. Approval for the design of the package by the Competent Authority is required for Excepted package
    a. Type A package
    b. Type B package
    c. Industrial package

180. Generally, the radiopharmaceuticals used in nuclear medicine examinations are transported as
    a. Type A package
    b. Type B package
    c. An accepted package

181. The method used for Overexposure exceeding 100 mSv calls for
    a. Assessment of NVD symptoms
    b. Assessment of C.B.C. in peripheral blood
    c. Chromosomal aberration analysis
    d. Urine analysis creatinine

182. In India, the present limit of ^{131}I activity in the body of the patient for discharge from the hospital, under normal conditions, is less than
    a. 1100 MBq
    b. 500 MBq
    c. 50 MBq
    d. 30 MBq

183. Competent Authority for enforcement of radiation safety in India is
    a. The Chairman, Atomic Energy Commission

b. The Director, Bhabha Atomic Research Centre
c. The Chairman, Atomic Energy Regulatory Board

184. As per AERB directive 6/1999, the annual effective dose to individual radiation workers in any calendar year during the five-year block shall not exceed
a. 10 mSv
b. 20 mSv
c. 30 mSv
d. 50 mSv

185. Personnel monitoring dose records of a radiation worker are preserved at least for a period of
a. 75 years
b. 50 years
c. 30 years

186. In millimetres, which of the following measurements is taken?
a. Energy resolution
b. Spatial resolution
c. Uniformity
d. Sensitivity.

187. Basic guidelines for the safe disposal of ^{131}I is
a. Dilute and disperse
b. Decay and dispose
c. Concentrate and contain
d. a and b

188. Contaminated articles (e.g. vials and syringes contaminated with short-lived ^{99}Mo + ^{99m}Tc can be released into sanitary sewerage subject to the condition that the concentration of radioactivity is limited to
a. 22.2 MBq/m^3
b. 185 MBq/m^3
c. 74 MBq/m^3
d. 37 MBq/m3

189. The best method of disposal of combustible radioactive waste generated in nuclear medicine facility is
a. Concentrate and contain
b. Delay and decay
c. Incineration
d. b and c

190. The spatial resolution of images is better with a _____ energy isotope
a. high
b. low

191. Derived Working Limits (D.W.L.) for radioactive surface contamination in the monitoring area of a laboratory due to alpha-emitting radionuclide is
a. 0.37 Bq/cm^2
b. 3.7 Bq/cm^2
c. 37 Bq/cm^2
d. 370 Bq/cm^2

192. Which of the following transmission phantoms is required to test the gamma camera's spatial resolution?
a. Hine-Duley phantom
b. Four quadrant bar phantom
c. Equal space phantom with parallel lines
d. A phantom with orthogonal holes

193. An orthogonal hole phantom cannot be utilized to evaluate spatial linearity because it lacks lead bars
a. True
b. False

194. On Monday, a physicist does an extrinsic uniformity test, which indicates a nonuniformity of 3%, and on Tuesday, it is 8%. So, what's the next step?
a. Due to photopeak
b. Requiring machine service
c. Obtaining inherent uniformity flood
d. a or c

195. We utilize an asymmetric window around a photopeak for a reason:
a. To take into account other emissions
b. To eliminate the Compton ridge
c. To rule out Compton scatter
d. None of those above

196. There are 88 pixels between the activity peaks on the computed activity profile with a matrix size of 128 × 128 and two sources spaced 30 cm apart. What is the size of a pixel?
a. 6.8 millimetres per pixel
b. 3.4 millimetres per pixel
c. 1.4 millimetres per pixel
d. 0.14 centimetres per pixel

197. If the distance between two 2-point sources on the camera surface is 30 mm and the number of pixels between their activity profile peaks is 50 on the x-axis and 52 on the y-axis, then in the x and y directions, are the pixel dimensions within 0.5 mm of one another?
    a. Yes
    b. No

198. If the background is 1106 cpm, what is the sensitivity of a camera that registers 20,137 cpm from a 120 ci source?
    a. 80 counts per minute/microCurie
    b. 115 counts per minute/microCurie
    c. 158 counts per minute/microCurie
    d. 210 counts per minute/microCurie

199. Which Scintillation crystals are not used in P.E.T. scanners?
    a. Sodium iodide
    b. Lutetium oxyorthosilicate
    c. Bismuth germinate
    d. Lead sulphate

200. Positron emission is
    a. A form of beta decay that involves the emission of a negatively charged beta particle
    b. Where two 511 keV gamma photons are emitted from an atom
    c. A form of beta decay that involves the emission of a positively charged beta particle (positron)

201. Positron emission tomography
    a. Is the same as SPECT but with 511 keV photons
    b. Uses any two 511 keV photons to help determine the position of the radioactive nucleus
    c. Uses the two 511 keV photons travelling 180° from each other that arise from the destruction of positron and electron to determine the position of the radioactive nucleus

202. Possible types of coincidences that occur are
    a. True, Random, Multiple and Scatter
    b. True, Random and Scatter
    c. True and Scatter

203. Random coincidences can be reduced by:
    a. Only scanning slim patients

    b. Decreasing activity and decreasing coincidence window width
    c. Changing the position, the patient is scanned

204. Scattered coincidences can be reduced by:
    a. Reducing energy window width
    b. Applying scatter correction
    c. Scanning only in 2D mode

205. The radionuclide used in P.E.T. imaging can affect:
    a. Sensitivity
    b. Spatial Resolution and Sensitivity
    c. Spatial resolution

206. A block detector is
    a. A solid-state detector system used specifically in P.E.T.
    b. Uses a system of several photomultiplier tubes and a scored light guide
    c. Identical to the detection system used in a gamma camera

207. A scintillator with a short decay time can
    a. Have a narrower coincidence window width which helps reduce the number of scattered events registered
    b. Have a narrower coincidence window width which helps reduce the number of random events registered
    c. Have a wider coincidence window width which makes the system more sensitive

208. The difference between 2D and 3D P.E.T. is
    a. 2D P.E.T. has septa in front of the detectors, limiting coincidences to those within a plane or between neighbouring planes.
    b. 2D P.E.T. produces a 2D (planar) dataset
    c. 3D P.E.T. contains fewer scattered events than 2D PET

209. The sensitivity of 3D P.E.T. systems is:
    a. 4–5 times higher than 2D P.E.T
    b. 4–5 times lower than 2D P.E.T
    c. The same as 2D PET

210. Time of Flight:
    a. Makes no difference to image quality
    b. Allows the position of the positron and electron annihilated to be determined exactly
    c. Uses the time difference between detected events to determine the where-

abouts on the line of response the positron and electron annihilated

211. Random events are:
    a. Distributed depending on the source distribution
    b. Dependent on the imaging tracer used
    c. Uniformly distributed across the field of view

212. Deadtime correction is
    a. Applied in P.E.T., though it is not often relevant in clinical P.E.T
    b. Applied in P.E.T. and is very relevant in clinical P.E.T
    c. Not applied in P.E.T

213. Scatter events are:
    a. Independent of the activity distributions and scatter correction methods use this information to correct for scattered events
    b. Dependent on the imaging tracer used
    c. Dependent on the activity distributions and scatter correction methods use this information to correct for scattered events

214. Normalization correction is applied to:
    a. Remove ring artefacts in the image
    b. Correct for differences in detector response over all lines of response
    c. Make the images look normal

215. The attenuation of a disintegration event is
    a. Dependent on the total path length of the line of response
    b. Dependent on the longest path length of a 511 keV photon
    c. Dependent on the shortest path length of a 511 keV photon

216. The use of C.T. for attenuation maps:
    a. Allows the same energy photons to be used as that used in P.E.T
    b. Is quicker than acquiring attenuation maps with transmission sources
    c. Is less prone to produce P.E.T. artefacts than transmission source attenuation correction

217. Standardized Uptake Values (S.U.V.) are:
    a. A way of determining glucose metabolism within tissue
    b. Comparing uptake between different imaging centres

c. A measure of activity concentration within a tissue

218. In FDG PET, Standardized Uptake Values are:
    a. Affected by the time post-injection that they are imaged/measured
    b. Are not affected by the patient's plasma glucose level
    c. Are affected by changes in height/weight
    d. All of the above

219. Spatial Resolution is:
    a. Affected by the isotope used
    b. Not affected by the diameter of the detector ring
    c. The same at all points in the field of view

220. The positron fraction is:
    a. The fraction of positrons imaged in a study
    b. The percentage of disintegrations in a radioactive nucleus that leads to the emission of a positron
    c. The fraction of positrons that annihilate with an electron

221. Noise Equivalent Count Rate is
    a. The count rate that would give rise to the observed noise, if Randoms and Scatters were removed
    b. The count rate you get once the noise is removed
    c. The count rate you achieve when Randoms and Scatters have been removed

222. The ability of a P.E.T. system to differentiate between two spots after image reconstruction is referred to as
    a. Contrast
    b. Resolution and
    c. Attenuation

223. The data that is used alongside transmission data in the derivation of attenuation correction factors in the P.E.T. scanner quality control operation is called
    a. Attenuation correction
    b. Calibration
    c. Blank scan

224. The annihilation photons' non-collinearity and the finite positron range are inherent

features of positron emission tomography, resulting in:

a. Artefacts of attenuation
b. Positional inaccuracy
c. Scatter
d. Truncation

225. Daily P.E.T. scanner quality control checks should be performed:
a. After the last procedure
b. During the uptake phase
c. At the end of the day and
d. Before the patient is injected

226. The following scanner-related characteristics must be corrected in quantitative P.E.T. imaging, such as S.U.V. calculation. EXCEPT for
a. Random coincidence correction and
b. Scattering coincidence correction
c. Adjustment for attenuation effects
d. Table speed correction

227. The event positioned on a line of response (L.O.R.), joined by two detectors, are called
a. The annihilation event
b. Scatter event
c. Coincidence event
d. Random event

228. The data plotted on Y-axis in the P.E.T. sinogram represents
a. The L.O.R.'s angle of orientation
b. The L.O.R.'s shift from the gantry's centre
c. The L.O.R.'s window of coincidence and
d. The L.O.R.'s displacement from the F.O.V.'s centre

229. The relative differences in count densities between adjacent areas in an object's image are referred to as
a. Contrast
b. Background
c. Noise or
d. Shadow

230. A PET scanner parameter defined as the coincident count rate in a measurement that does not include scattered or random coincidences is called:
a. Noise equivalent count rate (NECR)
b. Contrast

c. Signal-to-noise ratio (SNR)
d. Sensitivity

231. A medical event (misadministration) is:
a. A diagnostic dose that differs 50% or more than the prescribed dose
b. A therapeutic dose that differs 20% or more than the prescribed dose
c. A dose that results in more than 50 mSv (5 rem) effective dose equivalent to the prescribed dosage
d. a and b only
e. a, b and c

232. Once the medical event is established:
i. The R.S.O. is informed immediately
ii. The referring physician and individual should be informed within 24 hours
iii. Records are maintained for 10 years
a. (i) only
b. (i) and (ii)
c. All the three
d. (i) and (iii)

233. The kinetic energy received by an electron when a 1 volt potential difference accelerates it is known as
a. Electron volt
b. Kilo-electron volt
c. Mega-electron volt
d. None of the above

234. In the medium, the charged particles lose energy through collision with
a. The electrons of the atoms
b. Nuclei of the atoms
c. Electrons and nuclei of the atoms
d. None of the above

235. The amount of specific ionization is proportional to
a. The particle's mass and charge are inversely proportional to its velocity
b. The particle's mass and charge are inversely proportional to its charge
c. The particle's charge and velocity, and inversely to its mass
d. The particle's velocity is inversely proportional to its mass and charge

236. L.E.T. (Linear Energy Transfer) is usually expressed in the units of
a. $eV/\mu m$
b. $keV/\mu m$

c. keV/m

d. eV/m

237. The major interactions in soft tissues are 100 keV to 10 MeV in the energy range.

  a. Photoelectric effect

  b. Compton scattering

  c. Pair production

  d. Both a and c

238. In the case of lightweight nuclei, like hydrogen, the interaction of neutron

  a. Radiative capture

  b. Inelastic scattering

  c. Elastic scattering

  d. Neutron capture producing other particles

239. The Atomic Energy (Safe Disposal of Radioactive Waste) rules (G.S.R-125) came in the year

  a. 1962

  b. 1987

  c. 2004

  d. None of the above

240. The equilibrium exists in the relationship $^{68}Ge \rightarrow {}^{68}Ga$.

  a. Transient

  b. Secular

  c. No equilibrium

  d. Equilibrium changes with time

241. The Transportation Regulations do not apply to the following:

  a. Radioactive substance that is a necessary component of the mode of transportation (e.g. aircraft wings)

  b. Radioactive material transferred within an enterprise that is subject to the establishment's relevant safety rules and that does not involve public highways or railways (e.g. from hot lab to imaging room for administration)

  c. Radioactive material implanted or incorporated into a living person or animal for diagnostic or therapeutic purposes. (For example, treating the eye with eye plaque)

  d. All of the above

242. ^{131}I sealed in the capsule for the treatment of Thyroid disorder is termed as

  a. The normal form of radioactive material

b. A special form of radioactive material

c. Depends upon the activity of ^{131}I in the capsule

d. None of the above

243. For ^{99m}Tc radionuclide A_1 and A_2 values

  a. A_2 is greater than A_1

  b. A_1 is greater than A_2

  c. A_1 is the same as A_2

  d. None of the above

244. The role of stannous ion in the preparation of ^{99m}Tc radiopharmaceuticals is

  a. To reduce the valence state of ^{99m}Tc

  b. To increase the valence state from +3 to +7

  c. To reduce the amount of Moly breakthrough

  d. In helping to reduce the radiation burden on the patients

245. The limit for Al^{3+} in ^{99m}Tc eluate is:

  a. 5 µg/ml

  b. 10 µg/ml

  c. 15 µg/ml

  d. No limit

246. Chemical impurities in a ^{99m}Tc eluate:

  a. Degrades the image quality due to poor labelling

  b. Increases radiation burden on the patient

  c. Image quality does not affect

  d. Changes the valence state of ^{99m}Tc

247. Which among the following is radiochemical impurity?

  a. The presence of free ^{99m}Tc in radiopharmaceutical preparation

  b. The presence of ^{99}Mo in $^{99m}TcO_4^-$ solution

  c. The presence of aluminium ions in preparation

  d. None of the above

248. The presence of liver uptake in a bone scan is due to:

  a. Residual radioactivity from previous colloid scan

  b. Excessive aluminium ion from the generator

  c. Excessive reduced-hydrolysed ^{99m}Tc-MDP forming radiocolloids

  d. All of the above

249. Digital pocket dosimeters work on the principle of:

  a. Ionization chamber-based detectors

b. G.M.-based detectors

c. Scintillation-based detectors

d. Semiconductor-based detectors

250. Argon is the choice of gas filled in dose calibrators due to:

a. It being non-reactive in most physical conditions

b. It being the cheapest among all noble gases

c. It having relatively higher Z and lowering the first ionization potential

d. All of the above

24.2 Section B

24.2.1 State True or False

1. The mean free path of primary ionization in a medium is smaller for low L.E.T. radiation.

2. A rise in temperature due to a lethal dose of radiation such as > 6–8 Gy is significantly higher and can inactivate many biological molecules.

3. Most of the damage by low L.E.T. radiation is created by indirect interactions such as free radical formation and interactions.

4. The number of neutrons is always equal to the number of protons in any stable isotope.

5. The majority of the mass of the atom is distributed outside the nucleus.

6. Binding energy per nucleon in a nucleus is maximum for iron, and it reduces as mass number increases.

7. Successive decay of radioactive isotopes leads to decay series and eventually ends with stable isotopes.

8. In some isotopes, electron capture is a competing process for $\beta+$ emission.

9. It is easier to obtain large specific activity sources if the half-life is also larger.

10. K-shell electrons have the largest binding energy compared to any other shells.

11. Annual Limit on Intake (A.L.I.) is that amount of a radionuclide, if inhaled or ingested in a year, would lead to a committed effective dose of 20 mSv.

12. A.L.I. (inhalation) and D.A.C. are interrelated.

13. The A.L.I. value of ^{131}I is more than ^{99m}Tc.

14. Exposure can be measured in Roentgen or Coulombs/kg.

15. Photons have more radiation weighting factors than alpha particles.

16. The annual effective dose due to natural radiation sources is 2.36 mSv.

17. D.A.C. stands for Derived Air concentration of radioactive material.

18. The additive projection model assumes uniform risk throughout the lifetime after exposure + latency period.

19. The multiplicative model addresses many confounding factors that are ignored in the additive model, such as an age-related increase in cancer, a specific type of cancer occurring in high frequency in a population.

20. Dose and dose rate effectiveness factor refers to reduced risk at a low dose and low dose rates compared to A.B.S. evaluated risks.

21. Estimation of detriment accounts for fatal, non-fatal, curability factors associated with a non-fatal and average length of life lost due to cancers associated with a tissue.

22. The justification principle in radiation protection promotes radiation technology despite other technologies that may provide better results.

23. The optimization principle in radiation protection is to achieve maximum profit and cost-effectiveness of radiation technology.

24. ALARA principle of radiation technology helps lower the occupational and public dose well below the dose limits.

25. The dose rate at a given point decreased with the inverse of the square of the distance since fluence gets distributed in all directions in spherical geometry.

26. Dose limits applicable for pregnant women are 1 mSv for the calendar year of the end of pregnancy or childbirth.

27. Dose limits for apprentices or students aged 16–18 years are the same as the general public.

28. Any radioactive waste that has passed ten half-lives can be disposed of in the public domain.

29. Compared to medical exposure in a population, natural background radiation levels are negligible.

30. Biologically incorporated radionuclides in soft tissues are harder to eliminate by biological means.

31. When dealing with short-lived radionuclides, the effective half-life is dominated by physical half-life and vice versa.

32. Effective half-life is always more than either physical or biological half-life.

33. It is considered safe to use the radioactivity administering area for general occupancy and/or consumption of food during lunch breaks.

34. Dragging or kicking method to transport a consignment from one room to another or from a place of delivery to the administering room will help reduce the personnel dose.

35. In case of death of a patient after radioactivity is administered, the corpse can be released for cremation after 8 h.

36. Analysing personnel exposure to nuclear medicine professionals reveals the highest personnel dose compared to any other radiation technology.

37. Inhalation and ingestion exposure can be minimized by avoiding mouth pipetting and using masks and fume hoods.

38. Tissue weighting factors are based on the mass of each organ.

39. The cell membrane constitutes the critical target for cell killing by radiation.

40. An exposed individual does not suffer from genetic effects.

41. Dose and dose-rate effectiveness factor (DDREF) is applied while calculating risk for low dose and low dose rate exposure conditions to compensate for the decreased biological risk.

42. Stochastic effects do not appear in the majority of the radiation exposed individuals.

43. A whole-body dose of 6 Gy is not lethal to a human being.

44. People exposed with higher radiation have been found to have a statistically significant increase in the incidence of solid malignancies.

45. No significant stochastic effects are observed among radiation workers exposed within limits.

46. Neutrons are as efficient in inducing chromosomal aberrations as ^{60}Co gamma rays.

47. Partial body irradiation is less dangerous than whole-body irradiation.

48. The time of onset of radiation sickness and its persistence is dose-dependent.

49. There is no human evidence of increased risk of genetic effects among Hiroshima and Nagasaki atomic bomb survivors.

50. Epidemiological studies of high background areas and occupationally workers exposed to low doses do not show any significant increase in cancer incidence.

51. Cyclotron-produced radionuclides for diagnosis have energy higher than other routinely used radionuclides.

52. When constructing a high-dose therapeutic facility, there shall be no impact on the general population or the environment.

53. Nuclear Medicine facility with High Dose Therapy can be at the entrance of the hospital block.

54. ^{177}Lu for therapy purposes requires an isolation ward as per current regulations in India.

55. The fume hood is required for handling ^{32}P or ^{89}Sr radioisotope.

56. The outlet of the decay and delay tank should be at a higher level than the inlet pipe of the sewage line.

57. ^{131}I capsules containing 100 μCi of activity cannot be used to treat thyrotoxicosis disease.

58. Radiation exposure to the patient is NIL in the case of R.I.A. examination.

59. ^{131}I is used for imaging thyroid and treatment of certain thyroid disorders.

60. TREMCARD details the actions to be taken in case of a normal situation during the transport of the radioactive package.

61. Radioactive waste containing a millicurie level of ^{131}I used for the diagnostic procedure is by dilute and disperse technique.

62. The most commonly used radionuclide in R.I.A. is ^{123}I.

63. The Nuclear Medicine Committee is the Competent Authority for Approving plans for a nuclear medicine laboratory.

64. The Atomic Energy Act of 1962 regulates the use of radioactive materials and equipment that generate radiation.

65. The disposal limit for ground burial for ^{131}I is 37 MBq.

66. The basis of control of occupational exposure to the female who is not pregnant is different from male.

67. LD$_{50}$(60) for the humans is 3–5 Gy.

68. If the radiation exposure level at a 1 cm distance from a 1 MBq point source of radioactive material is 3.1 mGy/hr, the radioactive source is ^{89}Sr.

69. A PET Scanner works on the principle of anti-coincidence circuitry.

70. The TLD badges used for personnel monitoring use LiF: Mg, Cu, P.

71. During decontamination, the surface should be swiped from inside to outside.

72. The absorbed fraction from the source to the target organ for beta radiation is either 0 or 1.

73. If the radiation level at 1 m distance from a point source of radioactive material is 1 Gy/h, the radiation level at 10 meters is equal to 1 mGy.

74. The radiation dose delivered to the thyroid in a therapeutic procedure using ^{131}I is mainly from beta particles.

75. Alumina column cannot be directly used to separate ^{99m}Tc from low specific activity ^{99}Mo.

76. X-rays are more efficient in inducing chromosomal aberrations than ^{60}Co gamma rays.

77. The photoelectric process involves bound electrons.

78. NEMA specifications give specific limits of Q.C. values.

79. Q.C. protocol should be consistent to monitor the long-term performance of the Gamma Camera.

80. Disposal of radioactive waste containing a millicurie level of ^{99m}Tc is diluting and dispersing.

81. Radiation with low L.E.T. is preferred in nuclear medicine diagnostic procedures.

82. The permissible limit of surface contamination on the active worktable for all radionuclides used in nuclear medicine is 40 Bq/cm^2. In the case of ^{14}C and ^{3}H, it is 100 times that of the given limit.

83. The dilute and disperse principle always involves the disposal of radioactive waste generated in nuclear medicine laboratories.

84. In nuclear medicine practice, external exposure is more likely than internal exposure.

85. If I$_1$ and I$_2$ are intensities at distances d$_1$ and d$_2$, the law can be represented by I$_1$/ I$_2$ = (d$_2$)^2/(d$_1$)^2.

86. R.B.E. of radiation is independent of energy.

87. Personnel Monitoring Badge protects the radiation worker from radiation effects.

88. Characteristic X-rays of energy 28 keV emitted in the decay of ^{125}I are considered non-penetrating.

89. ^{186}Re is a cyclotron-produced radioisotope.

90. In the case of beta-emitting radionuclides, the risk of internal contamination is less than the external exposure.

91. Long-lived radioisotopes are produced by bombarding high-energy charged particles in accelerators.

92. Quality Assurance in nuclear medicine also refers to routine quality control of the imaging systems used in nuclear medicine.

93. In a radioactive decay by electron capture process, the atomic number of the radionuclide increases.

94. Beta has more radiation weighting factor than alpha particles.

95. The $T_{1/2}$ of ^{11}C is 2.04 min.
96. In the use of unsealed sources, the external hazard is less than the hazard due to the intake of radioactivity.
97. H.V.T. for 140 keV photons in the lead is 0.3 mm.
98. The radiation energy emitted by a radionuclide will reduce to half after a lapse of one physical half-life.
99. ^{131}I capsules containing 100 mCi of activity cannot be used to treat thyrotoxicosis disease.
100. The effective doses of 10 mSv and above recorded by a badge in a monitoring period are considered overexposure.

24.3   Section C

24.3.1 Fill in the blanks

1. sievert (Sv) is the unit of _____
2. An equivalent dose is a product of _____ and _____
3. Person sievert (Sv) is the unit of _____
4. The specific activity of a radiolabeled product is the activity per unit _____
5. Shielding material used for handling 370 MBq of ^{32}P radioactivity is _____
6. After six half-lives, the radionuclide activity reduces by a factor of _____
7. Physical Half-Life T_p of ^{13}N is _____
8. The gamma-ray constant (K-factor) for ^{131}I is 2.2 R-cm^2/mCi-h; the exposure rate from 1 mCi of ^{131}I at 1 cm will be _____
9. In case of internal contamination, radionuclides _____ are administered for quick elimination of the contaminant.
10. Perspex + lead is a good shield for high energy _____ particles.
11. In nuclear medicine, the types of packages used for the transport of radioactive materials are mainly _____
12. As per AERB Safety Code for nuclear medicine laboratories, patients normally with more than _____ MBq of ^{131}I activity are hospitalized, and radiation level at a distance of 1 m from the external surface at the time of discharge should be less than _____

13. The new Radiation Protection Rule (R.P.R.) promulgated under Atomic Energy Act 1962 came in year _____
14. The minimum distance from the medical cyclotron and any residential premises is _____
15. Medical Cyclotron is categorized under two types _____ and _____
16. Production of bremsstrahlung radiation increases with an increase in _____ of material and decreases with the increase in _____ of the charged particle.
17. External radiation hazards can easily be controlled by the three fundamental procedures, namely, _____, _____ and _____
18. The relation between Half Value Thickness (H.V.T.) and Tenth Value Thickness (T.V.T.) is T.V.T. = _____ x H.V.T.
19. Radionuclide generators are possible if the half-life of the parent nuclide is _____ than that of the daughter, and the two can exist in equilibrium.
20. The dose rate at 1 meter at the time of discharge of ^{131}I administered patient from isolation ward should be less than _____ mR/h.
21. The atomic number of lead used as a shielding material for gamma-ray sources is _____ than the aluminium and is equal to _____
22. The maximum energy of beta rays emitted by ^{89}Sr is _____
23. The equation used to evaluate the absorbed dose for complete decay of radioactivity is $D_\infty = 1.44 \times D_0 \times$_____
24. If λ_p is physical decay constant and λ_b is biological decay constant, effective decay constant λ_{eff} will be _____
25. The best methods for reducing exposure from the patient in nuclear medicine studies are _____ and _____

26. A cataract is an example of _____ effect of ionizing radiation.

27. The assessment of internal contamination due to beta emitters is normally done by _____

28. The threshold energy required for pair production is _____ MeV.

29. Genetic diseases are caused due to induction of _____ in germ cells.

30. Absorbed fraction (thyroid ← thyroid) due to beta particles from ^{131}I _____

31. In ^{131}I treatment of thyroid cancer patients the radiation dose to non-target tissues can be minimized by blocking the uptake with _____

32. R.I.A. kits are normally transported in _____ package.

33. In MIRD, the total dose calculated is the product of _____ and _____

34. The dosage of ^{99m}Tc labelled M.D.P. compound administered to the patient for bone scanning is about _____ MBq.

35. Radionuclides can be artificially produced by _____

36. The isotopes have the same atomic number but differ in the number of _____

37. The physical principle of radiation protection, which helps reduce the dose while working near the vicinity of radiation, is _____

38. Due to the divergence nature of radiation emission from isotope source, the intensity of radiation follows _____ with distance.

39. The threshold dose of a deterministic effect depends on physical factors of irradiation such as _____ and _____

40. The radiation weighting factor is used while estimating _____

41. The doubling dose used in genetic risk estimation is estimated to be _____

42. Separate limits are defined for extremities to avoid _____ type of biological effects.

43. Three components of stochastic effects for estimation of detriment are _____, _____ and _____

44. Estimated heritable risks for general public and occupation workers are _____ and _____ % Sv^{-1}

45. Genetic makeup and rate of division of cells in a tissue determine the _____ of the tissue towards radiation damage.

46. Stochastic risk of tissue also depends on _____ incidence of cancer associated with the tissue.

47. The total energy of the beta particle and anti-neutrino emitted is always the same _____

48. _____ is defined as radioactive decay per second.

49. An _____ particle is the same thing as a helium nucleus.

50. ^{201}Tl decay by _____ mechanism.

51. The stannous ion reduces the valence state of _____ in the preparation of ^{99m}Tc labelled radiopharmaceuticals.

52. The presence of ^{99}Mo in ^{99m}Tc eluate is an example of _____ impurity.

53. The volume to be withdrawn from a kit containing 140 mCi of ^{99m}Tc in 23 ml to obtain a dose of 5 mCi is _____

54. The presence of free ^{99m}Tc in preparation of ^{99m}Tc sulphur colloid is an example of _____ impurity.

55. A diphosphonate kit should be utilized within _____ hours of preparation.

56. The usual particle size of sulphur colloid is _____

57. Excessive aluminium in ^{99m}Tc eluate leads to the _____ uptake during the bone scan.

58. Limulus amebocyte lysate assay is carried out for testing for _____ in the sample.

59. 59. The least activity in the normal F.D.G. distribution would be in the _____

60. 18FDG-PET images depict a map of the _____ distribution throughout the body.

24.4   Section D

24.4.1 Match the following:

1. ALARA	$(\rightarrow)$	(A) Exposed cohort
2. Effective dose	$(\rightarrow)$	(B) Basic principle of radiation protection
3. DDREF	$(\rightarrow)$	(C) Solid cancer
4. Multiplicative model	$(\rightarrow)$	(D) Occupational type of exposure
5. Atomic bomb survivors	$(\rightarrow)$	(E) Tissue weighted equivalent dose
6. Genetic effects	$(\rightarrow)$	(F) Cellular level damage
7. DSB	$(\rightarrow)$	(F) Future generation
8. 1 in 1000 mortality	$(\rightarrow)$	(H) Detriment
9. Doubling dose	$(\rightarrow)$	(I) Maximum acceptable mortality
10. Morbidity	$(\rightarrow)$	(J) Genetic risk estimation

24.4.2 Indicate whether the following effects are Stochastic (S) or Deterministic (D).

1. Skin erythema	$(\rightarrow)$
2. Radiation sickness	$(\rightarrow)$
3. Leukaemia	$(\rightarrow)$
4. Sterility	$(\rightarrow)$
5. Genetic effects	$(\rightarrow)$
6. Prenatal effects	$(\rightarrow)$
7. Lung cancers	$(\rightarrow)$
8. Germ cell mutations	$(\rightarrow)$
9. Cataract	$(\rightarrow)$
10. Multiple myeloma	$(\rightarrow)$

24.4.3 Match the following

1. Radiation synovectomy	$(\rightarrow)$	(A) ^{125}I
2. Immunoradiometric assay	$(\rightarrow)$	(B) ^{131}I-mIBG
3. Painful bone metastases	$(\rightarrow)$	(C) T_3 and T_4
4. Neural crest imaging	$(\rightarrow)$	(D) ^{15}O
5. Renal function radiopharmaceutical	$(\rightarrow)$	(E) ^{90}Y-colloid
6. Cyclotron produced	$(\rightarrow)$	(F) 2.2 R-cm^2/h/mCi
7. Reduction of gall bladder dose	$(\rightarrow)$	(G) ^{99m}Tc-DTPA
8. Gamma ray dose constant for ^{131}I	$(\rightarrow)$	(H) Fatty meal
9. Effective dose limit for a pregnant NM staff	$(\rightarrow)$	(I) 1 mSv
10. Thyroid hormones	$(\rightarrow)$	(J) ^{153}Sm-EDTMP

24.4.4 Match the following:

1. Absorbed fraction $\phi(thyd \leftarrow thyd)$	$(\rightarrow)$	(A) HPGe detector
2. Radiation synovectomy	$(\rightarrow)$	(B) Geiger-Muller counter
3. Internal contamination of radionuclides	$(\rightarrow)$	(C) MIRD
4. Predominant genetic risk due to radiation	$(\rightarrow)$	(D) 500 mSv
5. Patient protection	$(\rightarrow)$	(E) Exponential phenomena
6. Biological elimination of radioactivity	$(\rightarrow)$	(F) Multi channel analyser
7. Annual equivalent dose to extremity	$(\rightarrow)$	(G) ^{90}Y-colloid
8. High energy resolution	$(\rightarrow)$	(H) Children and reproductive age group
9. Energy independent response	$(\rightarrow)$	(I) Chelating agent for GI
10. Energy discrimination	$(\rightarrow)$	(J) Guidance levels

24.5   Keys

24.5.1   Section A: Multiple Choice Questions

1	C	31	D	61	C	91	A
2	B	32	B	62	C	92	D
3	C	33	A	63	B	93	C
4	C	34	A	64	B	94	B
5	A	35	A	65	C	95	B
6	C	36	C	66	B	96	A
7	B	37	A	67	C	97	D
8	C	38	A	68	C	98	B
9	C	39	B	69	C	99	B
10	C	40	B	70	C	100	D
11	C	41	A	71	C	101	A
12	C	42	A	72	D	102	C
13	C	43	C	73	B	103	A
14	A	44	B	74	B	104	A
15	A	45	D	75	D	105	C
16	C	46	A	76	D	106	D
17	B	47	C	77	B	107	D
18	B	48	C	78	C	108	C
19	C	49	B	79	A	109	D
20	C	50	B	80	A	110	A
21	C	51	C	81	D	111	C
22	A	52	B	82	D	112	B
23	B	53	A	83	C	113	B
24	B	54	B	84	D	114	A
25	C	55	A	85	D	115	A
26	D	56	C	86	A	116	B
27	D	57	D	87	C	117	D
28	C	58	B	88	C	118	A
29	A	59	B	89	C	119	B
30	B	60	A	90	B	120	C

121	D	151	D	181	C	211	C
122	A	152	A	182	A	212	A
123	C	153	A	183	C	213	C
124	A	154	D	184	B	214	B
125	A	155	A	185	C	215	A
126	A	156	A	186	B	216	B
127	B	157	A	187	D	217	A
128	A	158	C	188	B	218	D
129	D	159	A	189	D	219	A
130	C	160	A	190	A	220	B
131	C	161	C	191	A	221	C
132	A	162	D	192	B	222	B
133	D	163	D	193	B	223	C
134	A	164	C	194	D	224	B
135	D	165	C	195	C	225	D
136	C	166	D	196	B	226	D

137	B	167	B	197	A	227	C
138	A	168	C	198	C	228	A
139	B	169	B	199	D	229	A
140	B	170	A	200	C	230	A
141	B	171	D	201	C	231	E
142	A	172	D	202	A	232	C
143	B	173	D	203	B	233	A
144	B	174	C	204	C	234	C
145	B	175	C	205	B	235	A
146	B	176	A	206	B	236	B
147	B	177	A	207	B	237	B
148	C	178	D	208	A	238	C
149	A	179	B	209	A	239	B
150	B	180	A	210	C	240	B

241	D	244	A	247	A	250	D
242	B	245	B	248	D		
243	B	246	A	249	D		

24.5.2   Section B: True/False

1	F	31	T	61	T	91	F
2	F	32	F	62	F	92	T
3	T	33	F	63	F	93	F
4	F	34	F	64	T	94	F
5	F	35	F	65	T	95	F
6	T	36	F	66	F	96	T
7	T	37	T	67	T	97	T
8	T	38	F	68	F	98	F
9	F	39	F	69	F	99	T
10	T	40	T	70	F	100	F
11	T	41	T	71	F		
12	T	42	T	72	T		
13	F	43	F	73	F		
14	T	44	T	74	T		
15	F	45	T	75	F		
16	T	46	F	76	T		
17	T	47	T	77	T		
18	T	48	T	78	F		
19	T	49	T	79	T		
20	T	50	T	80	T		
21	T	51	T	81	T		
22	F	52	T	82	F		
23	F	53	F	83	F		
24	T	54	T	84	F		
25	T	55	F	85	T		
26	F	56	T	86	F		
27	F	57	T	87	F		
28	F	58	T	88	T		
29	F	59	T	89	F		
30	T	60	T	90	F		

24.5.3 Section C: Fill in the Blanks

1	Effective dose & equivalent dose
2	Absorbed dose & radiation weighting factor
3	Collective effective dose
4	Mass
5	Perspex
6	2^6
7	10 min
8	2.2 R
9	Laxatives
10	Beta
11	Type-A
12	1100 & 50 µSv/h
13	2004
14	30 m
15	Self-shielded & non-self-shielded
16	Atomic number & energy
17	Time, distance & shielding
18	3.3
19	More
20	5
21	More, 82
22	1.463 MeV
23	T_{eff}
24	$\lambda p + \lambda b$
25	Time & distance
26	Late deterministic
27	Bioassay
28	1.022
29	Mutation
30	1
31	Potassium iodate (KIO_3)
32	Excepted packages
33	Time-integrated activity & S-factor
34	740
35	Reactor
36	Mass
37	Shielding
38	Inverse square law
39	Dose rate and LET
40	Equivalent dose
41	1 Gy
42	Deterministic
43	Fatal cancer, non-fatal cancer and heritable effects
44	0.2 and 0.1
45	Sensitivity
46	Background or natural
47	Same
48	Becquerel
49	Alpha
50	Electron capture
51	^{99m}Tc
52	Radionuclidic
53	0.8 mL
54	Radiochemical
55	4 h
56	0.3–1.0 µm
57	Liver
58	Pyrogens
59	Bone
60	Glucose

24.5.4 Section D: Match the Following

I		II		III		IV	
1	B	1	D	1	E	1	C
2	E	2	D	2	A	2	G
3	D	3	S	3	J	3	I
4	C	4	D	4	B	4	H
5	A	5	S	5	G	5	J
6	G	6	D	6	D	6	E
7	F	7	S	7	H	7	D
8	I	8	S	8	F	8	A
9	J	9	D	9	I	9	B
10	H	10	S	10	C	10	F

Milton Keynes UK
Ingram Content Group UK Ltd.
UKHW051017201123
432904UK00003B/62